# The Coronaviridae

# THE VIRUSES



---

THE VIRUSES: Catalogue, Characterization, and Classification
Heinz Fraenkel-Conrat

*Recent volumes in the series:*

THE ARENAVIRIDAE
Edited by Maria S. Salvato

THE BACTERIOPHAGES
Volumes 1 and 2 • Edited by Richard Calendar

THE CORONAVIRIDAE
Edited by Stuart G. Siddell

THE HERPESVIRUSES
Volumes 1–3 • Edited by Bernard Roizman
Volume 4 • Edited by Bernard Roizman and Carlos Lopez

THE INFLUENZA VIRUSES
Edited by Robert M. Krug

THE PAPOVAVIRIDAE
Volume 1 • Edited by Norman P. Salzman
Volume 2 • Edited by Norman P. Salzman and Peter M. Howley

THE PARAMYXOVIRUSES
Edited by David W. Kingsbury

THE PARVOVIRUSES
Edited by Kenneth I. Berns

THE PLANT VIRUSES
Volume 1 • Edited by R. I. B. Francki
Volume 2 • Edited by M. H. V. Van Regenmortel and Heinz Fraenkel-Conrat
Volume 3 • Edited by Renate Koenig
Volume 4 • Edited by R. G. Milne

THE REOVIRIDAE
Edited by Wolfgang K. Joklik

THE RETROVIRIDAE
Volumes 1–4 • Edited by Jay A. Levy

THE RHABDOVIRUSES
Edited by Robert R. Wagner

THE TOGAVIRIDAE AND FLAVIVIRIDAE
Edited by Sondra Schlesinger and Milton J. Schlesinger

THE VIROIDS
Edited by T. O. Diener

# The Coronaviridae

Edited by

STUART G. SIDDELL

*Institute of Virology*
*University of Würzburg*
*Würzburg, Germany*

PLENUM PRESS • NEW YORK AND LONDON

**Library of Congress Cataloging-in-Publication Data**

The coronaviridae / edited by Stuart G. Siddell.
p. cm. -- (The viruses)
Includes bibliographical references and index.
ISBN 0-306-44972-2
1. Coronaviruses. 2. Coronavirus infections. I. Siddell, S.
II. Series.
[DNLM: 1. Coronaviridae. 2. Coronaviridae Infections. QW
168.5.C8 C8217 1995]
QR399.C66 1995
576'.6484--dc20
DNLM/DLC
for Library of Congress 95-32957
CIP

ISBN 0-306-44972-2


A Division of Plenum Publishing Corporation
233 Spring Street, New York, N.Y. 10013

10 9 8 7 6 5 4 3 2 1



Printed in the United States of America

# Contributors

**Robert Anderson**, Department of Microbiology and Immunology, Dalhousie University, Halifax, Nova Scotia, B3H 4H7 Canada

**David A. Brian**, Department of Microbiology, University of Tennessee, Knoxville, Tennessee 37996-0845

**I. Brierley**, Virology Division, Department of Pathology, Cambridge University, Cambridge CB2 1QP, England

**T. D. K. Brown**, Virology Division, Department of Pathology, Cambridge University, Cambridge CB2 1QP, England

**David Cavanagh**, Institute for Animal Health, Division of Molecular Biology, Compton Laboratory, Compton, Newbury, Berkshire RG20 7NN, England

**Susan R. Compton**, Section of Comparative Medicine, Yale University, New Haven, Connecticut 06520

**Jane K. A. Cook**, Intervet UK Ltd., Huntingdon, Cambridgeshire PE17 2BQ, England

**Samuel Dales**, Cytobiology Group, Department of Microbiology and Immunology, The University of Western Ontario, London, Ontario, N6A 5C1 Canada

**Raoul J. de Groot**, Virology Division, Department of Infectious Diseases and Immunology, University of Utrecht, 3584 CL Utrecht, The Netherlands

**Luis Enjuanes**, Centro Nacional de Biotecnología, CSIC, Campus Universidad Autonóma, Cantoblanco, 28049 Madrid, Spain

**David J. Garwes**, Ministry of Agriculture, Fisheries & Food, London SW1P 3JR, England

**Brenda G. Hogue**, Department of Microbiology and Immunology, Division of Molecular Virology, Baylor College of Medicine, Houston, Texas 77030

**Kathryn V. Holmes**, Department of Pathology, Uniformed Services University of the Health Sciences, Bethesda, Maryland 20892

**Marian C. Horzinek**, Virology Division, Department of Infectious Diseases and Immunology, University of Utrecht, 3584 CL Utrecht, The Netherlands

**Thomas E. Kienzle**, John L. McClellan Memorial Veterans Administration Medical Research Service, Little Rock, Arkansas 72205; *present address*: Department of Ophthalmology, Baylor College of Medicine, Houston, Texas 77030

**Marion Koopmans**, Virology Section, National Institute of Public Health and Environmental Protection, 3720 BA Bilthoven, The Netherlands

**Hubert Laude**, Laboratoire de Virologie et Immunologie Moléculaires, INRA, Centre de Recherches de Jouy-en-Josas, 78350 Jouy-en-Josas, France

**Willem Luytjes**, Department of Virology, Institute of Medical Microbiology, Leiden University, 2300 AH Leiden, The Netherlands

**Paul S. Masters**, Wadsworth Center for Laboratories and Research, New York State Department of Health, Albany, New York 12201-0509

**A. P. A. Mockett**, Intervet, Inc., Millsboro, Delaware 19966

**Steven H. Myint**, Department of Microbiology, University of Leicester, Leicester LE1 9HN, England

**Peter J. M. Rottier**, Institute of Virology, Department of Infectious Diseases and Immunology, University of Utrecht, 3584 CL Utrecht, The Netherlands

**Stuart G. Siddell**, Institute of Virology and Immunobiology, University of Würzburg, 97078 Würzburg, Germany

**Eric J. Snijder**, Department of Virology, Institute of Medical Microbiology, Leiden University, 2300 AH Leiden, The Netherlands

**Willy J. M. Spaan**, Department of Virology, Institute of Medical Microbiology, Leiden University, 2300 AH Leiden, The Netherlands

**Robbert G. van der Most**, Department of Virology, Institute of Medical Microbiology, Leiden University, 2300 AH Leiden, The Netherlands

**Bernard A. M. Van der Zeijst**, Institute of Infectious Diseases and Immunology, School of Veterinary Medicine, University of Utrecht, 3508 TD Utrecht, The Netherlands

# Preface

Coronaviruses were recognized as a group of enveloped, RNA viruses in 1968 and accepted by the International Committee on the Taxonomy of Viruses as a separate family, the *Coronaviridae*, in 1975. By 1978, it had become evident that the coronavirus genomic RNA was infectious (i.e., positive strand), and by 1983, at least the framework of the coronavirus replication strategy had been perceived. Subsequently, with the application of recombinant DNA techniques, there have been remarkable advances in our understanding of the molecular biology of coronaviruses, and a mass of structural data concerning coronavirus genomes, mRNAs, and proteins now exists. More recently, attention has been focused on the role of essential and accessory gene products in the coronavirus replication cycle and a molecular analysis of the structure–function relationships of coronavirus proteins. Nevertheless, there are still large gaps in our knowledge, for instance, in areas such as the genesis of coronavirus subgenomic mRNAs or the function of the coronavirus RNA-dependent RNA polymerase.

The diseases caused by coronaviruses have been known for much longer than the agents themselves. Possibly the first coronavirus-related disease to be recorded was feline infectious peritonitis, as early as 1912. The diseases associated with infectious bronchitis virus, transmissible gastroenteritis virus, and murine hepatitis virus were all well known before 1950. However, it was only the realization in the late 1960s that a coronavirus was responsible for a human disease, the common cold, that brought these viruses to the attention of academic virologists. The advances of the last few years, for example, in the molecular analysis of coronavirus protein immunogenicity or the identification of coronavirus receptors, have undoubtedly given an impetus to studies on coronavirus pathogenesis. At the same time, however, they have also begun to reveal the complexities of the coronavirus–host relationship.

In parallel with the studies on coronaviruses, recent insights into the genome organization, replication strategy, and nucleotide sequences of two other groups of viruses, the toroviruses and the arteriviruses, have led to the idea of a "coronaviruslike" superfamily. This idea, which is based essentially on a molecular view of evolution, led, in 1992, to the inclusion of the toroviruses as a second genus in the *Coronaviridae*. The question of how the evolutionary links between the *Coronaviridae* and the arteriviruses should be reflected in their taxonomic status is still open to discussion.

This volume consists essentially of two parts. The first 12 chapters discuss the molecular aspects of coronavirus and torovirus biology. First, the unique features of coronavirus replication, including the genome organization, transcription, and translation strategies, and the importance of RNA recombination are reviewed. This is followed by a chapter on coronavirus receptors. Then each of the virus structural and nonstructural proteins is described in detail. Finally, the molecular biology of toroviruses and the idea of a "coronaviruslike" superfamily is discussed. In the second part of the volume, which consists of seven chapters, selected biological aspects of the most important coronavirus and torovirus infections are covered. These range from the experimental study of murine coronavirus infections to the problems of coronavirus infection in poultry and swine.

I hope that this volume will be useful to both academic researchers and their students, as well as clinicians and veterinarians with an interest in coronavirus-related diseases. I would like to thank the contributors, the staff of Plenum Press, and my wife for their patience during the preparation of this volume.

Stuart G. Siddell

*University of Würzburg*
*Würzburg, Germany*

# Contents

*Chapter 1*

**The *Coronaviridae*: An Introduction**

*Stuart G. Siddell*



*Chapter 2*

**Coronavirus Replication, Transcription, and RNA Recombination**

*Robbert G. van der Most and Willy J. M. Spaan*

*Chapter 3*

**Coronavirus Gene Expression: Genome Organization and Protein Synthesis**

*Willem Luytjes*



*Chapter 4*

**Coronavirus Receptors**

*Kathryn V. Holmes and Susan R. Compton*



*Chapter 5*

**The Coronavirus Surface Glycoprotein**

*David Cavanagh*

*Chapter 11*

**The Molecular Biology of Toroviruses**

*Eric J. Snijder and Marian C. Horzinek*

*Chapter 12*

**The Coronaviruslike Superfamily**

*Eric J. Snijder and Willy J. M. Spaan*



*Chapter 13*

**Pathogenesis and Diseases of the Central Nervous System Caused by Murine Coronaviruses**

*Samuel Dales and Robert Anderson*

CHAPTER 1

# The *Coronaviridae*

## An Introduction

STUART G. SIDDELL

## I. INTRODUCTION

The *Coronaviridae* are a group of enveloped, positive-strand RNA viruses with nonsegmented genomes of about 30,000 nucleotides. The family comprises two genera, coronavirus and torovirus, which share similarities in the organization and expression of their genomes and the structure of the viral gene products. Viruses in the two genera have, however, only limited sequence similarity and display; for example, distinctly different nucleocapsid morphologies.

In this introductory chapter, I shall briefly discuss the classification of the *Coronaviridae*, the organization of corona- and torovirus genomes, and the morphology and physicochemical properties of the virus particles. The molecular aspects of coronavirus and torovirus biology are discussed extensively in Chapters 2 to 12, and Chapters 13 to 19 deal with the biological aspects of coronavirus and torovirus infections.

## II. CLASSIFICATION

Viruses that are classified in the *Coronaviridae* fulfill the following criteria:

- A positive-strand, nonsegmented genome of about 30,000 nucleotides. The genome contains an unusually large RNA polymerase gene (about

STUART G. SIDDELL • Institute of Virology and Immunobiology, University of Würzburg, 97078 Würzburg, Germany.

*The Coronaviridae*, edited by Stuart G. Siddell, Plenum Press, New York, 1995.

20,000 nucleotides), a large surface glycoprotein gene, an integral membrane protein gene, and a nucleocapsid protein gene. These genes are arranged on the genome in a specific order (as described above in the 5′ → 3′ direction) but may be interspersed with additional genes encoding further structural or nonstructural proteins.

- A virion envelope bearing pronounced surface projections. These projections are composed of the large surface glycoprotein ($M_r$ = ~200 kDa) that characteristically exhibits a coiled-coil structure in the carboxy-terminal, membrane-anchoring half.
- An integral membrane protein ($M_r$ = ~25 kDa). This protein characteristically has three membrane-spanning regions in the amino-terminal half.
- A 3′ coterminal set of four or more intracellular subgenomic mRNAs. Only the open reading frames (ORFs) contained within the 5′ unique region of each mRNA (i.e., the region not found in the next smallest mRNA) are expressed as protein.
- An RNA polymerase gene composed of two overlapping ORFs. Expression of the downstream ORF is mediated by (−1) ribosomal frameshifting.

The features that distinguish coronaviruses and toroviruses are:

- The size of the nucleocapsid protein ($M_r$ = ~60 kDa and ~18 kDa, respectively) and the shape of the helical nucleocapsid structure (extended or tubular, respectively).
- The absence of a leader sequence at the 5′ end of torovirus mRNAs.
- With some limited exceptions, a lack of sequence similarity in the coronavirus and torovirus gene products.

At the present time 13 viruses are classified as species in the *Coronaviridae* (Cavanagh *et al.*, 1994). These are listed with their acronyms and natural hosts in Table I.

The prototype of the coronavirus genus is avian infectious bronchitis virus. The name "coronavirus" is derived from the solar corona-like (L. *corona* = crown) appearance of virus particles in negatively stained electron micrographs. The prototype of the torovirus genus is Berne virus (BEV) and the name "torovirus" is derived from the curved tubular (L. *torus* = lowest convex molding in the base of a column) morphology of the nucleocapsid structure.

In addition to the viruses listed in Table I, two viruses, rat coronavirus (RCV) and rabbit coronavirus (RbCV), are considered as tentative species of the coronavirus genus. Three more viruses, feline enteric coronavirus (FECV), sialoacryoadenitis virus (SADV), and porcine respiratory coronavirus (PRCV), are often discussed in the literature as coronavirus species, but the available data suggest that they could be equally well considered as variants of feline infectious peritonitis virus (FIPV), RCV, and transmissible gastroenteritis virus (TGEV), respectively.

In relation to the torovirus genus, there are reports of an enveloped virus in the stools of patients with gastroenteritis that resembles Breda viruses (Beards *et al.*, 1984). However, this virus is not yet recognized as a member of the genus.

TABLE I. *Coronaviridae*

| Natural host | Virus | Acronym |
|---|---|---|
| Coronavirus | | |
| Chicken | Avian infectious bronchitis virus | IBV |
| Cattle | Bovine coronavirus | BCV |
| Dog | Canine coronavirus | CCV |
| Man | Human coronavirus 229E | HCV 229E |
| Man | Human coronavirus OC43 | HCV OC43 |
| Cat | Feline infectious peritonitis virus | FIPV |
| Mouse | Murine hepatitis virus | MHV |
| Pig | Porcine epidemic diarrhea virus | PEDV |
| Pig | Porcine hemagglutinating encephalomyelitis virus | HEV |
| Pig | Porcine transmissible gastroenteritis virus | TGEV |
| Turkey | Turkey coronavirus | TCV |
| Torovirus | | |
| Horse | Berne virus | BEV |
| Cattle | Breda virus | BRV |

## A. Serological Relationships

A variety of serological assays have been used to determine the antigenic relationships among coronaviruses. They have included neutralization, hemagglutination-inhibition, immunofluorescence, immunoblotting, enzyme-linked immunosorbent assay (ELISA), and radioimmunoprecipitation using polyvalent and monovalent antisera and monoclonal antibodies (Spaan *et al.*, 1990; Dea *et al.*, 1990). The results of these studies can be summarized by a taxonomy placing the 11 coronavirus species in three antigenic groups. These groups are shown in Table II.

The inclusion of human coronavirus (HCV) 229E in antigenic group 1 has been questioned, essentially because no cross-reactivity could be demonstrated between HCV 229E and a large panel of TGEV-specific monoclonal antibodies (Sanchez *et al.*, 1990). However, the sequence relationships of coronavirus proteins appear to support the inclusion of HCV 229E in this group.

TABLE II. Antigenic Relationships of Coronaviruses

| Group 1 | Group 2 | Group 3 |
|---|---|---|
| HCV 229E | HCV OC43 | IBV |
| TGEV | MHV | |
| PEDV | BCV | |
| CCV | HEV | |
| FIPV | TCV | |

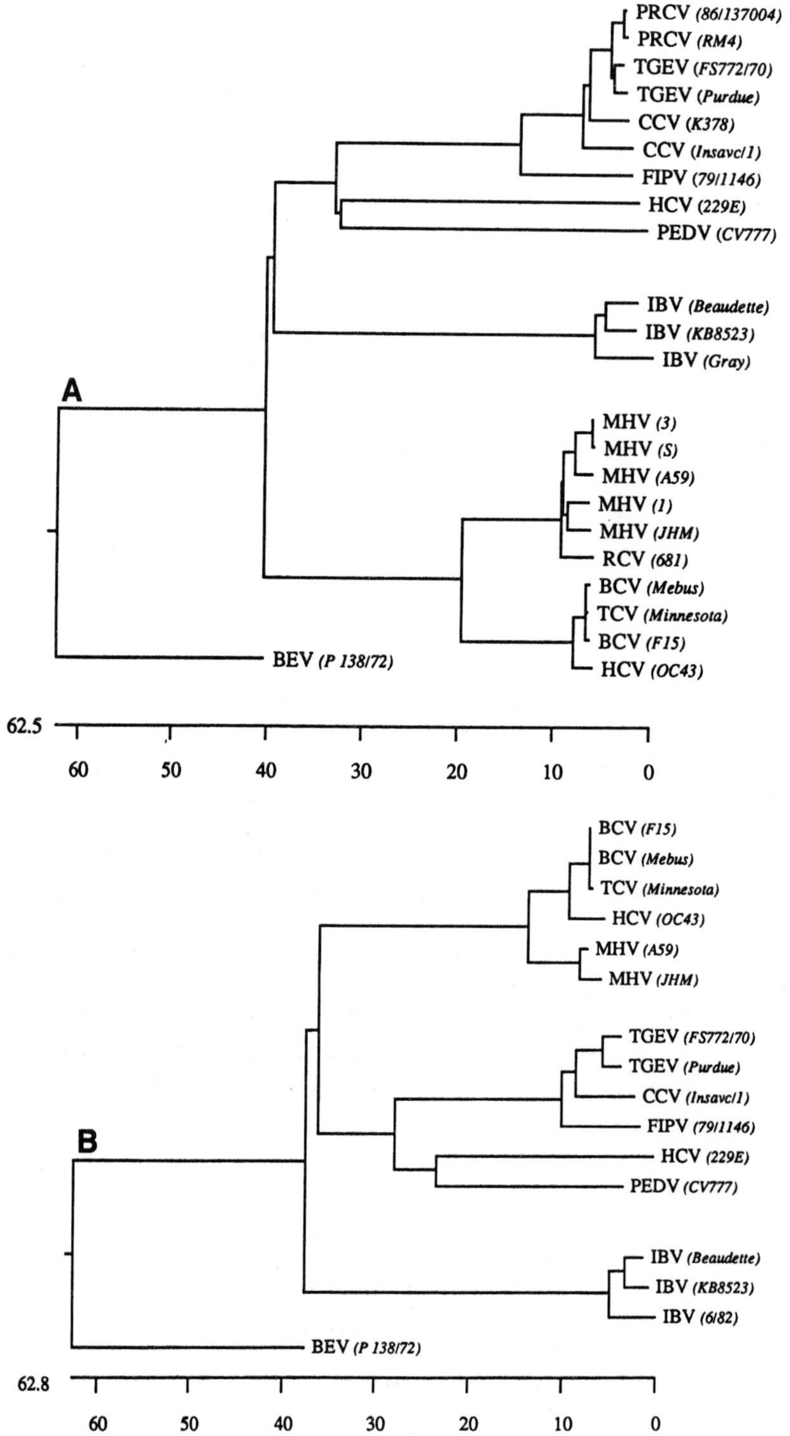
PRCV (86/137004)
PRCV (RM4)
TGEV (FS772/70)
TGEV (Purdue)
CCV (K378)
CCV (Insavc/1)
FIPV (79/1146)
HCV (229E)
PEDV (CV777)
IBV (Beaudette)
IBV (KB8523)
IBV (Gray)
A
MHV (3)
MHV (S)
MHV (A59)
MHV (1)
MHV (JHM)
RCV (681)
BCV (Mebus)
TCV (Minnesota)
BCV (F15)
HCV (OC43)
BEV (P 138/72)
62.5
60
50
40
30
20
10
0
BCV (F15)
BCV (Mebus)
TCV (Minnesota)
HCV (OC43)
MHV (A59)
MHV (JHM)
TGEV (FS772/70)
TGEV (Purdue)
CCV (Insavc/1)
FIPV (79/1146)
HCV (229E)
PEDV (CV777)
B
IBV (Beaudette)
IBV (KB8523)
IBV (6/82)
BEV (P 138/72)
62.8
60
50
40
30
20
10
0

FIGURE 1.

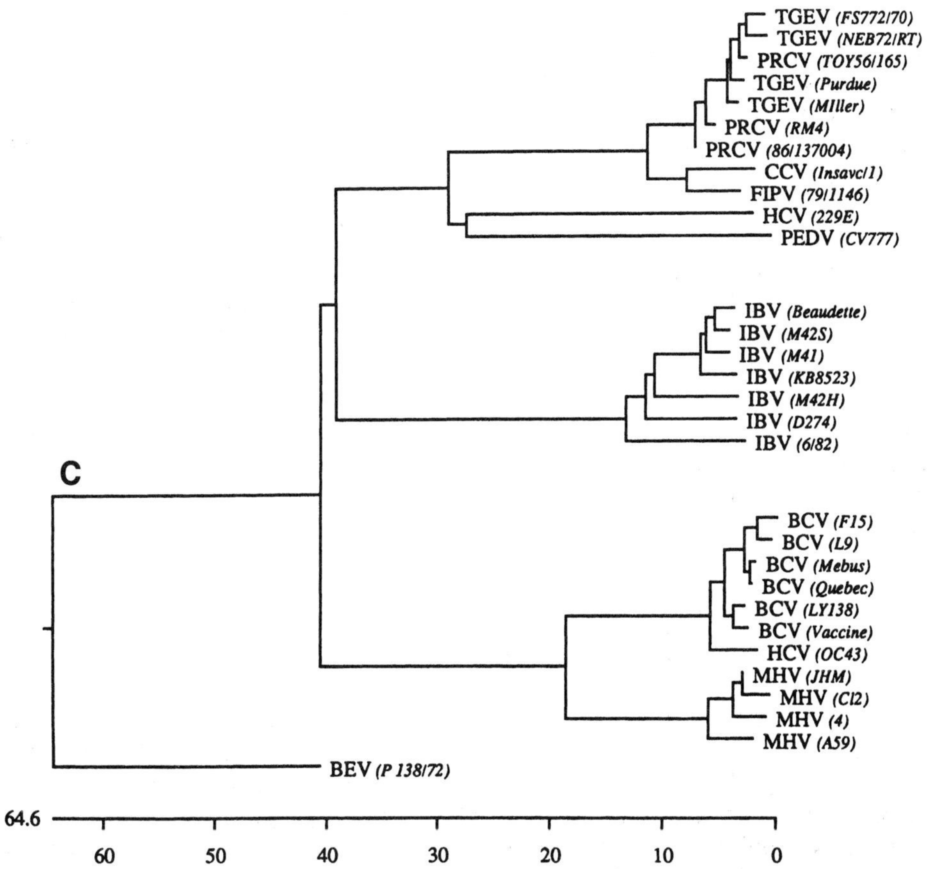


FIGURE 1. Phylogenetic relationships of the (A) nucleocapsid protein, (B) integral membrane protein, and (C) large-surface glycoprotein of coronaviruses. Amino acid sequences were aligned using the "clustal" method (Higgins and Sharp, 1989) and phylogenetic trees were constructed using the neighborhood-joining method (Saitou and Nei, 1987). The analyses were done using the MegAlign module of the Lasergene software suite (DNASTAR Ltd). The phylogenies are rooted assuming a biological clock.

## B. Sequence Relationships

The sequence analysis of a number of coronavirus genes, including the complete genomes of avian infectious bronchitis virus (IBV), murine hepatitis virus (MHV), and HCV 229E, provides a database for the analysis of possible evolutionary relationships between coronaviruses. This type of analysis is shown for the coronavirus large-surface glycoprotein, the integral membrane protein, and the nucleocapsid protein in Fig. 1. By and large, the nucleotide sequence data provide a striking confirmation of the conclusions derived from serological analysis. A closely related genetic cluster composed of HCV OC43, MHV, bovine coronavirus (BCV), and turkey coronavirus (TCV) is evident in the cladograms derived for all three proteins. The viruses comprising antigenic group 1 also form a genetic cluster. However, these viruses are clearly less closely related, and an evolutionary divergence of, in particular, the human coronavirus HCV 229E and the porcine coronavirus porcine epidemic diarrhea virus (PEDV) from the remainder of the group is suggested by the data. It should be remembered that the analyses presented here do not account for the possible role of recombination in the evolution of coronavirus genes (see Chapters 2 and 12).

As evidenced by immunofluorescence, seroneutralization, ELISA, and radioimmunoprecipitation, Berne virus and Breda virus are antigenically related to each other but there is no cross-reactivity with antisera specific for other animal viruses, including coronaviruses. The available data suggest that the cross-reactive torovirus antigen is predominantly the surface glycoprotein (Weiss and Horzinek, 1987). The evidence for antigenic cross-reactivity with human toroviruses is based mainly on immunoelectron microscopy (Beards *et al.*, 1986). At the present time, there is insufficient sequence data to evaluate the phylogeny of toroviruses. Limited but convincing sequence similarities in some of the gene products of toroviruses and coronaviruses (see Chapter 11) support the inclusion of these two genera in one family.

## III. GENOME ORGANIZATION

At the present time, the available data suggest that the genomes of coronaviruses and toroviruses contain between 6 and 11 functional ORFs (Spaan *et al.*, 1988) (Fig. 2). The easiest to define are those encoding the structural proteins, in particular, the large surface glycoprotein (S), the integral membrane protein (M), and the nucleocapsid protein (N), which have a specific size and/or location toward the 3′ end of the genome (for an explanation of nomenclature, see Table III). All coronaviruses contain a fourth structural protein ORF encoding the sM protein that is invariably located 5′ proximal to the M protein ORF. Additionally, the genomes of BEV and coronaviruses of the antigenic group 2 (e.g., HCV OC43, MHV, BCV) contain a fifth structural protein ORF, the hemagglutinin-esterase (HE) gene. The location of this ORF in coronaviruses is 5′ proximal to the S protein gene, and in BEV it lies between the M and N protein genes.

The HE gene of coronaviruses and BEV appears to encode a protein with an accessory function that is not required for replication in cultured cells. Thus, in some coronavirus isolates (e.g., MHV A59) and the single BEV isolate, mutation, possibly in cell culture, has resulted in the loss of a functional HE ORF and conversion of these sequences to pseudogenes (Yokomori *et al.*, 1991; Snijder and Horzinek, 1993).

In relation to the ORFs that encode nonstructural proteins, the coronavirus and torovirus genome is dominated by the RNA polymerase locus which consists of two large ORFs encompassing approximately 20,000 nucleotides toward the 5′ end of the genome. The remainder of the coronavirus nonstructural protein ORFs present a diverse and complex pattern both in their number, size, and arrangement. The reasons for the complexity are probably twofold. First, as

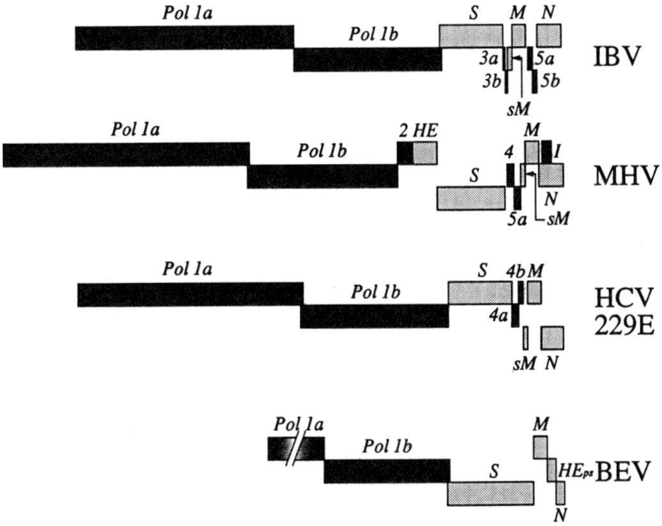


FIGURE 2. The genomic organization of a torovirus (BEV), a group 1 coronavirus (HCV 229E), a group 2 coronavirus (MHV), and a group 3 coronavirus (IBV). The genomic ORFs encoding structural proteins are lightly shaded; those encoding nonstructural proteins are filled. The ORFs are drawn to scale in the correct relative reading frames. ORF Pol 1a is defined as reading frame zero.

mentioned above, many coronavirus genes appear to encode proteins with accessory functions. Thus, under certain conditions (e.g., adaptation and propagation in cultured cells), mutations accumulate that lead to the inactivation and even deletion of these genes. Second, not only divergence from a common ancestor but also RNA recombination appear to be a major driving force in the evolution of coronavirus and torovirus genomes (see Chapter 12). This appears to have introduced a plasticity to the corona/torovirus genome that is exceptional, even among RNA viruses.

The genome maps shown in Fig. 2 should, therefore, be taken as an over-

TABLE III. *Coronaviridae* Structural Proteins[a]

| Gene product | Coronavirus | | Torovirus | |
|---|---|---|---|---|
| Nucleocapsid protein | **N** | **Nucleocapsid** | **N** | **Nucleocapsid** |
| Large-surface glycoprotein | **S** | **Surface** | **S** | **Surface** |
| | P | Peplomer | P | Peplomer |
| | E2 | Envelope 2 | | |
| Integral membrane protein | **M** | **Membrane** | **M** | **Membrane** |
| | E1 | Envelope 1 | E | Envelope |
| Small-membrane protein | **sM** | **Small membrane** | | — |
| Hemagglutinin-esterase glycoprotein | **HE** | **Hemagglutinin-esterase** | **HEps** | **Hemagglutinin-esterase (pseudogene)** |
| | E3 | Envelope 3 | ORF 4 product | |

[a]The nomenclature shown in bold type is recommended.

view, and isolates that display differences occur. The reader is also cautioned that the nomenclature of the ORFs encoding the nonstructural proteins of coronaviruses is not logical. In some cases, for example, the ORF 3a of canine coronavirus (CCV) and TGEV, the gene products are homologous, while others carrying the same designation, for example, the ORF 3a of IBV, are not.

## IV. MORPHOLOGY AND PHYSICOCHEMICAL PROPERTIES

Coronaviruses are described as pleiomorphic but roughly spherical enveloped particles, approximately 60 to 200 nm in diameter with a characteristic "fringe" of 20-nm-long surface projections. An "inner fringe" of short surface projections is sometimes seen on MHV, BCV, TCV, and HCV OC43 particles. Toroviruses are also pleiomorphic enveloped particles, although only 120 to 140 nm in diameter. Toroviruses are disk-, kidney-, or rod-shaped and are likewise decorated with 20-nm surface projections. The nucleocapsid of both coronaviruses and toroviruses has a helical symmetry, but differs remarkably in its morphology. Coronavirus nucleocapsids are extended while those of toroviruses are tubular. These morphological features and the assignment of struc-

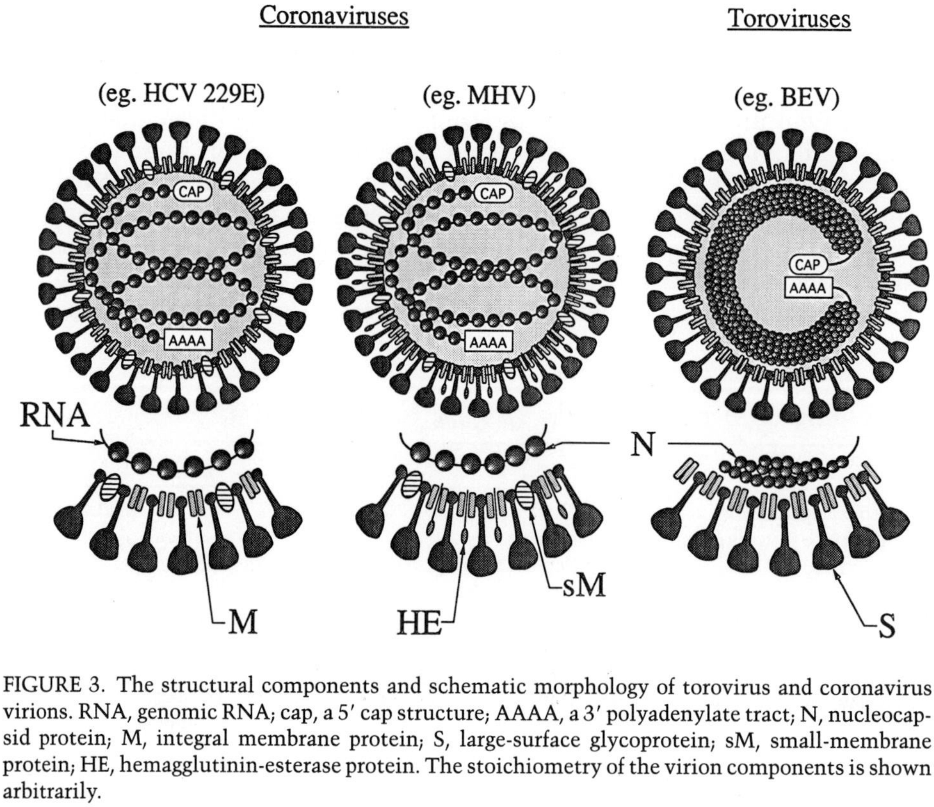


FIGURE 3. The structural components and schematic morphology of torovirus and coronavirus virions. RNA, genomic RNA; cap, a 5′ cap structure; AAAA, a 3′ polyadenylate tract; N, nucleocapsid protein; M, integral membrane protein; S, large-surface glycoprotein; sM, small-membrane protein; HE, hemagglutinin-esterase protein. The stoichiometry of the virion components is shown arbitrarily.

tural proteins to the virus particles are illustrated for a group 1 coronavirus, a group 2 coronavirus, and a torovirus in Fig. 3. In this diagram, the sM protein is depicted as a structural protein of group 2 coronaviruses, although this has not yet been shown experimentally.

Coronaviruses have an estimated molecular mass of $400 \times 10^6$ and a buoyant density in sucrose of 1.15 to 1.19 g/cm$^3$. The buoyant density in CsCl is 1.23 to 1.24 g/cm$^3$ and the sedimentation coefficient $s_{20,w}$ is 300 to 500. Coronaviruses are sensitive to heat, lipid solvents, nonionic detergents, formaldehyde, and oxidizing agents. Toroviruses have a buoyant density of 1.16 to 1.17 g/cm$^3$ in sucrose and an estimated sedimentation coefficient $s_{20,w}$ of 400 to 500. Virus infectivity is stable between pH 2.5 and 9.7 but rapidly inactivated by heat, organic solvents, and radiation. There is no inactivation of infectivity by phospholipase C or sodium deoxycholate.

In the last 15 to 20 years many articles that describe the structural, functional, and antigenic properties of the coronavirus and torovirus proteins have been published. However, it was only in 1990 that a standard nomenclature for the structural proteins of coronaviruses was introduced (Cavanagh *et al.*, 1990). This nomenclature has been adopted throughout this book but, to help the reader in the literature, a list of synonyms is presented in Table III. Unfortunately, there is, as yet, no standard nomenclature for the nonstructural proteins of coronaviruses or toroviruses. This is partly because there is insufficient information on the expression, processing, and function of these gene products, but also because of the aforementioned variability in the genomes of corona- and toroviruses. In this respect, the reader is advised to consider each virus species individually.

## V. CONCLUSION

The readers of this book will, I hope, appreciate that considerable progress has been made in elucidating many aspects of coronavirus and torovirus biology in a relatively short time. However, there remain many unanswered questions, particularly in areas such as the genesis of corona- and torovirus subgenomic mRNAs, the structure and function of the viral nonstructural proteins, and the pathogenesis of natural infections. I hope that reading this book may stimulate young scientists to address some of these questions.

## VI. REFERENCES

Beards, G. M., Hall, C., Green, J., Flewett, T. H., Lamouliatte, F., and Du Pasquier, P., 1984, An enveloped virus in stools of children and adults with gastroenteritis that resembles the Breda virus of calves, *Lancet* **2:**1050.

Beards, G. M., Brown, D. W. G., Green, J., and Flewett, T. H., 1986, Preliminary characterisation of torovirus-like particles of humans: Comparison with Berne virus of horses and Breda viruses of calves, *J. Med. Virol.* **20:**67.

Cavanagh, D., Brian, D. A., Enjuanes, L., Holmes, K. V., Lai, M. M. C., Laude, H., Siddell, S. G., Spaan, W. J. M., Taguchi, F., and Talbot, P. J., 1990, Recommendations of the coronavirus study group for the nomenclature of the structural proteins, mRNAs and genes of coronaviruses, *Virology* **176**:306.

Cavanagh, D., Brian, D. A., Brinton, M. A., Enjuanes, L., Holmes, K. V., Horzinek, M. C., Lai, M. M. C., Laude, H., Plagemann, P. W., Siddell, S. G., Spaan, W. J. M., Taguchi, F., and Talbot, P. J., 1995, *Coronaviridae*, in: *Virus Taxonomy, The Classification and Nomenclature of Viruses. Sixth Report of the ICTV* (D. H. L. Bishop *et al.*, eds.), Springer-Verlag, Wien, New York.

Dea, S., Verbeek, A. J., and Tijssen, P., 1990, Antigenic and genomic relationships among turkey and bovine enteric coronaviruses, *J. Virol.* **64**:3112.

Higgins, D. G., and Sharp, P. M., 1989, Fast and sensitive multiple sequence alignments on a microcomputer, *CABIOS* **5**:151.

Saitou, N., and Nei, M., 1987, A new method for reconstructing phylogenetic trees, *Mol. Biol. Evol.* **4**:406.

Sanchez, C. M., Jiminez, G., Laviada, M. D., Correa, I., Sune, C., Bullido, M. J., Gebauer, F., Smerdou, C., Callebaut, P., Escribano, J. M., and Enjuanes, L., 1990, Antigenic homology among coronaviruses related to transmissible gastroenteritis virus, *Virology* **174**:410.

Snijder, E., and Horzinek, M. C., 1993, Toroviruses: Replication, evolution and comparison with other members of the coronavirus-like superfamily, *J. Gen. Virol.* **74**:2305.

Spaan, W., Cavanagh, D., and Horzinek, M. C., 1988, Coronaviruses: Structure and genome expression, *J. Gen. Virol.* **69**:2939.

Spaan, W. J. M., Cavanagh, D., and Horzinek, M. C., 1990, Coronaviruses, in: *Immunochemistry of Viruses*, Vol. 2, *The Basis for Serodiagnosis and Vaccines* (M. H. V. Regenmortel and A. R. Neurath, eds.), pp. 359–379, Elsevier, Amsterdam.

Weiss, M., and Horzinek, M. C., 1987, The proposed family *Toroviridae*: Agents of enteric infections, *Arch. Virol.* **92**:1.

Yokomori, K., Banner, L. R., and Lai, M. M. C., 1991, Heterogeneity of gene expression of the haemagglutinin-esterase (HE) protein of murine coronaviruses, *Virology* **183**:647.

CHAPTER 2

# Coronavirus Replication, Transcription, and RNA Recombination

ROBBERT G. VAN DER MOST AND WILLY J. M. SPAAN

## I. INTRODUCTION

With the advent of recombinant DNA technology, our knowledge of the transcription and replication processes of positive-strand RNA viruses has increased profoundly. A major breakthrough in this field has been the development of full-length cDNA clones from which infectious RNA transcripts can be made. Such clones have now been constructed for members of the *Picorna-* (Racaniello and Baltimore, 1981), *Bromo-* (Ahlquist *et al.*, 1984), *Alpha-* (Rice *et al.*, 1987) and *Flaviviridae* (Lai *et al.*, 1991; Rice *et al.*, 1989). Unfortunately, the enormous length of the coronavirus genome has hampered the construction of a full-length cDNA clone so far.

In this chapter we will address new data that have been published in the field of coronavirus RNA replication, transcription, and recombination. In addition, we will discuss how cDNA clones of defective interfering (DI) RNAs may provide a promising alternative to study the replication and transcription of coronaviruses. Previous reviews that dealt with coronavirus transcription and replication appeared in 1988 (Spaan *et al.*, 1988) and 1990 (Lai, 1990).

ROBBERT G. VAN DER MOST AND WILLY J. M. SPAAN • Department of Virology, Institute of Medical Microbiology, Leiden University, 2300 AH Leiden, The Netherlands.

*The Coronaviridae*, edited by Stuart G. Siddell, Plenum Press, New York, 1995.

## II. REPLICATION

Coronaviruses are positive-strand RNA viruses. Thus, upon infection, the genomic RNA is translated to produce the RNA-dependent RNA polymerase. The RNA polymerase transcribes the viral genome into a complementary RNA. The negative-stranded RNA in turn serves as a template for the synthesis of genomic RNA. The viral and/or host enzymes involved in the RNA replication process of coronaviruses are unknown. The translation product(s) of the huge polymerase gene are proteolytically processed; however, the processing pathway is not well characterized (Baker *et al.*, 1989, 1993; Denison and Perlman, 1987; Denison *et al.*, 1991, 1992; Weiss *et al.*, 1994). By analogy with other positive-strand RNA viruses (Andino *et al.*, 1993; Lemm and Rice, 1993a,b), it is thought that processed end-products as well as precursor polypeptides and host factors may be involved in the replication process.

The RNA polymerase must recognize specific signals that control the initiation of plus- and minus-strand RNA synthesis. Mapping of these *cis*-acting signals is a prerequisite for our understanding of coronavirus replication. Because a full-length cDNA clone of a coronavirus is not available, DI RNAs are the tools of choice to study these signals. DI RNAs are replicated in virus-infected cells, and must therefore contain the essential *cis*-acting replication signals.

During the last few years several natural DI RNAs of murine hepatitis virus (MHV) strains JHM (Makino *et al.*, 1985, 1988a, 1990; Makino and Lai, 1989a) and A59 (van der Most *et al.*, 1991) have been characterized and used as templates to construct full-length cDNA clones. Synthetic DI RNAs can be transcribed *in vitro* from these clones and are replicated on transfection into MHV-infected cells. The structures of the JHM DIs DIssE and DIssF and the A59 DI MIDI are schematically shown in Fig. 1. All three DI RNAs contain the 5′ and 3′ terminal sequences of the viral genome. MIDI is a synthetic copy of the natural DI RNA DI-a and consists of three noncontiguous fragments of the viral genome, i.e., the 5′ terminal 3889 nucleotides (nt), 799 nt derived from the 3′ end of open reading frame 1b (ORF1b), and the 3′ terminal 806 nt (van der Most *et al.*, 1991). Both DIssF and MIDI contain sequences from the 3′ end of ORF1b, which are required for RNA packaging (Makino *et al.*, 1990; van der Most *et al.*, 1991). This packaging signal was mapped as a 61 nt sequence that could be folded into a bulged stem-loop structure (Fosmire *et al.*, 1992). It is not yet known whether this sequence alone is sufficient for RNA packaging.

The minimal sequence requirements for replication can be mapped by comparing the sequences of the three natural DI RNAs and by deletion mutagenesis. Deletion mapping of DIssE, DIssF, and MIDI has demonstrated that the 3′ replication signal comprises the 3′ terminal 447 nt (Kim *et al.*, 1993a; Lin and Lai, 1993; van der Most and Spaan, unpublished results) (Fig. 1). It has been suggested that a short conserved sequence element in the 3′ nontranslated region of the genomic RNA (5′-CGAAGAGC-3′) acts as a replication signal for minus-strand synthesis (Lai, 1990). Clearly, the results from the deletion mapping suggest that this sequence alone is not sufficient and that the signal(s) comprise a larger stretch of sequences. However, very recently Lin *et al.* have provided evidence that the 3′ terminal 55 nucleotides of the positive strand RNA plus the poly(A) tail comprise the cis-acting signal for the initiation of

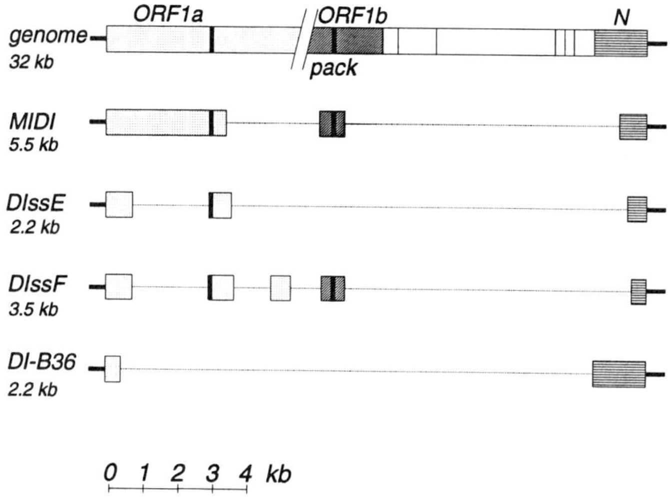


FIGURE 1. Schematic structure of the genomes of MIDI, DIssE, DIssF, and B36. The sequences derived from ORF1a, ORF1b, and the 3′ end of the genome (N) are indicated by different shadings. The putative enhancer sequence in ORF1a and the RNA packaging signal in ORF1b (pack) are indicated by black bars.

minus-strand synthesis (Lin *et al.*, 1994). This would imply that part of the 3′ terminal 447 nt that were identified as the 3′ replication signal, are in fact required for *positive strand* synthesis (Lin *et al.*, 1994). Thus, initiation of positive strand synthesis could involve interactions between 5′- and 3′-terminal signals.

Mapping of the 5′ replication signal has yielded conflicting results on two points. First, it is unclear which sequences at the very 5′ terminus of the genome are needed. An artificial MHV-A59 DI RNA, DI-B36, which consists of the 5′ terminal 467 nt and the 3′ terminal 1671 nt (Fig. 1), is replicated in MHV-A59-infected cells (Masters *et al.*, 1994). DI-B36 was originally constructed on the basis of a bovine coronavirus (BCV) DI RNA (Hofmann *et al.*, 1990; Masters *et al.*, 1994), which contains the 5′ terminal 498 nt and which is also replicated on transfection into BCV-infected cells (Brian *et al.*, 1994). These data suggest that the 5′ 467 nt comprise the 5′ replication signal. Accordingly, the 5′ terminal 474 nt were found to be sufficient for replication of DIssE (Kim *et al.*, 1993a). In contrast, the 5′ terminal 482 nt do not support replication of DIssF (Lin and Lai, 1993). Deletion of the sequence between positions 482 and 706 abrogates replication of DIssF (Lin and Lai, 1993), whereas deletion of the same fragment from DIssE only reduces replication efficiency slightly (Kim *et al.*, 1993a). Preliminary evidence from our laboratory suggests that for (efficient) replication of the MIDI-derivatives the 5′ 1.1 kb is essential (Luytjes, Gerritsma, and Spaan, unpublished results). It is possible that these differences reflect the different experimental protocols used. The deletion mapping experiments with DIssF and MIDI have involved *in vivo* transcription of DI RNA by a vaccinia virus-encoded T7 RNA polymerase (Fuerst *et al.*, 1986), in combination with DNA transfection (Lin and Lai, 1993; van der Most *et al.*, 1992). Replication of the DIssE mutants has been assessed in MHV-infected and RNA-transfected cells (Kim *et al.*, 1993a) and the analysis of DI-B36 has been done with a yet different protocol, which involves seeding of RNA-transfected L cells onto a monolayer of 17ClI cells (Masters *et al.*, 1994).

The second issue on which the current data conflict is the finding that replication of the JHM DIs DIssE and DIssF requires an internal sequence element of 135 nt (Kim *et al.*, 1993a; Lin and Lai, 1993) (Fig. 1). This sequence element is located 3 kb from the 5′ terminus of the genome and is contained in MIDI, but is not present in most of its derivatives [e.g., ΔH-*in* (de Groot *et al.*, 1992)], nor in DI-B36 (Masters *et al.*, 1994). These DI RNAs are, however, replicated, indicating that this presumptive signal is not an absolute requirement for MHV DI replication. It has been suggested that this signal functions as a replication enhancer (Kim *et al.*, 1993a), which would explain why RNA replication is readily detected in DIssE-transfected and virus-infected cells. Clearly, the precise sequence of the *cis*-acting signals, the extent to which these data can be extrapolated to the viral genome, and the role of the putative enhancer remain to be elucidated. The sequence elements that have been mapped comprise the 5′ and 3′ nontranslated regions of the genome, as well as parts of ORF1a and the N gene. To date, it is unclear which proteins interact with these replication signals. Several host factors have been shown to bind to the 5′ terminus of the plus-strand and to the 3′ terminus of the minus-strand RNA (Furuya and Lai, 1993), and it is possible that these proteins play a role in the initiation of RNA synthesis. Further characterization of the enzymology of replication will await the reconstruction of a functional polymerase gene or the development of an *in vitro* RNA replication system.

A peculiar feature of the MHV DI RNAs is that they contain large ORFs. For instance, the three fragments of DIssE as well as the three fragments of MIDI are joined in-frame (Makino *et al.*, 1988a; van der Most *et al.*, 1991). There is now substantial evidence that the presence of large ORFs in the genomes of MHV DI particles is biologically significant (de Groot *et al.*, 1992; Kim *et al.*, 1993b). For DIssE it appears that the ORF is not required in DI-transfected and MHV-infected cells, but is important after one passage (Kim *et al.*, 1993b). In principle, the requirement for a large ORF could be related to RNA stability (de Groot *et al.*, 1992), replication, packaging, or uncoating. The results obtained for DIssE may suggest that the large ORF is not required for RNA replication. However, these data have not yet been confirmed for MIDI and its derivatives (van der Most and Spaan, unpublished results). Furthermore, it is not known if these data can be extrapolated to the viral genomic RNA.

## III. TRANSCRIPTION

### A. Introduction

Coronavirus gene expression involves the synthesis of five to eight subgenomic mRNAs (Lai, 1990; Spaan *et al.*, 1988). These mRNAs are synthesized in different but constant molar ratios in the infected cell. As an illustration,

TABLE I. MHV Subgenomic mRNAs, Coding Assignments, and Relative Molarities

| MHV mRNA | Protein | Relative molarity (percent)[a] |
|---|---|---|
| mRNA1 (genome) | RNA-dependent RNA polymerase | 10.4 |
| mRNA2 | Nonstructural protein 2a | 4.7 |
| mRNA2-1 (not all strains) | HE glycoprotein | |
| mRNA3 | Spike glycoprotein | 4.2 |
| mRNA4 | Nonstructural protein | 6.5 |
| mRNA5 | ORF5a: nostructural protein | 13.1 |
| | ORF5b: small membrane protein | |
| mRNA6 | Membrane glycoprotein | 15.7 |
| mRNA7 | Nucleocapsid protein | 45.5 |

[a]From Jacobs *et al.* (1981).

Table I summarizes the MHV mRNAs, their translation products, and their molar ratios. The mRNAs form a 3′ coterminal nested set, i.e., they have identical 3′ ends but extend for different lengths in the 5′ direction (illustrated for MHV in Fig. 2). Only the 5′ most gene(s), which is not present in the next smaller mRNA, is translated (Lai, 1990; Spaan *et al.*, 1988) (see Chapter 3). On the genome, the transcription units are preceded by short sequence elements, termed *intergenic regions*. The complementary intergenic regions on the negative-stranded RNA are thought to function as promoters and will therefore be referred to as the "intergenic promoter sequences" or "promoters" in this chapter.

The subgenomic mRNAs not only form a 3′ coterminal nested set, but are also 5′ coterminal: all mRNAs contain a common 5′ leader sequence (Lai *et al.*, 1984; Spaan *et al.*, 1983) (Fig. 2). For different coronaviruses, the length of the leader ranges from 65 to 98 nt. The leader sequence is only found at the very 5′ terminus of the genome, which implies that the synthesis of subgenomic mRNAs involves fusion of noncontiguous sequences. To explain the synthesis of leader-containing subgenomic mRNAs, several models have been put forward. For instance, it has been suggested that the mRNAs arise via splicing of larger precursor RNAs or by "looping-out" of intervening sequences. The so-called leader-primed transcription model proposes that short leader RNAs are transcribed from the 3′ end of the genomic negative strand and act as primers for subgenomic mRNA synthesis (Lai *et al.*, 1984; Spaan *et al.*, 1983). Because in earlier experiments, only genome-length negative-strands were found in MHV-infected cells (Lai *et al.*, 1982), it seemed that these molecules served as the exclusive templates for transcription. However, the recent discovery of subgenomic negative-strands and double-stranded subgenomic RNAs in coronavirus-infected cells has provoked renewed discussion (Hofmann *et al.*, 1990; Sawicki and Sawicki, 1990; Sethna *et al.*, 1989): these findings imply that the subgenomic mRNAs could also be transcribed from their negative-stranded counterparts.

Much of our knowledge of coronavirus transcription has been obtained by

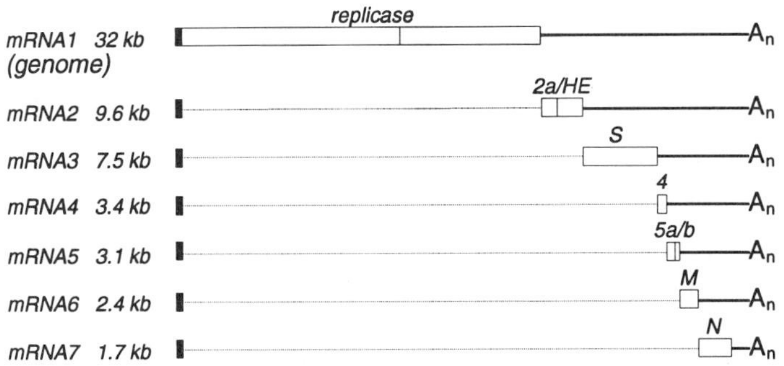


FIGURE 2. Coding assignments and the nested set structure of the MHV subgenomic mRNAs. The common 5′ leader sequence is indicated by a black box. The 5′ most unique ORF(s) of each mRNA are indicated by open boxes. The sequences downstream of the leader of the subgenomic mRNAs will be referred to as "body" sequences.

sequencing subgenomic mRNAs and intergenic regions and by analyzing negative-stranded RNAs. Furthermore, an experimental system has been set up based on synthetic DI RNAs carrying intergenic promoter sequences (Makino *et al.*, 1991; van der Most *et al.*, 1994). In the present situation, two basic questions can be asked. How are the leader sequences attached to the downstream "body" sequences and which are the templates for mRNA synthesis? The next sections will address the experimental evidence that has accumulated on these issues. We will focus on the discovery of the subgenomic negative strands and its consequences for the current coronavirus transcription models. We will also discuss the role and the sequence requirements of the intergenic regions and the role of basepairing in determining mRNA abundance.

## B. Negative-Stranded RNAs

The actual double-stranded structure that "produces" RNA is designated the replicative intermediate (RI). Visualization of double-stranded RNAs by gel electrophoresis and fluorography usually involves digestion of the excess single-stranded RNA by RNase A. The remaining double-stranded RNase-resistant molecules are referred to as the replicative forms (RFs). The RFs do not always correspond to the RIs because the RIs could also have been digested at certain RNase-sensitive sites to yield smaller RFs.

In 1989, Sethna and co-workers reported the discovery of genomic and subgenomic negative-stranded RNAs in transmissible gastroenteritis virus (TGEV)-infected cells. They identified the negative-stranded RNAs by using strand-specific oligonucleotide probes in Northern (RNA) blot analyses. This observation was confirmed for BCV (Hofmann *et al.*, 1990). It was subsequently demonstrated that the subgenomic negative strands contained an anti-leader sequence (Sethna *et al.*, 1991).

RNase digestion of cytoplasmic RNA isolated from TGEV-infected cells yielded a set of RFs corresponding to the subgenomic mRNAs (Sethna *et al.*,

1989). The RFs contained very small quantities of full-length positive-stranded RNAs, whereas full-length negative-stranded RNAs were abundant. This suggested that positive-stranded RNAs were being synthesized from a negative-strand template. However, these authors did not provide evidence that the subgenomic RFs did in fact correspond to subgenomic RIs. In 1990, Sawicki and Sawicki reported the existence of both genomic and subgenomic RFs in MHV-infected cells. They showed that each subgenomic RF was derived from a corresponding subgenomic RI. The RIs were transcriptionally active: following a very short pulse (2 min) with [$^3$H]uridine, the label was predominantly incorporated into the RIs and to a much lesser extent into the subgenomic RNAs. After longer pulse periods (up to 300 min), only labeled subgenomic mRNAs could be detected. This was interpreted as an indication that the label "flows through" the RIs into the single-stranded RNAs produced from them (Sawicki and Sawicki, 1990).

The discovery of subgenomic negative strands has led to two different models of how coronaviruses produce their mRNAs, which differ in the nature of the template(s) for mRNA synthesis. Sethna and co-workers speculated that the subgenomic mRNAs might in fact be amplified as independent replicons (Fig. 3A). The initial subgenomic mRNA templates for replication could either be present in the virion or could be synthesized from the antigenome via leader-primed transcription (Sethna *et al.*, 1989). This model implies that, at least early in infection, the genomic RI would yield subgenomic RFs upon RNase digestion (Sawicki and Sawicki, 1990). Sawicki and Sawicki could not detect such RFs and therefore proposed an alternative model. They suggested that subgenomic negative strands transcribed from the genome serve as templates for the synthesis of the corresponding mRNAs and not vice versa (Fig. 3B). To date, all attempts to obtain direct evidence for mRNA replication by transfecting synthetic subgenomic RNAs into coronavirus-infected cells failed to demonstrate any such replication (Brian *et al.*, 1994; Makino *et al.*, 1991; Luytjes and Spaan, unpublished results). Thus, several lines of evidence seem to suggest that the mRNAs may not undergo replication; instead, the hypothesis that the subgenomic mRNAs are transcribed from the subgenomic negative strands has gained some momentum (Sawicki and Sawicki, 1990; Hofmann *et al.*, 1993). Alternatively, it has been argued that the negative strands are synthesized from the subgenomic mRNAs as dead-end products (Jeong and Makino, 1992). It should be noted, however, that these conclusions are based on negative results. For instance, the failure to directly detect mRNA replication could also be explained by assuming that transfected subgenomic RNAs are, for some unknown reason, not suitable templates for replication.

Very recently, Schaad and Baric (1994) have provided strong genetic evidence indicating that, at least late in infection, the subgenomic negative strand RNAs are the functional templates for subgenomic mRNA synthesis. They studied a temperature-sensitive MHV-mutant, defective in negative strand synthesis, and found that the subgenomic RFs still incorporated radiolabel after a shift to the restrictive temperature. Because minus-strand synthesis was blocked, these results strongly suggest that the transcription indeed occurs on subgenomic negative strands (Schaad and Baric, 1994).

A

B

C

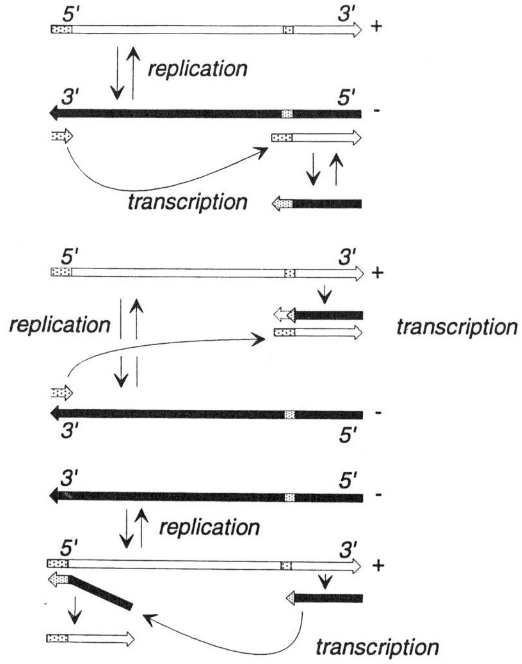


FIGURE 3. Schematic representation of three transcription models. Plus- and minus-strands are indicated by open and by black bars, respectively. On the plus-strands, the leader sequences and intergenic regions are indicated by shading; on the minus-strand, the intergenic promoter is shown by shading. (A) A genomic minus-strand is the template for subgenomic mRNA synthesis. The subgenomic mRNAs are first synthesized by leader-primed transcription and are then amplified. (B) Subgenomic minus-strands are synthesized by premature termination of transcription and serve as templates for leader-primed transcription. (C) A subgenomic minus-strand is synthesized by discontinuous transcription, i.e., the template jump occurs during minus-strand synthesis. The subgenomic minus-strand are the templates for subgenomic mRNA synthesis.

## C. Leader-Primed Transcription

The experiments described above do not address the question of how the leader sequences are attached to the body sequences. According to the original model of leader-primed transcription, short leader RNAs are transcribed from the 3′ end of the genomic negative strand, translocated to the intergenic promoters on the negative strand(s), and then extended (Lai, 1990; Spaan *et al.*, 1988). A key observation that has been interpreted in favor of this model is that the leader contains a complementary promoter sequence (CPS) near its undetermined 3′ end. This would allow basepairing to occur between the 3′ end of the leader and the promoter, thus facilitating the priming event (Budzilowicz *et al.*, 1985; Spaan *et al.*, 1983).

Before the discovery of subgenomic negative strands, alternative transcription models, e.g., *cis*-splicing of genome-length precursor RNAs, could be excluded on the basis of the UV inactivation kinetics of mRNAs (Jacobs *et al.*,

1981; Stern and Sefton, 1982). UV inactivation studies demonstrated that, late in infection, the UV target size of each mRNA corresponds to its physical size. This indicated that transcription of the mRNAs was independent of the synthesis of a genome length precursor. Based on the conclusion that subgenomic negative strands did not exist (Lai *et al.*, 1982), these data provided strong support for the leader-primed transcription model. The discovery of the subgenomic negative strands radically changed the picture, because alternative models could no longer be ruled out. For instance, any model in which mRNAs are transcribed from subgenomic templates is consistent with the UV transcription mapping.

One variant leader-primed transcription model, based on the existence of transcriptionally active subgenomic RIs, is that a nested set of subgenomic negative strands is synthesized first (Fig. 3b). The intergenic promoter sequences act as transcription attenuators in this model. These RNAs then serve as templates for leader-primed transcription (Sawicki and Sawicki, 1990). Thus, leader priming and transcription from subgenomic negative strands are not mutually exclusive. However, this model still assumes the existence of free leader RNAs. There are some indications that free leader transcripts do indeed exist. Baric and co-workers (1985, 1987) detected short RNA molecules containing leader sequences in MHV-infected cells. The biological significance of the small RNAs is not clear, though. Baker and Lai (1990) subsequently demonstrated that synthetic leader transcripts are incorporated into subgenomic mRNAs when added to an *in vitro* transcription system. An important observation was that the CPS at the 3′ of the leader was essential. When the CPS was deleted, the exogenous leaders were not incorporated. Although these data are suggestive, it cannot completely be ruled out that the synthetic leader RNAs did not actually prime transcription but merely functioned as templates for RNA recombination (Liao and Lai, 1992). The high-frequency leader reassortment that occurs during mixed infection of MHV strains has also been interpreted in support of the leader-primed transcription model (Jeong and Makino, 1994; Makino *et al.*, 1986a; Makino and Lai, 1989a; Zhang *et al.*, 1991). However, leader reassortment does not prove leader priming, but rather indicates that there is at least one step *in trans* in mRNA synthesis. If, for instance, the discontinuous transcription step occurs during negative-strand synthesis (and not by leader priming) (Fig. 3c), one would still observe "leader reassortment" when the (reassorted) negative strands are transcribed to produce mRNAs. Interestingly, leader reassortment appears to depend on a 9 nt sequence directly downstream of the leader; deletion of this sequence results in an exclusive use of the helper virus leader, which is provided *in trans* (Makino and Lai, 1989a; Zhang *et al.*, 1994). The function of the 9 nt sequence is unknown, but it has been proposed that it functions as a transcriptional terminator, controlling the synthesis of free leader transcripts (Zhang *et al.*, 1994).

As discussed in Section B, there is now genetic evidence strongly suggesting that subgenomic minus-strands are the functional templates for subgenomic mRNA synthesis (Schaad and Baric, 1994). Indeed, one experimental approach to determine whether leader-primed transcription occurs on subgenomic negative strands would be to transfect subgenomic negative strands and determine whether subgenomic mRNAs are transcribed from these templates.

## D. The Intergenic Promoter Sequence

The intergenic regions are short sequence elements upstream of the transcription units. Because the leader–mRNA junction occurs within the intergenic region, this sequence or its minus-sense counterpart, the intergenic promoter sequence, is considered to be crucial for mRNA synthesis. The potential basepairing between the leader CPS and the promoter is one of the cornerstones of the leader-primed transcription model. The intergenic promoter sequences for different genes in one coronavirus may differ slightly in sequence and in the extent of potential basepairing with the leader CPS. For MHV, the extent of potential basepairing ranges from 9 to 18 basepairs; every promoter contains the sequence 3′ UUAGAUUUG 5′, or a closely related sequence in the case of the promoters for mRNA2, mRNA2-1, and mRNA6 (Table II). The consensus sequences of different coronaviruses are quite similar, though different in length (Table III). Avian infectious bronchitis virus (IBV) has the most divergent sequence, whereas the consensus sequences of the coronaviruses belonging to clusters I and II (Chapter 3) share homology.

Until recently, many basic features of transcription remained unclear due to the lack of an appropriate experimental system. Such issues included the minimal sequence requirements for transcription initiation, the role of basepairing in promoter recognition, and the position(s) at which transcription starts. An important development has been the use of synthetic MHV DI RNAs containing additional intergenic promoter sequences (Makino *et al.*, 1991; van der Most *et al.*, 1994) (Fig. 4). Makino and co-workers (1991) have inserted the mRNA7 promoter and 0.54 kilobase (kb) of its flanking sequences into the genome of DIssF. Transfection of this DI RNA into MHV-infected cells resulted in the synthesis of the DI-derived subgenomic RNA. We have cloned short oligonucleotides (10–18 nt), comprising only the intergenic promoter sequences

TABLE II. Sequences of MHV Intergenic Promoter Sequences

| Promoter | | Sequence (on negative strand) |
|---|---|---|
| mRNA2 | | 3′-UUAGAAUAG-5′ |
| mRNA2-1 | A59: | 3′-AUUA– –UUCGAA-5′ |
| | JHM: | 3′-AUUA– –UUUGAA-5′ |
| mRNA3 | | 3′-AUUAGAUUUG-5′ |
| mRNA4 | | 3′-UUAGAUUUG-5′ |
| mRNA5 | | 3′-GAUUAGAUUUG-5′ |
| mRNA6 | | 3′-UAGAUUAGGUUUGA-5′ |
| mRNA7 | | 3′-UUAGAUUAGAUUUGAAAU-5′ |
| Leader (+ strand) | | 5′-UGUAGUUUAAAUCUAAUCUAAACUUUAUAAA-3′ |

TABLE III. Consensus Sequences for Several Coronaviruses from the Three Clusters[a]

| Virus | Antigen cluster | Consensus sequence |
|---|---|---|
| MHV-A59 | I | 3′-UUAG (A/G)U(U/A)UG-5′ |
| BCV | I | 3′-AG(A/G)UUUG-5′ |
| FIPV | II | 3′-UUGAUUUG-5′ |
| TGEV | II | 3′-UUGAUUUG-5′ |
| IBV | III | 3′-GAAUUGUU-5′ |

[a]From Spaan *et al.* (1988).

and no flanking sequences, and found that these sequences alone suffice to direct subgenomic DI RNA synthesis (van der Most *et al.*, 1994).

In this system, promoter strength has been defined as the ratio between the amounts of subgenomic and genomic DI RNAs synthesized in the cell. Insertion of the mRNA7 promoter with 0.54 kb of flanking sequences into DIssF gave a promoter strength of 0.8. Extension or deletion of the non-basepaired flanking sequences did not affect promoter strength (Makino and Joo, 1993). Deletions within the intergenic region reduced promoter strength: efficient

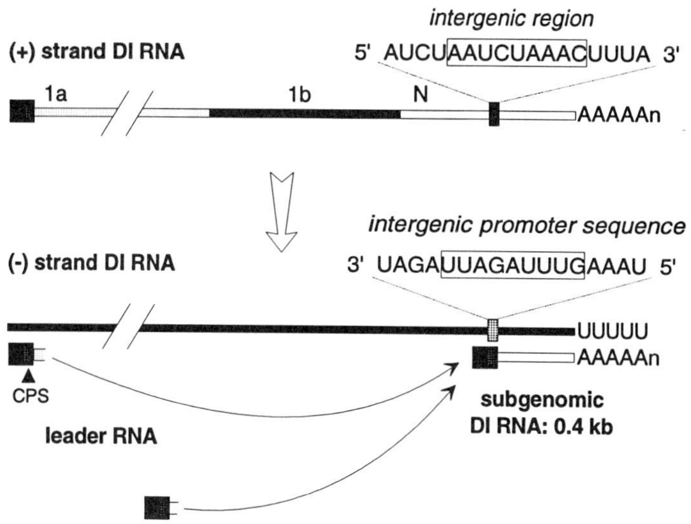


FIGURE 4. Schematic representation of the synthesis of a subgenomic RNA from a DI template. The sequences of the DI that are derived from ORF1a (1a), ORF1b (1b), and the 3′ end of the genome (N) are indicated by different shadings. The DI minus-strand is indicated by a black bar. A 17-nucleotide promoter sequence is inserted downstream of the N ORF and is indicated as a dotted box on both the plus- and minus-strands. The complementary promoter sequence (CPS) in the leader is indicated. The 3′ UUAGAUUUG 5′ consensus sequence and its plus-sense counterpart are boxed. Also indicated are the leader RNAs that originate either from the DI or from the viral genome.

subgenomic RNA synthesis requires a promoter larger than the partial consensus sequence 3′ AGAUUUG 5′ (Joo and Makino, 1992; Makino and Joo, 1993; Makino *et al.*, 1991; van der Most *et al.*, 1994). Thus, the intergenic promoter sequence is required for transcription, whereas the flanking sequences are not important.

## E. The Control of Subgenomic mRNA Abundance

The coronavirus subgenomic mRNAs are synthesized in different quantities. For instance, in MHV-infected cells, mRNA7 is synthesized in much greater quantities than mRNA3 (Table I). An important question is whether the determinants that control mRNA abundance reside in the intergenic promoter sequence. Both the sequence of the promoter and the extent of potential basepairing with the leader CPS could determine mRNA levels. For MHV there is a rough correlation between mRNA abundance and the length of the potential duplex. Based on this correlation, it has been proposed that the extent of basepairing controls mRNA abundance (Shieh *et al.*, 1987). However, no such correlation exists for IBV (Konings *et al.*, 1988) and BCV (Hofmann *et al.*, 1993). The relation between the extent of basepairing and mRNA abundance has been tested in our laboratory by inserting the 10 nt mRNA3 and 18 nt mRNA7 promoters into the MIDI genome (van der Most *et al.*, 1994). The "basepairing hypothesis" would predict that the mRNA7 promoter would be stronger than the mRNA3 promoter in this system. On the contrary, we found that the mRNA3 and mRNA7 promoters were of similar strength. In fact, promoter activity was even slightly higher for the mRNA3 promoter. Thus, these data clearly demonstrate that the extent of basepairing does not control mRNA levels.

The role of the specific nucleotide sequence has been studied by mutating single nucleotides in different promoters. Introduction of single substitutions into a 13 nt promoter demonstrated that the sequence of the intergenic promoter can be very flexible (Joo and Makino, 1992). Results from our laboratory show that promoter strength is affected only slightly when a single nucleotide in a 17 nt promoter or in the 18 nt mRNA7 promoter is mutated (van der Most *et al.*, 1994). In contrast, introduction of the same substitutions into the 10 nt mRNA3 promoter results in a more than tenfold reduction of promoter activity. This could suggest that transcription initiation requires a duplex of a certain minimal stability. Once this condition is met, extending the basepairing does not increase promoter strength. However, some of the substitutions in the 17 or 18 nt promoters did reduce promoter strength, without interrupting the leader–promoter duplex because G-U basepairs could be formed. This could indicate that recognition of the promoter sequence does not depend on basepairing only. Instead, direct interactions between the polymerase and the promoter could play a role in promoter recognition. Hofmann *et al.* (1993) came to the same conclusion based on their finding that the leader–body junction of one BCV mRNA had occurred on the unusual sequence 3′-CCAUCUG-5′ instead of the expected sequence 3′-AGGUUUG-5′.

The fact that some mutations reduce promoter strength opens the possibility that the levels of certain mRNAs are downregulated because their promoters contain point mutations (e.g., the MHV mRNA6 promoter) (Table II). However, this does not seem to be a general principle. Coronavirus mRNAs are, in general, synthesized at a rate inversely related to their length (Hofmann *et al.*, 1993; Konings *et al.*, 1988; Sethna *et al.*, 1989). This is similar to observations made by French and Ahlquist (1988) who, upon introduction of multiple promoter elements into a synthetic RNA3 of brome mosaic virus, found a gradient of accumulation favoring the smaller mRNAs. Thus, for coronaviruses the transcriptional activity of a promoter may also be (in part) determined by its position on the negative-stranded template. This could be accomplished by premature termination during plus-strand synthesis, i.e., larger RNA molecules are more prone to premature termination of transcription (Konings *et al.*, 1988). Alternatively, premature termination could occur during negative-strand synthesis (Sawicki and Sawicki, 1990). This latter idea is consistent with the transcription model in which a nested set of subgenomic minus strands is synthesized first: larger negative strands are produced in lower quantities because they encounter more transcription attenuators (see Section C) during their synthesis. However, any such model should take into account that, for feline infectious peritonitis virus (FIPV) and TGEV, the shortest mRNAs (RNAs 7 and 6, respectively) are produced in much lower quantities than the next larger mRNA encoding the nucleocapsid protein (de Groot *et al.*, 1987; Sethna *et al.*, 1989). Direct experimental evidence that mRNA abundance is influenced by the presence of additional up- or downstream promoters is not yet available.

## F. Heterogeneity of Subgenomic mRNAs

For MHV, every intergenic promoter contains the sequence 3′ UUAGAUUUG 5′, or a closely related sequence (Table II). The MHV-A59 leader contains two repeats of the 5′ UCUAA 3′ sequence (Makino *et al.*, 1988b). As mentioned above, the fusion between leader and body sequences occurs within the leader–promoter duplex. For most promoters, a perfect duplex is formed; it is therefore impossible to map the site(s) at which transcription starts. However, some promoters from MHV (e.g., the mRNA6 promoter) and BCV contain mutations, and these mutations are transcribed into the mRNAs (Armstrong *et al.*, 1984; Hofmann *et al.*, 1993). This indicates that transcription started upstream of the mutations. The same phenomenon was observed in an *in vitro* transcription system (Baker and Lai, 1990). To explain this, Baker and Lai (1990) proposed that 3′ terminal sequences of the leader RNA that do not basepair are trimmed and that a hypothetical nuclease recognizes mismatches within the duplex. By using the substitutions in the DI promoters as markers, it was subsequently found that not all mutations were transcribed into the subgenomic mRNAs (Joo and Makino, 1992; van der Most *et al.*, 1994). Mutations at the 5′ end of the promoter were always transcribed, but mutations near the 3′ end gave rise to two populations of mRNAs: mRNAs that had transcribed the mutation and mRNAs that contained the leader sequence (Fig. 5). This sug-

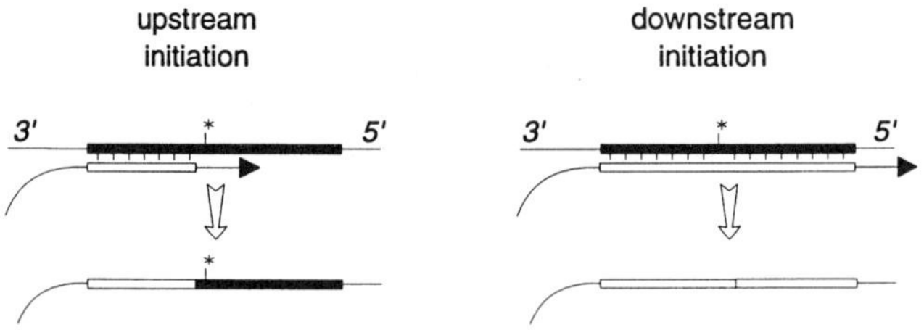


FIGURE 5. The use of markers to map the position(s) of transcription reinitiation. The marker is indicated by an asterisk. Open bars depict leader sequences. Black bars represent the negative-stranded template and the sequences that are transcribed from the template.

gested that transcription initiates at multiple—possibly random—sites, presumably resulting from trimming of the leader; mismatch-specific repair does not occur (van der Most *et al.*, 1994).

A different type of heterogeneity has been described for MHV-A59 mRNA6 and MHV–JHM mRNAs 6 and 7 (Armstrong *et al.*, 1984; Baker and Lai, 1990; Makino *et al.*, 1988b; Shieh *et al.*, 1987). For these mRNAs, the number of UCUAA repeats at the leader–body junction is variable. This was also observed for the mRNA7 promoter that had been inserted into MIDI (van der Most *et al.*, 1994). In this system, the leader, which was derived from MHV-A59, contains two pentanucleotide repeats. Two populations of mRNAs were detected: the majority of mRNAs contained two repeats, whereas a minor population had three repeats (Fig. 6). The heterogeneity has been explained by alternative basepairing between these promoters and the leader CPS (Makino *et al.*, 1988b). However, as exemplified in Fig. 6, these promoters contain two (imperfect) repeats of the promoter consensus sequence. Each of these individual "promoter domains" could give rise to a transcript, and these transcripts would differ in the number of repeats. This notion was supported by analyzing mutant promoters in which the ratio between the two transcripts was altered (van der Most *et al.*, 1994).

The number of pentanucleotide repeats in the leader CPS differs for the various MHV strains, e.g., MHV-A59 has two repeats in the leader, whereas the MHV-2c leader contains four repeats (Makino *et al.*, 1988b). During passage of MHV–JHM, the number of pentanucleotide repeats changes from three to two (Makino and Lai, 1989b). The differences in the number of UCUAA repeats of the JHM leader affect the transcription of certain subgenomic mRNAs. For example, transcriptional activity of the mRNA2-1 promoter of MHV–JHM (which encodes the hemagglutinin esterase gene) depends on the number of UCUAA repeats in the leader CPS: mRNA2-1 is only synthesized when the leader CPS contains two UCUAA repeats and not when three repeats are present (Shieh *et al.*, 1989). Additional examples of differential transcription have been described recently (La Monica *et al.*, 1992). There is as yet no explanation for the differential activity of these promoters. This phenomenon has been exploited to demonstrate that the transcription mechanism acts *in trans*. The mRNA2-1 promoter was inserted into a DI RNA with a three-repeat leader; this

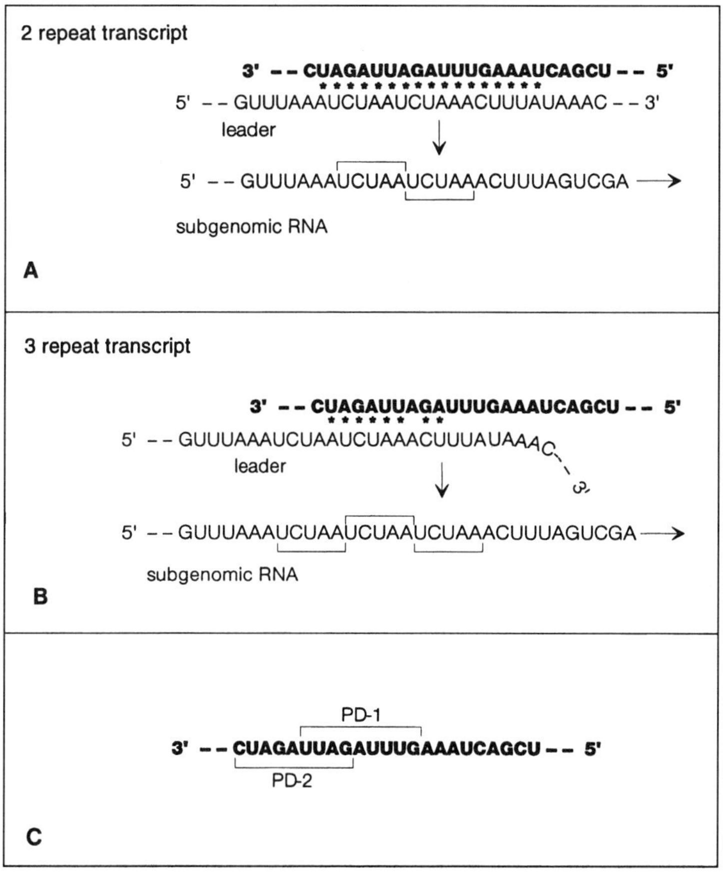


FIGURE 6. Heterogeneity at the leader–body fusion sites of subgenomic RNAs, illustrated for the 17 nt promoter inserted in MIDI. The structure of the DI genome is shown in Fig. 3. The UCUAA repeats in the leader are indicated. Basepairing is indicated by asterisks. (A,B) Synthesis of subgenomic RNAs containing two or three UCUAA repeats. (C) Schematic view of the overlapping promoter domains PD-1 and PD-2.

leader does not support transcription from the inserted promoter. When the helper virus contained a two-repeat leader, a subgenomic DI RNA was synthesized. Because RNA recombination could be excluded, these data indicate that the transcription process occurs *in trans* (Jeong and Makino, 1994).

## IV. RNA RECOMBINATION

Homologous RNA recombination has been demonstrated for picornaviruses (Cooper, 1968; King *et al.*, 1982), bromoviruses (Allison *et al.*, 1990; Bujarski and Kaesberg, 1986) and coronaviruses (Lai *et al.*, 1985). During a mixed infection of two MHV strains, RNA recombination occurs at a high frequency (Lai *et al.*, 1985; Makino *et al.*, 1986b). High-frequency RNA recombination does not only occur in tissue culture but also in infected mouse brains (Keck *et al.*, 1988a). Based on the analysis of a set of temperature sensitive (*ts*) mutants, the frequency of recombination for the entire MHV genome has been estimated to be 25% (Baric *et al.*, 1990). Recombinant murine coronaviruses have been isolated by employing different selection schemes, such as rescue of *ts* lesions (Baric *et al.*, 1990; Lai *et al.*, 1985; Makino *et al.*, 1986b), resistance to neutralizing monoclonal antibodies (Makino *et al.*, 1987), and rescue of a fusion-negative MHV variant (Keck *et al.*, 1988b). Many of these recombinant viruses had undergone additional recombination events outside the selection markers, again illustrating the high frequency of recombination (reviewed in Lai, 1992). At present, it is not known how widespread homologous RNA recombination is among other coronaviruses. The only other coronavirus for which recombination *in vivo* has been observed is IBV (Kusters *et al.*, 1990).

Although coronavirus RNA recombination was first thought to take place at certain "hot spots" (Banner *et al.*, 1990), it was later recognized that recombination is almost random and that these previously identified hot spots resulted from selection of recombinant viruses (Banner and Lai, 1991). Coronavirus RNA recombination most likely occurs via template switching: the polymerase and the nascent RNA strand dissociate from the original template and anneal to a different viral RNA molecule after which RNA synthesis continues. Experimental evidence for this copy-choice mechanism has only been obtained for poliovirus recombination (Kirkegaard and Baltimore, 1986), but it is assumed that this mechanism accounts for all examples of homologous RNA recombination (reviewed in Jarvis and Kirkegaard, 1991; Lai, 1992).

The high-frequency RNA recombination that occurs in MHV-infected cells has been exploited to introduce mutations into the viral genome (Koetzner *et al.*, 1992; Masters *et al.*, 1994; van der Most *et al.*, 1992). Recombination between the viral genome and co-replicating synthetic DI RNAs yielded recombinant viruses carrying DI sequences. Recombinant viruses were isolated in the absence of any selection or by rescuing a MHV-A59 *ts* mutant, Albany-4 (van der Most *et al.*, 1992). The *ts* phenotype of Albany-4 results from a deletion in the N gene which is repaired by recombination with DI-borne N sequences. The synthetic RNAs need not be replicating: Albany-4 was also rescued by using a synthetic mRNA7 transcript (Koetzner *et al.*, 1992), which is most probably not amplified on transfection (Brian *et al.*, 1994; Makino *et al.*, 1991; Luytjes and Spaan, unpublished results). However, when a replicating RNA was used, the frequency with which recombinants are isolated is more than 100-fold increased (Masters *et al.*, 1994).

The copy-choice model of recombination proposes that transcription initiates on one template (the "donor" template) and jumps to another template (the "acceptor" template) after which RNA synthesis proceeds (Jarvis and Kirkegaard, 1992). The number of recombinants that are generated depends on the concentrations of both parental RNAs: the concentration of donor template determines the number of initiation events, whereas the frequency of template switching depends on the concentration of acceptor template. This explains why the recombination frequency was increased more than 100-fold when a replicating DI RNA was used instead of the mRNA7 transcript (Masters *et al.*, 1994).

As long as a full-length "infectious" cDNA clone is not available, the only way to introduce genetic changes into a coronavirus genome will be through homologous RNA recombination. A more general applicability of this approach depends on several factors. It will be important to optimize transfection protocols and to investigate which sequences can be cloned into a DI RNA without destroying its capacity to be replicated. Recent results from our laboratory show that a large part of the MHV S gene can be inserted into a DI genome, although the resulting DI RNA replicates less efficiently than the parental DI RNA. One potential limitation of this approach is that introduction of mutations into more internal regions of the genome would require double recombination events. So far, mutations have only been introduced at the 5′ and 3′ regions of the genome. Also, in the case of mutations that decrease viral fitness, screening for recombinant viruses will be tedious. Presumably, such difficulties may be solved by developing efficient screening protocols and by applying selection, e.g., via rescue of *ts* lesions or by using monoclonal antibodies. Studies to address these issues are undoubtedly in progress.

ACKNOWLEDGMENTS. The authors gratefully acknowledge Dr. Raoul de Groot for stimulating discussions and Dr. Stanley Sawicki for his advice with regard to the role of negative-strand RNAs in transcription.

## V. REFERENCES

Ahlquist, P., French, R., Janda, M., and Loesch-Fries, L. S., 1984, Multicomponent RNA plant virus infection derived from cloned viral cDNA, *Proc. Natl. Acad. Sci. USA* **81:**7066.

Allison, R., Thompson, C., and Ahlquist, P., 1990, Regeneration of a functional RNA virus genome by recombination between deletion mutants and requirement for cowpea chlorotic mottle virus 3a and coat proteins for systemic infection, *Proc. Natl. Acad. Sci. USA* **87:**1820.

Andino, R., Rieckhof, G. E., Achacoso, P. L., and Baltimore, D., 1993, Poliovirus RNA synthesis utilizes an RNP complex formed around the 5′-end of viral RNA, *EMBO J.* **12:**3587.

Armstrong, J., Niemann, H., Smeekens, S., Rottier, P., and Warren, G., 1984, Sequence and topology of a model intracellular membrane protein, E1 glycoprotein, from a coronavirus, *Nature* **308:**751.

Baker, S. C., and Lai, M. M., 1990, An *in vitro* system for the leader-primed transcription of coronavirus mRNAs, *EMBO J.* **9:**4173.

Baker, S. C., Shieh, C. K., Soe, L. H., Chang, M. F., Vannier, D. M., and Lai, M. M., 1989, Identification of a domain required for autoproteolytic cleavage of murine coronavirus gene A polyprotein, *J. Virol.* **63:**3693.

Baker, S. C., Yokomori, K., Dong, S., Carlisle, R., Gorbalenya, A. E., Koonin, E. V., and Lai, M. M. C., 1993, Identification of the catalytic sites of a papain-like cysteine proteinase of murine coronavirus, *J. Virol.* **67:**6056.

Banner, L. R., and Lai, M. M., 1991, Random nature of coronavirus RNA recombination in the absence of selection pressure, *Virology* **185:**441.

Banner, L. R., Keck, J. G., and Lai, M. M. C., 1990, A clustering of RNA recombination sites adjacent to a hypervariable region of the peplomer gene of murine coronavirus, *Virology* **175:**548.

Baric, R. S., Stohlman, S. A., Razavi, M. K., and Lai, M. M., 1985, Characterization of leader-related small RNAs in coronavirus-infected cells: Further evidence for leader-primed mechanism of transcription, *Virus. Res.* **3**:19.

Baric, R. S., Shieh, C. K., Stohlman, S. A., and Lai, M. M., 1987, Analysis of intracellular small RNAs of mouse hepatitis virus: evidence for discontinuous transcription, *Virology* **156**:342.

Baric, R. S., Fu, K., Schaad, M. C., and Stohlman, S. A., 1990, Establishing a genetic recombination map for murine coronavirus strain A59 complementation groups, *Virology* **177**:646.

Brian, D. A., Chang, R.-Y., Hofmann, M. A., and Sethna, P. B., 1994, Role of subgenomic minus-strand RNA in coronavirus replication, *Arch. Virol.* (Suppl.) **9**:173.

Budzilowicz, C. J., Wilczynski, S. P., and Weiss, S. R., 1985, Three intergenic regions of coronavirus mouse hepatitis virus strain A59 genome RNA contain a common nucleotide sequence that is homologous to the 3′ end of the viral mRNA leader sequence, *J. Virol.* **53**:834.

Bujarski, J. J., and Kaesberg, P., 1986, Genetic recombination in a multipartite plant virus, *Nature* **321**:528.

Cooper, P. D., 1968, A genetic map of poliovirus temperature-sensitive mutants, *Virology* **35**:584.

de Groot, R. J., ter Haar, R. J., Horzinek, M. C., and van der Zeijst, B. A. M., 1987, Intracellular RNAs of the feline infectious peritonitis coronavirus strain 79-1146, *J. Gen. Virol.* **68**:995.

de Groot, R. J., van der Most, R. G., and Spaan, W. J. M., 1992, The fitness of defective interfering murine coronavirus DI-a and its derivatives is decreased by nonsense and frameshift mutations, *J. Virol.* **66**:5898.

Denison, M., and Perlman, S., 1987, Identification of putative polymerase gene product in cells infected with murine coronavirus A59, *Virology* **157**:565.

Denison, M. R., Zoltick, P. W., Leibowitz, J. L., Pachuk, C. J., and Weiss, S. R., 1991, Identification of polypeptides encoded in open reading frame-1b of the putative polymerase gene of the murine coronavirus mouse hepatitis virus-a59, *J. Virol.* **65**:3076.

Denison, M. R., Zoltick, P. W., Hughes, S. A., Giangreco, B., Olson, A. L., Perlman, S., Leibowitz, J. L., and Weiss, S. R., 1992, Intracellular processing of the N-terminal ORF 1a proteins of the coronavirus MHV-A59 requires multiple proteolytic events, *Virology* **189**:274.

Fosmire, J. A., Hwang, K., and Makino, S., 1992, Identification and characterization of a coronavirus packaging signal, *J. Virol.* **66**:3522.

French, R., and Ahlquist, P., 1988, Characterization and engineering of sequences controlling *in vitro* synthesis of brome mosaic virus subgenomic RNA, *J. Virol.* **62**:2411.

Fuerst, T. R., Niles, E. G., Studier, F. W., and Moss, B., 1986, Eukaryotic transient-expression system based on recombinant vaccinia virus that synthesizes bacteriophage T7 RNA polymerase, *Proc. Natl. Acad. Sci. USA* **83**:8122.

Furuya, T., and Lai, M. M. C., 1993, Three different cellular proteins bind to complementary sites of the 5′-end-positive and 3′-end-negative strands of mouse hepatitis virus RNA, *J. Virol.* **67**:7215.

Hofmann, M. A., Sethna, P. B., and Brian, D. A., 1990, Bovine coronavirus mRNA replication continues throughout persistent infection in cell culture, *J. Virol.* **64**:4108.

Hofmann, M. A., Chang, R.-Y., Ku, S., and Brian, D. A., 1993, Leader–mRNA junction sequences are unique for each subgenomic mRNA species in the bovine coronavirus and remain so throughout persistent infection, *Virology* **196**:163.

Jacobs, L., Spaan, W. J. M., Horzinek, M. C., and van der Zeijst, B. A. M., 1981, Synthesis of subgenomic mRNAs of mouse hepatitis virus is initiated independently: Evidence from UV transcription mapping, *J. Virol.* **39**:401.

Jarvis, T. C., and Kirkegaard, K., 1991, The polymerase in its labyrinth: Mechanisms and implications of RNA recombination, *Trends Genet.* **7**(6):186.

Jarvis, T. C., and Kirkegaard, K., 1992, Poliovirus RNA recombination: Mechanistic studies in the absence of selection, *EMBO J.* **11**:3135.

Jeong, Y. S., and Makino, S., 1992, Mechanism of coronavirus transcription: Duration of primary transcription initiation activity and effects of subgenomic RNA transcription on RNA replication, *J. Virol.* **66**:3339.

Jeong, Y. S., and Makino, S., 1994, Evidence for coronavirus discontinuous transcription, *J. Virol.* **68**:2615.

Joo, M., and Makino, S., 1992, Mutagenic analysis of the coronavirus intergenic consensus sequence, *J. Virol.* **66**:6330.

Keck, J. G., Matsushima, G. K., Makino, S., Fleming, J. O., Vannier, D. M., Stohlman, S. A., and Lai,

M. M., 1988a, *In vivo* RNA-RNA recombination of coronavirus in mouse brain, *J. Virol.* **62:**1810.

Keck, J. G., Soe, L. H., Makino, S., Stohlman, S. A., and Lai, M. M., 1988b, RNA recombination of murine coronaviruses: recombination between fusion-positive mouse hepatitis virus A59 and fusion-negative mouse hepatitis virus 2, *J. Virol.* **62:**1989.

Kim, Y.-N., Jeong, Y. S., and Makino, S., 1993a, Analysis of *cis*-acting sequences essential for coronavirus defective interfering RNA replication, *Virology* **197:**53.

Kim, Y.-N., Lai, M. M. C., and Makino, S., 1993b, Generation and selection of coronavirus defective interfering RNA with large open reading frame by RNA recombination and possible editing, *Virology* **194:**244.

King, A. M., McCahon, D., Slade, W. R., and Newman, J. W., 1982, Recombination in RNA, *Cell* **29:**921.

Kirkegaard, K., and Baltimore, D., 1986, The mechanism of RNA recombination in poliovirus, *Cell* **47:**433.

Koetzner, C. A., Parker, M. M., Ricard, C. S., Sturman, L. S., and Masters, P. S., 1992, Repair and mutagenesis of the genome of a deletion mutant of the coronavirus mouse hepatitis virus by targeted RNA recombination, *J. Virol.* **66:**1841.

Konings, D. A. M., Bredenbeek, P. J., Noten, J. F. H., Hogeweg, P., and Spaan, W. J. M., 1988, Differential premature termination of transcription as a proposed mechanism for the regulation of coronavirus gene expression, *Nucleic Acids Res.* **16:**10849.

Kusters, J. G., Jager, E. J., Niesters, H. G., and van der Zeijst, B. A., 1990, Sequence evidence for RNA recombination in field isolates of avian coronavirus infectious bronchitis virus, *Vaccine* **8:**605.

Lai, C.-J., Zhao, B., Hori, H., and Bray, M., 1991, Infectious RNA transcribed from stably cloned full-length cDNA of dengue type 4 virus, *Proc. Natl. Acad. Sci. USA* **88:**5139.

Lai, M. M. C., 1990, Coronavirus—organization, replication and expression of genome, *Annu. Rev. Microbiol.* **44:**303.

Lai, M. M. C., 1992, RNA recombination in animal and plant viruses, *Microbiol. Rev.* **56:**61.

Lai, M. M. C., Patton, C. D., and Stohlman, S. A., 1982, Replication of mouse hepatitis virus: Negative-stranded RNA and replicative form RNA are of genome length, *J. Virol.* **44:**487.

Lai, M. M., Baric, R. S., Brayton, P. R., and Stohlman, S. A., 1984, Characterization of leader RNA sequences on the virion and mRNAs of mouse hepatitis virus, a cytoplasmic RNA virus, *Proc. Natl. Acad. Sci. USA* **81:**3626.

Lai, M. M., Baric, R. S., Makino, S., Keck, J. G., Egbert, J., Leibowitz, J. L., and Stohlman, S. A., 1985, Recombination between nonsegmented RNA genomes of murine coronaviruses, *J. Virol.* **56:**449.

La Monica, N., Yokomori, K., and Lai, M. M., 1992, Coronavirus mRNA synthesis: Identification of novel transcription initiation signals which are differentially regulated by different leader sequences, *Virology* **188:**402.

Lemm, J. A., and Rice, C. M., 1993a, Assembly of functional sindbis virus RNA replication complexes: Requirement for coexpression of p123 and p34, *J. Virol.* **67:**1905.

Lemm, J. A., and Rice, C. M., 1993b, Roles of nonstructural polyproteins and cleavage products in regulating sindbis virus RNA replication and transcription, *J. Virol.* **67:**1916.

Liao, C. L., and Lai, M. M., 1992, RNA recombination in a coronavirus: Recombination between viral genomic RNA and transfected RNA fragments, *J. Virol.* **66:**6117.

Lin, Y.-J., and Lai, M. M. C., 1993, Deletion mapping of a mouse hepatitis virus defective interfering RNA reveals the requirement of an internal and discontinuous sequence for replication, *J. Virol.* **67:**6110.

Lin, Y.-J., Liao, C.-L., and Lai, M. M. C., 1994, Identification of the cis-acting signal for minus-strand RNA synthesis of a murine coronavirus: Implications for the role of minus-strand RNA in RNA replication and transcription, *J. Virol.* **68:**8131.

Makino, S., and Joo, M., 1993, Effect of intergenic consensus sequence flanking sequences on coronavirus transcription, *J. Virol.* **67:**3304.

Makino, S., and Lai, M. M. C., 1989a, High-frequency leader sequence switching during coronavirus defective interfering RNA replication, *J. Virol.* **63:**5285.

Makino, S., and Lai, M. M. C., 1989b, Evolution of the 5′-end of genomic RNA of murine coronaviruses during passages *in vitro*, *Virology* **169:**227.

Makino, S., Fujioka, N., and Fujiwara, K., 1985, Structure of the intracellular defective viral RNAs of defective interfering particles of mouse hepatitis virus, *J. Virol.* **54:**329.

Makino, S., Stohlman, S. A., and Lai, M. M., 1986a, Leader sequences of murine coronavirus mRNAs can be freely reassorted: Evidence for the role of free leader RNA in transcription, *Proc. Natl. Acad. Sci. USA* **83:**4204.

Makino, S., Keck, J. G., Stohlman, S. A., and Lai, M. M., 1986b, High-frequency RNA recombination of murine coronaviruses, *J. Virol.* **57:**729.

Makino, S., Fleming, J. O., Keck, J. G., Stohlman, S. A., and Lai, M. M., 1987, RNA recombination of coronaviruses: Localization of neutralizing epitopes and neuropathogenic determinants on the carboxyl terminus of peplomers, *Proc. Natl. Acad. Sci. USA* **84:**6567.

Makino, S., Shieh, C. K., Soe, L. H., Baker, S. C., and Lai, M. M., 1988a, Primary structure and translation of a defective interfering RNA of murine coronavirus, *Virology* **166:**550.

Makino, S., Soe, L. H., Shieh, C., and Lai, M. M. C., 1988b, Discontinuous transcription generates heterogeneity at the leader fusion sites of coronavirus mRNAs, *J. Virol.* **62:**3870.

Makino, S., Yokomori, K., and Lai, M. M., 1990, Analysis of efficiently packaged defective interfering RNAs of murine coronavirus: Localization of a possible RNA-packaging signal, *J. Virol.* **64:**6045.

Makino, S., Joo, M., and Makino, J. K., 1991, A system for study of coronavirus mRNA synthesis: A regulated, expressed subgenomic defective interfering RNA results from intergenic site insertion, *J. Virol.* **65:**6031.

Masters, P. S., Koetzner, C. A., Kerr, C. A., and Heo, Y., 1994, Optimization of targeted RNA recombination and mapping of a novel nucleocapsid gene mutation in the coronavirus mouse hepatitis virus, *J. Virol.* **68:**328.

Racaniello, V. R., and Baltimore, D., 1981, Cloned poliovirus complementary DNA is infectious in mammalian cells, *Science* **214:**916.

Rice, C. M., Levis, R., Strauss, J. H., and Huang, H. V., 1987, Production of infectious RNA transcripts from Sindbis virus cDNA clones: Mapping of lethal mutations, rescue of a temperature-sensitive marker, and *in vitro* mutagenesis to generate defined mutants, *J. Virol.* **61:**3809.

Rice, C. M., Grakoui, A., Galler, R., and Chambers, T. J., 1989, Transcription of infectious yellow fever virus RNA from full-length cDNA templates produced by *in vitro* ligation, *New Biol.* **1:**285.

Sawicki, S. G., and Sawicki, D. L., 1990, Coronavirus transcription: Subgenomic mouse hepatitis virus replicative intermediates function in RNA synthesis, *J. Virol.* **64:**1050.

Schaad, M. C., and Baric, R. S., 1994, Genetics of mouse hepatitis virus transcription: Evidence that subgenomic negative strands are functional templates, *J. Virol.* **68:**8169.

Sethna, B. P., Hung, S.-L., and Brian, D. A., 1989, Coronavirus subgenomic minus-strand RNAs and the potential for mRNA replicons, *Proc. Natl. Acad. Sci. USA* **86:**5626.

Sethna, P. B., Hofmann, M. A., and Brian, D. A., 1991, Minus-strand copies of replicating coronavirus messenger RNAs contain antileaders, *J. Virol.* **65:**320.

Shieh, C. K., Soe, L. H., Makino, S., Chang, M. F., Stohlman, S. A., and Lai, M. M., 1987, The 5′-end sequence of the murine coronavirus genome: Implications for multiple fusion sites in leader-primed transcription, *Virology* **156:**321.

Shieh, C. K., Lee, H. J., Yokomori, K., La Monica, N., Makino, S., and Lai, M. M., 1989, Identification of a new transcriptional initiation site and the corresponding functional gene 2b in the murine coronavirus RNA genome, *J. Virol.* **63:**3729.

Spaan, W., Delius, H., Skinner, M., Armstrong, J., Rottier, P., Smeekens, S., van der Zeijst, B. A. M., and Siddell, S. G., 1983, Coronavirus mRNA synthesis involves fusion of non-contiguous sequences, *EMBO J.* **2:**1839.

Spaan, W., Cavanagh, D., and Horzinek, M. C., 1988, Coronaviruses: Structure and genome expression, *J. Gen. Virol.* **69:**2939.

Stern, D. F., and Sefton, B. M., 1982, Synthesis of coronavirus mRNAs: Kinetics of inactivation of infectious bronchitis virus RNA synthesis by UV light, *J. Virol.* **42:**755.

van der Most, R. G., Bredenbeek, P. J., and Spaan, W. J., 1991, A domain at the 3′ end of the polymerase gene is essential for encapsidation of coronavirus defective interfering RNAs, *J. Virol.* **65:**3219.

van der Most, R. G., Heijnen, L., Spaan, W. J. M., and de Groot, R. J., 1992, Homologous RNA recombination allows efficient introduction of site-specific mutations into the genome of coronavirus MHV-A59 via synthetic co-replicating RNAs, *Nucleic Acids Res.* **20:**3375.

van der Most, R. G., de Groot, R. J., and Spaan, W. J. M., 1994, Subgenomic RNA synthesis directed by a synthetic defective interfering RNA of mouse hepatitis virus: A study of coronavirus transcription initiation, *J. Virol.* **68:**3656–3666.

Weiss, S. R., Hughes, S. A., Bonilla, P. J., Turner, J. D., Leibowitz, J. L., and Denison, M. R., 1994, Coronavirus polyprotein processing, *Arch. Virol.* (Suppl.) **9:**349.

Zhang, X., Liao, C.-L. and Lai, M. M. C., 1994, Coronavirus leader RNA regulates and initiates subgenomic mRNA transcription both in trans and in cis, *J. Virol.* **68:**4738.

CHAPTER 3

# Coronavirus Gene Expression

## Genome Organization and Protein Synthesis

WILLEM LUYTJES

## I. INTRODUCTION

Recent studies on coronaviruses have provided detailed insights into the arrangement of open reading frames (ORFs) in the coronavirus genome and the strategies used to translate these ORFs. This chapter will present an updated and detailed overview of coronavirus genomic organization as well as of the coding assignments of the coronavirus ORFs (genes) and their translation strategies (see also the general reviews on coronaviruses by Spaan *et al.*, 1988; Lai, 1990; Holmes, 1991).

Early studies, based on gel electrophoresis of RNA, estimated the length of the coronavirus genome to be about 18 kilobases (kb). It was not until a complete cDNA library was cloned and sequenced for coronavirus avian infectious bronchitis virus (IBV) (Boursnell *et al.*, 1987) that the size of the genome could be established at around 30 kb, the largest among RNA viruses. The genomic RNA of three other coronaviruses has since been completely sequenced, and for several other coronaviruses partial sequence data have been obtained.

From this wealth of sequence information a general picture of coronavirus genome structure emerges, but many significant differences exist. Based on these differences, three groups of structurally related coronaviruses (named after the boldfaced member) can be distinguished: (1) murine hepatitis virus (**MHV**), bovine coronavirus (BCV), human coronavirus OC43 (HCV-OC43), and

---

WILLEM LUYTJES • Department of Virology, Institute of Medical Microbiology, Leiden University, 2300 AH Leiden, The Netherlands.

*The Coronaviridae*, edited by Stuart G. Siddell, Plenum Press, New York, 1995.

turkey coronavirus (TCV); (2) transmissible gastroenteritis virus (**TGEV**), feline infectious peritonitis virus (FIPV), canine coronavirus (CCV), porcine epidemic diarrhea virus (PEDV), and HCV-229E; and (3) **IBV**.

Molecular studies have also provided new insights into translation strategies of coronavirus genes. This has led to the realization that coronaviruses possibly possess as wide a variety of translation strategies as can be found in the entire virus kingdom.

In this chapter, we will refer to coronavirus ORFs and mRNAs according to the recommendations made by the coronavirus study group in 1989 (Cavanagh *et al.*, 1990). Thus, mRNAs are designated by a number according to decreasing size, genomic RNA being 1, and ORFs are numbered corresponding to the mRNA from which they are expressed. In the literature variants, isolates, and serotypes of coronaviruses are described without any clear consensus. To avoid confusion we will refer to such viruses as variants.

## A. General Genome Structure

Positive-strand coronavirus RNAs (genome and messenger RNAs) are capped and polyadenylated. At both ends of the genome, regions of nontranslated sequences (NTR) are present that are most likely involved in replication (see Chapter 2). By virtue of their nested set structure, the 3′-NTR [200–500 nucleotides (nt)] is common to all coronavirus positive-strand RNAs. Colinearity between positive-strand RNAs is also present at the 5′-NTR but is restricted to the leader sequence.

Leader sequences and 3′-NTR sequences are highly similar between coronaviruses from the same structural cluster, but diverge in sequence and in length between these clusters. In fact, only one short stretch of nucleotides, 5′-GGAAGAGC-3′, is completely conserved in the 3′-NTR of all coronaviruses. This region, first detected by Lapps *et al.* (1987), is believed to be important for minus-strand RNA synthesis.

The arrangement of the elementary genes (those that are essential for replication and virion structure) in the genomic RNA is the same for all coronaviruses studied to date (Fig. 1). The polymerase/replicase and associated functions are encoded by the 5′-two thirds of all coronavirus genomes. The major structural proteins, i.e., those that are found in all coronaviruses, are encoded by genes located in the order: 5′-S (large surface protein gene), -sM (small membrane protein gene), -M (membrane protein gene), -N (nucleocapsid protein gene), -3′. Depending on the coronavirus, additional ORFs are present at different positions in the genome. The functions of the proteins encoded by these ORFs are mostly unknown.

Coronaviral protein synthesis requires the synthesis of mRNAs by a mechanism described in Chapter 2. Briefly, short conserved stretches of nucleotides, previously called intergenic sequences or junction sequences but which in fact are mRNA promoters on the minus strands, separate coding regions on the genome. On the minus strands, the promoter upstream of a coding region initiates transcription of a mRNA containing this coding region and all se-

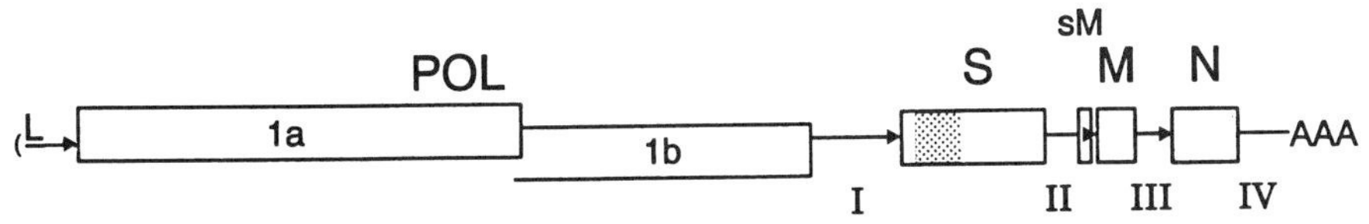


FIGURE 1. Overview of the coronavirus genome organization. The open reading frames of the five common elementary genes (see text) are shown. The shaded area in the S ORF represents the hypervariable region in this gene. Abbreviations: POL, polymerase gene; S, spike or large-surface protein gene; sM, small-membrane protein gene; M, membrane protein gene; N, nucleocapsid protein gene; 1a, ORF 1a of the POL gene; 1b, ORF 1b of the POL gene; L, leader RNA; (, cap; AAA, poly-A tail; >, promoter sequence (also called intergenic or junction sequence). Only promoter sequences common to all coronavirus genomes are shown; I–IV, regions with additional ORFs, described in Figs. 2–5 and in the text.

quences downstream. However, only the first coding region, which may comprise several ORFs on the mRNA, is translated.

The evolution of coronavirus genomes appears to involve extensive deletions or insertions in the genomic RNA. This can be deduced from the fact that differences between coronavirus genomes are characterized by the number and position of nonstructural genes. Also, differences between related, present-day coronaviruses mainly involves insertions or deletions in the genome. In those cases where ORFs are involved, these genomic changes can be associated with different pathogenic properties of the viruses. This is especially true for the large-surface protein gene.

## B. General Translation Strategy

Coronavirus mRNAs are similar to host mRNAs insofar as they supply all the structural prerequisites for recognition by the eukaryotic host translation machinery. The initiation of translation in eukaryotes is generally accepted to occur through ribosomal scanning, starting at the 5′-cap structure and proceeding until an AUG codon in the preferred context is found (Kozak, 1989). Protein synthesis then continues until a termination codon is encountered and the ribosome complex detaches from the RNA and dissociates.

All coronavirus mRNAs possess a common leader sequence that provides a cap structure. Therefore, by means of ribosomal scanning, an ORF encountered in the mRNA coding region can be translated. Indeed, early *in vitro* translation studies on coronavirus mRNAs that encode readily detectable structural proteins confirmed that only the 5′-extreme ORF, which on these mRNAs comprised the entire 5′ coding region, was translated (Rottier *et al.*, 1981; Siddell, 1983). It was thus concluded that coronavirus mRNAs were translated according to the ribosomal scanning model and, although containing more than one coding region (as a consequence of the discontinuous transcription of coronavirus mRNA, see Chapter 2), were functionally monocistronic. However, it soon became apparent that some coronavirus mRNAs were quite different. In these cases, the 5′ coding region contained more than one translationally active

ORF. It has thus become clear, as will be described in this chapter, that coronavirus mRNAs can be functionally monocistronic or polycistronic.

Little is known about translational regulation of coronavirus mRNAs. For the functionally monocistronic mRNAs, the levels of protein expressed will probably mainly depend on the amount of mRNA, which is transcriptionally regulated (Chapter 2). However, there are some indications that the leader sequence may play, at least to some extent, a role in determining translation levels. Hofmann *et al.* (1993) have found that during persistent infection of BCV in tissue culture cells an intraleader ORF is selected. This may downregulate the expression of the POL gene, resulting in persistency. Recently, Tahara *et al.* (1994) have shown that the MHV leader sequence is able to stimulate viral translation *in cis*, probably in conjunction with a virus-specified or virus-induced factor. Structural features such as hairpins, often implicated in translational control, can be observed in the leader sequence of coronavirus mRNAs; but since these are common to all mRNAs, they will not play a role in determining relative protein levels. Whether the length of the 5′-NTRs on the different mRNAs, which can vary considerably, especially between coronaviruses, plays any role in determining translation efficiency remains to be defined.

Translation regulation mechanisms are more evident for the functionally polycistronic mRNAs. The best studied of these mechanisms is ribosomal frameshifting, occurring in the POL mRNA. Also, leaky scanning and internal entry by ribosomes have been proposed for the translation of other coronavirus ORFs and these mechanisms will be discussed in more detail.

## II. CORONAVIRUS GENOME: CODING ASSIGNMENTS AND TRANSLATION STRATEGY OF CONSERVED GENES

### A. The Polymerase (POL) Coding Region

The coronavirus genome is infectious; thus the proteins responsible for replication and mRNA transcription must be translated from the incoming genomic RNA. The polymerase coding region is at the 5′-end of every coronavirus genome. In IBV, MHV-JHM, MHV-A59, and HCV-229E, it comprises two overlapping reading frames, ORF 1a and 1b. In other coronaviruses, this is probably also true, but sequence data are only available for the viruses mentioned. The POL gene of IBV is 19,887 nt (Boursnell *et al.*, 1987), that of MHV-JHM 21,798 nt (Lee *et al.*, 1991), that of MHV-A59 21,530 nt (Bredenbeek *et al.*, 1990b; Bonilla *et al.*, 1994), and that of HCV-229E is 20,274 nt in length (Herold *et al.*, 1993).

Soon after the sequence of ORF 1a and 1b of IBV was determined it was clear that translation of ORF 1b could not be explained by ribosomal scanning, which would lead to translation of ORF 1a only. It was noticed, however, that the configuration around the 42 nt overlap of the two polymerase ORFs resembled that in retroviruses where ribosomal frameshifting occurs. Indeed, Brierley *et al.* (1987) could show *in vitro*, using SP6 transcripts of IBV cDNA covering the frameshift region and flanking reporter genes, that two overlapping reading frames could be translated to a transframe protein by (−1) ribosomal frameshifting. It was later established that the same is true for MHV-A59 (Bredenbeek *et al.*, 1990b) and HCV-229E (Herold *et al.*, 1993). This frameshifting event is mediated by a RNA-pseudoknot structure which causes slippage of the ribosomes on a stretch of adenines and uridines (UUUAAAC) just upstream of the pseudoknot structure (Brierley *et al.*, 1989). The pseudoknot is essential for this event: it could not be replaced by other stem-loop structures and mutations were only allowed when they were compensated on the complementary arms of the pseudoknot stems (Brierley *et al.*, 1991). From these experiments it can be inferred that POL ORF 1b is translated *in vivo* by ribosomal frameshifting (see Chapter 10).

The POL ORFs have a coding capacity for an extremely large replicase/transcriptase protein, which is most likely extensively cleaved. Comparison of the complete sequences of the POL coding regions of IBV, MHV, and HCV-229E and short stretches of 3′-ORF 1b sequence obtained for other coronaviruses revealed that ORF 1b is highly conserved. However, between coronaviruses, ORF 1a is only moderately conserved at the 3′-end. Probably, ORF 1b encodes general coronavirus polymerase functions, while ORF 1a may contain more specific functions, for example, those related to replication in the specific host cell.

*In vitro* translation studies of purified genomic RNA of MHV produced a 28-kDa and a 220-kDa protein from a 250-kDa precursor. This cleavage was sensitive to the protease inhibitor leupeptin (Denison and Perlman, 1986, 1987). The 28-kDa protein could be identified as the N-terminal cleavage product of ORF 1a (Soe *et al.*, 1987). Baker *et al.* (1989, 1993) subsequently showed that this protein is the product of a *cis*-acting papainlike cysteine protease activity, with Cys-1137 and His-1288 as catalytic residues.

Using a panel of antibodies raised against bacterial fusion proteins and antipeptide sera derived from the 5′-6.5 kb of ORF 1a of MHV-A59, a number of processing products were detected in infected cells (Denison *et al.*, 1992; Weiss *et al.*, 1994). These proteins were mapped to ORF 1a in the order, p28-p65-p50-p240, the latter two being cleavage products of a p290 precursor. Cleavage of p28 occurred early and was insensitive to the protease inhibitor leupeptin, which is in contrast to the *in vitro* data. The p290 cleavage is sensitive to leupeptin, suggesting that two distinct proteases are involved in processing this part of the ORF 1a product.

At the other end of the POL gene, Bredenbeek *et al.* (1990c) identified a 33-kDa protein among the *in vitro* translation products of the 3′-most 1.85 kb of the MHV-A59 ORF 1b. They used antibodies raised against a peptide consisting of the C-terminal amino acids of ORF 1b. Denison *et al.* (1991) identified several products of ORF 1b of MHV-A59 using antibodies raised against fusion proteins containing sequences from the 3′-2 kb of this ORF. Among the *in vitro* translation products of purified genomic RNA, five different polypeptides could be detected: p90, p74, p53, p44, and p32. The latter possibly corresponds to the 33-kDa protein found by Bredenbeek *et al.* (1990a). As for the *in vitro* data on ORF 1a products, the fact that protease inhibitor did not affect cleavage of the ORF 1b *in vitro* products may indicate that care should be taken in extrapolation to the *in vivo* situation. For IBV, Brierley *et al.* (1990) have found several POL products in virus-infected cell lysates using antibodies directed to fusion proteins containing parts of the POL ORFs.

As yet, no function can be assigned to any of the polypeptides detected, nor is it known how the ORF 1a/1b products relate to the large polymerase precursor. As was shown in the one case for MHV by Baker *et al.* (1989, 1993), virus-encoded proteases, encoded somewhere in the POL ORFs, are believed to be responsible for the processing of the POL precursors. Computer analyses of the polymerase protein sequences have led to the prediction of several putative protease domains in the POL protein, as well as potential protease cleavage sites (Gorbalenya *et al.*, 1989, 1991; Lee *et al.*, 1991; see Chapter 10).

## B. The Large-Surface (S) Protein Gene

The largest structural gene common to coronaviruses is the large-surface protein or spike protein gene. The S protein is the most prominent protein on coronavirions. Anchored in the viral envelope, it mediates recognition of and binding to the receptor on the host cell and fusion with its membrane. In the TGEV and IBV clusters, S is translated from mRNA 2, and in the MHV cluster from mRNA 3. The S ORF encodes a signal sequence, and its translation will, thus, direct ribosomes to the endoplasmic reticulum, allowing direct insertion into the membrane. The S gene has been cloned for most coronaviruses and expressed using a variety of systems. Functions and biological properties of the S protein will be discussed in Chapter 5.

The S gene of coronaviruses from the different clusters varies in size from 3675 to 4350 nt. Size differences have been mapped to the 5′-part of the gene (De Groot *et al.*, 1987). Large size differences have also been located in the 5′-region of the S genes encoded by closely related coronaviruses (TGEV/PRCV: Rasschaert *et al.*, 1990; Britton *et al.*, 1991; Wesley *et al.*, 1991), by related strains (MHV-A59/JHM: Luytjes *et al.*, 1987; Schmidt *et al.*, 1987; other MHV strains, Parker *et al.*, 1989b; BCV: Zhang *et al.*, 1991a) and by variants (La Monica *et al.*, 1991). Apparently, this part of the coronavirus genome is highly variable, probably since it encodes parts of the S protein associated with host cell recognition and the immune response. The 3′ part of the S gene is more conserved between coronaviruses and contains information necessary for the structure of the spike oligomers (De Groot *et al.*, 1987).

## C. The Small-Membrane (sM) Protein Gene

The small-membrane protein gene of coronaviruses has only recently been recognized as a structural protein gene. In 1990, Smith *et al.* reported the first observation that a small protein encoded by IBV mRNA 3 was membrane-associated. Subsequently, it was established that this protein (then designated

3c) is associated with the viral envelope, and thus represents a new structural (membrane) protein of coronaviruses (Liu and Inglis, 1991). In TGEV, a protein with similar properties was detected (Godet *et al.*, 1992) and the name "small-membrane protein" was suggested. Recently, Yu *et al.* (1994) have identified a homologous protein in the virus particles of MHV-A59. The sM protein will be discussed in Chapter 9.

It now seems clear that the sM ORF will be present in all coronaviruses. The ORF is invariably located directly upstream of the M ORF. Amino acid and nucleotide homology between the different sM ORFs is limited, but the general predicted structure is highly similar. The strategy used for the translation of the sM ORF is very different between the members of the three coronavirus clusters. In IBV, the sM ORF is the third ORF in the 5′ coding region of the tricistronic mRNA 3. Using synthetic mRNAs, in which the influenza virus nucleoprotein (NP) ORF was placed in front of the three IBV mRNA 3 ORFs, only NP and sM proteins were translated *in vitro*. When the region containing the 3a and 3b ORFs was deleted, no translation of sM was observed (Liu and Inglis, 1992b). Thus, it was concluded that the sM ORF is translated by internal ribosomal entry in the region containing the 3a and 3b ORFs. Internal ribosome entry sites (IRES) are characterized by a specific RNA structure: the ribosomal landing pad (RLP). Such an RLP structure has been predicted by Liu and Inglis (1992b), and recently Le *et al.* (1994) have suggested an alternative folding of the upstream RNA into an RLP structure, similar to that found in picornaviruses. Interestingly, unlike picornaviruses, the IBV RLP structure comprises functional ORFs (3a and 3b, see Section III.B.). Also, the coronavirus IRES would be unique in that it occurs on a polycistronic RNA.

The sM ORF in MHV is located as the second, downstream ORF on mRNA 5 (see Fig. 3). The sM ORF product was detected in infected cells and in virions (Leibowitz *et al.*, 1988; Yu *et al.*, 1994). *In vitro*, the sM ORF is translated much less efficiently than the upstream 5a ORF. However, recent *in vitro* translation data indicate that expression of the MHV sM ORF may also be mediated by internal ribosome entry (Thiel and Siddell, 1994). This would be similar to the situation in IBV.

In BCV and OC43, sM is expressed from monocistronic mRNAs 5-1 and 5, respectively. The sM ORF product could be detected in BCV-infected cells by immunofluorescence using antiserum raised against the MHV-A59 sM protein. The BCV sM is similar, 50–60% at the amino acid level, to its MHV counterpart (Abraham *et al.*, 1990; Woloszyn *et al.*, 1990). It will be interesting to establish whether BCV harbors an IRES in the mRNA 5 coding region. If not, the sequence differences between these viruses might reflect the position of essential domains in the MHV RLP.

In the members of the TGEV cluster, the sM gene is expressed from monocistronic mRNA 4 (equivalent to HCV-229E mRNA 5). The size of the gene is similar to those found in the members of the MHV cluster. An interesting topic for future research will be to investigate why only some coronaviruses require such an elaborate mechanism of translation for the sM protein. It may be that adaptation to the host is involved.

## D. The Membrane (M) Protein Gene

The membrane-associated protein is translated from the M gene. The M protein assumes a tightly membrane-associated conformation which is believed to be crucial for virion structure formation. The properties of the M protein will be discussed in Chapter 6. The predicted membrane topology of the M protein is highly conserved between coronaviruses, and the gene does not show any large insertions or deletions. Yet, M protein sequences between viruses from different clusters are only moderately similar. The M gene ranges in size from 675 to 786 nt. The main area of divergence is at the 5′-end, where the M genes of the TGEV cluster encode signal sequences that target M to the ER (Kapke *et al.*, 1988). In some variants of IBV and some enteric strains of MHV differences were also observed in this region (Cavanagh and Davis, 1988; Homberger, 1994).

Surprisingly, the M gene sequence of the avian coronavirus TCV is almost identical (only two nts are different) to that of mammalian coronavirus BCV (Verbeek and Tijssen, 1991), a level of sequence similarity not found between any other two coronaviruses.

## E. The Nucleocapsid (N) Protein Gene

The nucleocapsid gene encodes the protein that encapsidates the genomic RNA. It is translated from the most abundant viral mRNA, which in most coronaviruses is the smallest: mRNA 7 in viruses from the MHV cluster and mRNA 6 in those from the IBV cluster. Coronaviruses from the TGEV cluster express the N gene from mRNA 6, but these viruses, excepting HCV-229E and PEDV, also express a smaller mRNA (see Section III.D.).

The amino acid similarity of the N proteins between coronaviruses from the different clusters is low. The N genes of coronaviruses from a single cluster are generally of similar size. However, the PEDV N gene is 135 nt larger than that of the other members of this cluster (Bridgen *et al.*, 1993). The authors suggest that the extra sequence may be the evolutionary result of polymerase stuttering, since it contains a 36 nt (12 amino acid) periodicity. From an evolutionary point of view, TCV is again noticeable: its N gene has merely a single nt difference (including the 3′-NTR) from the same area in BCV (Verbeek and Tijssen, 1991).

Interestingly, in coronaviruses from the MHV cluster, i.e., BCV (where it was first recognized by Lapps *et al.*, 1987), TCV, SDAV (Kunita *et al.*, 1993), and MHV strains A59, S, 1, and 3, but not JHM (Parker and Masters, 1990), and in those from the TGEV cluster, i.e., PEDV and HCV-229E (Schreiber *et al.*, 1989; Bridgen *et al.*, 1993), an internal ORF of considerable length (612–660 nt in the MHV cluster; 336 nt in the TGEV cluster) is present in the N gene with an AUG codon in a favorable context. It has been shown by Senanayake *et al.* (1992) that this ORF is expressed in BCV-infected cells and encodes a membrane-associated protein. Thus, BCV mRNA7 is functionally bicistronic, probably by means of a ribosomal leaky scanning mechanism. In HCV-OC43, the homologue of the internal ORF of BCV is shorter because of a point mutation (Kamahora *et al.*, 1989). Recently it has been suggested by Schaad and Baric (1993) that additional, short internal ORFs in the N gene of MHV might exist which are preceded by promoter-similar sequences that would produce subgenomic mRNAs. This observation remains to be confirmed as it could reflect aberrant initiation events. More detailed information on the coronavirus N protein is presented in Chapter 7.

## III. CORONAVIRUS GENOME: CODING ASSIGNMENTS AND EXPRESSION OF NONCONSERVED GENES

### A. ORFs between the POL and S Genes: Region I

In the MHV cluster of coronaviruses, two ORFs, which are not present in members of the TGEV and the IBV clusters, are located between the polymerase and the large-surface protein gene (Fig. 2). The first of these, ORF 2a, is translated from mRNA 2 and encodes a small protein designated ns2a or 30 kDa. This protein has been identified as a phosphoprotein in BCV (Cox *et al.*, 1991). Early *in vitro* translation experiments showed a product of MHV mRNA 2 of approximately 30,000 molecular weight (Leibowitz *et al.*, 1982; Siddell, 1983). Using antibodies raised against a fusion product of the cloned gene of MHV-A59, the protein could be detected in cells infected with either MHV-A59 or MHV-JHM but not in purified virions, suggesting a nonstructural nature (Bredenbeek *et al.*, 1990a; Zoltick *et al.*, 1990). ORF 2a has been sequenced for MHV strains A59 (Luytjes *et al.*, 1988), JHM (Shieh *et al.*, 1989), and for BCV (Cox *et al.*, 1989). In the MHV-JHM variant Wb1 a large deletion has eliminated the entire 5′-end of ORF 2a, including the mRNA2 promoter. Consequently no 30-kDa protein is synthesized (Schwarz *et al.*, 1990). This deletion does not affect growth in tissue culture, suggesting that the 30-kDa protein may only be important to the virus in its natural host.

The second ORF in this region, exclusive to viruses of the MHV cluster, encodes a protein that constitutes the major difference in virion structure between the MHV, TGEV, and IBV groups. It has been designated HE (hemagglutinin esterase) in analogy with the influenza C virus hemagglutinin- esterase-fusion (HEF), with which it shows a high degree of structural and functional homology (Luytjes *et al.*, 1988; Vlasak *et al.*, 1988a,b). The HE gene is translated from mRNA 2-1 and encodes a 65,000 molecular weight protein, which in its dimeric form is the second spike protein for several viruses from the MHV cluster. HE translation is signal sequence mediated, and the protein is inserted into the ER membrane through a transmembrane domain which may be the homologue of the HEF fusion domain. It has been hypothesized that the HE gene has been acquired by a RNA-recombination event between ancestral coronaviruses and influenza C viruses (Luytjes *et al.*, 1988).

The HE locus in MHV is highly heterogeneous between strains, but also within a single strain. In MHV-A59, where it was first detected, the HE gene is not expressed, since it lacks an initiation codon (Luytjes *et al.*, 1988). More

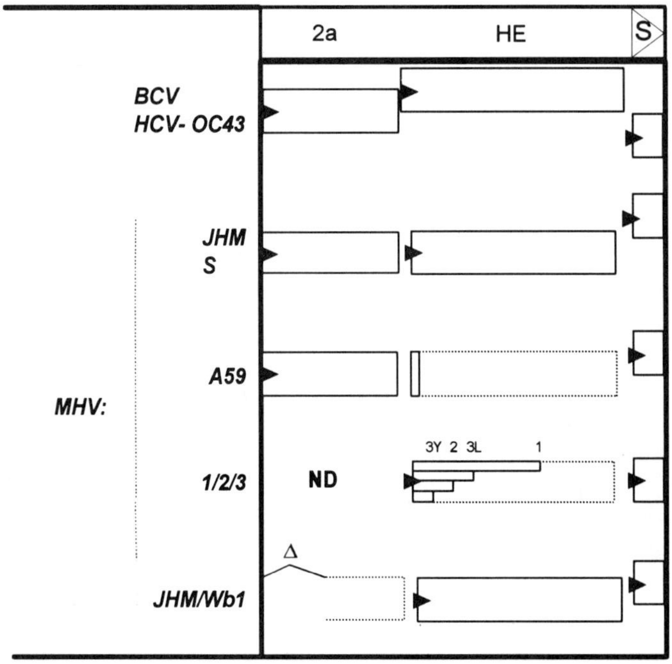


FIGURE 2. Genetic structure of coronavirus RNA in region I between the POL and S genes. The left panel indicates the coronavirus; the right panel gives a diagram of the genomic structure. Boxes represent reading frames. Functional open reading frames are in white boxes, stippled boxes show the outline of nontranslated reading frames. The positions of the ORFs are represented relative to that of the upstream flanking, conserved ORF (names shown in triangles). Deletions are indicated by Δ. ND = not determined. In the upper, shaded box the nomenclature of the ORFs is displayed (see text). Variants of MHV-1/2/3 are indicated above their respective reading frames. Other legends are as described in Fig. 1.

important, mutations have destroyed the upstream promoter that determines the synthesis of mRNA 2-1. In other MHV strains the HE gene is truncated at the 3′-end by mutations and insertions/deletions that have introduced translation termination codons and thus preclude membrane insertion. The gene for a full-length HE protein is present in MHV strains S and JHM, although the HE protein was only detected at low levels in MHV-JHM virions (Shieh *et al.*, 1989; Yokomori *et al.*, 1991). Single-strain heterogeneity is indicated by the observation that MHV-JHM(2) variants, which contained HE regions truncated to different extents, could be isolated over a period of time from the brains of infected animals (Yokomori *et al.*, 1993).

The HE gene has been cloned and sequenced for four BCV strains (Quebec: Parker *et al.*, 1989a; Mebus: Kienzle *et al.*, 1990; L9 and LY138: Zhang *et al.*, 1991b) and one HCV strain (OC43: Zhang *et al.*, 1992). In these viruses the gene is highly conserved and shows none of the variability in structure observed for MHV. The functional HE gene ranges in size from 1275 to 1320 nt. Chapter 8 will present a more detailed analysis of the function and structure of the HE protein.

## B. ORFs between the S and sM Genes: Region II

There is great variability between coronaviruses in the area of the genome between the S and the sM genes, in the size and in the number of ORFs present (Fig. 3). In IBV, two ORFs (3a and 3b) are present, each of which is expressed from mRNA 3. IBV mRNA 3 also encodes the sM protein (see Section II.C.). The corresponding proteins (3a: 6700 molecular weight; 3b: 7400 molecular weight) have been detected in IBV-infected cells, and both ORFs can be translated from T7/SP6 transcripts of cloned cDNA of mRNA 3 (Smith *et al.*, 1990; Liu *et al.*, 1991). IBV mRNA 3 is thus functionally poly(tri)cistronic.

Smith *et al.* (1990) proposed that translation of the IBV 3a and 3b ORFs is governed by the "leaky scanning" model of Kozak (1987, 1989). In this model the strength of the initiation codon of a downstream reading frame on a polycistronic mRNA has to be higher than that of the ORF that is located upstream. Skipping of the preceding, weaker AUGs by a portion of the ribosomes enables production of the downstream encoded proteins. Consistent with this model, the AUG codon of ORF 3a is in a weak context and no additional AUG codons are present in these ORFs. Note that ORFs 3a and 3b are in the region of mRNA 3 considered to fold into the RLP for translation of the sM ORF (see Section II.C.).

In MHV, 2 or 3 ORFs are located between genes S and sM. In MHV-JHM, a single ORF 4 is translated from mRNA 4 and encodes a 15,000 molecular weight protein, which has been detected in infected cells (Ebner *et al.*, 1988). MHV-A59 contains two ORFs in this region: 4a (predicting a protein of 2200 molecular weight) and 4b (predicting a protein of 11,700 molecular weight; Weiss *et al.*, 1993). Alignment of the sequences of MHV-JHM and A59 from this area indicates that two single nt insertions/deletions have caused the difference. The short ORF 4a of MHV A59 is in a different reading frame and ORF 4b is the shorter homologue of the MHV-JHM ORF 4. *In vitro* translation of T7-transcribed RNA containing ORF 4a and 4b produced only a 2.2-kDa protein. Also, antibodies against the 15-kDa ORF 4 product of MHV-JHM did not detect any 11.7-kDa protein in MHV-A59-infected cells (Weiss *et al.*, 1993). Thus, early assumptions that the 14-kDa protein observed in analyses of MHV-A59 encoded proteins (Rottier *et al.*, 1981) and cell-free lysates (Leibowitz *et al.*, 1982) is a product of the coding region of mRNA 4 may be incorrect.

The sM ORF of MHV mRNA 5 (ORF 5b) is expressed preferentially to ORF 5a in *in vitro* translation (Budzilowicz and Weiss, 1987). Since the product of ORF 5a has not yet been identified, it remains to be seen whether this mRNA is, like mRNA 3 of IBV, functionally polycistronic. If so, it would be the only coronavirus mRNA that is functionally monocistronic from a downstream ORF.

Nothing is known about the function of the ORF 4 (a,b) and ORF 5a gene products. However, in MHV strain S, which is easily propagated in tissue culture, ORF 5a is deleted and ORF 4 is not expressed because synthesis of mRNA 4 has been eliminated by mutations in its promoter. It thus appears that these proteins are not required for replication, at least not *in vitro* and for this strain of MHV (Yokomori and Lai, 1991).

In BCV, the region between the S and sM genes contains 3 ORFs (Abraham

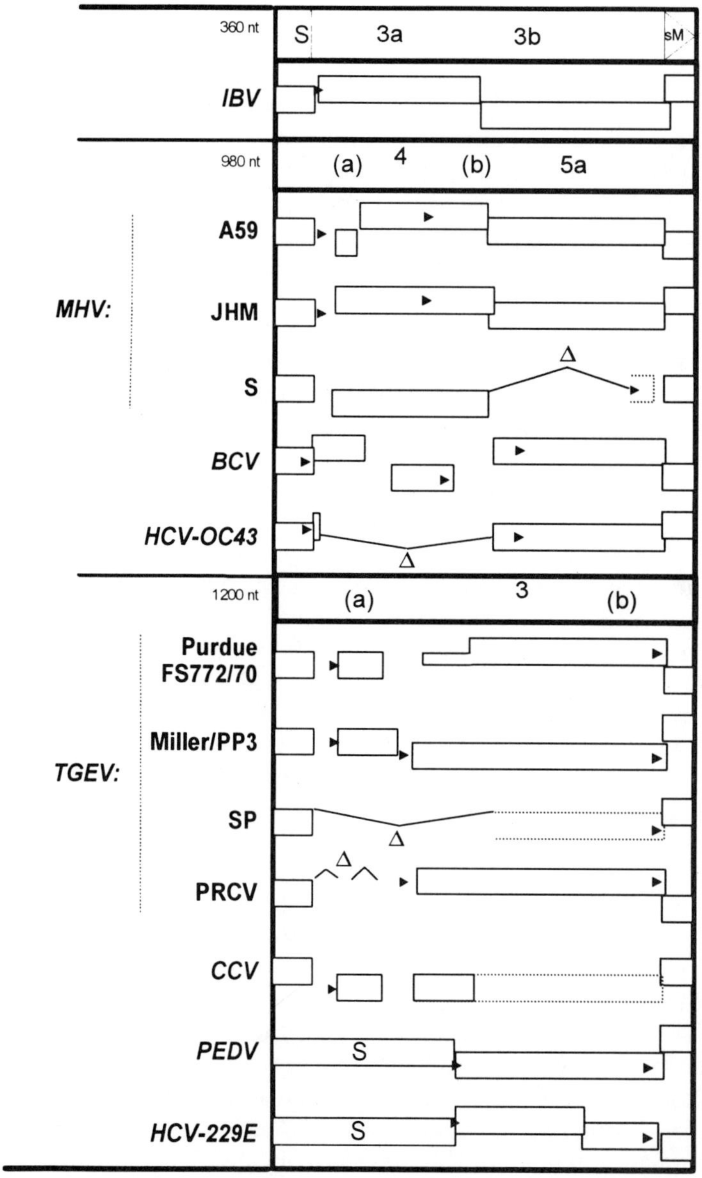


FIGURE 3. Genetic structure of coronavirus RNA in region II between the S and sM genes. The three structural clusters are outlined in separate panels. The numbers in the upper corners of the left panels represent the approximate lengths of the region in the respective clusters. Other legends are as in Figs. 1 and 2.

*et al.*, 1990). The first two of these (ORFs 4a and 4b) are located in the coding region of mRNA 4, are homologous to MHV-JHM ORF 4, and may have arisen from a single base deletion in an ancestral BCV ORF 4. ORF 4a encodes a protein of 4900 molecular weight and can be translated *in vitro* from mRNA 4. The putative product of ORF 4b would be a protein of 4800 molecular weight, but it has not been detected after *in vitro* translation (Abraham *et al.*, 1990). Unlike MHV ORFs 4/4a, BCV ORF 4a overlaps briefly with the end of the S gene. Accordingly, the promoter responsible for the generation of mRNA 4 is located some 330 nt inside the upstream gene. The third BCV ORF in this region is located in the coding region of mRNA 5, which has its promoter located internally in ORF 4b. ORF 5 encodes a protein of 12,700 molecular weight (Abraham *et al.*, 1990; Woloszyn *et al.*, 1990).

The HCV-OC43 strain reported by Mounir and Talbot (1993) also shows considerable variation in this area compared to BCV, although otherwise both viruses are highly homologous at the nucleotide and amino acid levels. The difference is due to a large deletion in the mRNA 4 coding region of HCV-OC43. This deletion has left only 11 amino acids from ORF 4a. The authors found that this truncated ORF is followed by a copy of a large part of the HCV-OC43 leader sequence. This finding is possibly an artifact, but it is not unique because Taguchi *et al.* (1994) have found a similar arrangement between the M and N genes of MHV strain S. The partial leader copy is succeeded by the homologue of the BCV ORF 5, although for HCV-OC43 it should be designated ORF 4b because it is located in the coding region of mRNA 4. The deletion event also removed the promoter that would give rise to the OC43 version of BCV mRNA 5. Whether this configuration occurs in other OC43 variants remains to be determined.

In the coronaviruses of the TGEV cluster, the region between the S and sM genes is extremely variable. Generally, two ORFs are present. In two TGEV strains, FS772/70 and Purdue-115, the ORFs are located on mRNA 3 (Britton *et al.*, 1989; Rasschaert *et al.*, 1990). Therefore, the ORFs have been designated ORF 3a and 3b. However, Wesley *et al.* (1989) reported for the Miller/PP3 strain of TGEV that these ORFs were each translated from a separate RNA, designated mRNAs 3 and 4, respectively. Consequently, these ORFs were named ORF 3 and 4. Following the proposed coronavirus nomenclature and to preserve analogy with other coronaviruses, these mRNAs are called mRNAs 3 and 3-1, encoding ORFs 3a and 3b.

In TGEV-related strains porcine respiratory coronavirus (PRCV) 86/137004 and 86/135308, the promoter responsible for the synthesis of mRNA 3 has been deleted. A new one has been generated by mutations approximately 300 nt downstream, just in front of the AUG of ORF 3b (Page *et al.*, 1991). Thus, PRCV synthesizes a different mRNA 3 (related to mRNA 3-1 of TGEV Miller/PP3), containing ORF 3. To make the picture even more complicated, a TGEV Miller/PP3 variant, SP, contains a large deletion stretching from just upstream of the promoter for mRNA 3 into ORF 3b and this virus produces no mRNA 3 at all (Wesley *et al.*, 1990).

Coronavirus CCV is closely related to TGEV strains FS772/70 and Purdue. As in these strains, two CCV ORFs are present, 3a and 3b, both in the coding region of mRNA3. While the nt sequences are highly similar to TGEV, ORF 3b is truncated by 20% compared to TGEV 3b, because of an acquired termination codon (Horsburgh *et al.*, 1992).

In FIPV, the counterparts for the TGEV ORFs 3a and b are reportedly present, but mutations have split ORF 3b into two separate ORFs and an extra ORF, starting in 3a, can be observed, bringing the number of FIPV ORFs in this region to 4 (Vennema, 1991). These four ORFs are apparently located in the coding region of mRNA 3. There are no data on expression of these ORFs.

Two other members of the TGEV cluster, PEDV and HCV-229E, are somewhat divergent in region II, since both seem to lack a homologue of ORF 3a. In PEDV, one ORF is present between the S and sM ORFs (Duarte *et al.*, 1994). This ORF 3 is related to ORF 3b in the TGEV strains and is located in the coding region of mRNA 3. ORF 3 appears to be polymorphic, as sequence variation was observed between independent clones and between two variants, which included deletions leading to truncation of the ORF. HCV-229E contains two ORFs in this area, which together are homologous to PEDV ORF 3. They briefly overlap and are located in the coding region of a single mRNA (Raabe and Siddell, 1989; Raabe *et al.*, 1990). The nomenclature for the HCV-229E mRNAs used by the authors is different from that for the other members of this cluster. They have found an extra RNA with a coding region starting internally in the S gene, not only by hybridization but also by metabolic labeling (Schreiber *et al.*, 1989). Therefore, the next smaller mRNA is numbered 4, rather than 3. Consequently, the ORFs located between the S and sM genes are named 4a and 4b. For the sake of clarity, it may be better to number these ORFs 3a and 3b, on the coding region of mRNA 3. The upstream promoter in S would then produce RNA 2-1. There is no information on the expression and biological relevance of these ORFs.

## C. ORFs between the M and N Genes: Region III

In IBV, extra ORFs can be found between the M and the N genes (Fig. 4). Two ORFs, 5a and 5b, encoding proteins of 7400 and 9500 molecular weight, respectively, are present. The ORFs are both expressed from mRNA 5 and can be detected in infected cells, implying that the IBV mRNA 5 is also functionally polycistronic (Liu and Inglis, 1992a). The "leaky scanning" model does not seem to apply here since the AUG codon of ORF 5a is in a stronger context than that of ORF 5b. The translation mechanism of ORF 5b therefore remains to be determined.

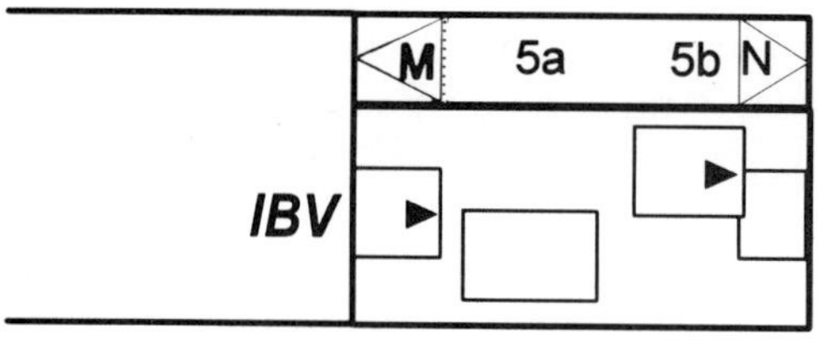


FIGURE 4. Genetic structure of coronavirus RNA in region III between the M and N genes. Legends are as in Figs. 1 and 2.

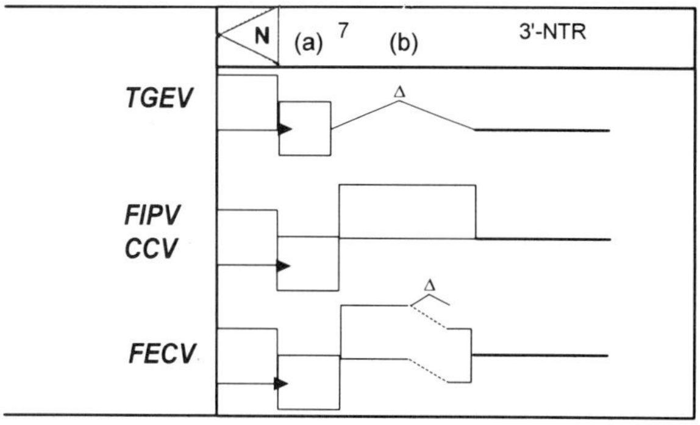


FIGURE 5. Genetic structure of coronavirus RNA in region IV downstream of the N gene. Legends are as in Figs. 1 and 2.

## D. ORFs Downstream of the N Gene: Region IV

Unlike viruses from the MHV and IBV clusters, coronaviruses from the TGEV cluster encode proteins downstream of the N gene (Fig. 5). As was determined by De Groot *et al.* (1988), two ORFs are present in this region of the FIPV genome: ORF 7a and 7b, located on mRNA 7, which is generated from a promoter immediately downstream of the N gene. In TGEV, only one ORF is found (ORF 7), apparently a deletion product of the homologue of ORF 7a of FIPV (Kapke and Brian, 1986; Rasschaert *et al.*, 1987; De Groot *et al.*, 1988; Britton *et al.*, 1988). Coronaviruses CCV and FECV are similar to FIPV. The CCV ORFs are colinear with those of FIPV but only 60–80% homologous at the amino acid level. In contrast, the FECV ORFs are 89–99% homologous, but ORF 7b has been truncated by a point mutation (Vennema *et al.*, 1992a; Horsburgh *et al.*, 1992). The 9100-molecular-weight translation product of ORF 7 of TGEV (Garwes *et al.*, 1989; Tung *et al.*, 1992) and a 14,000-molecular-weight product of ORF 7b of FIPV, CCV, and FECV (Vennema, 1991) have been detected *in vivo* by immunoprecipitation. The ORF 7 protein of TGEV is endoplasmic reticulum (ER) membrane associated (Garwes *et al.*, 1989; Tung *et al.*, 1992), leading to the suggestion that it may play a role during virus assembly or budding. The ORF 7b protein of FIPV is a soluble protein containing the amino acid sequence KTEL, an ER-retention signal (Vennema *et al.*, 1992b). This suggests that it can be retrieved into the ER lumen after transport. Its role in virus replication is not yet clear. If a product of ORF 7a can be detected in FIPV-infected cells, the FIPV mRNA 7 will be another example of a coronavirus mRNA that is functionally polycistronic. The separate position held by HCV-229E and PEDV in the TGEV cluster is again illustrated by the fact that these viruses do not contain any ORFs or promoters in this region (Schreiber *et al.*, 1989; Bridgen *et al.*, 1993).

# IV. CONCLUSION

The genomes of coronaviruses apparently contain a relatively flexible set of conserved reading frames that are essential and variable or nonconserved reading frames that often appear to be dispensable *in vitro*. It should be noted, however, that many viruses described in the literature are tissue culture variants, or at least adapted to growth in tissue culture. They may not be infectious to the natural host in their mutated form and dispensable ORFs may indeed prove essential to the virus in the field.

Nevertheless, the data do emphasize the adaptability of coronaviruses and the flexibility of the genome. The implications for evolution are apparent. The

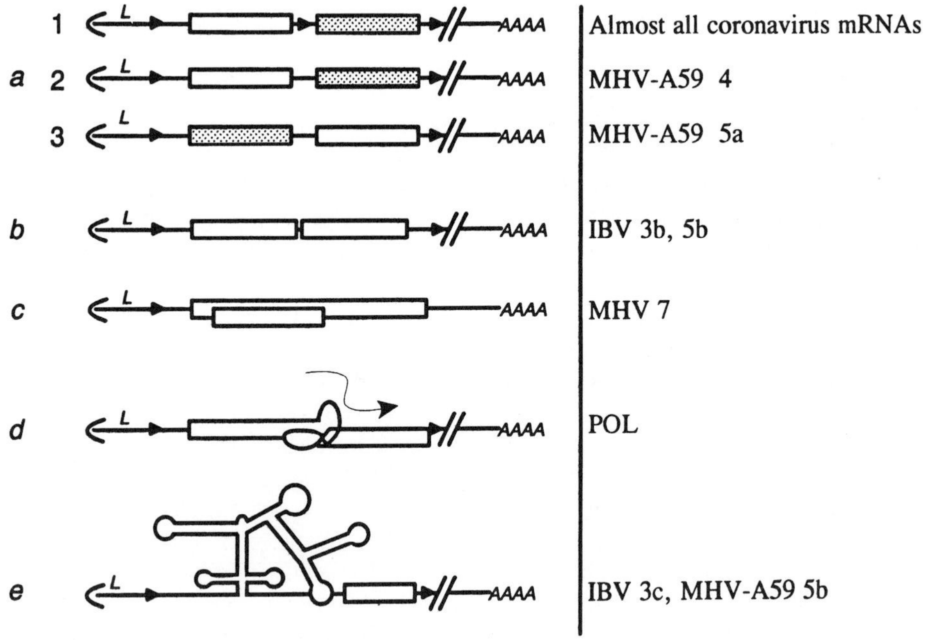


FIGURE 6. Summary of translation strategies used by coronaviruses. Diagrams of coronavirus RNAs are represented with a cap (C), the leader (L), promoters (>) bordering the coding region of the RNA, its functional ORF in white, and nontranslated ORFs in gray. The nontranslated ORFs are generally downstream of the second promoter on the RNA and therefore not in the coding region (a-1). In one case (MHV-A59 mRNA 4), a nontranslated ORF occurs inside the coding region (a-2) and in another case (MHV-A59 mRNA 5) only the downstream ORF is translated (a-3). In b through e, only the coding region is shown and it is indicated how the additional downstream ORFs in coding regions are translated. See the text for additional information. (a) The ribosomal 40S subunit associates with the mRNA cap via cap binding and initiation factors, scans the RNA, finds the initiating AUG codon, whereupon the 60S subunit associates and translation follows. After reaching the termination codon, the complex dissociates and no further downstream ORFs are translated. (b,c) A second downstream ORF (b) or internal ORF (c) is translated by leaky scanning. Thus, a portion of the ribosomal 40S subunits fail to recognize the AUG codon of the first ORF and initiation at the AUG codon of the downstream ORF occurs. (d) A large, downstream ORF is translated by (−1) ribosomal frameshifting, (mediated by a slippery sequence and RNA-pseudoknot) into this ORF. (e) A ribosomal landing pad, which contains functional ORFs, serves as an internal ribosomal entry site to allow translation of the downstream ORF indicated. In the right panel examples of the ORFs for which the particular translation strategy is used, or is likely to be used, are given.

potential of coronaviruses to evolve rapidly, by their ability to incorporate new genes, is most convincingly illustrated for the HE gene. For coronaviruses of the MHV cluster, the biological properties acquired with the HE gene are not fully understood, but the available data indicate that HE is involved in pathogenesis and the acquisition of the gene constitutes a significant evolutionary advantage to the viruses involved. The mechanism by which new genes are incorporated into the genome is very likely to be RNA recombination between heterologous RNAs. Obviously this will be a major subject of study on coronaviruses in the coming years.

Viruses have often developed alternative strategies for the expression of proteins essential for replication. Translational mechanisms different to the strategy of ribosomal scanning are also believed to allow for the regulation of protein levels. It now appears that coronaviruses employ a whole set of alternative translation strategies, including ribosomal frameshifting, internal ribosomal entry, and leaky scanning (Fig. 6). It is likely that ribosomal frameshifting in the POL gene provides a crucial means of regulating expression of the vital 1b ORF functions and is used by all coronaviruses. More puzzling is the expression of the sM ORF. Closely related coronaviruses use different strategies to express this ORF, which may reflect a fundamental difference between these viruses. It will be interesting to find out what these differences are. In particular, it will be especially important to find out whether the IRES elements of IBV and MHV function on genomic RNA, which would imply that the sM ORF in these viruses can be translated from the incoming genome.

Detailed analyses of coronavirus translation mechanisms, replication strategies, and the relative importance of the different ORFs, especially those of the polymerase coding region, are in progress. The analysis of the biological functions of these ORFs awaits the use of site-directed mutagenesis and the techniques of reverse genetics, which, hopefully, may soon become available to coronavirologists.

ACKNOWLEDGMENTS. The assistance and critical assessments of Dr. Willy Spaan and Dr. Raoul de Groot during the writing of this review are greatly acknowledged.

## V. REFERENCES

Abraham, S., Kienzle, T. E., Lapps, W. E., and Brian, D. A., 1990, Sequence and expression analysis of potential nonstructural proteins of 4.9, 4.8, 12.7, and 9.5 kDa encoded between the spike and membrane protein genes of the bovine coronavirus, *Virology* **177**:488.

Baker, S. C., Shieh, C.-K., Soe, L. H., Chang, M. F., Vannier, D. M., and Lai, M. M. C., 1989, Identification of a domain required for autoproteolytic cleavage of murine coronavirus gene A polyprotein, *J. Virol.* **63**:3693.

Baker, S. C., Yokomori, K., Dong, S., Carlisle, R., Gorbalenya, A. E., Koonin, E. V., and Lai, M. M., 1993, Identification of the catalytic sites of a papain-like cysteine proteinase of murine coronavirus, *J. Virol.* **67**:6056.

Bonilla, P. J., Gorbalenya, A. E., and Weiss, S. R., 1994, Mouse hepatitis virus strain A59 polymerase gene ORF 1a: Heterogeneity among MHV strains, *Virology* **198**:736.

Boursnell, M. E., Brown, T. D., Foulds, I. J., Green, P. F., Tomley, F. M., and Binns, M. M., 1987, Completion of the sequence of the genome of the coronavirus avian infectious bronchitis virus, *J. Gen. Virol.* **68**:57.

Bredenbeek, P. J., Noten, A. F. H., Horzinek, M. C., and Spaan, W. J. M., 1990a, Identification and stability of a 30-kDa non-structural protein encoded by mRNA 2 of mouse hepatitis virus in infected cells, *Virology* **175**:303.

Bredenbeek, P. J., Pachuk, C. J., Noten, A. F. H., Charité, J., Luytjes, W., Weiss, S. R., and Spaan, W. J. M., 1990b, The primary structure and expression of the 2nd open reading frame of the polymerase gene of the coronavirus MHV-A59—A highly conserved polymerase is expressed by an efficient ribosomal frameshifting mechanism, *Nucleic Acids Res.* **18**:1825.

Bredenbeek, P. J., Snijder, E. J., Noten, F. H., Den Boon, J. A., Schaaper, W. M., Horzinek, M. C., and Spaan, W. J., 1990c, The polymerase gene of corona- and toroviruses: Evidence for an evolutionary relationship, *Adv. Exp. Med. Biol.* **276**:307.

Bridgen, A., Duarte, M., Tobler, K., Laude, H., and Ackermann, M., 1993, Sequence determination of the nucleocapsid protein gene of the porcine epidemic diarrhoea virus confirms that this virus is a coronavirus related to human coronavirus 229E and porcine transmissible gastroenteritis virus, *J. Gen. Virol.* **74**:1795.

Brierley, I., Boursnell, M. E., Binns, M. M., Bilimoria, B., Blok, V. C., Brown, T. D., and Inglis, S. C., 1987, An efficient ribosomal frame-shifting signal in the polymerase-encoding region of the coronavirus IBV, *EMBO J.* **6**:3779.

Brierley, I., Digard, P., and Inglis, S. C., 1989, Characterization of an efficient coronavirus ribosomal frameshifting signal: Requirement for an RNA pseudoknot, *Cell* **57**:537.

Brierley, I., Boursnell, M. E. G., Binns, M. M., Bilimoria, B., Rolley, N. J., Brown, T. D. K., and Inglis, S. C., 1990, Products of the polymerase-encoding region of the coronavirus IBV, *Adv. Exp. Med. Biol.* **276**:275.

Brierley, I., Rolley, N. J., Jenner, A. J., and Inglis, S. C., 1991, Mutational analysis of the RNA pseudoknot component of a coronavirus ribosomal frameshifting signal, *J. Mol. Biol.* **220**:889.

Britton, P., Carmenes, R. S., Page, K. W., Garwes, D. J., and Parra, F., 1988, Sequence of the nucleoprotein gene from a virulent British field isolate of transmissible gastroenteritis virus and its expression in *Saccharomyces cerevisiae*, *Mol. Micro.* **2**:89.

Britton, P., Lopez Otin, C., Martin Alonso, J., and Parra, F., 1989, Sequence of the coding regions from the 3.0 kb and 3.9 kb mRNA. Subgenomic species from a virulent isolate of transmissible gastroenteritis virus, *Arch. Virol.* **105**:165.

Britton, P., Mawditt, K. L., and Page, K. W., 1991, The cloning and sequencing of the virion protein genes from a British isolate of porcine respiratory coronavirus: Comparison with transmissible gastroenteritis virus genes, *Virus Res.* **21**:181.

Budzilowicz, C. J., and Weiss, S. R., 1987, *In vitro* synthesis of two polypeptides from a nonstructural gene of coronavirus mouse hepatitis virus strain A59, *Virology* **157**:509.

Cavanagh, D., and Davis, P. J., 1988, Evolution of avian coronavirus IBV: Sequence of the matrix glycoprotein gene and intergenic region of several serotypes, *J. Gen. Virol.* **69**:621.

Cavanagh, D., Brian, D. A., Enjuanes, L., Holmes, K. V., Lai, M. M. C., Laude, H., Siddell, S. G., Spaan, W., Taguchi, F., and Talbot, P. J., 1990, Recommendations of the coronavirus study group for the nomenclature of the structural proteins, messenger RNAs, and genes of coronaviruses, *Virology* **176**:306.

Cox, G. J., Parker, M. D., and Babiuk, L. A., 1989, The sequence of cDNA of bovine coronavirus 32K nonstructural gene, *Nucleic Acids Res.* **17**:5847.

Cox, G. J., Parker, M. D., and Babiuk, L. A., 1991, Bovine coronavirus nonstructural protein ns2 is a phosphoprotein, *Virology* **185**:509.

De Groot, R. J., Luytjes, W., Horzinek, M. C., Van der Zeijst, B. A. M., Spaan, W. J., and Lenstra, J. A., 1987, Evidence for a coiled-coil structure in the spike proteins of coronaviruses, *J. Mol. Biol.* **196**:963.

De Groot, R. J., Andeweg, A. C., Horzinek, M. C., and Spaan, W. J. M., 1988, Sequence analysis of the 3′ end of the feline coronavirus FIPV 79-1146 genome: Comparison with the genome of porcine coronavirus TGEV reveals large insertions, *Virology* **167**:370.

Denison, M. R., and Perlman, S., 1986, Translation and processing of mouse hepatitis virus virion RNA in a cell-free system, *J. Virol.* **60**:12.

Denison, M. R., and Perlman, S., 1987, Identification of putative polymerase gene product in cells infected with murine coronavirus A59, *Virology* **157**:565.

Denison, M. R., Zoltick, P. W., Leibowitz, J. L., Pachuk, C. J., and Weiss, S. R., 1991, Identification of polypeptides encoded in open reading frame-1b of the putative polymerase gene of the murine coronavirus mouse hepatitis virus-A59, *J. Virol.* **65**:3076.

Denison, M. R., Zoltick, P. W., Hughes, S. A., Giangreco, B., Olson, A. L., Perlman, S., Leibowitz, J. L., and Weiss, S. R., 1992, Intracellular processing of the N-terminal ORF 1a proteins of the coronavirus MHV-A59 requires multiple proteolytic events, *Virology* **189**:274.

Duarte, M., Tobler, K., Bridgen, A., Rasschaert, D., Ackermann, M., and Laude, H., 1994, Sequence analysis of the porcine epidemic diarrhea virus genome between the nucleocapsid and spike protein genes reveals a polymorphic ORF, *Virology* **198**:466.

Ebner, D., Raabe, T., and Siddell, S. G., 1988, Identification of the coronavirus MHV-JHM mRNA 4 product, *J. Gen. Virol.* **69**:1041.

Garwes, D. J., Stewart, F., and Britton, P., 1989, The polypeptide of $M_r$ 14,000 of porcine transmissible gastroenteritis virus: Gene assignment and intracellular location, *J. Gen. Virol.* **70**:2495.

Godet, M., L'Haridon, R., Vautherot, J. F., and Laude, H., 1992, TGEV corona virus ORF4 encodes a membrane protein that is incorporated into virions, *Virology* **188**:666.

Gorbalenya, A. E., Koonin, E. V., Donchenko, A. P., and Blinov, V. M., 1989, Coronavirus genome: Prediction of putative functional domains in the non-structural polyprotein by comparative amino acid sequence analysis, *Nucleic Acids Res.* **17**:4847.

Gorbalenya, A. E., Koonin, E. V., and Lai, M. M., 1991, Putative papain-related thiol proteases of positive-strand RNA viruses. Identification of rubi- and aphthovirus proteases and delineation of a novel conserved domain associated with proteases of rubi-, alpha- and coronaviruses, *FEBS Lett.* **288**:201.

Herold, J., Raabe, T., Schelle-Prinz, B., and Siddell, S. G., 1993, Nucleotide sequence of the human coronavirus 229E RNA polymerase locus, *Virology* **195**:680.

Hofmann, M. A., Senanayake, S. D., and Brian, D. A., 1993, A translation-attenuating intraleader open reading frame is selected on coronavirus messenger RNAs during persistent infection, *Proc. Natl. Acad. Sci. USA* **90**:11733.

Holmes, K. V., 1991, Coronaviridae and their replication, in: *Fundamental Virology* (B. N. Fields and D. M. Knipe, eds.), pp. 471–488, Raven Press, New York.

Homberger, F. R., 1994, Nucleotide sequence comparison of the membrane protein genes of three enterotropic strains of mouse hepatitis virus, *Virus Res.* **31**:49.

Horsburgh, B. C., Brierley, I., and Brown, T. D., 1992, Analysis of a 9.6 kb sequence from the 3′ end of canine coronavirus genomic RNA, *J. Gen. Virol.* **73**:2849.

Kamahora, T., Soe, L. H., and Lai, M. M. C., 1989, Sequence analysis of nucleocapsid gene and leader RNA of human coronavirus OC43, *Virus Res.* **12**:1.

Kapke, P. A., and Brian, D. A., 1986, Sequence analysis of the porcine transmissible gastroenteritis coronavirus nucleocapsid protein gene, *Virology* **151**:41.

Kapke, P. A., Tung, F. Y. T., Hogue, B. G., Brian, D. A., Woods, R. D., and Wesley, R., 1988, The amino-terminal signal peptide on the porcine transmissible gastroenteritis coronavirus matrix protein is not an absolute requirement for membrane translocation and glycosylation, *Virology* **165**:367.

Kienzle, T. E., Abraham, S., Hogue, B. G., and Brian, D. A., 1990, Structure and orientation of expressed bovine coronavirus hemagglutinin-esterase protein, *J. Virol.* **64**:1834.

Kozak, M., 1987, An analysis of 5′-noncoding sequences from 699 vertebrate messenger RNAs, *Nucleic Acids Res.* **15**:8125.

Kozak, M., 1989, The scanning model for translation: An update, *J. Cell Biol.* **108**:229.

Kunita, S., Mori, M., and Terada, E., 1993, Sequence analysis of the nucleocapsid protein gene of rat coronavirus SDAV-681, *Virology* **193**:520.

Lai, M. M. C., 1990, Coronavirus—organization, replication and expression of genome, *Annu. Rev. Microbiol.* **44**:303.

La Monica, N., Banner, L. R., Morris, V. L., and Lai, M. M., 1991, Localization of extensive deletions in the structural genes of two neurotropic variants of murine coronavirus JHM, *Virology* **182**:883.

Lapps, W., Hogue, B. G., and Brian, D. A., 1987, Sequence analysis of the bovine coronavirus nucleocapsid and matrix protein genes, *Virology* **157**:47.

Le, S-Y., Sonenberg, N., and Maizel, J. V., Jr., 1994, Distinct structural elements and internal entry of ribosomes in mRNA3 encoded by infectious bronchitis virus, *Virology* **198:**405.

Lee, H. J., Shieh, C. K., Gorbalenya, A. E., Koonin, E. V., Lamonica, N., Tuler, J., Bagdzhadzhyan, A., and Lai, M. M. C., 1991, The complete sequence (22 kilobases) of murine coronavirus gene-1 encoding the putative proteases and RNA polymerase, *Virology* **180:**567.

Leibowitz, J. L., Weiss, S. R., Paavola, E., and Bond, C. W., 1982, Cell-free translation of murine coronavirus RNA, *J. Virol.* **43:**905.

Leibowitz, J. L., Perlman, S., Weinstock, G. M., De Vries, J. R., Budzilowicz, C. J., Weissemann, J. M., and Weiss, S. R., 1988, Detection of a murine coronavirus nonstructural protein encoded in a downstream open reading frame, *Virology* **164:**156.

Liu, D. X., and Inglis, S. C., 1991, Association of the infectious bronchitis virus 3c protein with the virion envelope, *Virology* **185:**911.

Liu, D. X., and Inglis, S. C., 1992a, Identification of two new polypeptides encoded by mRNA5 of the coronavirus infectious bronchitis virus, *Virology* **186:**342.

Liu, D. X., and Inglis, S. C., 1992b, Internal entry of ribosomes on a tricistronic mRNA encoded by infectious bronchitis virus [published erratum appears in *J. Virol.* 1992 **66**(11):6840], *J. Virol.* **66:**6143.

Liu, D. X., Cavanagh, D., Green, P., and Inglis, S. C., 1991, A polycistronic mRNA specified by the coronavirus infectious bronchitis virus, *Virology* **184:**531.

Luytjes, W., Sturman, L. S., Bredenbeek, P. J., Charité, J., Van der Zeijst, B. A. M., Horzinek, M. C., and Spaan, W. J., 1987, Primary structure of the glycoprotein E2 of coronavirus MHV-A59 and identification of the trypsin cleavage site, *Virology* **161:**479.

Luytjes, W., Bredenbeek, P. J., Noten, A. F. H., Horzinek, M. C., and Spaan, W. J. M., 1988, Sequence of mouse hepatitis virus A59 mRNA2: Indications for RNA-recombination between coronaviruses and influenza C virus, *Virology* **166:**415.

Mounir, S., and Talbot, P. J., 1993, Human coronavirus OC43 RNA 4 lacks two open reading frames located downstream of the S gene of bovine coronavirus, *Virology* **192:**355.

Page, K. W., Mawditt, K. L., and Britton, P., 1991, Sequence comparison of the 5′ end of messenger RNA 3 from transmissible gastroenteritis virus and porcine respiratory coronavirus, *J. Gen. Virol.* **72:**579.

Parker, M. M., and Masters, P. S, 1990, Sequence comparison of the N genes of five strains of the coronavirus mouse hepatitis virus suggests a three domain structure for the nucleocapsid protein, *Virology* **179:**463.

Parker, M. D., Cox, G. J., Deregt, D., Fitzpatrick, D. R., and Babiuk, L. A., 1989a, Cloning and *in vitro* expression of the gene for the E3 haemagglutinin glycoprotein of bovine coronavirus, *J. Gen. Virol.* **70:**155.

Parker, S. E., Gallagher, T. M., and Buchmeier, M. J., 1989b, Sequence analysis reveals extensive polymorphism and evidence of deletions within the E2 glycoprotein gene of several strains of murine hepatitis virus, *Virology* **173:**664.

Raabe, T., and Siddell, S., 1989, Nucleotide sequence of the human coronavirus HCV 229E mRNA 4 and mRNA 5 unique regions, *Nucleic Acids Res.* **17:**6387.

Raabe, T., Schelleprinz, B., and Siddell, S. G., 1990, Nucleotide sequence of the gene encoding the spike glycoprotein of human coronavirus HCV-229E, *J. Gen. Virol.* **71:**1065.

Rasschaert, D., Gelfi, J., and Laude, H., 1987, Enteric coronavirus TGEV: Partial sequence of the genomic RNA, its organization and expression, *Biochimie* **69:**591.

Rasschaert, D., Duarte, M., and Laude, H., 1990, Porcine respiratory coronavirus differs from transmissible gastroenteritis virus by a few genomic deletions, *J. Gen. Virol.* **71:**2599.

Rottier, P. J. M., Spaan, W. J. M., Horzinek, M. C., and van der Zeijst, B. A. M., 1981, Translation of three mouse hepatitis virus strain A59 subgenomic RNAs in Xenopus laevis oocytes, *J. Virol.* **38:**20.

Schaad, M. C., and Baric, R. S., 1993, Evidence for new transcriptional units encoded at the 3′ end of the mouse hepatitis virus genome, *Virology* **196:**190.

Schmidt, I., Skinner, M., and Siddell, S., 1987, Nucleotide sequence of the gene encoding the surface projection glycoprotein of coronavirus MHV-JHM, *J. Gen. Virol.* **68:**47.

Schreiber, S. S., Kamahora, T., and Lai, M. M. C., 1989, Sequence analysis of the nucleocapsid protein gene of human coronavirus 229E, *Virology* **169:**142.

Schwarz, B., Routledge, E., and Siddell, S. G., 1990, Murine coronavirus nonstructural protein NS2 is not essential for virus replication in transformed cells, *J. Virol.* **64:**4784.

Senanayake, S. D., Hofmann, M. A., Maki, J. L., and Brian, D. A., 1992, The nucleocapsid protein gene of bovine coronavirus is bicistronic, *J. Virol.* **66:**5277.

Shieh, C. K., Lee, H. J., Yokomori, K., La Monica, N., Makino, S., and Lai, M. M., 1989, Identification of a new transcriptional initiation site and the corresponding functional gene 2b in the murine coronavirus RNA genome, *J. Virol.* **63:**3729.

Siddell, S., 1983, Coronavirus JHM: Coding assignments of subgenomic mRNAs, *J. Gen. Virol.* **64:**113.

Smith, A. R., Boursnell, M. E. G., Binns, M. M., Brown, T. D. K., and Inglis, S. C., 1990, Identification of a new membrane-associated polypeptide specified by the coronavirus infectious bronchitis virus, *J. Gen. Virol.* **71:**3.

Soe, L. H., Shieh, C. K., Baker, S. C., Chang, M. F., and Lai, M. M., 1987, Sequence and translation of the murine coronavirus 5′-end genomic RNA reveals the N-terminal structure of the putative RNA polymerase, *J. Virol.* **61:**3968.

Spaan, W., Cavanagh, D., and Horzinek, M. C., 1988, Coronaviruses: Structure and genome expression, *J. Gen. Virol.* **69:**2939.

Taguchi, F., Ikeda, T., Makino, S., and Yoshikura, H., 1994, A murine coronavirus MHV-S isolate from persistently infected cells has a leader and two consensus sequences between the M and N genes, *Virology* **198:**355.

Tahara, S. M., Dietlin, T. A., Bergmann, C. C., Nelson, G. W., Kyuwa, S., Anthony, R. P., and Stohlman, S. A., 1994, Coronavirus translational regulation: Leader affects mRNA efficiency, *Virology* **202:**621.

Thiel, V., and Siddell, S. G., 1994, An internal ribosome entry site in the coding region of murine hepatitis virus mRNA 5, *J. Gen. Virol.* **75:**3041.

Tung, F. Y. T., Abraham, S., Sethna, M., Hung, S.-L., Sethna, P., Hogue, B., and Brian, D. A., 1992, The 9-kDa hydrophobic protein encoded at the 3′ end of the porcine transmissible gastroenteritis virus genome is membrane-associated, *Virology* **186:**676.

Vennema, H., 1991, *The Proteins of Feline Infectious Peritonitis Coronavirus: Their Biosynthesis and Involvement in Pathogenesis*, thesis, State University of Utrecht.

Vennema, H., Heijnen, L., Rottier, P. J., Horzinek, M. C., and Spaan, W. J., 1992a, A novel glycoprotein of feline infectious peritonitis coronavirus contains a KDEL-like endoplasmic reticulum retention signal, *J. Virol.* **66:**4951.

Vennema, H., Rossen, J. W., Wesseling, J., Horzinek, M. C., and Rottier, P. J., 1992b, Genomic organization and expression of the 3′ end of the canine and feline enteric coronaviruses, *Virology* **191:**134.

Verbeek, A., and Tijssen, P., 1991, Sequence analysis of the turkey enteric coronavirus nucleocapsid and membrane protein genes: A close genomic relationship with bovine coronavirus, *J. Gen. Virol.* **72:**1659.

Vlasak, R., Luytjes, W., Leider, J., Spaan, W., and Palese, P., 1988a, The E3 protein of bovine coronavirus is a receptor-destroying enzyme with acetylesterase activity, *J. Virol.* **62:**4686.

Vlasak, R., Luytjes, W., Spaan, W., and Palese, P., 1988b, Human and bovine coronaviruses recognize sialic acid-containing receptors similar to those of influenza C viruses, *Proc. Natl. Acad. Sci. USA* **85:** 4526.

Weiss, S. R., Zoltick, P. W., and Leibowitz, J. L., 1993, The ns 4 gene of mouse hepatitis virus (MHV), strain A 59 contains two ORFs and thus differs from ns 4 of the JHM and S strains, *Arch. Virol.* **129:**301.

Weiss, S. R., Hughes, S. A., Bonilla, P. J., Turner, J. D., Leibowitz, J. L., and Denison, M. R., 1994, Coronavirus polyprotein processing, *Arch. Virol. Suppl.* **9:**349.

Wesley, R. D., Cheung, A. K., Michael, D. D., and Woods, R. D., 1989, Nucleotide sequence of coronavirus TGEV genomic RNA: Evidence for 3 mRNA species between the peplomer and matrix protein genes, *Virus Res.* **13:**87.

Wesley, R. D., Woods, R. D., and Cheung, A. K., 1990, Genetic basis for the pathogenesis of transmissible gastroenteritis virus, *J. Virol.* **64:**4761.

Wesley, R. D., Woods, R. D., and Cheung, A. K., 1991, Genetic analysis of porcine respiratory coronavirus, an attenuated variant of transmissible gastroenteritis virus, *J. Virol.* **65:**3369.

Woloszyn, N., Boireau, P., and Laporte, J., 1990, Nucleotide sequence of the bovine enteric coronavirus BECV F15 mRNA5 and mRNA 6 unique regions, *Nucleic Acids Res.* **18:**1303.

Yokomori, K., and Lai, M. M., 1991, Mouse hepatitis virus S RNA sequence reveals that nonstructural proteins ns4 and ns5a are not essential for murine coronavirus replication, *J. Virol.* **65:**5605.

Yokomori, K., Banner, L. R., and Lai, M. M. C., 1991, Heterogeneity of gene expression of the hemagglutinin-esterase (HE) protein of murine coronaviruses, *Virology* **183:**647.

Yokomori, K., Stohlman, S. A., and Lai, M. M., 1993, The detection and characterization of multiple hemagglutinin-esterase (HE)-defective viruses in the mouse brain during subacute demyelination induced by mouse hepatitis virus, *Virology* **192:**170.

Yu, X., Bi, W., Weiss, S. R., and Leibowitz, J. L., 1994, Mouse hepatitis virus gene 5b protein is a new virion envelope protein, *Virology* **202:**1018.

Zhang, X. M., Kousoulas, K. G., and Storz, J., 1991a, Comparison of the nucleotide and deduced amino acid sequences of the S-genes specified by virulent and avirulent strains of bovine coronaviruses, *Virology* **183:**397.

Zhang, X. M., Kousoulas, K. G., and Storz, J., 1991b, The hemagglutinin/esterase glycoprotein of bovine coronaviruses: Sequence and functional comparisons between virulent and avirulent strains, *Virology* **185:**847.

Zhang, X., Kousoulas, K. G., and Storz, J., 1992, The hemagglutinin/esterase gene of human coronavirus strain OC43: Phylogenetic relationships to bovine and murine coronaviruses and influenza C virus, *Virology* **186:**318.

Zoltick, P. W., Leibowitz, J. L., Oleszak, E. L., and Weiss, S. R., 1990, Mouse hepatitis virus ORF-2A is expressed in the cytosol of infected mouse fibroblasts, *Virology* **174:**605.

CHAPTER 4

# Coronavirus Receptors

KATHRYN V. HOLMES AND SUSAN R. COMPTON

## I. CORONAVIRUS SPECIES AND TISSUE TROPISMS

Coronaviruses are highly species-specific in that they generally cause disease in only one host species (Möstl, 1990; Wege *et al.*, 1982). However, experimental inoculation of other species with several coronaviruses, either by artificial routes such as intracerebral inoculation or in the highly susceptible neonatal period, can result in mild or asymptomatic infection as shown in Table I. In general, coronaviruses only infect cells from their normal host species or from species that are susceptible to infection with an antigenically related coronavirus (Table II). Host-dependent differences in susceptibility to coronavirus infection can be demonstrated within a species. For example, different strains of inbred mice vary greatly in their susceptibility to infection with various murine hepatitis virus (MHV) strains (Bang and Warwick, 1960; Stohlman *et al.*, 1980; Wege *et al.*, 1982).

Coronaviruses exhibit strong tissue tropisms *in vivo*. While coronavirus infections are usually initiated in the respiratory and/or enteric epithelium and some coronavirus infections are limited to these tissues, several coronaviruses can spread to specific other organs such as the liver, lymphoid organs, brain, peritoneum, or kidney. Different isolates of the same or closely related coronaviruses may exhibit differing degrees of tropism for enteric or respiratory epithelium. For example, more than 20 isolates of MHV are divided into two biotypes based on their initial site of replication in either the respiratory or enteric tract (Barthold, 1986).

In this chapter, we will discuss the role of virus receptors as determinants of the species specificity, host strain specificity, and tissue tropism of corona-

KATHRYN V. HOLMES • Department of Pathology, Uniformed Services University of the Health Sciences, Bethesda, Maryland 20892. SUSAN R. COMPTON • Section of Comparative Medicine, Yale University, New Haven, Connecticut 06520.

*The Coronaviridae*, edited by Stuart G. Siddell, Plenum Press, New York, 1995.

TABLE I. Infection of Animals with Coronavirus

| Coronavirus | Mouse | Rat | Cow | Turkey | Human | Pig | Dog | Cat | References |
|---|---|---|---|---|---|---|---|---|---|
| MHV | **d, D**[a] | b[b] | | | | | | | a |
| SDAV | d, A | **d, D** | | | | | | | b, c |
| BCV | d | b | **d, A** | a[c] | A | | | | b, d, e, f, g |
| TCV | | | | **d, D** | | | | | |
| HEV | b | | | | | **d, A** | | | h |
| HCV-OC43 | d | | | | **d, D** | | | | b, i |
| HCV-229E | | | | | **d, D** | | | A | j |
| TGEV | | | | | | **d, D** | a | A | k, l, m |
| CCV | | | | | | a, A | **d, A** | a, A | n, o, p, q, r |
| FIPV | | | | | | d, A | | **d, D** | n, s |

[a]Boldface indicates infection of natural host; nonboldface indicates infection of a foreign species.
[b]Lower case letters indicate neonatal animal; upper case letters indicate weanling/adult animal.
[c]a or A indicates asymptomatic infection following natural route of exposure; b or B indicates disease following intracerebral inoculation only; d or D indicates disease following natural route of exposure.
References: a, Cheever *et al.*, 1949; b, Barthold *et al.*, 1990; c, Bhatt *et al.*, 1977; d, Kaye *et al.*, 1975; e, Akashi *et al.*, 1981; f, Dea *et al.*, 1991; g, Storz and Rott, 1981; h, Hirahara *et al.*, 1992; i, McIntosh *et al.*, 1967; j, Barlough *et al.*, 1985; k, Larson *et al.*, 1979; l, McClurkin *et al.*, 1970; m, Reynolds *et al.*, 1979; n, Woods *et al.*, 1981; o, Woods and Wesley, 1986; p, Barlough *et al.*, 1984; q, Stoddart *et al.*, 1988; r, McArdle *et al.*, 1992; s, Woods and Pederson, 1979.

virus infection. It is important to note, however, that susceptibility to coronavirus infection and disease depends on many other host-dependent factors in addition to receptor availability and specificity, including intracellular determinants of virus replication and immunological responses to virus infection.

## II. CORONAVIRUS ENVELOPE GLYCOPROTEINS THAT INTERACT WITH RECEPTORS

Coronavirus envelopes may exhibit either one or two envelope glycoproteins that interact with different cellular receptors. The virus attachment proteins are the spike glycoprotein (S), which forms the large peplomers characteristic of coronaviruses, and the hemagglutinin esterase (HE) glycoprotein, which forms short spikes in some coronaviruses in the MHV/OC43/BCV (bovine coronavirus) serogroup (Spaan *et al.*, 1988; Sturman and Holmes, 1985; Vlasak *et al.*, 1988a,b). Aspects of the S and HE glycoproteins that may be important for receptor interactions are summarized briefly here. These envelope glycoproteins will be discussed in detail in other chapters of this volume.

The S glycoprotein forms a trimeric spike that is responsible for virus attachment to specific receptor glycoproteins, mediates virus-induced membrane fusion, and induces neutralizing antibody and cell-mediated immune responses (Daniel *et al.*, 1993; Delmas and Laude, 1990; Rasschaert *et al.*, 1987). The S glycoprotein of BCV, like its HE glycoprotein, can bind to 9-*O*-acetylated sialic acid residues on host cell macromolecules (Schultze *et al.*, 1991a). Monoclonal antibodies to S inhibit virus infection and/or virus-induced cell fusion, and virions that lack S protein are not infectious (Daniel *et al.*, 1993; Holmes *et al.*, 1981; Rasschaert *et al.*, 1987). Sequencing of the genes encoding the S

TABLE II. Infection of Cultured Cells with Coronaviruses

| Coronavirus | Species from which cell lines were derived | | | | | | | | | References |
|---|---|---|---|---|---|---|---|---|---|---|
| | Mouse | Rat | Cow | Turkey | Human | Pig | Dog | Cat | Monkey | |
| MHV | **+**[a] | + | | | + | | | | | a, b |
| SDAV | + | **+** | | | | | | | | c |
| BCV | | | **+** | | + | + | + | | + | d, e, f, g, h |
| TCV | | | | **+** | + | | | | | i |
| HEV | | | | | + | **+** | + | | | h, j |
| HCV-OC43 | | | | | **+** | | | | + | k |
| HCV-229E | | | | | **+** | | | | | |
| TGEV | | | | | | **+** | + | + | + | l, m, n, o, p |
| CCV | | | | | + | | **+** | + | | p, q, r |
| FIPV | | | | | | | | **+** | | |

[a]Boldface indicates infection of cells from normal host species; nonboldface indicates infection of cells from a foreign species. References: a, Lucas *et al.*, 1977; b, McIntosh *et al.*, 1969; c, Percy *et al.*, 1989; d, Gerna *et al.*, 1981; e, King and Brian, 1982; f, Schultze and Herrler, 1992; g, Dea *et al.*, 1980; h, Schultze *et al.*, 1990; i, Dea *et al.*, 1989; j, Vautherot *et al.*, 1992; k, Kapikian *et al.*, 1972; l, Ishii *et al.*, 1992; m, Welter, 1965; n, Barlough *et al.*, 1983; o, Horzinek *et al.*, 1982; p, Woods, 1982; q, LaPorte *et al.*, 1980; r, Evermann *et al.*, 1981.

glycoproteins of many coronaviruses indicates that there is a high degree of variability between different viruses and between different strains of the same virus (Cavanagh *et al.*, 1988). Newly synthesized S proteins of many coronaviruses can be cleaved at a cluster of basic amino acids near the center of the protein by trypsin or related host proteases associated with the Golgi apparatus to yield two subunits called S1 and S2 (Lai, 1990; Spaan *et al.*, 1988). S1, the amino-terminal subunit located on the tip of the spike, exhibits a high degree of variability. S2, the carboxy-terminal subunit, is predicted to include the stalk of the spike, the transmembrane, and intracytoplasmic domains. The extent of S cleavage depends on the type of coronavirus, the virus strain, and the host cell used to grow the virus. For several coronaviruses including BCV and rat coronavirus (RCV), addition of trypsin to the culture medium increases cell-fusing activity, infectivity, and/or plaquing ability (Gaertner *et al.*, 1991; Storz *et al.*, 1981). Mutations at the cleavage site of the S glycoprotein of MHV are associated with reduced virulence, delayed cell fusion, and virus persistence (Gallagher *et al.*, 1991; Kant *et al.*, 1992). Large insertions or deletions that occur near specific sites in S1 of some coronaviruses have been associated with altered receptor interactions, different tissue tropism, and altered virulence (Gallagher *et al.*, 1990; La Monica *et al.*, 1991; Wang *et al.*, 1992). For example, two closely related porcine coronavirus strains, transmissible gastroenteritis virus (TGEV) and porcine respiratory coronavirus (PRCV), that differ markedly in the lengths of their S1 glycoproteins cause either enteric or respiratory disease, respectively (Laude *et al.*, 1993).

The HE glycoprotein found on some coronavirus envelopes is a 120 to 140-kDa disulfide-linked dimer of 60 to 65-kDa monomers that forms a fringe of short spikes (Herrler *et al.*, 1991; Vlasak *et al.*, 1988a). HE is encoded in the genomes of coronaviruses in the serogroup that includes MHV, BCV, OC43, hemagglutinating encephalitis virus (HEV), and turkey coronavirus (TCV), but not in the genomes of the avian infectious bronchitis virus (IBV) or HCV229E groups of coronaviruses (Dea *et al.*, 1989; Hogue and Brian, 1986; Schultze *et al.*, 1990; Vautherot *et al.*, 1992). Whether or not full-length HE protein is expressed in cells infected with a particular virus strain is determined by the availability of an mRNA with a 5′-terminal HE gene and by initiation and termination codons in the gene. Thus, HE protein is found on envelopes of some MHV strains but not others (Yokomori *et al.*, 1991). Coronavirus HE is closely related in amino acid sequence to the HE glycoprotein of influenza C (Herrler *et al.*, 1991; Luytjes *et al.*, 1988). These HE proteins bind to 9-*O*-acetylated sialic acid receptor moieties, Neu5,9$Ac_2$, that are found on glycoproteins, glycolipids, and gangliosides on membranes of erythrocytes, enterocytes, brain, and other cell types *in vivo* and on cell lines from several species (Herrler *et al.*, 1987, 1991; Varki, 1992, 1993). Acetylation at the 9 position of sialic acid is developmentally regulated in various cell types of different species. The acetyl-esterase of HE releases the acetyl groups from the neuraminic acid, destroying virus-binding activity (Schultze *et al.*, 1991b). Treatment of virions with pronase selectively removes HE and inhibits hemagglutination (Dea *et al.*, 1989; Hogue and Brian, 1986; King *et al.*, 1985). Monoclonal antibodies to the HE glycoproteins can inhibit hemagglutination, neutralize virus infectivity, and inhibit esterase activity (Parker *et al.*, 1989, 1990; Storz *et al.*, 1991; Vautherot *et al.*, 1992; Yoo *et al.*, 1992).

## III. CORONAVIRUS BINDING AND PENETRATION

Coronavirus envelope glycoproteins may interact with specific cell membrane receptors at several steps during the virus replicative cycle. To initiate infection, virions bind to cell membrane receptors by either the S or HE glycoprotein, or possibly by sequential binding of HE followed by binding of S. Next, conformational changes in either the viral attachment glycoprotein, S, or the receptor, or both may be induced by virus binding and lead to fusion of the viral envelope with host cell membranes. This membrane fusion event has similar properties to the cell–cell fusion that occurs in cell cultures or tissues of animals infected with some coronaviruses (Barthold, 1986). MHV-A59-induced fusion is optimal at neutral or mildly alkaline pH (Sturman *et al.*, 1990), like that induced by Sendai and HIV, which fuse at the plasma membrane (White, 1990). For other coronaviruses, such as HCV-229E, IBV, or some MHV variants, infected cells generally do not fuse at neutral pH, and the viral envelope may fuse with endosomal membranes rather than the plasma membrane (Gallagher *et al.*, 1991). After synthesis of viral genomic RNA and proteins, virions mature by budding at intracellular membranes in a special pre-Golgi compartment and are transported through the Golgi apparatus into post-Golgi vesicles that probably fuse with the plasma membrane to release virions from infected cells (Griffiths and Rottier, 1992; Tooze and Tooze, 1985). Newly synthesized cellular glycoproteins that serve as virus receptors are probably also present in the endoplasmic reticulum and Golgi membranes, so that interactions of envelope glycoproteins with nascent receptors could occur in these compartments. Coronavirus virions tend to adhere to the plasma membrane of the host cell following release, forming dense mats of adsorbed virions (Oshiro, 1973), which suggests that elution of virions from receptors is inefficient. The viral spike glycoprotein, S, which is not incorporated into virions at intracellular membranes, is transported to the plasma membrane where it can bind to specific virus receptors on adjacent uninfected cells and initiate cell–cell fusion.

Attachment and penetration are complex processes that have not yet been studied in great detail for most coronaviruses. Initial binding of a single HE or S glycoprotein to a receptor may be quickly followed by the recruitment of additional receptors beneath the virion. Cooperative interactions of many envelope glycoproteins with receptors could prolong adherence of the virion to the plasma membrane. Subsequent fusion of the viral envelope with the plasma membrane or endosomal membrane may depend on the affinity of S for its receptor, the fusing activity of the S glycoprotein, and the lipid composition of the cell membrane. The early events in coronavirus–cell interactions are technically difficult to study due to the high ratio of noninfectious to infectious virions in coronavirus preparations and the tendency of S1 to detach from the virions (Boursnell *et al.*, 1989; Cavanagh *et al.*, 1986; Sturman *et al.*, 1990). Studies on the interactions of purified virus attachment proteins with purified recombinant receptor proteins will provide additional useful information about receptor affinities.

## IV. CARCINOEMBRYONIC ANTIGEN-RELATED RECEPTORS FOR MHV

As shown in Table I, MHV naturally infects only mice, although young rats can be infected following intracerebral inoculation. Inbred mouse strains show varying degrees of susceptibility to different MHV strains. Princeton mice (PRI) are fully susceptible and die from MHV-2 infection (Bang and Warwick, 1960). Similarly, BALB/c, C57Bl/6, and DBA mice are fully susceptible to fatal hepatitis from MHV-3 infection (Le Prevost *et al.*, 1975). In contrast, C3H mice are semiresistant to MHV-2, which produces lower virus titers and only mild or inapparent disease, and to MHV-3, which gives adult C3H mice chronic neurological disease but not hepatitis (Virelizier *et al.*, 1975). A/J mice, which are also semiresistant to MHV-3, do not develop disease, although virus titers in the liver are comparable to those of BALB/c mice (Dindzans *et al.*, 1986; Le Prevost *et al.*, 1975). Of all inbred mouse strains tested, only SJL mice are resistant to MHV-JHM-induced encephalomyelitis, and peritoneal macrophages from SJL but not other mouse strains are resistant to infection with MHV-JHM and MHV-A59 (Knobler *et al.*, 1984; Stohlman and Frelinger, 1978).

To determine whether receptors play a role in the mouse strain specificity and tissue tropism of MHV infections, Boyle *et al.*, (1987) studied binding of MHV virions to membranes from murine enterocytes or hepatocytes, which are the natural target tissues of MHV. In a solid-phase assay, MHV-A59 bound to membranes from MHV-susceptible BALB/c mice, but not to membranes from MHV-resistant SJL mice. This assay also showed that binding of MHV-A59 to intestinal brush border membranes is highly species specific, and no virus binding could be detected on membranes from the rat, cat, dog, pig, human, rabbit, or chicken (Compton *et al.*, 1992). A virus-overlay protein blot assay showed that in BALB/c liver and intestine only one 110 to 120-kDa BALB/c membrane glycoprotein, called the mouse hepatitis virus receptor (MHVR), bound MHV-A59 virions (Boyle *et al.*, 1987). Virus binding to proteins from SJL membranes was not detected. An anti-receptor monoclonal antibody, MAb-CC1, was generated by immunization of SJL mice with BALB/c intestinal brush border membranes. MAb-CC1 bound to the 100 to 120-kDa MHVR and to a 55 to 58-kDa glycoprotein in immunoblots of BALB/c liver and intestine membranes, blocked binding of MHV to cultured BALB/c or C3H fibroblasts, and protected them from infection with MHV-A59 and other strains of MHV (Dveksler *et al.*, 1991, 1993a; Williams *et al.*, 1990). Binding of radiolabeled MAb-CC1 to membrane preparations from different mouse tissues showed that the highest levels of MHVR protein were on colon, small intestine, and liver (Williams *et al.*, 1991). MHVR was not detected on other BALB/c tissues or on any SJL/J tissues by this MAb-CC1 binding assay. Thus, the small intestine and liver, which are major target tissues for MHV infection *in vivo*, express the highest levels of

MHVR. The mouse strain-dependent expression of MHVR also correlates with the susceptibility of different mouse strains to MHV infection.

Immunoaffinity chromatography of Swiss-Webster mouse liver using MAb-CC1 yielded a mixture of a 100- to 120-kDa and a 55- to 58-kDa glycoprotein (Williams *et al.*, 1990). Comparison of the N-terminal amino acid sequence of the 100- to 120-kDa glycoprotein, called MHVR, with known sequences revealed strong homology with members of the carcinoembryonic antigen (CEA) family of glycoproteins in the immunoglobulin superfamily (Williams *et al.*, 1991). The sequence of a full-length cDNA clone encoding MHVR predicted a mature MHVR protein of 424 amino acids with 4 immunoglobulinlike domains, a transmembrane domain, and a short intracytoplasmic domain (Dveksler *et al.*, 1991). The protein is highly glycosylated, with 16 predicted N-linked glycosylation sites. The coding sequence of MHVR is identical to that of mmCGM1 (BgpA), a murine CEA-related glycoprotein in the biliary glycoprotein (BGP) subgroup (Beauchemin *et al.*, 1989a). Expression of MHVR in a vaccinia virus vector yielded a recombinant glycoprotein that binds MHV and MAb-CC1. Human cells and hamster cells are resistant to infection with MHV-A59 virions, but when transiently or stably transfected with cDNA encoding MHVR, they become susceptible to MHV-A59 and all other strains of MHV tested. Infection of these MHVR-transfected cells can be blocked by pretreatment of the cells with MAb-CC1 (Dveksler *et al.*, 1991, 1993a). These data show conclusively that MHVR is a functional receptor for MHV and suggest that the block in MHV replication in cells of non-murine species is in receptor-dependent virus attachment and entry.

On the MHVR glycoprotein, the binding sites for the S glycoprotein of MHV and for MAb-CC1 are located on the N-terminal Ig-like domain (Dveksler *et al.*, 1993b). Several other members of the immunoglobulin superfamily of glycoproteins serve as receptors for viruses, including ICAM-1 for human rhinoviruses, CD4 for HIV, and poliovirus receptor for polioviruses (White and Littman, 1989). For each of these receptors, as for MHVR, the virus binds to determinants on the N-terminal immunoglobulinlike domain.

In addition to MHVR, liver and enterocyte membranes from BALB/c mice and other MHV-susceptible strains contain a 55- to 58-kDa glycoprotein that is recognized by MAb-CC1 (Williams *et al.*, 1990). From these tissues, cDNAs derived from several alternatively spliced transcripts of MHVR were cloned and sequenced. The 55 to 58-kDa glycoprotein is encoded by one of these transcripts which includes only the N-terminal and fourth Ig-like domain and transmembrane and cytoplasmic domains (Dveksler *et al.*, 1993a). The second and third Ig-like domains were deleted by splicing. Therefore, this protein was called MHVR(2d) to distinguish it from the original receptor, which is now called MHVR(4d) (Dveksler *et al.*, 1993a). Additional splice variants generated 2 and 4 domain proteins with longer cytoplasmic domains. When transfected into hamster fibroblasts, each of these new MHVR-related cDNAs encodes an MHVR glycoprotein isoform that is recognized by MAb-CC1 and that makes the cells susceptible to infection with MHV-A59 and other strains of MHV, as evaluated by development of viral antigens in the cytoplasm of cells, or by synthesis and release of infectious virions (Dveksler *et al.*, 1993a; Yokomori and

Lai, 1992a,b). Thus, in MHV-susceptible strains of mice there are at least four alternatively spliced transcripts of the *Bgp* gene that encode functional MHV receptors.

MHV-resistant adult SJL/J mice were studied to determine whether they express proteins homologous to MHVR. A polyclonal rabbit antibody directed against the 15 amino acids at the amino-terminus of MHVR and a polyclonal antibody directed against human CEA both recognized 55-kDa and 105- to 115-kDa glycoproteins in SJL/J intestinal brush border membranes (Williams *et al.*, 1990). In addition, reverse transcriptase polymerase chain reaction (RTPCR), Northern blot, and RNase protection assays were used to characterize cDNAs from SJL/J mice and outbred CD-1 mice that encode glycoproteins homologous to MHVR (Dveksler *et al.*, 1993a; Yokomori and Lai, 1992a,b). Sequencing of these cDNAs revealed major differences in the amino acid sequence of the N-terminal domain between MHVR and its SJL/J homologue, as well as minor differences in other domains. One of three potential N-linked glycosylation sites from the N-terminal Ig domain of MHVR was missing in the SJL homologue, which could result in the slightly smaller size of the resulting glycoproteins. The coding sequence of the SJL homologue is identical to that of mmCGM2 derived from outbred CD-1 mice (Beauchemin *et al.*, 1989b; McCuaig *et al.*, 1993). Both 2 domain and 4 domain isoforms of mmCGM2 were identified, and a transcript encoding an mmCGM2 with a longer cytoplasmic domain was cloned. No MHVR transcripts or genes could be found in SJL tissues by RTPCR or RNase protection assays, which indicates that MHVR and mmCGM2 are alleles of the same murine *Bgp* gene (Dveksler *et al.*, 1993a; Nedellec *et al.*, 1993). None of the mmCGM2 isoforms was recognized by MAb-CC1, and none bound MHV-A59 virions in solid phase binding assays or virus overlay protein blot assays (Dveksler *et al.*, 1993a; Williams *et al.*, 1990). Therefore, it was surprising that each of the recombinant mmCGM2 isoforms when expressed in hamster fibroblasts made the cells susceptible to infection with MHV-A59 (Dveksler *et al.*, 1993a). Differences in the characteristics of the N-terminal Ig domains of the BALB/c and SJL isoforms could account for the observed differences in virus and MAb-CC1 binding and receptor activity between MHVR and mmCGM2 isoforms. Transfection of mmCGM2 cDNA into SJL cell lines made them susceptible to MHV-A59 infection (Dveksler *et al.*, 1993a), suggesting that the block in coronavirus replication in these cells is at an early step in virus attachment or entry. The Bgp glycoproteins expressed in adult SJL mice may have weaker MHV-binding activity than the MHVR glycoprotein, and overexpression of a recombinant SJL protein may at least partially compensate for its weak binding activity. Direct measurement of affinity between MHV spike glycoproteins and receptor isoforms will be required to test this hypothesis.

Several cellular functions of the Bgp protein isoforms of mice and rats have been identified. These proteins may bind calmodulin and can serve as calcium-independent or calcium-dependent homophilic cell adhesion molecules (Culic *et al.*, 1992; Turbide *et al.*, 1991). The rat Bgp protein has ecto-ATPase activity (Lin and Guidotti, 1989), and Bgp proteins act as bile acid transporters in the liver (Sippel *et al.*, 1993). The expression of Bgps and other CEA-related glycoproteins is regulated in a tissue-dependent manner during development, and is often elevated in transformed cells (Beauchemin *et al.*, 1987). Consequently, a single cell type from an outbred CD-1 mouse or an SJL X BALB/c hybrid mouse may express several isoforms of MHVR as well as several isoforms of mmCGM2. How the expression of each of these related glycoproteins is regulated, how the proteins are targeted, and how they interact with each other or with other cell membrane molecules are not yet known. Immunolabeling of BALB/c tissues with MAb-CC1 shows that large amounts of MHVR antigen are expressed in BALB/c mice on the intestinal brush borders membranes, respiratory epithelium, bile canaliculi, endothelial cells, and proximal tubules of the kidney (Godfraind, in preparation). MHVR has also been detected by fluorescence activated cell sorting (FACS) analysis on B cells, macrophages, and fibroblast cell lines (Coutelier *et al.*, 1994). Thus, large amounts of receptor glycoproteins are expressed on the membranes of the respiratory and enteric tracts where virus enters the body and on hepatocytes which are a major target tissue for MHV infection, and receptor is also expressed on macrophages and endothelial cells which could play a role in spread of infection by viremia. It is not yet clear whether the expression of different levels of receptor isoforms in different tissues affect the tissue tropisms of diverse MHV strains.

## V. AMINOPEPTIDASE N RECEPTORS FOR TGEV AND HCV- 229E

Coronaviruses serologically related to HCV-229E include canine coronavirus (CCV), feline infectious peritonitis virus (FIPV) and feline enteric coronavirus (FECV), and TGEV and PRCV. Unlike MHV, which binds only to intestinal brush border membranes from MHV-susceptible strains of mice, HCV-229E and serologically related coronaviruses bind to receptors on brush border membranes of several unrelated species (Table III) (Compton, 1988; Compton *et al.*, 1992). The species specificity of infection with these viruses appears to be determined by a step subsequent to virus binding.

The strategy used to identify cellular receptors for members of the HCV-229E group of coronaviruses was similar to that used to identify the MHV receptors. Monoclonal antibodies directed against membranes of virus-susceptible cells were prepared and tested for the ability to block virus infection, and the protein precipitated by that antibody was identified, cloned, and expressed in foreign species to be tested for virus receptor activity (Delmas *et al.*, 1992; Yeager *et al.*, 1992). Several MAbs directed against porcine cell membrane proteins were found to prevent infection with TGEV (Delmas *et al.*, 1992). These antibodies immunoprecipitated a 130- to 150-kDa membrane glycoprotein. N-terminal amino acid sequencing of the immunoprecipitated protein identified it as porcine aminopeptidase N (APN), a metalloprotease that is a class 2 glycoprotein (Ashmun and Look, 1990; Look *et al.*, 1986). Direct binding of TGEV to purified porcine APN (pAPN) was demonstrated (Delmas *et al.*, 1992). When the cDNA encoding pAPN was cloned and expressed in TGEV-resistant canine cells, the cells became susceptible to TGEV infection (Delmas *et al.*, 1992). Thus, pAPN is a functional receptor for TGEV.

TABLE III. Binding of Coronaviruses to Intestinal Brush Border Membranes of Different Species

| | Source of brush border membranes | | | | |
|---|---|---|---|---|---|
| Virus | Mouse | Human | Dog | Cat | Pig |
| MHV | **++**[a] | — | — | — | — |
| HCV-229E | ± | **++** | + | + | + |
| CCV | — | ++ | **++** | ++ | ++ |
| FIPV | — | ++ | ++ | **++** | ++ |
| TGEV | — | ++ | ++ | + | **++** |

[a]Boldface indicates binding to membranes from normal host; non-boldface indicates binding to membranes from a foreign species. ++, Strong binding; +, moderate binding; ±, weak binding; —, no detectable binding.

Similarly, a monoclonal antibody, MAb-RBS, directed against membrane glycoproteins on human cell lines was found to block HCV-229E infection of human WI-38 fibroblasts and HL-60 cells (Yeager *et al.*, 1992). This MAb also immunoprecipitates a 150-kDa glycoprotein that is recognized by MAbs directed against human APN (hAPN), which is also called CD13 (Look *et al.*, 1989). Murine fibroblasts are normally resistant to infection by HCV-229E, but they become susceptible to the virus following transfection with a cDNA clone that encodes hAPN, and infection is blocked by MAb-RBS (Yeager *et al.*, 1992). Interestingly, although hAPN is a functional receptor for HCV-229E, it is not a good receptor for TGEV which is serologically related (Holmes, in preparation). This observation suggests that other species-specific host factors may block critical steps in virus replication that follow virus binding to APN to determine the species specificity of infection with this group of coronaviruses.

The amino acid sequences and domain structures of APN and Bgp are unrelated, although both cellular glycoproteins can serve as receptors for one or more coronaviruses. APN is a zinc-binding protein that is found on the plasma membrane as a homodimer with its N-terminal domain anchored in the cytoplasm (Ashmun and Look, 1990). A short transmembrane region is followed by a stalk domain, and the catalytic site of the enzyme is on its globular carboxy-terminal domain near a HELAH consensus sequence.

*In vivo*, APN is found on enterocyte brush border membranes where it removes the N-terminal amino acids from oligopeptides generated in the gut lumen by endopeptidases. APN is expressed on myeloid tumor cells and may play a role in metastasis of melanoma, and APN at synaptic junctions helps to degrade neuroactive peptides (Ashmun and Look, 1990; Hersh, 1985). Because the S glycoproteins of some coronaviruses require protease cleavage to activate cell fusion and/or viral infectivity, it was important to determine whether the protease activity of APN plays a role in its receptor function. Several lines of evidence suggest that the S glycoprotein of HCV-229E may bind at or near the catalytic site of hAPN. MAb-RBS, which blocks virus binding to hAPN, also inhibits enzyme activity (Yeager *et al.*, 1992). A cDNA that encodes hAPN containing a 39 amino acid deletion that includes the HELAH sequence encoded a protein that was not recognized by MAb-RBS and could not serve as a receptor for HCV-229E when expressed in mouse fibroblasts (Ashmun *et al.*, 1992; Yeager *et al.*, 1992). Finally, chelation of zinc inactivated both the enzymatic activity of hAPN and its HCV-229E-receptor activity (Yeager *et al.*, 1992). Other evidence strongly suggests that the protease activity of hAPN or pAPN is not required for its coronavirus receptor function. Small inhibitors of the protease, actinonin and bestatin, did not block HCV-229E or TGEV infection (Delmas *et al.*, 1992; Yeager *et al.*, 1992), and point mutations that abolish enzyme activity did not reduce virus receptor activity (Yeager *et al.*, 1992; Ashmun and Holmes, in preparation). Recently, it has been shown that determinants that are essential for the TGEV-receptor interaction reside within a domain of pAPN that is distinct from the enzymatic site (Delmas *et al.*, 1994). In summary, the serologically related coronaviruses TGEV and HCV-229E, which cause disease in porcine intestine and human respiratory tract, respectively, both utilize APN glycoproteins of their normal host species as receptors. The APN glycoproteins appear to be used as receptors because of their expression on respiratory and intestinal epithelial cell membranes, and their enzymatic activity does not appear to play a role in coronavirus infection.

## VI. CARBOHYDRATE RECEPTORS FOR CORONAVIRUSES

The observation that the HE glycoproteins of several coronaviruses in the MHV/BCV/OC43 serogroup bind to Neu5,9Ac$_2$ residues on erythrocytes like the HE glycoprotein of influenza C, suggested that this carbohydrate moiety might serve as a receptor determinant for these coronaviruses (Schultze *et al.*, 1990, 1991b; Vlasak *et al.*, 1988a,b). The acetyl esterase activities of the HE glycoproteins of BCV, HCV-OC43, and HEV, like that of influenza C virus, cleaves acetyl groups from Neu5,9Ac$_2$, rendering the erythrocytes resistant to agglutination by these viruses. Because 9-*O*-acetylated neuraminic acid moieties are found on a variety of cell types in many species (Varki, 1992, 1993), any coronaviruses that use this carbohydrate as their only receptor would be expected to have broader host range and tissue tropism than those of coronaviruses that use species-specific determinants of a particular glycoprotein such as murine CEA-related glycoproteins or APN. Several observations on BCV support this hypothesis. BCV, which expresses high levels of HE and esterase activity (Schultze and Herrler, 1992; Vlasak *et al.*, 1988a,b), infects cultured cells and tissues of more species than most other coronaviruses (Tables I and II). Pretreatment of several BCV-susceptible cell lines with neuraminidase or acetylesterase rendered the cells resistant to BCV infection, and resialylation with Neu5,9Ac$_2$ but not with nonacetylated sialic acids, restored susceptibility to BCV infection (Schultze and Herrler, 1992). Treatment of BCV virions with diisopropyl fluorophosphate, which inhibits the esterase activity of the HE glycoprotein, drastically reduces BCV infectivity (Vlasak *et al.*, 1988b).

Although coronavirus HE may facilitate virus binding to cells, interaction of the viral S glycoprotein with a specific glycoprotein receptor may be required for virus penetration. This hypothesis has not yet been tested directly because viruses that express HE but not S glycoprotein are not available. However, MHV variants that express S but not HE are infectious *in vitro* and *in vivo*, indicating that HE is not essential for MHV infectivity (La Monica *et al.*, 1991; Yokomori *et al.*, 1991). The role of HE expression in determining coronavirus virulence is not yet understood. Because the purified S glycoproteins of BCV and HCV-OC43, in addition to the HE glycoproteins, bind Neu5,9$Ac_2$ (Kunkel and Herrler, 1993; Schultze *et al.*, 1991a), the experiments on modulation of virus susceptibility by removal or replacement of Neu5,9$Ac_2$ must be interpreted with caution. It is possible that some coronavirus virions first bind to cells by interaction of HE with 9-*O*-acetylated neuraminic acid residues on any macromolecule, and then their S glycoproteins bind to Neu5,9$Ac_2$ or another moiety on a specific receptor glycoprotein, leading to virus penetration and uncoating.

The expression of HE glycoprotein could affect the entry of coronaviruses into cells in several ways. HE might mediate initial attachment to cells and stabilize virus binding until S interacts with protein receptors to initiate membrane fusion. If the esterase activity of HE can cleave Neu5,9$Ac_2$ on mucins or on cells such as erythrocytes that lack the capability to internalize virus, then expression of HE would increase the probability that a virion would reach receptors on susceptible cells. HE in the Golgi apparatus of infected cells might cleave acetyl residues from Neu5,9$Ac_2$ on nascent cell membrane molecules, resulting in cells deficient in Neu5,9$Ac_2$. This might facilitate release of virions from the infected cells.

The coronaviruses, like picornaviruses and retroviruses, have evolved to utilize a variety of different receptor determinants for entry into susceptible cells. The characterization of receptor-binding sites on viral envelope glycoproteins and the selection and characterization of mutant viruses that utilize different receptors may elucidate the molecular mechanisms that determine coronavirus tissue tropism and species specificity.

## VII. REFERENCES

Akashi, H., Inaba, Y., Miura, Y., Sato, K., Tokuhisa, S., Asagi, M., and Hayashi, Y., 1981, Propagation of the Kakegawa strain of bovine coronavirus in suckling mice, rats and hamsters, *Arch. Virol.* **67**:367.

Ashmun, R. A., and Look, A. T., 1990, Metalloprotease activity of CD13/aminopeptidase N on the surface of human myeloid cells, *Blood* **75**:462.

Ashmun, R. A., Shapiro, L. H., and Look, A. T., 1992, Deletion of the zinc-binding motif of CD13/aminopeptidase N molecules results in loss of epitopes that mediate binding of inhibitory antibodies, *Blood* **79**:3344.

Bang, F. B., and Warwick, A., 1960, Mouse macrophages as host cells for the mouse hepatitis virus and the genetic basis of their susceptibility, *Proc. Natl. Acad. Sci. USA* **46**:1065.

Barlough, J. E., Jacobson, R. H., and Scott, F. W., 1983, Macrotiter assay for coronavirus neutralizing activity in cats using a canine continuous cell line (A-72), *Lab. Anim. Sci.* **33**:567.

Barlough, J. E., Stoddart, C. A., Soresso, G. P., Jacobson, R. H., and Scott, F. W., 1984, Experimental inoculation of cats with canine coronavirus and subsequent challenge with feline infectious peritonitis virus, *Lab. Anim. Sci.* **34**:592.

Barlough, J. E., Johnson-Lussenburg, C. M., Stoddart, C. A., Jacobson, R. H., and Scott, F. W., 1985,

Experimental inoculation of cats with human coronavirus 229E and subsequent challenge with feline infectious peritonitis virus, *Can. J. Comp. Med.* **49**:303.

Barthold, S. W., 1986, Mouse hepatitis virus biology and epizootiology, in: *Viral and Mycoplasmal Infections of Laboratory Rodents. Effects on Biomedical Research* (P.N. Bhatt, R.O. Jacoby, H.C. Morse, III, and A.E. New, eds.), p. 571, Academic Press, Orlando, FL.

Barthold, S. W., de Souza, M. S., and Smith, A. L., 1990, Susceptibility of laboratory mice to intranasal and contact infection with coronaviruses of other species, *Lab. Anim. Sci.* **40**:481.

Beauchemin, N. S., Benchimol, S., Cournoyer, D., Fuks, A., and Stanners, C. P., 1987, Isolation and characterization of full-length functional cDNA clones for human carcinoembyronic antigen, *Mol. Cell Biol.* **7**(9):3221.

Beauchemin, N., Turbide, C., Afar, D., Bell, J., Raymond, M., Stanners, C. P., and Fuks, A., 1989a, A mouse analogue of the human carcinoembryonic antigen, *Cancer Res.* **49**:2017.

Beauchemin, N., Turbide, C., Huang, J. Q., Benchimol, S., Jothy, S., Shirota, K., Fuks, A., and Stanners, C. P., 1989b, Studies on the function of carcinoembryonic antigen, in: *The Carcinoembryonic Antigen Gene Family* (A. Yachi and J.E. Shively, eds.), p. 49, Elsevier Science Publishers BV (Biomedical Division), New York.

Bhatt, P.N., Jacoby, R. O., and Jonas, A. M., 1977, Respiratory infection of mice with sialodacryoadenitis virus, a coronavirus of rats, *Infect. Immun.* **18**:823.

Boursnell, M. E., Binns, M. M., Brown, T. D., Cavanagh, D., and Tomley, F. M., 1989, Molecular biology of avian infectious bronchitis virus, *Prog. Vet. Microbiol. Immunol.* **5**:65.

Boyle, J. F., Weismiller, D. G., and Holmes, K. V., 1987, Genetic resistance to mouse hepatitis virus correlates with absence of virus-binding activity on target tissues, *J. Virol.* **61**:185.

Cavanagh, D., Davis, P. J., Darbyshire, J. H., and Peters, R. W., 1986, Coronavirus IBV: Virus retaining spike glycopolypeptide S2 but not S1 is unable to induce virus-neutralizing or haemagglutination-inhibiting antibody, or induce chicken tracheal protection, *J. Gen. Virol.* **67**:1435.

Cavanagh, D., Davis, P. J., and Mockett, A. P., 1988, Amino acids within hypervariable region 1 of avian coronavirus IBV (Massachusetts serotype) spike glycoprotein are associated with neutralization epitopes, *Virus Res.* **11**:141.

Cheever, F. S., Daniels, J. B., Pappenheimer, A. M., and Bailey, O. T., 1949, A murine virus (JHM) causing disseminated encephalomyelitis with extensive destruction of myelin. I. Isolation and biological properties of the virus, *J. Exp. Med.* **90**:181.

Compton, S. R., 1988, *Coronavirus Attachment and Replication*, Ph.D. dissertation, The Uniformed Services University of the Health Sciences, Bethesda, MD.

Compton, S. R., Stephensen, C. B., Snyder, S. W., Weismiller, D. G., and Holmes, K. V., 1992, Coronavirus species specificity: Murine coronavirus binds to a mouse-specific epitope on its carcinoembryonic antigen-related receptor glycoprotein, *J. Virol.* **66**:7420.

Coutelier, J.-P., Godfraind, C., Dveksler, G. S., Wysocka, M., Cardellichio, C. B., Noel, H., and Holmes, K. V., 1994, B lymphocyte and macrophage expression of carcinoembryonic antigen-related adhesion molecules that serve as receptors for murine coronavirus, *Eur. J. Immunol.* **24**:1383.

Culic, O., Huang, Q. H., Flanagan, D., Hixson, D., and Lin, S. H., 1992, Molecular cloning and expression of a new rat liver cell-CAM105 isoform. Differential phosphorylation of isoforms, *Biochem. J.* **285**:47.

Daniel, C., Anderson, R., Buchmeier, M. J., Fleming, J. O., Spaan, W. J., Wege, H., and Talbot, P. J., 1993, Identification of an immunodominant linear neutralization domain on the S2 portion of the murine coronavirus spike glycoprotein and evidence that it forms part of complex tridimensional structure, *J. Virol.* **67**:1185.

Dea, S., Roy, R. S., and Begin, M. E., 1980, Bovine coronavirus isolation and cultivation in continuous cell lines, *Am. J. Vet. Res.* **41**:30.

Dea, S., Garzon, S., and Tijssen, P., 1989, Isolation and trypsin-enhanced propagation of turkey enteric (bluecomb) coronaviruses in a continuous human rectal adenocarcinoma cell line, *Am. J. Vet. Res.* **50**:1310.

Dea, S., Verbeek, A., and Tijssen, P., 1991, Transmissible enteritis of turkeys: Experimental inoculation studies with tissue-culture-adapted turkey and bovine coronaviruses, *Avian Dis.* **35**:767.

Delmas, B., and Laude, H., 1990, Assembly of coronavirus spike protein into trimers and its role in epitope expression, *J. Virol.* **64:**5367.

Delmas, B., Gelfi, J., L'Haridon, R., Vogel, L. K., Sjöström, H., Norën, O., and Laude, H., 1992, Aminopeptidase N is a major receptor for the entero-pathogenic coronavirus TGEV, *Nature* **357:**417.

Delmas, B., Gelfi, J., Kut, E., Sjöström, H., Noren, O., and Laude, H., 1994, Determinants essential for the transmissible gastroenteritis virus–receptor interaction reside within a domain of aminopeptidase-N that is distinct from the enzymatic site, *J. Virol.* **68:**5216.

Dindzans, V. J., Skamene, E., and Levy, G. A., 1986, Susceptibility/resistance to mouse hepatitis virus strain 3 and macrophage procoagulant activity are genetically linked and controlled by two non-H2-linked genes, *J. Immunol.* **137:**2355.

Dveksler, G. S., Pensiero, M. N., Cardellichio, C. B., Williams, R. K., Jiang, G. S., Holmes, K. V., and Dieffenbach, C. W., 1991, Cloning of the mouse hepatitis virus (MHV) receptor: Expression in human and hamster cell lines confers susceptibility to MHV, *J. Virol.* **65:**6881.

Dveksler, G. S., Dieffenbach, C. W., Cardellichio, C. B., McCuaig, K., Pensiero, M. N., Jiang, G. S., Beauchemin, N., and Holmes, K. V., 1993a, Several members of the mouse carcinoembryonic antigen-related glycoprotein family are functional receptors for the coronavirus mouse hepatitis virus- A59, *J. Virol.* **67:**1.

Dveksler, G. S., Pensiero, M. N., Dieffenbach, C. W., Cardellichio, C. B., Basile, A. A., Elia, P. E., and Holmes, K. V., 1993b, Mouse hepatitis virus strain A59 and blocking antireceptor monoclonal antibody bind to the N-terminal domain of cellular receptor, *Proc. Natl. Acad. Sci. USA* **90:**1716.

Evermann, J. F., Baumgartener, L., Ott, R. L., Davis, E. V., and McKeirnan, A. J., 1981, Characterization of a feline infectious peritonitis virus isolate, *Vet. Pathol.* **18:**256.

Gaertner, D. J., Smith, A. L., Paturzo, F. X., and Jacoby, R. O., 1991, Susceptibility of rodent cell lines to rat coronaviruses and differential enhancement by trypsin or DEAE-dextran, *Arch. Virol.* **118:**57.

Gallagher, T. M., Parker, S. E., and Buchmeier, M. J., 1990, Neutralization-resistant variants of a neurotropic coronavirus are generated by deletions within the amino-terminal half of the spike glycoprotein, *J. Virol.* **64:**731.

Gallagher, T. M., Escarmis, C., and Buchmeier, M. J., 1991, Alteration of the pH dependence of coronavirus-induced cell fusion: Effect of mutations in the spike glycoprotein, *J. Virol.* **65:**1916.

Gerna, G., Cereda, P. M., Cattaneo, M. G., Battaglia, M., and Gerna, M. T., 1981, Antigenic and biological relationships between human coronavirus OC43 and neonatal calf diarrhoea coronavirus, *J. Gen. Virol.* **54:**91.

Griffiths, G., and Rottier, P., 1992, Cell biology of viruses that assemble along the biosynthetic pathway, *Semin. Cell Biol.* **3:**367.

Herrler, G., Reuter, G., Rott, R., Klenk, H. D., and Schauer, R., 1987, *N*-acetyl-9-*O*-acetylneuraminic acid, the receptor determinant for influenza C virus, is a differentiation marker on chicken erythrocytes, *Biol. Chem. Hoppe Seyler* **368:**451.

Herrler, G., Szepanski, S., and Schultze, B., 1991, 9-*O*- acetylated sialic acid, a receptor determinant for influenza C virus and coronaviruses, *Behring. Inst. Mitt.* **89:**177.

Hersh, L. B., 1985, Characterization of membrane-bound aminopeptidases from rat brain: Identification of the enkephalin-degrading aminopeptidase, *J. Neurochem.* **44:**1427.

Hirahara, T., Yasuhara, H., Yamanaka, M., Matsui, O., Kimura, Y., Izumida, A., Yoshiki, K., Sato, K., Kodama, K., Sasaki, N., *et al.*, 1992, Pathogenicity of porcine hemagglutinating encephalomyelitis virus for mouse and guinea pig, *J. Vet. Med. Sci.* **54:**163.

Hogue, B. G., and Brian, D. A., 1986, Structural proteins of human respiratory coronavirus OC43, *Virus Res.* **5:**131.

Holmes, K. V., Doller, E. W., and Sturman, L. S., 1981, Tunicamycin resistant glycosylation of coronavirus glycoprotein: Demonstration of a novel type of viral glycoprotein, *Virology* **115:**334.

Horzinek, M. C., Lutz, H., and Pedersen, N. C., 1982, Antigenic relationships among homologous structural polypeptides of porcine, feline, and canine coronaviruses, *Infect. Immun.* **37:**1148.

Ishii, H., Watanabe, I., Mukamoto, M., Kobayashi, Y., and Kodama Y., 1992, Adaption of transmissible gastroenteritis virus to growth in non-permissive Vero cells, *Arch. Virol.* **122:**201.

Kant, A., Koch, G., van Roozelaar, D. J., Kusters, J. G., Poelwijk, F. A., and van der Zeijst, B. A., 1992, Location of antigenic sites defined by neutralizing monoclonal antibodies on the S1 avian infectious bronchitis virus glycopolypeptide, *J. Gen. Virol.* **73:**591.

Kapikian, A. Z., James, Jr., H. D., Kelly, S. J., King, L. M., Vaughn, A. L., and Chanock, R. M., 1972, Hemadsorption by coronavirus strain OC43, *Proc. Soc. Exp. Biol. Med.* **139:**179.

Kaye, H.S., Yarbough, W. B., and Reed, C. J., 1975, Calf diarrhoea coronavirus, *Lancet* **2:**509.

King, B., and Brian, D. A., 1982, Bovine coronavirus structural proteins, *J. Virol.* **42:**700.

King, B., Potts, B. J., and Brian, D. A., 1985, Bovine coronavirus hemagglutinin protein, *Virus Res.* **2:**53.

Knobler, R. L., Tunison, L. A., and Oldstone, M. B., 1984, Host genetic control of mouse hepatitis virus type 4 (JHM strain) replication. I. Restriction of virus amplification and spread in macrophages from resistant mice, *J. Gen. Virol.* **65:**1543.

Kunkel, F., and Herrler, G., 1993, Structural and functional analysis of the surface protein of human coronavirus OC43, *Virology* **195:**195.

Lai, M. M., 1990, Coronavirus: Organization, replication and expression of genome, *Annu. Rev. Microbiol.* **44:**303.

La Monica, N., Banner, L. R., Morris, V. L., and Lai, M. M., 1991, Localization of extensive deletions in the structural genes of two neurotropic variants of murine coronavirus JHM, *Virology* **182:**883.

LaPorte, J., Bobulesco, P., and Rossi, F., 1980, Une lignee cellulaire particulierememt sensible a la replication du coronavirus enterique bovine: les cellules HRT18. *Compt. Rend. Hebdomadaire Sances Acad Sci* (Serie D) **290:**623.

Larson, D. J., Morehouse, L. G., Solorzano, R. F., and Kinden, D. A., 1979, Transmissible gastroenteritis in neonatal dogs: Experimental intestinal infection with transmissible gastroenteritis virus, *Am. J. Vet. Res.* **40:**477.

Laude, H., Van Reeth, K., and Pensaert, M., 1993, Porcine respiratory coronavirus: molecular features and virus–host interactions, *Vet. Res.* **24:**125.

Le Prevost, C., Levy-Leblond, E., Virelizier, J. L., and Dupuy, J. M., 1975, Immunopathology of mouse hepatitis virus type 3 infection. Role of humoral and cell-mediated immunity in resistance mechanisms, *J. Immunol.* **114:**221.

Lin, S. H., and Guidotti, G., 1989, Cloning and expression of a cDNA coding for a rat liver plasma membrane ecto-ATPase. The primary structure of the ecto-ATPase is similar to that of the human biliary glycoprotein I, *J. Biol. Chem.* **264:**14408.

Look, A. T., Peiper, S. C., Rebentisch, M. B., Ashmun, R. A., Roussel, M. F., Lemons, R. S., Le Beau, M. M., Rubin, C. M., and Sherr, C. J., 1986, Molecular cloning, expression, and chromosomal localization of the gene encoding a human myeloid membrane antigen (gp150), *J. Clin. Invest.* **78:**914.

Look, A. T., Ashmun, R. A., Shapiro, L. H., and Peiper, S. C., 1989, Human myeloid plasma membrane glycoprotein CD13 (gp150) is identical to aminopeptidase N, *J. Clin. Invest.* **83:**1299.

Lucas, A., Flintoff, W., Anderson, R., Percy, D., Coulter, M., and Dales, S., 1977, *In vivo* and *in vitro* models of demyelinating diseases: Tropism of the JHM strain of murine hepatitis virus for cells of glial origin, *Cell* **12:**553.

Luytjes, W., Bredenbeek, P. J., Noten, A. F., Horzinek, M. C., and Spaan, W. J., 1988, Sequence of mouse hepatitis virus A59 mRNA 2: Indications for RNA recombination between coronaviruses and influenza C virus, *Virology* **166:**415.

McArdle, F., Bennett, M., Gaskell, R. M., Tennant, B., Kelly, D. F., and Gaskell, C. J., 1992, Induction and enhancement of feline infectious peritonitis by canine coronavirus, *Am. J. Vet. Res.* **53:**1500.

McClurkin, A. W., Stark, S. L., and Norman, J. O., 1970, Transmissible gastroenteritis (TGE) of swine: The possible role of dogs in the epizootiology of TGE, *Can. J. Comp. Med.* **34:**347.

McCuaig, K., Rosenberg, M., Nedellec, P., Turbide, C., and Beauchemin, N., 1993, Expression of the *Bgp* gene and characterization of mouse colon biliary glycoprotein isoforms, *Gene* **127:**173.

McIntosh, K., Dees, J. H., Becker, W. B., Kapikian, A. Z., and Chanock, R. M., 1967, Recovery in tracheal organ cultures of novel viruses from patients with respiratory disease, *Proc. Natl. Acad. Sci. USA* **57:**933.

McIntosh, K., Kapikian, A. Z., Hardison, K. A., Hartley, J. W., and Chanock, R. M., 1969, Antigenic relationships among the coronaviruses of man between human and animal coronaviruses, *J. Immunol.* **102:**1109.

Möstl, K., 1990, Coronaviridae, pathogenetic and clinical aspects: An update, *Comp. Immunol. Microbiol. Infect. Dis.* **13**:169.

Nedellec, P., Dveksler, G. S., Daniels, E., Turbide, C., Chow, B., Basile, A. A., Holmes, K. V., and Beauchemin, N., 1993, Bgp2, a new member of the carcinoembryonic antigen-related gene family, encodes an alternative receptor for mouse hepatitis viruses, *J. Virol.* **68**:4525.

Oshiro, L. S., 1973, Coronaviruses, in: *Ultrastructure of Animal Viruses and Bacteriophages: An Atlas* (A. J. Dalton and F. Haguenau, eds.), p. 331, Academic Press, New York.

Parker, M. D., Cox, G. J., Deregt, D., Fitzpatrick, D. R., and Babiuk, L. A., 1989, Cloning and *in vitro* expression of the gene for the E3 haemagglutinin glycoprotein of bovine coronavirus, *J. Gen. Virol.* **70**:155.

Parker, M. D., Yoo, D., and Babiuk, L. A., 1990, Expression and secretion of the bovine coronavirus hemagglutinin-esterase glycoprotein by insect cells infected with recombinant baculoviruses, *J. Virol.* **64**:1625.

Percy, D., Bond, S., and MacInnes, J., 1989, Replication of sialodacryoadenitis virus in mouse L-2 cells, *Arch. Virol.* **104**:323.

Rasschaert, D., Delmas, B., Charley, B., Grosclaude, J., Gelfi, J., and Laude, H., 1987, Surface glycoproteins of transmissible gastroenteritis virus: Functions and gene sequence, *Adv. Exp. Med. Biol.* **218**:109.

Reynolds, D. J., and Garwes, D. J., 1979, Virus isolation and serum antibody responses after infection of cats with transmissible gastroenteritis virus, *Arch. Virol.* **60**:161.

Schultze, B., and Herrler, G., 1992, Bovine coronavirus uses *N*-acetyl-9-*O*-acetylneuraminic acid as a receptor determinant to initiate the infection of cultured cells, *J. Gen. Virol.* **73**:901.

Schultze, B., Gross, H. J., Brossmer, R., Klenk, H. D., and Herrler, G., 1990, Hemagglutinating encephalomyelitis virus attaches to *N*-acetyl-9-*O*-acetylneuraminic acid-containing receptors on erythrocytes: Comparison with bovine coronavirus and influenza C virus, *Virus Res.* **16**:185.

Schultze, B., Gross, H. J., Brossmer, R., and Herrler, G., 1991a, The S protein of bovine coronavirus is a hemagglutinin recognizing 9-*O*-acetylated sialic acid as a receptor determinant, *J. Virol.* **65**:6232.

Schultze, B., Wahn, K., Klenk, H. D., and Herrler, G., 1991b, Isolated HE-protein from hemagglutinating encephalomyelitis virus and bovine coronavirus has receptor-destroying and receptor-binding activity, *Virology* **180**:221.

Sippel, C. J., Suchy, F. J., Ananthanarayanan, M., and Perlmutter, D. H., 1993, The rat liver ecto-ATPase is also a canalicular bile acid transport protein, *J. Biol. Chem.* **268**:2083.

Spaan, W., Cavanagh, D., and Horzinek, M. C., 1988, Coronaviruses: Structure and genome expression, *J. Gen. Virol.* **69**:2939.

Stoddart, C. A., Barlough, J. E., Baldwin, C. A., and Scott, F. W., 1988, Attempted immunisation of cats against feline infectious peritonitis using canine coronavirus, *Res. Vet. Sci.* **45**:383.

Stohlman, S. A., and Frelinger, J. A., 1978, Resistance to fatal central nervous system disease by mouse hepatitis virus strain JHM. I. Genetic analysis, *Immunogenetics* **6**:277.

Stohlman, S. A., Frelinger, J. A., and Weiner, L. P., 1980, Resistance to fatal central nervous system disease by mouse hepatitis virus, strain JHM. II. Adherent cell-mediated protection, *J. Immunol.* **124**:1733.

Storz, J., and Rott, R., 1981, Reactivity of antibodies in human serum with antigens of an enteropathogenic bovine coronavirus, *Med. Microbiol. Immunol.* **169**:169.

Storz, J., Rott, R., and Kaluza, G., 1981, Enhancement of plaque formation and cell fusion of an enteropathogenic coronavirus by trypsin treatment, *Infect. Immun.* **31**:1214.

Storz, J., Herrler, G., Snodgrass, D. R., Hussain, K. A., Zhang, X. M., Clark, M. A., and Rott, R., 1991, Monoclonal antibodies differentiate between the haemagglutinating and the receptor-destroying activities of bovine coronavirus, *J. Gen. Virol.* **72**:2817.

Sturman, L., and Holmes, K. V., 1985, The novel glycoproteins of coronaviruses, *Trends Biochem. Sci.* **10**:17.

Sturman, L. S., Ricard, C. S., and Holmes, K. V., 1990, Conformational change of the coronavirus peplomer glycoprotein at pH 8.0 and 37 degrees C correlates with virus aggregation and virus-induced cell fusion, *J. Virol.* **64**:3042.

Tooze, J., and Tooze, S. A., 1985, Infection of AtT20 murine pituitary tumour cells by mouse hepatitis virus strain A59: Virus budding is restricted to the Golgi region, *Eur. J. Cell Biol.* **37**:203.

Turbide, C., Rojas, M., Stanners, C. P., and Beauchemin, N., 1991, A mouse carcinoembryonic antigen gene family member is a calcium-dependent cell adhesion molecule, *J. Biol. Chem.* **266**:309.

Varki, A., 1992, Diversity in the sialic acids, *Glycobiology* **2**:25.

Varki, A., 1993, Biological roles of oligosaccharides: All of the theories are correct, *Glycobiology* **3**:97.

Vautherot, J. F., Madelaine, M. F., Boireau, P., and Laporte, J., 1992, Bovine coronavirus peplomer glycoproteins: Detailed antigenic analyses of S1, S2 and HE, *J. Gen. Virol.* **73**:1725.

Virelizier, J. L., Dayan, A. D., and Allison, A. C., 1975, Neuropathological effects of persistent infection of mice by mouse hepatitis virus, *Infect. Immun.* **12**:1127.

Vlasak, R., Luytjes, W., Leider, J., Spaan, W., and Palese, P., 1988a, The E3 protein of bovine coronavirus is a receptor-destroying enzyme with acetylesterase activity, *J. Virol.* **62**:4686.

Vlasak, R., Luytjes, W., Spaan, W., and Palese, P., 1988b, Human and bovine coronaviruses recognize sialic acid-containing receptors similar to those of influenza C viruses, *Proc. Natl. Acad. Sci. USA* **85**:4526.

Wang, F. I., Fleming, J. O., and Lai, M. M., 1992, Sequence analysis of the spike protein gene of murine coronavirus variants: Study of genetic sites affecting neuropathogenicity, *Virology* **186**:742.

Wege, H., Siddell, S., and ter Meulen, V., 1982, The biology and pathogenesis of coronaviruses, *Curr. Top. Microbiol. Immunol.* **99**:165.

Welter, C.J., 1965, TGE of swine I. Propagation of a virus in cell culture and development of a vaccine, *Vet. Me./Small Anim. Clin.* **60**:1054.

White, J. M., 1990, Viral and cellular membrane fusion proteins, *Annu. Rev. Physiol.* **52**:675.

White, J. M., and Littman, D. R., 1989, Viral receptors of the immunoglobulin superfamily, *Cell* **56**:725.

Williams, R. K., Jiang, G.-S., Snyder, S. W., Frana, M. F., and Holmes, K. V., 1990, Purification of the 110-kilodalton glycoprotein receptor for mouse hepatitis virus (MHV)-A59 from mouse liver and identification of a nonfunctional, homologous protein in MHV-resistant SJL/J mice, *J. Virol.* **64**:3817.

Williams, R. K., Jiang, G. S., and Holmes, K. V., 1991, Receptor for mouse hepatitis virus is a member of the carcinoembryonic antigen family of glycoproteins, *Proc. Natl. Acad. Sci. USA* **88**:5533.

Woods, R. D., 1982, Studies on enteric coronaviruses in a feline cell line, *Vet. Microbiol.* **7**:427.

Woods, R. D., and Pedersen, N.C., 1979, Cross-protection studies between feline infectious peritonitis and transmissible gastroenteritis viruses, *Vet. Microbiol.* **4**:11.

Woods, R. D., and Wesley, R. D., 1986, Immune response in sows given transmissible gastroenteritis virus or canine coronavirus, *Am. J. Vet. Res.* **47**:1239.

Woods, R. D., Cheville, N. F., and Gallagher, J. E., 1981, Lesions in the small intestine of newborn pigs inoculated with porcine, feline, and canine coronaviruses, *Am. J. Vet. Res.* **42**:1163.

Yeager, C. L., Ashmun, R. A., Williams, R. K., Cardellichio, C. B., Shapiro, L. H., Look, A. T., and Holmes, K. V., 1992, Human aminopeptidase N is a receptor for human coronavirus 229E, *Nature* **357**:420.

Yokomori, K., and Lai, M. M., 1992a, The receptor for mouse hepatitis virus in the resistant mouse strain SJL is functional: Implications for the requirement of a second factor for viral infection, *J. Virol.* **66**:6931.

Yokomori, K., and Lai, M. M. C., 1992b, Mouse hepatitis virus utilizes two carcinoembryonic antigens as alternative receptors, *J. Virol.* **66**:6194.

Yokomori, K., Banner, L. R., and Lai, M. M., 1991, Heterogeneity of gene expression of the hemagglutinin-esterase (HE) protein of murine coronaviruses, *Virology* **183**:647.

Yoo, D., Graham, F. L., Prevec, L., Parker, M. D., Benko, M., Zamb, T., and Babiuk, L. A., 1992, Synthesis and processing of the haemagglutinin-esterase glycoprotein of bovine coronavirus encoded in the E3 region of adenovirus, *J. Gen. Virol.* **73**:2591.

CHAPTER 5

# The Coronavirus Surface Glycoprotein

David Cavanagh

## I. PHYSICOCHEMICAL PROPERTIES

### A. Electron Microscope Observations

Coronaviruses are frequently claimed to have a characteristic morphology, including the possession of a "club-shaped" surface projection or spike (S) glycoprotein. However, in common with other aspects of the coronaviruses, the group exhibits variation with respect to the shape, size, and distribution of the S protein on the virion surface. Davies and Macnaughton (1979) described the spikes of infectious bronchitis virus (IBV) and human coronavirus (HCV) 229E as being "tear-drop" shaped and widely spaced, whereas those of murine hepatitis virus (MHV) type 3 were mostly "cone-shaped" and closely spaced, although in some MHV-3 preparations the spikes were more bulbous. Dimensions of S vary not only among the coronaviruses but also depending on the staining procedure; following potassium phosphotungstate staining all three viruses had spikes approximately 20 nm long and 10 nm wide at the bulbous end, except for the cone-shaped spikes of MHV, which had a diameter of only 5 nm (Davies and Macnaughton, 1979). The entire S protein has been observed after solubilization and purification (Sturman *et al.*, 1980; Cavanagh, 1983c). The nonenvelope-associated S1 subunit of the IBV S protein can become detached from the virion (Stern and Sefton, 1982a; Cavanagh and Davis, 1986).

DAVID CAVANAGH • Institute for Animal Health, Division of Molecular Biology, Compton Laboratory, Compton, Newbury, Berkshire RG20 7NN, England.

*The Coronaviridae*, edited by Stuart G. Siddell, Plenum Press, New York, 1995.

## B. Sedimentation Characteristics

Purification of the S protein of MHV, IBV, HCV strain 229E, bovine coronavirus (BCV), and porcine hemagglutinating encephalomyelitis virus (HEV) has been achieved using a combination of nonionic detergent and sucrose gradient centrifugation (Sturman *et al.*, 1980; Hasony and Macnaughton, 1981; Cavanagh, 1983b; Schultze *et al.*, 1990, 1991). When milligram quantities of IBV were used, it was necessary to dissociate and sediment the virus proteins in the presence of 1 M NaCl; otherwise the M protein cosedimented with S (Cavanagh, 1983b). Spike has also been purified by affinity chromatography (Mockett, 1985; Daniel and Talbot, 1990). Sedimentation studies have been variously interpreted as indicating that S from virions is a homodimer or homotrimer (IBV: Cavanagh, 1983c), homodimer (MHV: Vennema *et al.*, 1990b), or homotrimer (TGEV: Delmas and Laude, 1990).

## C. Electrophoretic Analysis

It cannot be said that the S protein, of the coronaviruses as a group, exhibits characteristic migration in sodium dodecyl sulfate–polyacrylamide gel electrophoresis (SDS-PAGE). The apparent molecular weight of uncleaved S is in the range 170–220,000 Da, depending on the virus. The extent of cleavage into amino-(N)terminal S1 and carboxy-(C)terminal S2 glycopolypeptides depends on the virus type, virus strain, and cell type used, and ranges from 0 to 100%. The two cleavage products also vary in molecular weight, from 84 to 135,000, and may be separated or comigrate in polyacrylamide gels. Examples of electrophoretic analysis include BCV (Hogue *et al.*, 1984; Deregt *et al.*, 1987; Cyr-Coats *et al.*, 1988; Vautherot *et al.*, 1992a), MHV-JHM (Siddell, 1982), MHV-A59 (Frana *et al.*, 1985), HCV-OC43 (Hogue *et al.*, 1984), HEV (Callebaut and Pensaert, 1980), IBV (Cavanagh *et al.*, 1986a–c), turkey coronavirus (TCV) (Dea *et al.*, 1989a), canine coronavirus (CCV), and porcine transmissible gastroenteritis virus (TGEV) (Garwes and Reynolds, 1981), and porcine epidemic diarrhea virus (PEDV) (Egberink *et al.*, 1988). Additional references are to be found in the review by Siddell *et al.* (1983). S2 of IBV may aggregate (Stern and Sefton, 1982a) and stain poorly with Coomassie Brilliant Blue (Cavanagh, 1983b), while others have reported high-molecular-weight forms of S that are irregularly obtained (e.g., Hogue *et al.*, 1984). Care must be exercised, therefore, in the interpretation of polyacrylamide gel polypeptide profiles of coronaviruses that have not been extensively studied.

# II. PRIMARY STRUCTURE

## A. Overall Features

The S protein is a large glycoprotein that possesses an overall hydrophobic hydropathicity profile, a N-terminal signal sequence that is absent from the mature protein, a C-terminal hydrophilic sequence preceded by a membrane-spanning domain, and a large number (up to 35) of potential N-linked glycosylation sites, most of which appear to be occupied by glycans (Fig. 1). The entire S protein gene has been sequenced for many strains of IBV (Binns *et al.*, 1985, 1986; Niesters *et al.*, 1986; Sutou *et al.*, 1988; Jordi *et al.*, 1989; Kusters *et al.*, 1989a, 1990; Koch and Kant, 1990b; Cavanagh *et al.*, 1992b; Jia *et al.*, 1993a,b), TGEV (Jacobs *et al.*, 1987; Rasschaert and Laude, 1987; Britton and Page, 1990; Wesley, 1990), the TGEV variant porcine respiratory coronavirus (PRCV) (Rasschaert *et al.*, 1990; Britton *et al.*, 1991; Wesley *et al.*, 1991), the feline coronaviruses feline infectious peritonitis virus (FIPV) (de Groot *et al.*, 1987a) and feline enteric coronavirus (FECV) (Wesseling *et al.*, 1994), HCV-229E (Raabe *et al.*, 1990), CCV (Horsburgh *et al.*, 1992; Wesseling *et al.*, 1994), PEDV (Duarte and Laude, 1994), several strains of MHV (Luytjes *et al.*, 1987; Schmidt *et al.*, 1987; Parker *et al.*, 1989; Gallagher *et al.*, 1990a,b; La Monica *et al.*, 1991; Taguchi *et al.*, 1992), BCV (Abraham *et al.*, 1990; Boireau *et al.*, 1990; Parker *et al.*, 1990; Zhang *et al.*, 1991), and HCV-OC43 (Künkel and Herrler, 1993; Mounir and Talbot, 1993).

There is great variation among the coronaviruses with respect to the size of the S polypeptide. Currently the shortest S protein known is that of IBV, with approximately 1160 amino acids, while FIPV has the longest protein of 1452 amino acids (Fig. 1). The S protein of some, but not all, of the coronaviruses is cleaved to produce the N-terminal S1 and C-terminal S2 glycopolypeptides.

## B. Specific Features

### 1. Glycosylation, Transport, and Maturation

Another common feature of the spike glycoprotein of coronaviruses is the high degree of glycosylation, the number of potential glycosylation sites ranging from 21 (MHV) to 35 (FIPV) (Fig. 1) (see Section II.A for references). Both S1 and S2 are glycosylated, e.g., IBV-Beaudette has 16 and 12 potential glycosylation sites in S1 and S2, respectively. Comparison of the molecular weights after removal of the glycans and by comparison of the deduced molecular weight of the nascent polypeptide, obtained by sequencing, with that obtained from SDS-PAGE, indicates that most of these sites are occupied by glycans.

That the glycans, cotranslationally added, are of the N-linked (via asparagine residues) type has been shown by their susceptibility to endoglycosidase-H (Stern and Sefton, 1982b; Cavanagh, 1983a) and tunicamycin (Holmes *et al.*, 1981; Rottier *et al.*, 1981; Stern and Sefton, 1982b; van Berlo *et al.*, 1987; Dea *et al.*, 1989a,b). Indeed, following replication in the presence of tunicamycin, which prevents the attachment of N-linked glycans, virus particles lacking S protein are produced. Conversion of the initial high mannose or simple glycans to complex or hybrid molecules is a slow process, the half-life being one or several hours (Vennema *et al.*, 1990b,c). The MHV S protein is formed initially as a glycosylated monomer of molecular weight 150,000, which then oligomerizes slowly. The 150,000 glycopolypeptide is converted to a 180,000 form,

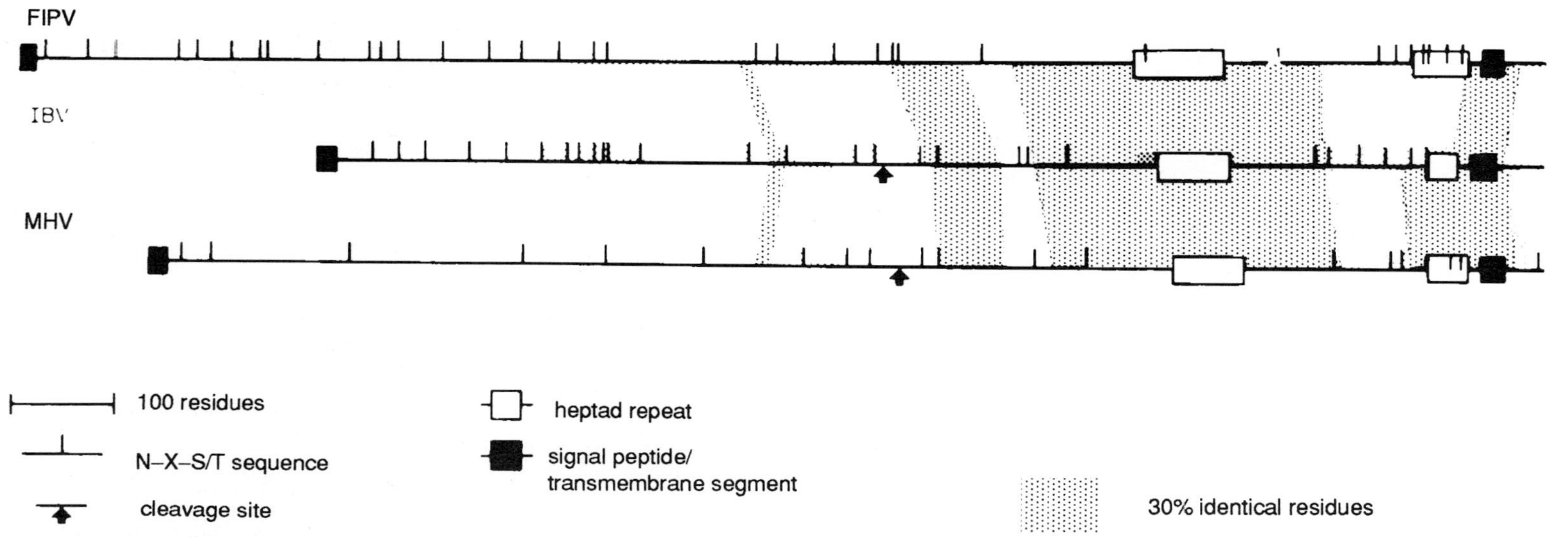


FIGURE 1. Diagrammatic comparison of the amino acid sequence of the spike glycoproteins of three coronaviruses: FIVP strain 79-1146, IBV strain M41, and MHV strain A59. Potential glycosylation sites (N-X-S/T) are shown by short verticle lines. Modified from de Groot *et al.* (1987b), with permission.

presumably involving modification of the simple glycans, after which cleavage into S1 and S2 occurs (Vennema *et al.*, 1990b). In FIPV-infected cells, the oligomerization process was observed to be slower (half-time of 40–60 min) than that of the G glycoprotein and hemagglutinin of vesicular stomatitis virus (VSV) and influenza virus, respectively, but faster than that of Rous sarcoma virus envelope glycoprotein (Vennema *et al.*, 1990b). The initial high mannose (simple) glycans become converted to complex ones, the half-time being about 1 hr in FIPV-infected cells (Vennema *et al.*, 1990b,c). When S was expressed using vaccinia virus recombinants, the half-time of this process was approximately 3 hr. Coronavirus budding occurs at pre-Golgi membranes (see Opstelten *et al.*, 1993, for references), the conversion of simple glycans to complex ones occurring after the budding. Thus the S protein would appear to move faster through the Golgi apparatus when it is in virions than when it is not (Vennema *et al.*, 1990c).

Similar observations had been made following infection of cells with TGEV (Delmas and Laude, 1990). Oligomers were formed before the simple glycans were converted to complex ones, the oligomerization being considered to have taken place not beyond the *cis* Golgi compartment. Many S monomers failed to oligomerize. When tunicamycin prevented glycosylation, oligomerization was prevented, indicating that the lack of glycans caused aberrant folding of the S polypeptide.

Although S1 and S2 are not linked by disulfide bonds, there are many intrachain disulfide bonds. The ectodomain of the S protein of MHV, for example, containes 42 cysteine residues. Opstelten *et al.* (1993) have studied the role of these disulfide bonds by incubating MHV-infected cells in the presence of dithiothreitol (DTT) to produce reducing conditions. The major conformational events involving disulfide bonds took about 20 min. The process would appear to start on the nascent polypeptide. The DTT did not interfere with the functioning of the endoplasmic reticulum. Indeed, the membrane protein (M) was transported normally. Opstelten and colleagues (1993) state in their discussion that M and S form complexes almost immediately after synthesis, which accumulate at the site of budding (pre-Golgi membranes). Thus it may be the interaction of S with M that prevents M progressing beyond the virus-budding site.

A small proportion of S does progress to the cell surface following coronavirus infection (Laviada *et al.*, 1990) and expression of S involving vaccinia virus recombinants (Vennema *et al.*, 1990c; Pulford and Britton, 1991; Oleszak *et al.*, 1992; Kubo and Taguchi, 1993) and baculovirus (Yoo *et al.*, 1990; Godet *et al.*, 1991). The S1 subunit, expressed in the absence of S2 (which forms the membrane-anchoring stalk of S) was excreted. A truncated TGEV S protein, which lacked the carboxy-terminal 292 amino acids, was retained in the endoplasmic reticulum and did not progress to the cell surface (Pulford and Britton, 1991). However, a mutant TGEV spike gene that lacked 70 amino acids from the carboxy-terminus was excreted from cells after expression from a baculovirus recombinant (Godet *et al.*, 1991). A mutant MHV S1 that contained the S2 transmembrane domain was transported to and retained at the cell surface, as was a mutant S2 to which had been added the S1 N-terminal signal peptide. When "normal" S1 and also S2 were transiently expressed in the same cells, S1 was retained at the plasma membrane, suggesting that S1 and S2 had associated in a manner similar to that when the intact S gene was expressed (Kubo and Taguchi, 1993).

## 2. Cleavage of the Propolypeptide

Although the S protein of some coronaviruses is cleavable by cell proteases, to yield S1 and S2, the extent to which this occurs varies, depending on the type and strain of virus and the host cell used. Cleaved spike has never been observed with two coronaviruses, TGEV and FIPV (Garwes and Pocock, 1975; Egberink *et al.*, 1988). In the case of IBV strains from chicken embryos and chick kidney cells (Cavanagh and Davis, 1987; Stern and Sefton, 1982a), PEDV from infected pigs (Egberink *et al.*, 1988), and TCV from human rectal carcinoma (HRT) cells (Dea *et al.*, 1989a), all or almost all of the virion-associated spike was in the form of S1 and S2. Virions of IBV-Beaudette from Vero cells (Cavanagh *et al.*, 1986b) and BCV-Mebus from HRT cells (Hogue *et al.*, 1984; Cyr-Coats *et al.*, 1988) had about 70% and 50% cleaved S, respectively. In contrast, almost none of the S of HCV-229E from human rhabdomyosarcoma cells (Schmidt and Kenny, 1982), MRC5 cells (Kemp *et al.*, 1984), and human embryonic lung cells (Arpin and Talbot, 1990) and of HCV-OC43 (Hogue and Brian, 1986) was cleaved, although the latter studies have shown that S of HCV-OC43 can be completely cleaved by trypsin. The role played by the cell type is well illustrated with MHV-A59, where the extent of S cleavage varies from 0–100%, depending on the host cell used (Frana *et al.*, 1985).

Notwithstanding the requirement that cells have appropriate proteases with which to cleave S, there are features of some spike molecules that make them potentially more readily cleavable than the S of other coronaviruses. Thus sequencing of the S gene of many IBV strains has shown that the cleavage site is adjacent to a basic amino acid sequence, RRFRR, RRSRR, RRHRR, or GRHRR (Binns *et al.*, 1985, 1986; Cavanagh *et al.*, 1986b, 1988; Niesters *et al.*, 1986; Kusters *et al.*, 1989a). MHV strains JHM and A59 have basic connecting peptides comprising RRARR and RRAHR, respectively (Schmidt *et al.*, 1987; Luytjes *et al.*, 1987), and BCV has the sequence KRRSRR or KRRSVR at the corresponding location (Abraham *et al.*, 1990; Boireau *et al.*, 1990; Parker *et al.*, 1990, Zhang *et al.*, 1991). These basic sequences, which would be situated at the surface of the protein, would be expected to make the spike sensitive to those cell proteases that hydrolyze adjacent to basic residues. In the case of influenza A viruses those strains that have the hemagglutinin (HA) protein cleaved in many cell types have an HA1-HA2 connecting peptide containing several basic residues, whereas those isolates in which cleavage occurs in fewer cell types have only a single basic residue at the cleavage site (Bosch *et al.*, 1981). In keeping with this observation, the S of TGEV and FIPV, which has not been observed in a cleaved form, and that of HCV-229E, which is hardly cleaved, does not have any pairs of basic residues in the region analogous to the cleavage site of IBV and MHV (Rasschaert and Laude, 1987; Jacobs *et al.*, 1987; de Groot *et al.*, 1987a; Raabe *et al.*, 1990). The capacity of the S protein of HCV-OC43 to be cleaved by trypsin is in keeping with the presence of the sequence RRSR, which is a potential cleavage site (Künkel and Herrler, 1993; Mounir and Talbot, 1993). The requirement for cleavage of S for the fusion process is discussed in Section V.3.

## 3. Structure of S2

The observation that urea removed S1 but no other polypeptide from IBV led to the suggestion that S was anchored in the envelope by S2 (Cavanagh, 1983c). Support for this view was provided by comparison of N-terminal and nucleotide sequencing data that showed that S2 was generated by cleavage of the S precursor and that near the carboxy-terminus of S2 there was a 44 residue, strongly hydrophobic sequence suggestive of a membrane-spanning domain (Binns *et al.*, 1985; Cavanagh *et al.*, 1986b). Analysis of the deduced amino acid sequences of the spike protein of IBV, MHV, and FIPV (de Groot *et al.*, 1987b), supported by Rasshaert and Laude's (1987) deductions based on TGEV data, has revealed the presence of two regions in S2 with a seven-residue periodicity, forming heptad repeats (Fig. 1). These are indicative of a coiled-coil structure. Sedimentation studies have indicated that S is either a homodimer or homotrimer. The results of a cross-linking study have favored the latter (Delmas and Laude, 1990). Currently, therefore, the oligomeric spike protein is envisaged as being anchored by an α-helical region near to the C-terminus of S2. Just beyond the outer membrane surface is the shorter (minor) repeat structure predicted to be an α-helix of 5 (IBV and MHV) or 7 nm (FIPV). The major repeat indicates a helix of some 10 (IBV and MHV) or 13 nm (FIPV) or more, which would span more than half of S (Fig. 1) (de Groot *et al.*, 1987b). Thus S is envisaged as a multimeric protein, possibly a homotrimer, the narrow stalk of S being a complex coiled-coil structure, somewhat analogous to the hemagglutinin trimer of influenza virus (Wilson *et al.*, 1981). Just before the membrane-spanning domain of S is the sequence KWP. Terminating 10 residues upstream of KWP is a leucine-zipper motif, the length of these varying from three to five heptad repeats (Britton, 1991).

S2 of IBV-M41 was susceptible to trypsin and other proteases, near its C-terminus, at residues close to the outer membrane surface (Cavanagh *et al.*, 1986a). The S1 subunit of IBV could be quantitatively released from S2 by treatment with urea (Cavanagh and Davis, 1986). Although urea did not have this effect on MHV-A59, incubation of MHV at pH 8.0 and 37 °C did result in the release of S1 (Sturman *et al.*, 1990; Weismuller *et al.*, 1990).

## 4. Conservation of S2 among Coronaviruses

Comparison of the amino acid sequences of the spike proteins from IBV, MHV, FIPV, and TGEV has shown that there is far greater sequence conservation within the C-terminal, S2 part of the molecule than in S1 (Schmidt *et al.*, 1987; de Groot *et al.*, 1987b; Rasschaert and Laude, 1987). Within S2 are regions exhibiting 30% identical residues (Fig. 1), while comparison of S2 of BCV strain F15 with that of MHV-A59 reveals about 70% homology (Fig. 4) (Boireau *et al.*, 1990). One notable feature, shared by the various coronaviruses, is that whereas the average cysteine content of S is about 3%, that in the C-terminal hydrophilic tail, which is probably within the lumen of the virus, is around 24%. Some of these cysteine residues may be associated with fatty acid chains, known to be present on S2 of MHV (Schmidt, 1982; Sturman *et al.*, 1985; van Berlo, 1987).

All the coronavirus S genes analyzed encode an highly conserved eight-residue sequence Lys-Trp-Pro-Trp-Trp/Tyr-Val-Trp-Leu. The leucine-zipper motifs described in the previous section ends 10 residues before the Lys-Trp-Pro of this conserved sequence, the remainder of which probably forms the first part of the transmembrane domain (see Britton, 1991).

### 5. Extensive Variation of S1 among Coronaviruses

In contrast to the modest conservation of S2 within the coronavirus genus, S1 shows very little amino acid identity when some pairs of viruses are compared, e.g., IBV with TGEV (Rasschaert and Laude, 1987), IBV with MHV (Schmidt *et al.*, 1987), and MHV with FIPV (de Groot *et al.*, 1987b) (Fig. 1). Indeed, the extensive heterogeneity exhibited by S1 extends to strains within some coronavirus species, notably IBV. In recent years there has been extensive analysis of the antibody-inducing epitopes of some coronaviruses, the S1 subunit being the major inducer of virus-neutralizing (VN) antibodies in some coronaviruses, e.g., IBV and TGEV. Since sequence differences in S1 are responsible for many of the antigenic differences between coronaviruses it is appropriate to discuss sequence and antigenic variation in the same section (Section III).

## III. STRUCTURAL AND ANTIGENIC VARIATION OF S

### A. S Protein Is a Major Inducer of Neutralizing Antibody

All three surface glycoproteins of coronaviruses, S, M, and, when present, HE, induce VN antibody (see Chapters 6 and 8, this volume, for M and HE, respectively). Although there are fewer S than M and N molecules in the virion (Cavanagh, 1981), the spike protein induces a good immune response following infection. That S induces VN antibody has been shown in three ways. First, by immunization with purified S protein of TGEV (Garwes *et al.*, 1978/79), MHV (Hasony and Macnaughton, 1981), and IBV (Cavanagh *et al.*, 1984). Second, by immunization with vectors expressing the S gene of IBV (Tomley *et al.*, 1987), FIPV (de Groot *et al.*, 1989), TGEV (Godet *et al.*, 1991), and MHV (Daniel and Talbot, 1990; Wesseling *et al.*, 1993). Third, by showing that many VN monoclonal antibodies (MAbs) bound to the S protein of MHV (references in subsequent sections).

### B. Location of Antigenic Sites

A number of approaches have been used to locate antigenic sites on S, all involving the use of MAbs: competition enzyme-linked immunosorbent assay

(ELISAs); production of polypeptides or peptides by proteolysis of S, prokaryotic expression vector or synthetically; and sequencing of MAb-resistant variants. No simple picture has emerged for the genus as a whole, except that S1 has more VN epitopes than S2 in some species.

Earlier, polyclonal antibody analyses (for references, see Spaan *et al.*, 1990) indicated that each coronavirus could be placed into one of four antigenic groups. Combining more recent biochemical, sequence, and MAb analyses would suggest three groups. In one group would be TGEV, FIPV, CCV, HCV-229E, and PEDV. A second group would contain MHV, BCV, HCV-OC43, and TCV, while IBV would constitute the sole member of the third group (Chapter 1, this volume). It is instructive to compare the S protein of coronavirus species within these groups. The S protein differences observed among species within a group may play a role in host range and tissue specificity and certainly do so with respect to antigenic differences of practical importance.

## 1. TGEV, FIPV, CCV, HCV-229E, and PEDV

Four or five major antigenic domains were detected on the S protein of TGEV using competition assays with MAbs (Delmas *et al.*, 1986; Jimenez *et al.*, 1986; Correa *et al.*, 1988; Simkins *et al.*, 1989; Enjuanes *et al.*, 1990). Subsequently, extensive analysis of the binding of MAbs to proteolytic fragments of S (Correa *et al.*, 1990; Delmas *et al.*, 1990a,b; Enjuanes *et al.*, 1990), to S polypeptides generated in *Escherichia coli* (Enjuanes *et al.*, 1990; Delmas *et al.*, 1990a,b), to synthetic peptides (Posthumus *et al.*, 1990a,b; Gebauer *et al.*, 1991), and sequencing of MAb-resistant (MAR) mutants (Delmas *et al.*, 1990a,b; Enjuanes *et al.*, 1990; Gebauer *et al.*, 1991) has revealed that the antigenic makeup of S is complex. In addition, some sites would appear to be composed of sequences that are linearly far apart, but which are brought into close proximity by the folding of the protein, with one sequence playing a more prominent role than the other.

The major research teams in this area have used different systems for naming antigenic domains (see Gebauer *et al.*, 1991, for a comparison) and a common nomenclature, using Roman numerals, has been proposed (Posthumus *et al.*, 1990a,b). I have used this nomenclature in Fig. 2 to show the approximate positions of the antigenic domains, taking into account data generated by all the approaches referenced above.

The S1 part of S is the major inducer of antibodies and many of the anti-S1 MAbs were VN. The major VN epitopes are in region I and are highly conserved among TGEV isolates. This, and the great difficulty experienced in obtaining MAR mutants to site I MAbs, has led to the conclusion that this region has an important function. Competition between MAbs and antibodies in sera raised following vaccination of pigs with TGEV confirmed the immunodominant nature of site A, in particular subsite antibody (Diego *et al.*, 1992). Sites II and IV are also associated with VN epitopes. Site III is exposed on denatured but not native S. Differentiation among strains of TGEV has been achieved by MAbs directed against epitopes at sites II, III, and IV (Welch and Saif, 1988; Sanchez *et al.*, 1990). Some epitopes may be formed by residues contributed from sites I and II (Correa *et al.*, 1990; Delmas *et al.*, 1990a,b), while residues in S2 (1176–1184)

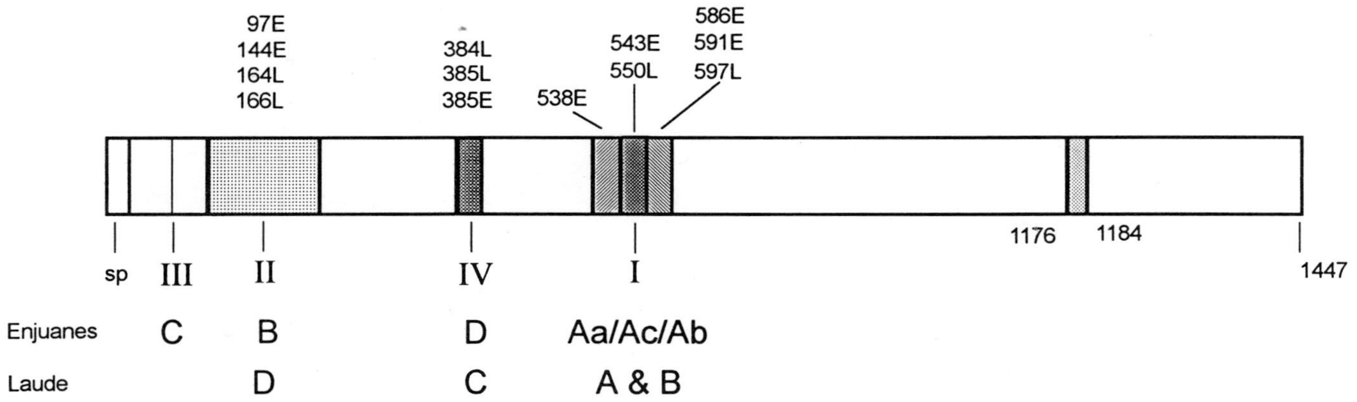


FIGURE 2. Antigenic sites on the S protein of TGEV. The rectangle represents the whole of S, including signal peptide (sp). Sites A–D as defined by the groups of Enjuanes and of Laude are shown, together with the alternative nomenclature I–IV (for references, see Section III. B. 1). The approximate locations of the sites are: I, residues 523–600; II, 90–227; III, 50–51; and IV, 378–390. The numbers above the rectangle show the location of amino acid differences among MAR mutants (L, E: data from the groups of Laude and Enjuanes, respectively). Residues 1176–1184 contribute in some way to the antigenicity of epitopes at site IV.

contribute in some way to the antigenicity of epitopes at site IV (Posthumus *et al.*, 1990a,b, 1991).

There were differences among the sites with respect to the influence of glycans. The binding of MAbs to site IV was little affected by the absence of glycans on S, whereas nonglycosylated/deglycosylated S did not bind site I and II MAbs, and site III was perhaps an intermediate case (Gebauer *et al.*, 1991; Delmas and Laude, 1991).

A very interesting variant of TGEV of economic importance was isolated in Europe in the mid-1980s (see Chapter 16, this volume). This variant was named porcine respiratory coronavirus because it replicated mainly in the respiratory tract of pigs and very little in the alimentary tract. Hence, it did not cause the enteric disease so characteristic of classical TGEV. Sequencing revealed mutations in gene 3 of PRCV, leading to loss of the first open reading frame, plus a large deletion of 672–681 nucleotides from near the beginning of the S gene (Rasschaert *et al.*, 1990; Britton *et al.*, 1991; Wesley *et al.*, 1991) (Fig. 3). Thus PRCV lacks antigenic sites II and III, and MAbs to these sites can be used to differentiate PRCV from TGEV (Garwes *et al.*, 1988; Callebaut *et al.*, 1988; Enjuanes *et al.*, 1990; Simkins *et al.*, 1992). The remainder of S of PRCV and TGEV have 98% identity and, correspondingly, have site I and IV epitopes in common (Sanchez *et al.*, 1990).

The S proteins of TGEV and FIPV are almost identical in length, and for residues 275-1447 there is 94% identity and some common site I and IV epitopes, including some that are VN and that mediate antibody-dependent enhancement of FIPV infectivity (see Chapter 14, this volume). In contrast, the first 274 residues have only 30% identity, and no common site II and III epitopes have been detected (Fig. 3) (Jacobs *et al.*, 1987; Sanchez *et al.*, 1990).

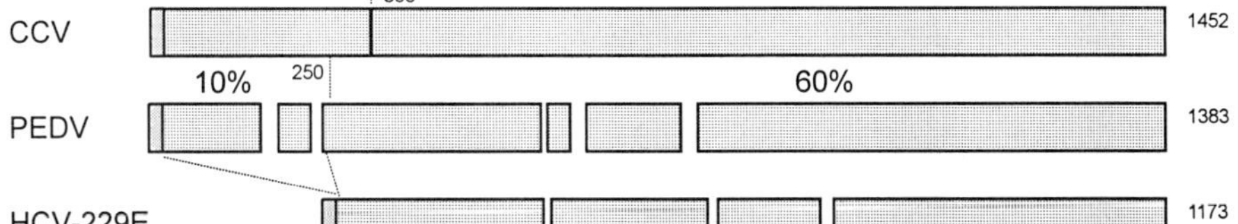

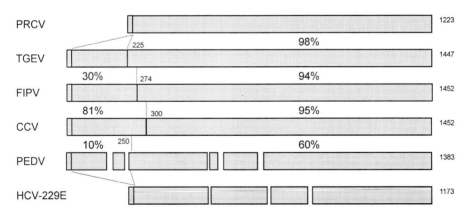


FIGURE 3. Amino acid sequence differences among the S proteins of members of the TGEV group of coronaviruses. The diagram shows the much greater variation among the amino-terminal 225–300 residues than among the remainder. The gaps between rectangles representing S of PEDV and HCV-229E show the approximate size and position of the largest deletions only. HCV-229E and the PRCV variant of TGEV lack the first 225 or so residues of the other species. The amino-terminal signal peptide is indicated by the lighter shading.

Feline coronaviruses could be divided into two antigenic groups on the basis of binding of anti-S MAbs (Fiscus and Teramoto, 1987a,b; Hohdatsu *et al.*, 1991). A feline coronavirus that did not cause FIP (FECV) had 90% and 97% amino acid identity within residues 1-300 and 301-1452, respectively, of a FIPV isolate (Wesseling *et al.*, 1994).

The CCV S protein is the same length as that of FIPV, and residues 300-1452 have 95% identity (Horsburgh *et al.*, 1992; Wesseling *et al.*, 1993) and some site I epitopes in common, including VN epitopes (Fig. 3) (Sanchez *et al.*, 1990; Hohdatsu *et al.*, 1991). When the first 300 residues are compared, the identity with FIPV is still high, at 81% (Wesseling *et al.*, 1993); many isolates of one species, IBV, differ from each other by this amount in the first 300 or so residues (see Section III.A.3). None of the TGEV site II and III MAbs of Sanchez *et al.* (1990) bound to CCV, in keeping with the low (38%) identity between the first 300 residues of these two viruses.

The sequence of several genes of another porcine coronavirus, PEDV, have been determined recently and have confirmed that this virus more closely resembles the TGEV group than the other coronavirus groups. Although somewhat shorter than that of TGEV, the 1383 residue PEDV S does include the N-terminal region that corresponds to the beginning of the TGEV S protein and is absent from PRCV (Fig. 3) (Duarte and Laude, 1994). However, there is only about 10% identity in this region when compared with TGEV, CCV, and FIPV. Not surprisingly, none of 25 anti-TGEV spike protein MAbs bound to PEDV (Sanchez *et al.*, 1990).

Finally, sequence analysis of HCV-229E has confirmed its relationship to the TGEV group, although, like PRCV, the 229E S protein lacks the first 225 or so amino acids (Fig. 3) (Raabe *et al.*, 1990). The extent of identity of the 229E S protein to that of TGEV, FIPV, and CCV is similar to that exhibited by PEDV. Thus, the first 550 or so amino acids have about 38% identity, the C-terminal half being greater (57%). In contrast, the identity with the first half of S of MHV and IBV is only 16–18% and for the second half 32–35% (Raabe *et al.*, 1990). Once again, none of the TGEV anti-S MAbs of Sanchez *et al.* (1990) bound to the 229E S protein.

### 2. MHV, BCV, HCV-OC43, HEV, and TCV

Recent sequencing and MAb studies of these viruses have confirmed their much greater similarity to each other than to other coronaviruses. Comparison has revealed a hypervariable region within S1 that is associated with some of the antigenic differences not only between these virus species but also among strains of a given species (Fig. 4). This hypervariable region may also be associated with host range and tissue tropism, although this is speculation. Sequence data are not available for the S protein of HEV and TCV but they are for the other members of this group; it is to these that the following discussion of sequence applies.

The lengths of the S proteins are very similar, ranging from 1351/53 for HCV-OC43 to 1376 for MHV-4 (see Section II.A for references). The BCV S protein is most closely related to that of HCV-OC43 (91% identity) and less so with MHV

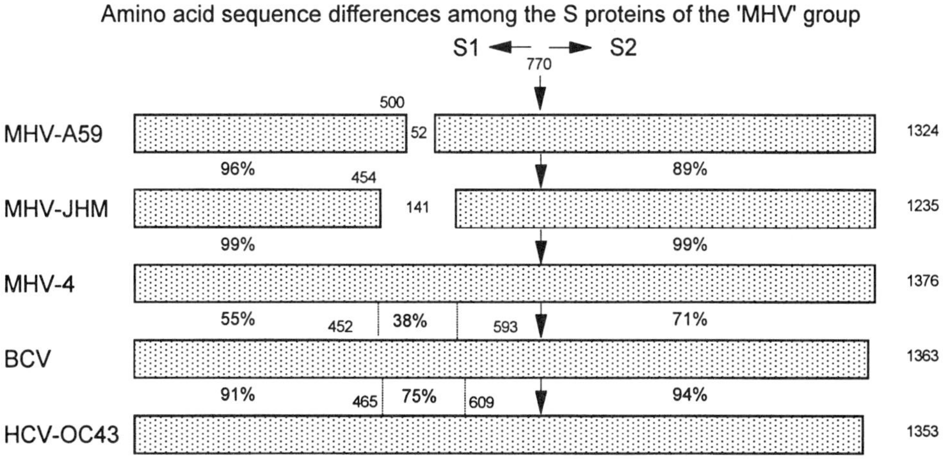


FIGURE 4. Amino acid sequence differences among the S proteins of the MHV group of coronaviruses. The rectangles represent the whole spike protein, the numbering having been normalized with respect to that of the longest known MHV spike protein, belonging to strain MHV4. The arrow at position 770 marks the cleavage site between S1 and S2. The numbers in the gaps show the size of deletions compared with MHV4. The percentage differences shown for the pairs MHV-A59/MHV-JHM and MHV-JHM/MHV-4 refer to comparisons of S1 and S2, excluding the deletions from S1. The percentage differences shown for the pairs MHV4/BCV and BCV/HCV-OC43 are for three areas, the middle one being the hypervariable region.

(about 60%). HCV-OC43 also has approximately 60% identity to the S protein of MHV. As observed with the other coronviruses, identity is greater in S2 (BCV:HCV, 95%; BCV:MHV, 71%) than in S1 (BCV:HCV, 88%; BCV:MHV, 55%).

The very high identity shown by BCV and HCV-OC43 is interrupted by a 135-residue sequence in S1 (residues 465-609 of HCV) in which the identity drops from >90% to 75% (Fig. 4). The similarly situated residues of BCV (residues 452–593) have <40% identity with the corresponding region of MHV-4 compared with >55% elsewhere in S (Fig. 4). Indeed, about half of the amino acid differences between S of MHV-4 and BCV occur within this region which accounts for only about 12% of S. It is in this same region that strains of MHV differ dramatically from each other. Whereas most of the S1 of MHV strains has about >90% identity, at this location there are deletions of a single stretch of amino acids as great as 159 residues (Taguchi *et al.*, 1985; Luytjes *et al.*, 1987; Morris *et al.*, 1989; Parker *et al.*, 1989; Taguchi and Fleming, 1989; Banner *et al.*, 1990; Gallagher *et al.*, 1990a,b; La Monica *et al.*, 1991). It is likely that MHV-4 is a prototypic MHV strain, the commonly used A59 and JHM strains having been selected during passage *in vitro*.

The S gene of six BCV strains, isolated in several countries in two continents, have been sequenced, revealing a very high conservation (98% overall) (Zhang *et al.*, 1991). Most of the differences (45–56 bases, 16 to 26 amino acids) were between virulent and avirulent strains, compared with only 6–14 base differences (<10 amino acids) among the four avirulent isolates. There were twofold more differences in S1 than S2, but the differences in S1 were not concentrated in one region. A BCV isolate from nasal swabs of a calf suffering from respiratory disorders had a S protein that had 98% amino acid identity with that of enteric BCV isolates (Zhang *et al.*, 1994).

The first coronavirus to be extensively studied with respect to the location of its antigenic sites was MHV. A complex picture has emerged and might be summarized as follows. Following infection with MHV, epitopes formed by the S1 subunit induce antibodies with the highest VN activity. Most of these epitopes are conformational. In these respects MHV resembles BCV, in addition to viruses of the TGEV group (Section III.B.1) and IBV (Section III.B.3). Some epitopes associated with the S2 subunit of MHV induce VN antibodies, but these have usually been of low titer. Several of these epitopes are continuous and have been mimicked by peptides. There is evidence from competition experiments with MAbs that some antigenic sites are formed or at least influenced by residues contributed by both S1 and S2.

Competition with MAbs revealed up to six epitope groups (Talbot *et al.*, 1984; Wege *et al.*, 1984; Talbot and Buchmeier, 1985) of which two (sites A and B) or three (sites A, B, and C) were associated with VN antibodies. Cell fusion was inhibited by site A MAbs. Direct evidence that the major VN epitopes reside on S1 has been derived from analysis of the binding of MAbs to S1 in the absence of S2. S1 expressed using a baculovirus was able to bind anti-S1 VN MAbs (Taguchi *et al.*, 1990; Takase-Yoden *et al.*, 1990). Using the cl-2 variant of MHV-JHM, the S1 and S2 glycopolypeptides of which can be distinguished in SDS-PAGE gels, Kubo *et al.* (1993) showed that of 15 MAbs raised against S, 14 immunoprecipitated S1 after this had been dissociated from S2 by urea and 2-mercaptoethanol. Nine of the 14 MAbs were VN, only one bound to S1 after boiling in the presence of SDS and DTT, and five of the VN MAbs failed to immunoprecipitate S1 after only mild denaturation (SDS at room temperature without DTT). Taken together, the MAb-binding studies of Takase-Yoden *et al.* (1990) and Kubo *et al.* (1993) show that some S1 residues form epitopes independently of S2 amino acids and that some of these epitopes are discontinuous in nature.

Neutralizing MAbs 11F and 30B were induced by MHV-JHM sequences in bacterial fusion proteins and mapped to S1 residues 33–40 and 395–406, respectively (Fig. 5c) (Routledge *et al.*, 1991). Competition experiments (Stuhler *et al.*, 1991) using purified MHV-JHM revealed that MAb 30B competed with highly neutralizing, discontinuous epitope MAbs of the site A described by Wege *et al.* (1984). This indicates that residues around amino acid position 400 of the MHV S1 protein form at least part of site A (Wege *et al.*, 1984) and play a major role in the induction of VN antibodies. This location is adjacent to a highly variable region of S1 (Fig. 5).

MAbs to sites B and C selected MHV-JHM variants that had large deletions in S1 (Talbot *et al.*, 1984; Dalziel *et al.*, 1986; Parker *et al.*, 1989; Gallagher and Buchmeier, 1990; Gallagher *et al.*, 1990; La Monica *et al.*, 1991). This per se does not prove that the epitopes for these MAbs were located in this part of S1; the epitopes might have been situated elsewhere in S1 or in S2 and their configuration disrupted by the large deletion in S1. In this regard it is noteworthy that neither of the MAbs of Talbot *et al.* (1984) used to select the variants with the deletions in S1 could react with denatured S in Western blots, indicating that

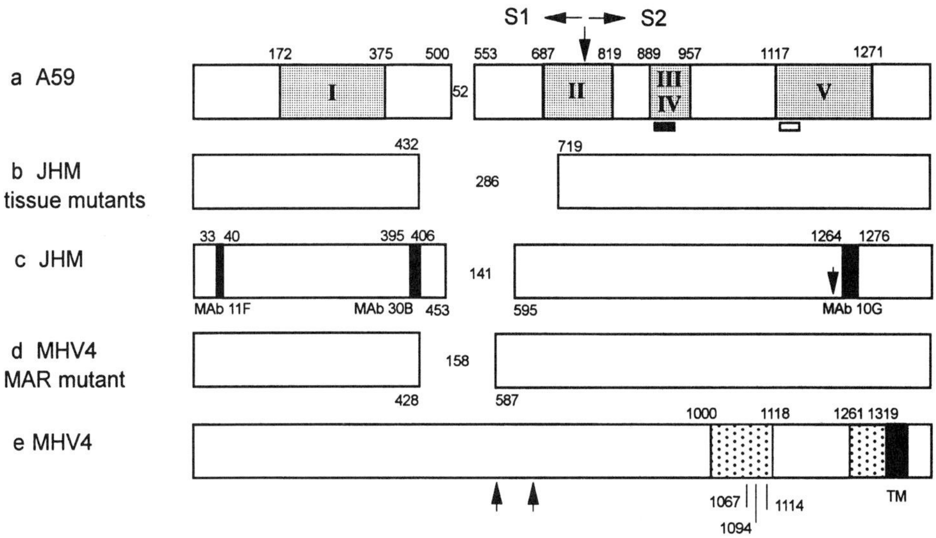


FIGURE 5. Antigenic and other sites within the S protein of strains of MHV. The amino acid numbering has been normalized with respect to that of the longest known MHV protein, that of MHV4. The numbers in the gaps show the size of deletions copared with MHV4. The arrow at position 770 marks the cleavage site between S1 and S2. (a) Shows the location of antigenic domains I–V identified by Daniel *et al.* (1993) using bacterially expressed fragments of S of MHV-A59. The small, filled rectangle represents the peptide of Luytjes *et al.* (1989) (residues 900–908), which bound the VN and anti-fusion Mab 19.2. The small, open rectangle represents the peptide of Talbot *et al.* (1988) (residues 1,135–1,144), which induced VN antibodies and protective immunity in mice. (b) S polypeptide of neurotropic variants of MHV-JHM isolated from the brain and spinal cord of JHM-infected rats (La Monica *et al.*, 1991). (c) S of MHV-JHM showing three sites identified by the binding of bacterial expression products with VN MAbs 11F and 30B (MAb 11F also inhibits fusion) and with VN and antifusion MAb 10G (Routlege *et al.*, 1991). The arrow indicates the positions (1250, 1251, and 1257) of amino acid substitutions in MHV-JHM MAR mutants selected by MAb 11F (Grosse and Siddell, 1994). (d) S of a MAR mutant of MHV4 selected with MAb V4B11.3 (Parker *et al.*, 1989). (e) S of MHV4 showing the heptad repeat regions (stippled) and membrane-spanning region (solid box). The locations are shown of three mutations present in mutants of MHV4 recovered from a persistently infected neural cell line, the mutants requiring a pH of 5.5–6.0 for membrane fusion (Gallagher *et al.*, 1991). The arrows show the position of two point mutations present in a MAR mutant selected with MAb 7.2-V-1 (Wang *et al.*, 1993).

the epitopes were of a discontinuous nature. Not all MAR mutants with mutations in S1 have deletions. For example, MAb 7.2-V-1 selected a MAR mutant of MHV-DL that had two point mutations, both in S1, at residues 588 and 650, the former within and the other very near the hypervariable region of MHV (Fig. 5e) (Wang *et al.*, 1992). The hypervariable region is even larger than that indicated by comparison of MAR mutants, as La Monica *et al.* (1991) have shown with neurotopic variants of JHM isolated from the brain and spinal cord of JHM-infected Wistar Furth rats. These variants lacked 147 amino acids in addition to those already absent from JHM when compared with MHV4 (Fig. 5b).

A number of studies have shown that the situation is more complex than the foregoing account would imply. MAb 11F, which was induced by residues

33–40 in a bacterially expressed product, was used to raise 10 MAR mutants. Sequencing of these showed that the single amino acid mutations were not located in residues 33–40 but at positions 1109, 1110, and 1116 of the JHM S protein (equivalent to residues 1250, 1251, and 1257 of MHV4; Fig. 5c) (Grosse and Siddell, 1994). These mutations did not affect the binding of many other S1 and S2 MAbs, showing that the mutations had not caused extensive alterations to the structure of S. This implies some specific interaction between amino acids in region 33–40 of S1 and those around residues 1109–1116 of S2.

Several other studies have shown the presence of epitopes on S2. A site A MAb that inhibited fusion reacted with an nine amino acid sequence, delineated using prokaryotic expression products and synthetic peptides, situated 131 residues downstream from the N-terminus of S2 (Fig. 5a) (Luytjes *et al.*, 1989). This region is particularly interesting because the MAb used, although neutralizing, could not select MAR mutants (Talbot and Buchmeier, 1985). Moreover, this site was highly conserved among MHV strains. Taken together, these observations suggest that this region of S may have a vital function for the virus. The MAb directed against this peptide was able to protect mice from lethal infection (Koolen *et al.*, 1990). The peptide has also been expressed as a chimeric protein and induced protection in mice (Brown *et al.*, 1994).

Further involvement of S2 in the induction of VN antibodies has been shown by the finding that a decapeptide, corresponding to the sequence starting at residue 365 from the N-terminus of S2 of MHV-JHM, not only induced VN antibodies but also induced in mice some protection from a lethal challenge of MHV (Fig. 5a) (Talbot *et al.*, 1988). Induction of a VN, and antimembrane fusion, MAb by MHV-JHM sequences in a bacterial fusion protein showed that S2 residues 1123–1137 of MHV-JHM (equivalent to residues 1264–1276 in the MHV4 strain) (Fig. 5c) induced antibodies with these properties (Routledge *et al.*, 1991). A VN MAb (2.2-V-1) selected a MAR mutant with a single amino acid substitution at residue 1113 in S2 of MHV-DL, a strain that has >99% identity with MHV-JHM (Fig. 5) (Wang *et al.*, 1992).

Recently Daniel *et al.* (1993) have investigated the antigenicity of bacterial fusion proteins containing overlapping fragments of the S protein of MHV-A59 (Luytjes *et al.*, 1989). This strain lacks many of the residues of the hypervariable region described above (Fig. 5a). The products were examined by immunoblotting, using mostly MAbs that had previously been shown to bind to denatured S (most likely binding to continuous epitopes) but also a few MAbs against discontinuous epitopes. The MAbs had been raised against native S protein. Five domains (I–V) of S, in both S1 and S2, bound 13 of these antibodies (Fig. 5), of which sites I, III, and IV bound VN antibodies. A synthetic 13-residue peptide, conserved among MHV strains, was located within domain III (Luytjes *et al.*, 1989). The combined data further support the view that sequences that are linearly situated far apart, e.g., domains II and V, are closely situated in the native protein, forming discontinuous epitopes.

In summary, for MHV, VN epitopes are formed largely by residues in S1, but there are some VN epitopes in S2. In some cases there may be interaction between S1 and S2 residues, brought into juxtaposition by the folding of the protein, to form VN epitopes, but there are also epitopes in S1 and S2 that are independent of each other. Some epitopes in both S1 and S2 are continuous in nature, while some VN epitopes in S1, at least, rely on conformation for antibody binding.

The S proteins of BCV, TCV, HEV, HCV-OC43, and MHV are antigenically related, as shown using both sera and MAbs (Hogue *et al.*, 1984; Tijssen *et al.*, 1990; Vautherot *et al.*, 1990, 1992a; Michaud and Dea, 1993).

Using five MAbs in competition experiments, Deregt and Babiuk (1987) detected two nonoverlapping sites, both associated with VN antibody, in S of BCV. Following interaction of the MAbs with fragments of S generated with proteases, Deregt *et al.* (1989a) tentatively concluded that these two VN antibody-inducing domains mapped within the S1 subunit. This was later confirmed using S1 that had been expressed independently of S2 using a baculovirus expression system (Yoo *et al.*, 1990). More detailed analysis has been performed by Vautherot *et al.* (1992a) who used 30 anti-S MAbs. Nineteen of these defined four independent antigenic sites on S1, one of which (S1B) was a major and the other a minor (S1-A) inducer of VN antibody. Site 1B could be subdivided. All the anti-S VN MAbs bound to S1. Some of the S1 VN MAbs failed to react with SDS-denatured polypeptides, while others did react provided that the polypeptides had not also been reduced, indicating that the epitopes were conformational. Although MAR mutants were obtained, no sequence data are available for these. However, Yoo *et al.* (1991b) have constructed a set of deletion mutants of S1, expressed them in a baculovirus system, and examined the binding of the A and B site VN MAbs of Deregt *et al.* (1989a). Only mutant S1 containing residues 324–720 bound the MAbs. The hypervariable region of this group of viruses is situated in the middle of this 324–720 sequence (Fig. 4). Two immunoreactive domains were further delineated, namely, one between residues 324 and 403 situated on the N-terminal side of the hypervariable region and one between residues 517–720, the N-terminal portion of which is part of the hypervariable region. Mapping of the 37-kDa trypsin fragment previously examined by Deregt *et al.* (1989a) further indicated that the second domain was formed by residues 517 to 621, most of which is in the hypervariable region, and that the first domain comprised residues 351–403. It was not possible to assign site A and B MAbs to one or other of these domains; both were required for binding of each MAb.

Six MAbs, all non-VN, defined two sites on S2 (Vautherot *et al.*, 1992a). Analysis of prokaryotic expression products and synthetic peptides have shown that MAbs to site S2A, and no other S2 or S1 MAbs, and polyclonal sera against BCV bound to epitopes formed by residues from within the first 20 N-terminal residues of S2 (Vautherot *et al.*, 1992b). The recognition of these epitopes by convalescent serum showed that these epitopes induced antibodies in the host animal. MAbs to these epitopes bind not only to BCV strains but also to isolates of MHV, HEV, and HCV-OC43 (Vautherot *et al.*, 1992a), even though there are amino acid differences in this region. A similarly situated continuous epitope has been identified at the N-terminus of S2 of IBV, although this induced weakly neuralizing antibody (Kusters *et al.*, 1989b; Kant *et al.*, 1992).

The MAbs of Vautherot *et al.* (1992a) to the spike VN sites A and B were specific to BCV, whereas antibody to the other two, non-VN S1 sites (C and D) bound to HCV-OC43 and one S1-D MAb bound to HEV and MHV. However, some BCV site B MAbs of Michaud and Dea (1993) did bind to HCV-OC43. These MAbs were highly VN and therefore probably correspond to the A and B sites of Vautherot *et al.* (1992a), Deregt and Babiuk (1987), and Deregt *et al.* (1989a,b), which are located within and to the N-terminal side of the hypervariable region (Yoo *et al.*, 1991b), as described above. It would be expected that some site A and/or B MAbs would bind to HCV-OC43 since the sequence of the BCV antigenic domain containing residues 351-403 is identical to that in HCV-OC43 (Künkel and Herrler, 1993; Mounir and Talbot, 1993). Some variation among BCV isolates has been detected using MAbs (El-Ghorr *et al.*, 1989; Hussain *et al.*, 1991; Michaud and Dea, 1993). Some of the anti-S MAbs raised against the cell-adapted, attenuated Mebus strain failed to bind to virulent field strains, which is in keeping with the sequence data that show that the field isolates resemble each other more than they do the Mebus strain. In addition, some anti-Mebus spike MAbs differentiated some Quebec isolates from each other. Michaud and Dea (1993) have shown that BCV and TCV had many spike VN epitopes in common.

In summary, the BCV VN epitopes identified to date reside in S1, probably located near and within the hypervariable region (Fig. 4). Few of these epitopes were common to other viruses of the group, whereas there was greater conservation with respect to the non-VN epitopes in S1 and even more so for S2 epitopes. These findings are consistent with our knowledge of the S sequences of these viruses.

### 3. IBV

Since its isolation in 1936, many serotypes of IBV have been defined by VN tests. This antigenic variation is of great practical importance, resulting in the production of vaccines to several serotypes and/or the frequent modification of prophylaxis protocols (King and Cavanagh, 1991).

The major VN antibody sites of IBV reside in S1. This has been deduced as (1) virus from which S1 had been removed by urea did not induce VN antibody (Cavanagh, 1983c; Cavanagh and Davis, 1986); (2) urea-released S1, although in monomeric form, did induce VN, and hemagglutination-inhibiting antibody (Cavanagh and Davis, 1986); and (3) strongly neutralizing MAbs bound to S1 (Mockett *et al.*, 1984; Koch *et al.*, 1990; Karaca *et al.*, 1992; Parr and Collisson, 1993) and selected MAR variants with mutations in S1 (Fig. 6) (Cavanagh *et al.*, 1988; Koch *et al.*, 1990; Kant *et al.*, 1992). The dependence on conformation for the antigenicity of the S1 sites has been shown not only by the failure of anti-S1 VN MAbs to react in Western blots but also by reduced antibody binding after deglycosylation of S with glycosidases (Koch and Kant, 1990a).

One site in S2 (site G) induced weakly neutralizing antibody (Kusters *et al.*, 1989b). Analysis with prokaryotic expression fragments and with synthetic peptides has shown that this S2 site, which is not denatured by SDS and mercaptoethanol (Koch and Kant, 1990a), is situated within the first 20 residues after the N-terminus of S2 (Fig. 6) (Lenstra *et al.*, 1989; Kusters *et al.*, 1989b; Posthumus *et al.*, 1990a,b). Part of this sequence is strongly conserved among isolates with otherwise very variable S sequences. A similarly situated se-

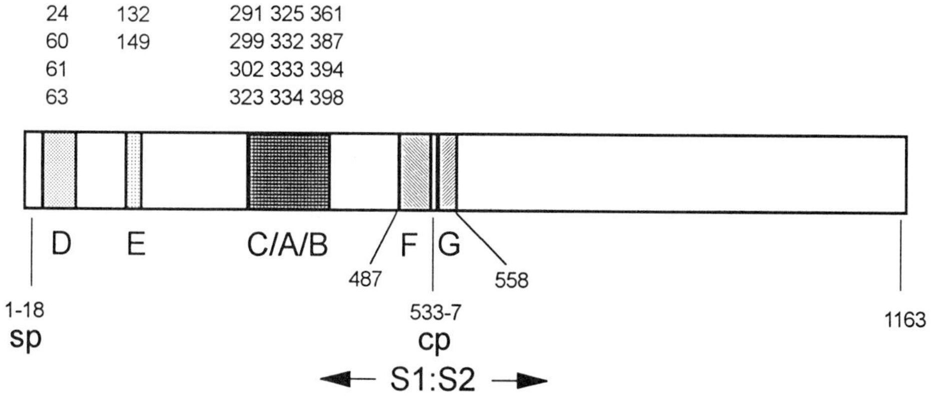


FIGURE 6. Location of some antigenic sites on the S protein of IBV. Conformation-dependent sites A to E are associated with VN epitopes and were defined by the sequencing of MAR mutants; amino acid position 63 was identified on strain M41 and most of the others on strains of the D207 group. The numbers refer to amino acid positions relative to the D207 isolate. The order of sites A, B, and C is uncertain. MAbs to site D also inhibited hemagglutination. Sites F and G contain epitopes that are not conformation-dependent and are defined using bacterial expression products and peptides and are associated with weakly neutralizing antibodies. sp, Signal peptide; cp, basic S1-S2 connecting peptide.

quence in the S protein of BCV is also very antigenic (Vautherot *et al.*, 1992b). Kusters *et al.* (1989b) have suggested that the beginning of S2 is a protruding protein segment with a high local mobility.

Competition assays with 27 anti-S MAbs resulted in the identification of six antigenic sites (A–F) in S1 of the D207 strain and two (G, H) in S2 (Fig. 6) (Koch and Kant, 1990a; Koch *et al.*, 1990). One site, F, was defined using bacterial expression proteins and one weakly neutralizing MAb (Fig. 6) (Kusters *et al.*, 1989b). The other five S1 sites bound VN antibody, two (D, E) and three (A, B, and C) sites being located in the first (region I) and third (region III) quarters of S, respectively (Figs. 6 and 7). MAR mutants selected with the VN MAbs exhibited amino acid changes at a total of 16 positions, 75% of these being in region I, the remainder in region III (Figs. 6 and 7). MAbs to site D, in addition to being VN, inhibited hemagglutination and selected MAR mutants of IBV-D207 (Kant *et al.*, 1992) and IBV-M41 (Cavanagh *et al.*, 1988) at positions 60-63 from the N-terminus of S1, including signal peptide.

The entire S protein gene of nine isolates of IBV has been sequenced, revealing that S1 can vary ≥ fivefold more than S2 (references in Section II.A). The amino acid differences in S1 are not scattered randomly but tend to be concentrated in regions I and III (Fig. 7). Among 13 strains of one serotype (Massachusetts) there were more than fivefold more variable positions in region I than in region II (Fig. 7b) (Cavanagh *et al.*, 1988, 1992a). Since all these strains were of the same serotype, this shows that the variation in region I did not make a great impact on serotype.

In a group of seven IBV isolates, including UK/6/82 and the D207 strains used by Koch and colleagues for MAR mutant studies, in which the maximum difference between any two strains was only 5% of S1 residues, 88% of the

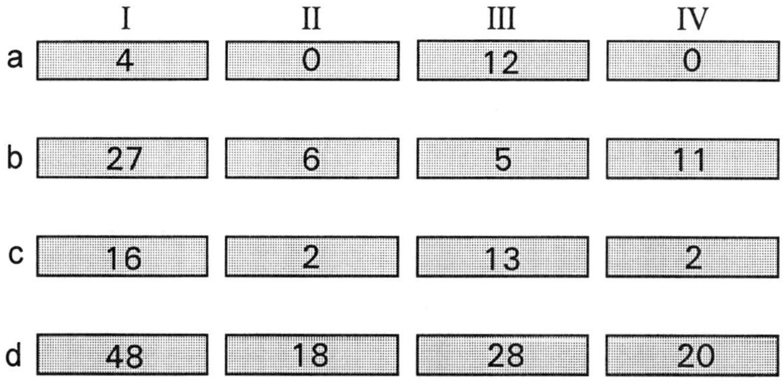


FIGURE 7. Amino acid variaton of the S1 sunbunit of the IBV spike protein. The rectangles represent the first (amino acid residues 1–130), second (131–260), third (261–391), and fourth (392 to approximately 520) quarters of S1, signal peptide excluded. The numbers show the number of amino acid differences in each quarter for the groups of isolates specified. (a) Location of amino acid changes among 16 MAR mutants of IBV, strain D207 (Kant *et al.*, 1992). (b) Location of amino acid differences among a group of 13 strains (excluding Beaudette) of the Massachussetts serotype (Cavanagh *et al.*, 1988, 1992b). (c) Location of amino acid differences among a group of seven isolates, including D207 and UK/6/82, where the maximum difference between two isolates is 5% (Cavanagh *et al.*, 1992a). (d) Number of amino acid differences between members of the UK/6/82 group and the Massachussetts M41 strain, the difference between M41 and individual members of the UK/6/82 group being approximately 20%.

differences were distributed approximately equally among regions I and III (Fig. 7c) (Cavanagh *et al.*, 1992b). Comparison of the MAR mutant and field isolate data shows that the location of changed residues in MAR mutants is very similar to that of the amino acid differences among field isolates, indicating that many of the differences in regions I and III of field isolates are associated with differences in antigenicity. Indeed, despite their overall high amino acid identity, these seven strains represented three serotypes, supporting the view that changes in these regions of S1 can be advantageous to the virus, in respect to avoidance, if only partial, of immune responses induced in chickens by prior infection. A comparison of Fig. 7a–c suggests that changes in region III may contribute more to the serotype than region I. The observation that isolates that differed by only a dozen or fewer amino acids in S1 exhibited very little cross-neutralization suggests that most of the VN antibody was induced by a small number of immunodominant epitopes.

As the differences between serotypes rise to about 20% of S1 residues, e.g., when the UK/6/82 group is compared with the Massachusetts M41 strain, the number of amino acid differences in regions II and IV does increase, but there is also a marked increase in the number of changes in regions I and III (Fig. 7d) (Cavanagh *et al.*, 1992a; Cavanagh and Davis, 1992). To what extent changes in regions II and IV contribute to antigenic variation is not known.

Recently, the sequence of S1 of several of the "classical" North American reference isolates has been determined, including the Arkansas 99 (Ark99) isolate (Wang *et al.*, 1993; Jia *et al.*, 1993a), SE17 and Pp14 (Wang *et al.*, 1993),

Gray (Kwon and Jackwood, 1993; Wang and Collisson, 1993), and JMK (Kwon and Jackwood, 1993), Holte (Wang and Collisson, 1993), CU-T2 (Jia *et al.*, 1993a), Iowa-609 (Wang *et al.*, 1994) and parts of S1 for the Connecticut-46 and Florida isolates and two strains recently isolated in Delaware (J. Gelb, C. Keeler, and W.A. Nix, personal communication). This has revealed that some of these strains differ substantially from each other (about 20% of S1 residues), while other isolates have high identity (>95%). The Gray and JMK strains share 98% identity, consistent with them being placed in the same serotype (Hopkins, 1974). However, Hopkins, unlike Johnson and Marquardt (1975), placed the Holte strain in the same serotype as JMK, whereas they differ by about 27% in the first 200 or so amino acids of S1. Similarly, Johnson and Marquardt considered that the Ark99 and Connecticut-46 were different serotypes, consistent with the finding that they differ by 29% in the first 200 residues, whereas Hopkins placed them in the same serotype. These results further indicate that serology is not a firm foundation on which to build a system for establishing relationships between IBV isolates.

The first 350 or so residues of S1 of many Australian strains, isolated over a 30-year period, have recently been determined (S. Sapats, P. J. Wright, and Y. Ignjatovic, personal communication). Comparisons show that many isolates obtained in the 1960s and 1970s had about 80% identity with each other and with several European and North American strains in this part of S1. Most interestingly, however, two strains isolated in different parts of Australia in the same year (N1/88 and Q3/88) had only 67% identity with each other over the whole of S1 and had percentage differences in the first 350 residues of about 55% (suggesting a possible difference of about 67% for the whole of S1) when compared with earlier Australian isolates and with non-Australian strains. Only vaccines based on Australian isolates have ever been permitted in Australia, and importation of live poultry was banned 50 years ago. Thus Australian strains of IBV may have been evolving independently, at least for the lifetime of the intensive poultry industry.

Another contributor to IBV variation is recombination during mixed infection, and there is sequence evidence that indicates that recombination has occurred within the S gene (Kusters *et al.*, 1989a, 1990; Cavanagh *et al.*, 1992a; Wang *et al.*, 1993; Jia *et al.*, 1993b). One consequence is that characterizing an IBV isolate by partial sequencing of S1 may not always result in the establishment of a more accurate relationship between two strains than has been achieved by serological analysis. Where it is important to establish the degree of relatedness of strains it would be advisable to sequence more than one part of S and also part of another gene.

## IV. INDUCTION OF PROTECTIVE IMMUNITY

There is strong evidence that the S protein is a major inducer of protective immune responses, although the other structural proteins undoubtedly also play a role (see Chapters 6 and 8, this volume). C57 BL/10 strain mice were protected from a lethal infection of MHV-3 by immunization with sucrose gradient-purified S (Hasony and Macnaughton, 1981), as were BALB/C mice immunized with affinity-purified S and challenged with MHV-A59 (Daniel and Talbot, 1990), although there were some clinical signs. Some anti-S MAbs also gave a measure of protection against MHV (Nakanaga *et al.*, 1986; Koolen *et al.*, 1990). When the MHV-A59 S protein was expressed in mice using an adenovirus vector, 50–90% of the mice, dependent on the challenge dose, were protected against the lethal effects of intracerebral A59 infection (Wesseling *et al.*, 1993). A recombinant adenovirus expressing the N protein also induced protective immunity, the protection being greater when N and S were jointly expressed compared with either protein alone. A peptide corresponding to a nine amino acid sequence located 131 residues from the amino-terminus of S2 (Fig. 5) (Luytjes *et al.*, 1989) induced protection in mice (Koolen *et al.*, 1990). This sequence has also been expressed as a chimeric protein including the VP2 capsid protein of a human parvovirus, expressed by a recombinant baculovirus (Brown *et al.*, 1994). Sixty percent of mice vaccinated with a chimeric protein and subsequently challenged with MHV A59 survived the challenge.

IBV from which the S1 subunit had been removed by urea was unable to induce immunity to respiratory infection in chickens (Cavanagh *et al.*, 1986c), indicating that the S protein, and possibly the S1 subunit in particular, was required for immunity. Chickens vaccinated with live vaccine of one IBV serotype are frequently poorly protected against challenge by other serotypes (Johnson and Marquardt, 1975). Since the S protein varies to a greater extent than the other virion proteins and given the experimental evidence for a major role in protection played by the S protein, it is likely that poor cross-protection is due in large measure to differences in the epitopes of the S protein.

Epitopes on the S protein of TGEV are more immunogenic in pigs than epitopes on the N and M proteins and are considered to be of major importance in the induction of lactogenic immunity (Diego *et al.*, 1992). Vaccination of cats against FIPV can be disadvantageous since antiviral antibodies lead to enhancement of infectivity of FIPV for monocytes and macrophages, leading to death of the animal (early death syndrome). That the S protein is involved has been shown by vaccinating cats with a vaccinia virus recombinant expressing the FIPV S protein; the cats succumbed earlier than controls after challenge with FIPV (Vennema *et al.*, 1990a,d).

## V. BIOLOGICAL FUNCTIONS

### A. Primary Attachment to Cells

This section concerns the virus components that mediate attachment to cells; cell receptors for coronaviruses are discussed in Chapter 4. It is generally assumed that the attachment of coronaviruses to cells is mediated by the S protein (Collins *et al.*, 1992), especially in the absence of the HE protein. Many, although not all, VN MAbs are directed against S (see Section III). Some coronaviruses that do not have a HE protein can cause hemagglutination (HA). IBV can cause HA, although this is inefficient. In the past, manifestation of the HA capacity of IBV has involved concentration of the virus and treatment with a crude preparation of *Clostridium welchii* phospholipase C type I (Bingham *et al.*, 1975; Cavanagh and Davis, 1986). A recent study has shown that treatment of IBV with purified neuraminidase (from *Vibrio cholerae* and from Newcastle disease virus) efficiently converts IBV into good HA material, indicating that it was the presence of small amounts of neuraminidase that had been effective in the crude bacterial extracts (Schultze *et al.*, 1992). Neuraminidase treatment of erythrocytes abolished HA with neuraminidase-treated IBV. This indicates that sialic acid (a2,6 linkage) on the erythrocytes was acting as a receptor. It is unclear why neuraminidase treatment of the virus enhanced HA. One reason may be that cellular material was attached to the S protein, via a sialic acid link, and that the neuraminidase "cleaned" this off. That it is the S protein that is responsible for attachment of IBV has been shown in a number of ways. Some MAbs reactive with S (Mockett *et al.*, 1984) and polyclonal antibody raised against purified S (Cavanagh *et al.*, 1984) inhibited HA. Moreover, the HA-inhibiting antibody was associated with the S1 subunit (Mockett *et al.*, 1984; Cavanagh *et al.*, 1986c), virus lacking S1 could not cause HA and was noninfectious (Cavanagh *et al.*, 1986c) and the HA-inhibiting MAbs also had VN activity. Porcine TGEV that, like IBV, does not have an HE protein can also cause HA (Noda *et al.*, 1987, 1988).

The HE protein, present in a subset of the coronaviruses, can also attach to erythrocytes (see Chapter 8, this volume). Schultze *et al.* (1991) have purified the S and HE proteins of BCV. Both proteins recognized the same type of sialic acid receptor, but the S protein caused HA more efficiently than did HE. They proposed that S was responsible for the primary attachment of the virus to cells.

## B. Membrane Fusion

Great care must be taken when generalizations are made about coronaviruses, not least where membrane fusion is concerned. It can be stated that S-induced cell–cell fusion (syncytium formation) does not have an absolute requirement for cleavage of S. Indeed, the S protein of members of the TGEV group is not cleaved but can induce syncytia. For those coronaviruses that do produce a cleaved S, prevention of cleavage does not necessarily result in the loss of fusion activity, although it results in less efficient fusion induction. No single sequence solely responsible for fusion has been identified. Indeed, the evidence to date is that several regions, in terms of the linear S molecule, influence fusogenic activity. With regard to pH, the spike protein of many strains of all coronavirus species so far examined will induce cell–cell fusion optimally at neutral pH and slightly above. For these viruses, fusion of an incoming virion with a cell membrane is affected little or not at all by lysosomotropic agents, since there is no detrimental effect on the successful infection of cells. Since lysosomotropic agents raise the pH in endosomes away from acidity, this indicates that virion–cell fusion either occurs at the cell surface or within endosomes prior to their acidification. However, there are some very interesting exceptions to this general observation.

### 1. Evidence for the Role of S in Fusion

Expression of the S gene of FIPV and MHV-JHM using virus vectors has provided unequivocal evidence that S alone among the virus proteins is required for syncytium formation (de Groot *et al.*, 1989; Pfleiderer *et al.*, 1990; Grosse and Siddell, 1994). The identification of a fusion domain has not proven possible, given that sequences in both S1 and S2 glycopolypeptides are involved and that the three-dimensional structure of S has not been deduced. Bacterial expression products, containing nonconformational epitopes, corresponding to both S1 and S2 sequences have induced MAbs with fusion-inhibiting activity (Routledge *et al.*, 1991). A nine-residue peptide, corresponding to the sequence beginning 131 residues from the N-terminus of S2 (strain A59), binds a fusion-inhibiting site A MAb (Luytjes *et al.*, 1989) (position 900–908 in Fig. 5a). Recombination experiments have also supported the view that S2 of MHV is involved in fusion (Keck *et al.*, 1988) and S2 of BCV expressed in insect cells caused cell fusion (Yoo *et al.*, 1991a). Amino acids in both of the hydrophobic regions of S2 that exhibit heptad repeats and are believed to form the coiled-coil structure of the oligomeric S protein have been associated with fusion activity (Fig. 5e). Gallagher *et al.* (1991) identified three residues, at positions 1067, 1094, and 1114 in the MHV4 S protein, which together were associated with the loss of fusion at neutral pH and a requirement of acidic pH, although it is not clear exactly how these changes cause such a marked effect on fusion (Fig. 5e). A peptide corresponding to residues 1264–1276 of MHV4 induced MAb 10G, which had both VN and antifusion activity (Fig. 5) (Routledge *et al.*, 1991). In the same study a peptide of residues 33–40, i.e., near the N-terminus of S1, also had both of these activities. The anti-S2 MAb 10G might destabilize the oligomeric structure of S and thereby interfere with the interaction of the fusogenic region with a cell membrane. A similar explanation may apply to MAb 19.2, which inhibits fusion and binds to a linear sequence within antigenic domain III (Lutyjes *et al.*, 1989; Daniel *et al.*, 1993), although this is not located directly within the hydrophobic areas (Fig. 5). It must be remembered, however, that S protein-induced cell–cell fusion is at least a two-step process, the fusion event being necessarily preceded by the attachment of the S protein, anchored in the plasma membrane of a cell, to the surface of an adjacent cell. Cells that were expressing the S protein of FIPV or MHV, following infection with a vector carrying the relevant S gene, fused with cocultivated cells when those cells were of a type that were susceptible to FIPV or MHV infection, respectively, but not to cells that were refractory to infection with these viruses (de Groot *et al.*, 1989; Vennema *et al.*, 1990b,c). This not only indicates that S must first attach to a receptor prior to fusion, but also that there is some specificity in that attachment. Taguchi *et al.* (1992) have shown, however, that vaccinia virus expression of MHV S resulted in fusion not only in murine cells but also in rabbit RK13 cells, which are refractory to infection but would appear to have a receptor for MHV, strain cl-2. It may be that MAbs that inhibit fusion might do so either by inhibiting attachment of S to cell membranes (where S can be either on virions or on the surface on an infected cell, adjacent to a noninfected one) or the fusion event. Some site A and B MAbs of MHV neutralize virus in addition to inhibiting fusion. The VN and antifusion MAb 11F that binds near the beginning of S1 (Fig. 5c) could interfere with the attachment process, thereby indirectly affecting fusion. Alternatively, residues 33–40 may be in close proximity to residues 1264–1319 in the native protein, this association being necessary for fusogenic activity; hence, this activity would be inhibited to antibodies against either linear location. A third possibility is suggested by the observation that MAR mutants selected by MAb 11F had mutations not in residues 33–40, which are known to have induced MAb 11F (Routledge *et al.*, 1991), but in S2 residues 1109, 1110, and 1116 (Grosse and Siddell, 1994). Sturman *et al.* (1990) have shown with MHV that mildly alkaline conditions result in the irreversible dissocation of S1 and S2. Furthermore, they have suggested that this event is necessary for fusion activity. It has been proposed (S. G. Siddell, personal communication) that MAb 11F might interfere with this dissociation process, given that the residues 33–40 would seem to be closely associated with residues 1109–1116 in the mature S protein. It should be remembered that the S proteins of some coronaviruses, e.g., those of the TGEV group, are not cleaved into S1 and S2 glycopolypeptides (Sections I.B.2 and IV.B.3). Therefore, if rearrangement of the N-terminal and C-terminal halves of S is a part of the fusion-activation process, this does not necessarily result in the total separation of S1 and S2. Knowledge of the three-dimensional structure of S would greatly facilitate interpretation of these data.

### 2. Role of pH in Fusion

There are two stages at which the fusogenic role of S may come into play, the first of which can be considered essential for natural infection. First, the virion membrane must fuse with a cell membrane to release the genome into the cytoplasm. Second, newly produced S protein may, on being transported to the plasma membrane, cause an infected cell to fuse to an adjacent, noninfected cell, thereby allowing spread of the infection without the absolute necessity for release of virions into the extracellular environment. Many investigations of the pH requirements for coronavirus fusion have focused on cell–cell fusion, to form syncytia, either by the addition of large amounts of virus to cells to give rapid syncytium formation [fusion from without (FFWO), a gross example of virion–cell fusion] or simply by observing syncytium formation after virus replication [fusion from within (FFWI)]. The virion–cell membrane fusion that results in genome uncoating has been studied by addition of lysosomotropic agents (ammonium chloride, chloroquine). These raise the pH in endosomes, the vesicles in which viruses become surrounded if they do not fuse at the cell surface, from <6.0 to almost 7.0. The rationale is that if infection is inhibited or greatly decreased by a lysosomotropic agent, this is indicative that an acidic environment is necessary for activation of the fusion protein.

Observation of fusion induction by the S protein of MHV, FIPV, BCV, and IBV have shown that for many strains fusion (virion–cell and cell–cell) occurs optimally at about pH 7.0 or slightly greater. This has been shown by experiments on FFWO with MHV-A59 (Sturman *et al.*, 1985; Frana *et al.*, 1985; Kooi *et al.*, 1991) and on FFWI following MHV4 replication (Gallagher *et al.*, 1991).

MHV-induced FFWI was inhibited when the culture medium had a pH $<7.0$ (Sawicki and Sawicki, 1986). Similar observations have been made with BCV (Payne and Storz, 1988). Neutral pH was sufficient for FFWI following expression of FIPV and MHV S protein from vaccinia virus vectors (de Groot *et al.*, 1989; Vennema *et al.*, 1990b,c; Pfleiderer *et al.*, 1990; Taguchi *et al.*, 1992; Stauber *et al.*, 1993). Lysosomotropic agents had little effect on the infection on cells by BCV and IBV strain UK/123/82 (Payne and Storz, 1988; Li and Cavanagh, 1990, 1992). Electron microscope observations showed fusion of BCV virions at the cell surface but not at internal membranes.

Kryzstyniak and Dupuy (1984) reported that infection by MHV3 was greatly diminished by lysosomotropic agents. Mizzen *et al.* (1985) found that ammonium chloride did not decrease the number of cells infected but did extend the eclipse phase. That is, in the presence of the agent, virus was transferred from the cell surface (as shown by resistance of proteinase-K treatment of the cells) and remained as infectious virus within the cells for an appreciably longer time than in the absence of ammonium chloride. However, even in the presence of ammonium chloride the cells ultimately became infected, and virus yields were the same as the controls. Kooi *et al.* (1991) have confirmed that ammonium chloride delays the appearance of new virus by about 4 hr and that the final titers were the same as in the controls. In parallel experiments the titer of vesicular stomatitis virus, a virus known to require an acidic pH for fusion, was greatly reduced in the presence of the lysosomotropic agent. These authors also showed that MHV-A59 was endocytosed by Vero and rat astrocytoma (C6) cells, but this did not lead to productive infection. The latter did occur after cells to which virus had attached were treated with polyethylene glycol to cause fusion at the cell surface. This agreed with the earlier work of Van Dinter and Flintoff (1987). The interpretation of these results is that although MHV-A59 can be endocytosed, this may result in transfer of virus to lysosomes, where it is destroyed. Rather, it is considered that the fusion of MHV-A59 virions with cells is not dependent on endocytosis and certainly not on an acidic environment, and occurs at the cell surface or possibly in endocytic vesicles and/or endosomes prior to acidification.

There are, however, exceptions to these observations. The productive infection of Vero and CK cells by IBV-Beaudette (in contrast to the UK/123/82 strain) was reduced by 80–95% when ammonium chloride was present early in infection, with a corresponding 0.9 $\log_{10}$ decrease in subsequent virus production by 10 hr postinfection, near the end of the first replication cycle (Li and Cavanagh, 1990, 1992). Syncytium formation by FFWI occurred optimally at pH 6.7. These results suggest that IBV-Beaudette virions fuse shortly after transfer to endosomes in a slightly acidic environment. Electron microscope investigations of the entry of IBV-Beaudette into chorioallantoic cells and CK cells did not reveal any fusion of the virus at the cell surface but did observe entry by endocytosis (Chasey and Alexander, 1976; Patterson and Bingham, 1976). In contrast, Payne *et al.* (1990) observed that BCV fused with plasma membrane but not with endosomal membranes, consistent with the observation that ammonium chloride had little inhibitory effect on BCV infection.

A more dramatic example has been described by Gallagher *et al.* (1991).

They showed that MHV4, in susceptible murine DBT and Sac− cells, caused syncytium formation by FFWI at pH 5.5–8.5, the optimum being at pH 7.0 and above. Lysosomotropic agents did not reduce replication, as assessed by the rate of viral RNA synthesis, although there was a delay in the time taken for peak production in the presence of chloroquine. However, variants of MHV4 obtained from a persistently infected neural cell line caused little or no fusion in DBT or Sac− cells at pH 7.0 and above. Instead, the optimum pH for fusion was 5.5–6.0. Correspondingly, productive infection by these mutants was inhibited substantially by lysosomotropic agents. The mutations corresponding to this changed phenotype were located in the larger of the heptad repeat areas of S2 (Fig. 5e). Passage of the acid pH-dependent variants in Sac− cells resulted in selection of neutral pH-fusing revertants. Ammonium chloride and chloroquine did not prevent replication of the revertants, although, as with MHV-A59 (Mizzen *et al.*, 1985; Kooi *et al.*, 1991), they delayed virus replication by a few hours.

The work of Gallagher *et al.* (1991) has shown that cell type was of paramount importance in the selection of MHV4 variants with greatly differing pH requirements for fusion. One suggestion by Gallagher *et al.* (1991) as to why low pH-dependent virus was selected in OBL21A cells is that it may be unable to fuse at the cell surface but can fuse after transfer to endosomes, because of some difference in the membranes. It has been shown that the composition of the membrane, for example, with respect to cholesterol, can effect membrane fusion (Daya *et al.*, 1988; Roos *et al.*, 1990; Kooi *et al.*, 1991). However, it does not necessarily follow that the spike protein gene should mutate to give a low pH-dependant protein. An alternative hypothesis (Gallagher *et al.*, 1991) is that sustained replication in OBL21A neural cells might require that the incoming nucleocapsid becomes available deep within the cell and not just beneath the surface. This would be more likely to happen if the S protein required a low pH for fusion activation, achieved when the endosomes had traveled away from the cell surface. They further speculate that Sac− cells selected neutral pH-requiring revertants because syncytium formation is advantageous to the spread of the virus.

In any event, the stage of nucleocapsid release from coronavirus virions is one that can be affected by cell type and may be a factor in determining host cell range and pathogenesis. Kooi *et al.* (1988) have further data that support the view that the degree to which cells are permissive to MHV can depend on the efficiency of the fusion process leading to the release of the nucleocapsid. Beushausen *et al.* (1987) have presented evidence that suggests that differentiated rat oligodendrocytes are resistant to MHV infection because normal processing of the phosphorylated nucleocapsid protein is inhibited in some manner.

### 3. Requirement of Cleavage for Fusion

Whether or not cleavage of the coronavirus S protein is required for its fusogenic activity has been studied in several ways, including the simple expedient of determining whether each coronavirus had cleaved or uncleaved S in its virions. Experiments involved FFWO and FFWI, with and without trypsin treatment, and expression of the S protein using virus vectors, with and without mutant S genes. Some coronaviruses produce virions with up to 100% cleaved S, while others have never been observed to have cleaved S (see Section I.C). Virions of both types are infectious, showing that cleavage is not essential for virus–cell fusion. However, several studies during the 1980s indicated that only cleaved S would induce fusion of one cell with another to produce syncytia.

MHV-A59 from Sac− cells, which had 100% of S in the cleaved form, was able to cause FFWO when added to L2 cells (Frana *et al.*, 1985). In contrast, MHV-A59 virus from 17Cl 1 cells, which had about 50% cleaved S (Frana *et al.*, 1985), was unable to cause FFWO unless it had previously been treated with trypsin to complete the cleavage process (Sturman *et al.*, 1985). Similarly, BCV-L9 only caused FFWI and plaque formation when trypsin was used to increase the proportion of cleaved S from 30–60% to almost 100% (Storz *et al.*, 1981; Payne and Storz, 1988; Cyr-Coats *et al.*, 1988). A variant of MHV-A59, which had no cleaved S, produced only small plaques and also did not cause cell fusion (Sawicki, 1987). IBV-Beaudette-induced FFWI of Vero cells was increased if trypsin was present, the effect being most marked with virus that had not been previously adapted to growth in Vero cells. Plaque development was also enhanced (D. Li and D. Cavanagh, unpublished observation). However, virions of both no, low, and high Vero cell-passaged virus had the same proportion (approximately 1:2) of uncleaved to cleaved S.

Initially these experiments were interpreted as showing that for viruses that had a spike that was capable of being cleaved into S1 and S2, cleavage was necessary for fusogenic activity leading to syncytium formation. An alternative interpretation is that, for these coronaviruses, both cleaved *and* uncleaved S has fusogenic activity but that cleaved S is more fusogenic. In those cases where addition of trypsin enhanced cell–cell fusion, it would then be concluded that prior to the addition of trypsin there was not enough S at the cell surface capable of efficiently forming sufficient microfusion events that would ultimately coalesce to form syncytia. Extraneous trypsin cleaved the uncleaved S molecules, thereby increasing the number and concentration of efficiently fusogenic S molecules to above a threshold value, gross fusion then occurring.

This alternative view arises in part from experiments in which S proteins have been expressed using virus vectors. A cleaved form of the spike of FIPV has never been observed. When the FIPV S protein was expressed using a vaccinia virus recombinant, syncytium formation resulted (de Groot *et al.*, 1989). To determine if uncleaved S of MHV-JHM would induce syncytia, Stauber *et al.* (1993) and Taguchi (1993) mutated the highly basic S1-S2 connecting peptide (Arg-Arg-Ala-Arg-Arg) to Ser-Val-Ser-Gly-Gly and Arg-Thr-Ala-Leu-Glu, respectively. When expressed, the mutant spikes were not cleaved, but they did induce cell–cell fusion. Taguchi (1993) observed a 2- to 4-hr delay in fusion in the cells expressing the mutant S, suggesting that although uncleaved S was fusogenic, it was perhaps less efficient than cleaved S.

Additional support for this conclusion was provided by Gombold *et al.* (1993), who examined mutant MHV-A59 isolated from productively but persistently infected primary mouse glial cell cultures. The mutants grew normally in L2 cells, but, whereas wild-type virus had caused almost 100% fusion of the monolayer by 10 hr after infection, the mutants had caused less than 10% fusion. Reduction of the pH had no enhancing effect on fusion, but trypsin did do so for several mutants, which in the absence of the enzyme produced only uncleaved S. Sequencing revealed that the mutants had the connecting peptide Arg-Arg-Ala-*Asp*-Arg instead of the wild-type Arg-Arg-Ala-*His*-Arg, i.e., a negatively charged aspartic acid residue in place of a weakly basic histidine. Analysis of revertants, which produced cleaved S and fused cells efficiently, showed that all the revertants had lost the aspartic acid residue in the connecting peptide, it having been replaced by the wild-type histidine in one case but by the small, uncharged residues alanine or glycine in others. These results showed that the presence of a negatively charged residue one position away from the C-terminal end of the S1-S2 connecting peptide prevented cleavage; that a basic residue at this position was not essential for cleavage; and that while cleaved S was not required for virus–cell fusion and was not essential for cell–cell fusion, uncleaved S was less efficient at inducing cell–cell fusion. The observation that virions of MHV and BCV with 50% or less cleaved S are infectious also shows that cleavage of all the S molecules is not necessary for the fusion between virion and cell membranes that results in genome uncoating.

The seemingly greater capacity of the S protein to fuse virions with cell membranes may be because the concentration of S molecules in the virion envelope may be greater than at the surface of an infected cell. If this is the case, then virions would be expected to cause more microfusion events, in close proximity to each other, resulting in fusion. Syncytium formation is a rather gross manifestation of S-induced membrane fusion, a process that is probably preceded by small, localized fusion events between adjacent cells. If this is the case, then it may be the progression from the localized to the extensive fusion that requires a particularly high density of cleaved S.

In conclusion, the process by which the coronavirus S protein induces membrane fusion would seem to be a complex, stepwise one. Interaction of the S protein with a cell surface may require primary and secondary receptors. This attachment process may induce conformational changes that ultimately result in gross changes in the juxtaposition of S1 and S2 sequences, followed by membrane fusion. Protonation is not required in most cases.

## VI. ROLE OF S IN PATHOGENICITY

It is likely that the coronavirus spike glycoprotein is a major determinant of cell tropism and hence of pathogenicity. However, in very few, if any, cases can it be definitively stated that a change in the S protein was responsible for a different biological behavior. In some mutants, genes encoding other proteins are known to have changes, while in all mutants there may well be mutations in other genes that have not been detected. Infectious DNA copies of coronavirus genomes will be required to obtain unequivocal answers about the role of the S protein in the varied pathogenicity of coronaviruses. However, it is worth highlighting those cases in which changes in virulence or tropism have been paralleled by changes, sometimes very marked ones, in S.

Several members of the TGEV group exhibit >90% amino acid identity in the S protein except for the first 200–300 residues, which are either missing or have identities as low as 30% (Fig. 3; Section III B 1). The most stiking example, perhaps, is that of TGEV compared with its mutant PRCV, with their tropisms for the alimentary and respiratory tracts, respectively. From the S gene to the poly-A tract there is 98% identity overall, except that PRCV lacks the first 224 amino acids of the TGEV spike. It is very tempting to believe that this deletion is responsible for the different tropisms of these viruses, one of which (PRCV) causes mild, sometimes inapparent, disease, while the other (TGEV) is frequently lethal for piglets. However, it is known that there are a few differences elsewhere in the genome, including some in gene 3 that have effectively resulted in the loss of the 3a open reading frame.

Variation within S of FIPV may be responsible for differences in pathogenicity. Avirulent strains of FIPV were internalized less rapidly than virulent strains, from which they could be distinguished by anti-S MAbs (Fiscus and Teramoto, 1987b).

Likewise, the hypothesis that many of the different tropisms exhibited by MHV strains and other similar coronaviruses are the result of differences in S is very attractive. In particular, the hypervariable region of S1 (Fig. 5) is a prime candidate for a determinant of pathogenicity. Strains of MHV cause a broad spectrum of CNS diseases in rodents as well as replicating in various non-CNS tissues (Matsubara *et al.*, 1991). Many of the neuroattenuated variants have arisen during passage in tissue culture. The MHV4 strain may be considered as a prototype MHV strain. Passage in tissue culture has resulted in the loss, for example, of 141 amino acids from the hypervariable region of S1 to generate the JHM strain (Fig. 5), which prior to sequencing had been considered to be identical to MHV4 (Baybutt *et al.*, 1984; Parker *et al.*, 1989; Gallagher *et al.*, 1990).

Interestingly, analysis of MHV-JHM mRNA obtained from the brains of Lewis rats at 5 to 7 days after intracerebral inoculation showed that the S mRNA was larger than that of the input virus (Taguchi *et al.*, 1985). Virus isolated at the same time produced the larger S mRNA during replication *in vitro*, and produced a correspondingly enlarged S protein. It is likely that the inoculum was a mixed population, a small proportion of which had a "full-size" S protein very similar to the MHV4 virus from which the JHM strain had been derived. Back-passage of the virus in rats then selected the virus with the full-length S.

Morris *et al.* (1989, 1990) isolated variants from the CNS of Wistar Furth rats after infection with MHV-JHM, and two of these variants have been sequenced by La Monica *et al.* (1991). One variant lacked the codons encoding the C-terminal 246 amino acids of the HE protein and had a 147 amino acid deletion in S1 (286 residues when compared with MHV4) (Fig. 5b). This variant caused demyelination in both 2- and 10-day-old rats, whereas the wild-type virus only produced this in the 10-day-olds, causing an acute encephalitis virus in 2-day-old rats. Another variant, which induced acute encephalitis in rats of both ages, did not have the deletion in S but did have a deletion in the HE gene.

A neural cell line persistently infected with MHV4 selected variants that had a dependence on low pH for membrane fusion, in contrast to wild-type virus (Gallagher *et al.*, 1991). Three mutations responsible for this were identified in S2 (Fig. 5e). The authors suggested that this would favor virus fusion in endosomes rather than at the cell surface, releasing the nucleocapsid deeper within the cell, this being required for some unknown reason. Revertants were selected when the mutants were back-passaged in Sac− cells; the revertants, unlike the mutants, caused extensive syncytia formation. This enables virus to spread without the necessity of being released from cells; this may be an advantage in some instances. Different, fusion-related mutations were identified by Gombold *et al.* (1993) after MHV-A59 had established productive persistent infection in mouse glial cells. The mutants had an altered S1-S2 connecting peptide and the spike was not cleaved and did not cause cell–cell fusion. Thus the spread of such mutants was limited to that attained by virions released from cells. A related example of a change in biological properties of MHV coincident with a change in S has been described by Fazakerley *et al.* (1992). A MAR mutant that had a 149 amino acid deletion in S1 had a neuroattenuated phenotype, considered to be a result of its reduced rate of spread in the CNS. Some other, similar deletion mutants have been selected from MHV4 stocks by VN MAbs. In four cell lines several such mutants had enhanced growth compared with the wild-type virus, attributed to delayed or diminished cytopathic effect, allowing cultures to support virus production for longer periods (Gallagher *et al.*, 1990).

Two MAb-selected variants of MHV-3 had a changed tissue tropism, the mutants accumulating more in the lung and kidneys rather than in the liver, the primary target organ of the wild-type virus (Martin *et al.*, 1990).

The S protein can affect the outcome of infection in other ways. MHV-JHM spike on the surface of infected cells can bind IgG via the Fc domain (Oleszak and Leibowitz, 1990). MHV-infected cells were lysed in the presence of spleen and lymph node cells from nonimmunized mice, leading to the proposal that S on the cell surface had, at neutral pH, caused fusion of the infected cell and those of the B cell lineage (Wysocka *et al.*, 1989).

ACKNOWLEDGMENTS. I would like to thank the following for providing me with data prior to publication: Jack Gelb, Jagoda Ignjatovic, Mark Jackwood, Hubert Laude, Syed Naqi, and Willy Spaan.

## VII. REFERENCES

Abraham, S., Kienzle, T. E., Lapps, W., and Brian, D. A., 1990, Deduced sequence of the bovine coronavirus spike protein and identification of the internal proteolytic cleavage site, *Virology* **176**:296.

Arpin, N., and Talbot, P. J., 1990, Molecular characterization of the 229E strain of human coronavirus, in: *Coronaviruses and Their Diseases* (D. Cavanagh and T. D. K. Brown, eds.), pp. 73–80, Plenum Press, New York.

Banner, L. R., Keck, J. G., and Lai, M. M. C., 1990, A clustering of RNA recombination sites adjacent to a hypervariable region of the peplomer gene of murine coronavirus, *Virology* **175**:548.

Baybutt, H. N., Wege, H., Carter, M. J., and ter Meulen, V., 1984, Adaptation of coronavirus JHM to persistent infection of murine Sac(−) cells, *J. Gen. Virol.* **65**:915.

Beushausen, S., Narindrasorasak, S., Sanwal, B. D., and Dales, S., 1987, *In vivo* and *in vitro* models of demyelinating disease: Activation of the adenylate cyclase system influences JHM virus expression in explanted rat oligodendrocytes, *J. Virol.* **61:**3795.
Bingham, R. W., Madge, M. H., and Tyrrell, D. A., 1975, Haemagglutination by avian infectious bronchitis virus—a coronavirus, *J. Gen. Virol.* **28:**381.
Binns, M. M., Boursnell, M. E. G., Cavanagh, D., Pappin, D. J. C., and Brown, T. D. K., 1985, Cloning and sequencing of the gene encoding the spike protein of the coronavirus IBV, *J. Gen. Virol.* **66:**719.
Binns, M. M., Boursnell, M. E. G., Tomley, F. M., and Brown, T. D. K., 1986, Comparison of the spike precursor sequences of coronavirus IBV strains M41 and 6/82 with that of IBV Beaudette, *J. Gen. Virol.* **67:**2825.
Boireau, P., Cruciere, C., and Laporte, J., 1990, Nucleotide sequence of the glycoprotein S (E2) gene of the bovine enteric coronavirus; comparison with mouse hepatitis virus, *J. Gen. Virol.* **71:**487.
Bosch, F. X., Garten, W., Klenk, H-D., and Rott, R., 1981, Proteolytic cleavage of influenza virus hemagglutinins: Primary structure of the connecting peptide between HA, and $HA_2$ determines proteolytic cleavability and pathogenicity of avian influenza viruses, *Virology* **113:**725.
Britton, P., 1991, Coronavirus motif, *Nature* **353:**394.
Britton, P., and Page, K. W., 1990, Sequence of the S gene from a virulent British field isolate of transmissible gastroenteritis virus, *Virus Res.* **18:**71.
Britton, P., Mawditt, K. L., and Page, K. W., 1991, The cloning and sequencing of the virion protein genes from a British isolate of porcine respiratory coronavirus: Comparison with transmissible gastroenteritis virus genes, *Virus Res.* **21:**181.
Brown, C. S., Welling-Wester, S., Feijlbrief, M., van Lent, J. W. M., and Spaan, W. J. M., 1994, Chimeric parvovirus B19 capsids for the presentation of foreign epitopes, *Virology* **198:**477.
Callebaut, P. E., and Pensaert, M. B., 1980, Characterisation and isolation of structural polypeptides in haemagglutinating encephalomyelitis virus, *J. Gen. Virol.* **48:**193.
Callebaut, P., Correa, I., Pensaert, M., Jiménez, G., and Enjuanes, L., 1988, Antigenic differentiation between transmissible gastroenteritis virus of swine and a related porcine respiratory coronavirus, *J. Gen. Virol.* **69:**1725.
Cavanagh, D., 1981, Structural polypeptides of coronavirus IBV, *J. Gen. Virol.* **53:**93.
Cavanagh, D., 1983a, Coronavirus IBV glycopolypeptides: Size of their polypeptide moieties and nature of their oligosaccharides, *J. Gen. Virol.* **64:**1187.
Cavanagh, D., 1983b, Coronavirus IBV: Further evidence that the surface projections are associated with two glycopolypeptides, *J. Gen. Virol.* **64:**1787.
Cavanagh, D., 1983c, Coronavirus IBV: Structural characterisation of the spike protein, *J. Gen. Virol.* **64:**2577.
Cavanagh, D., and Davis, P. J., 1986, Coronavirus IBV: Removal of spike glycoprotein S1 by urea abolishes infectivity and haemagglutination but not attachment to cells, *J. Gen. Virol.* **67:**1443.
Cavanagh, D., and Davis, P. J., 1987, Coronavirus IBV: Relationships among recent European isolates studied by limited proteolysis of the virion glycopolypeptides, *Avian Pathol.* **16:**1.
Cavanagh, D., and Davis, P. J., 1992. Sequence analysis of strains of avian infectious bronchitis coronavirus isolated during the 1960s in the UK, *Arch. Virol.* **130:**471.
Cavanagh, D., Darbyshire, J. H., Davis, P. J., and Peters, R. W., 1984, Induction of humoral neutralising and haemagglutination-inhibiting antibody by the spike protein of avian infectious bronchitis virus, *Avian Pathol.* **13:**573.
Cavanagh, D., Davis, P. J., and Pappin, D. J. C., 1986a, Coronavirus IBV glycopolypeptides: Locational studies using proteases and saponin, a membrane permeabilizer, *Virus Res.* **4:**145.
Cavanagh, D., Davis, P. J., Pappin, D. J. C., Binns, M. M., Boursnell, M. E. G., and Brown, T. D. K., 1986b, Coronavirus IBV: Partial amino terminal sequencing of spike polypeptide S2 identifies the sequence Arg-Arg-Phe-Arg-Arg at the cleavage site of the spike precursor propolypeptide of IBV strains Beaudette and M41, *Virus Res.* **4:**133.
Cavanagh, D., Davis, P. J., Darbyshire, J. H., and Peters, R. W., 1986c, Coronavirus IBV: Virus retaining spike glycopolypeptide S2 but not S1 is unable to induce virus-neutralizing or haemagglutination-inhibiting antibody or induce chicken tracheal protection, *J. Gen. Virol.* **67:**1435.
Cavanagh, D., Davis, P. J., and Mockett, A. P. A., 1988, Amino acids within hypervariable region 1 of

avian coronavirus IBV (Massachusetts serotype) spike glycoprotein are associated with neutralization epitopes, *Virus Res.* **11**:141.

Cavanagh, D., Davis, P. J., and Cook, J. K. A., 1992a, Infectious bronchitis virus: Evidence for recombination within the Massachusetts serotype, *Avian Pathol.* **21**:401.

Cavanagh, D., Davis, P. J., Cook, J. K. A., Li, D., Kant, A., and Koch, G., 1992b, Location of the amino acid differences in the S1 spike glycprotein subunit of closely related serotypes of infectious bronchitis virus, *Avian Pathol.* **21**:33.

Chasey, D., and Alexander, D. J., 1976, Morphogenesis of avian infectious bronchitis virus in primary chick kidney cells, *Arch. Virol.* **52**:101.

Collins, A. R., Knobler, R. L., Powell, H., and Buchmeier, M. J., 1982, Monoclonal antibodies to murine hepatitis virus-4 (strain JHM) define the viral glycoprotein responsible for attachment and cell–cell fusion, *Virology* **119**:358.

Correa, I., Jimenez, G., Sune, C., Bullido, M. J., and Enjuanes, L., 1988, Antigenic structure of the E2 glycoprotein from transmissible gastroenteritis coronavirus, *Virus Res.* **10**:77.

Correa, I., Gebauer, F., Bullido, M. J., Suñé, C., Baay, M. F. D., Zwaagstra, K. A., Posthumus, W. P. A., Lenstra, J. A., and Enjuanes, L., 1990, Localization of antigenic sites of the E2 glycoprotein of transmissible gastroenteritis coronavirus, *J. Gen. Virol.* **71**:271.

Cyr-Coats, K. ST., Storz, J., Hussain, K. A., and Schnorr, K. L., 1988, Structural proteins of bovine coronavirus strain L9: Effects of the host cell and trypsin treatment, *Arch. Virol.* **103**:35.

Dalziel, R. G., Lampert, P. W., Talbot, P. J., and Buchmeier, M. J., 1986, Site-specific alteration of murine hepatitis virus type 4 peplomer glycoprotein E2 results in reduced neurovirulence, *J. Virol.* **59**:463.

Daniel, C., and Talbot, P. J., 1990, Protection of mice from lethal coronavirus MHV-A59 infection by monoclonal affinity-purified spike glycoprotein, in: *Coronaviruses and Their Diseases* (D. Cavanagh and T. D. K. Brown, eds.), pp. 205–210. Plenum Press, New York.

Daniel, C., Anderson, R., Buchmeier, M. J., Fleming, J. O., Spaan, W. J. M., Wege, H., and Talbot, P. J., 1993, Identification of an immunodominant linear neutralization domain on the S2 portion of the murine coronavirus spike glycoprotein and evidence that it forms part of a complex tridimensional structure, *J. Virol.* **67**:1185.

Davies, H. A., and Macnaughton, M. R., 1979, Comparison of the morphology of three coronaviruses, *Arch. Virol.* **59**:25.

Daya, M., Cervin, M., and Anderson, R., 1988, Cholesterol enhances mouse hepatitis virus-mediated cell fusion, *Virology* **163**:276.

Dea, S., Garzon, S., and Tijssen, P., 1989a, Identification and location of the structural glycoproteins of a tissue culture-adapted turkey enteric coronavirus, *Arch. Virol.* **106**:221.

Dea, S., Garzon, S., and Tijssen, P., 1989b, Intracellular synthesis and processing of the structural glycoproteins of turkey enteric coronavirus, *Arch. Virol.* **106**:239.

De Groot, R. J., Maduro, J., Lenstra, J. A., Horzinek, M. C., van der Zeijst, B. A. M., and Spaan, W. J. M., 1987a, cDNA cloning and sequencing analysis of the gene encoding the peplomer protein of feline infectious peritonitis virus, *J. Gen. Virol.* **68**:2639.

De Groot, R. J., Luytjes, W., Horzinek, M. C., van der Zeijst, B. A. M., Spaan, W. J. M., and Lenstra, J. A., 1987b, Evidence for a coiled-coil structure in the spike proteins of coronaviruses, *J. Mol. Biol.* **196**:963.

De Groot, R. J., Van Leen, R. W., Dalderup, M. J. M., Vennema, H., Horzinek, M. C., and Spaan, W. J. M., 1989, Stably expressed FIPV peplomer protein induces cell fusion and elicits neutralizing antibodies in mice, *Virology* **171**:493.

Delmas, B., and Laude, H., 1990, Assembly of coronavirus spike protein into trimers and its role in epitope expression, *J. Virol.* **64**:5367.

Delmas, B., and Laude, H., 1991, Carbohydrate-induced conformational changes strongly modulate the antigenicity of coronavirus TGEV glycoproteins S and M, *Virus Res.* **20**:107.

Delmas, B., Gelfi, J., and Laude, H., 1986, Antigenic structure of transmissible gastroenteritis virus. II. Domains in the peplomer glycoprotein, *J. Gen. Virol.* **67**:1405.

Delmas, B., Godet, M., Gelfi, J., Rasschaert, D., and Laude, H., 1990a, Enteric coronavirus TGEV: Mapping of four major antigenic determinants in the amino half of peplomer protein E2, in: *Coronaviruses and Their Diseases* (D. Cavanagh and T. D. K. Brown, eds.), pp. 151–157, Plenum Press, New York.

Delmas, B., Rasschaert, D., Godet, M., Gelfi, J., and Laude, H., 1990b, Four major antigenic sites of the coronavirus transmissible gastroenteritis virus are located on the amino-terminal half of spike glycoprotein S, *J. Gen. Virol.* **71**:1313.

Deregt, D., and Babiuk, L. A., 1987, Monoclonal antibodies to bovine coronavirus: Characteristics and topographical mapping of neutralising epitopes on the E2 and E3 glycoproteins, *Virology* **161**:410.

Deregt, D., Sabra, M., and Babiuk, L. A., 1987, Structural proteins of bovine coronavirus and their intracellular processing, *J. Gen. Virol.* **68**:2863.

Deregt, D., Parker, M. D., Cox, G. C., and Babiuk, L. A., 1989a, Mapping of neutralizing epitopes to fragments of the bovine coronavirus E2 protein by proteolysis of antigen-antibody complexes, *J. Gen. Virol.* **70**:647.

Deregt, D., Gifford, G. A., Khalidijaz, M., Watts, T. C., Gilchrist, J. E., Haines, D. M., and Babiuk, L. A., 1989b, Monoclonal antibodies to bovine coronavirus glycoproteins E2 and E3: Demonstration of *in vivo* virus neutralizing activity, *J. Gen. Virol.* **70**:993.

Diego, M. de, Laviada, M. D., Enjuanes, L., and Escribano, J. M., 1992, Epitope specificity of protective lactogenic immunity against swine transmissible gastroenteritis virus, *J. Virol.* **66**: 6502.

Duarte, M., and Laude, H., 1994, Sequence of the spike protein of the coronavirus porcine epidemic diarrhoea virus, *J. Gen. Virol.* **75**:1195.

Egberink, H. F., Ederveen, J., Callebaut, P., and Horzinek, M. C., 1988, Characterization of the structural proteins of porcine epizootic diarrhea virus, strain CV777, *Am. J. Vet. Res.* **49**:1320.

El-Ghorr, A. A., Snodgrass, D. R., Scott, F. M. M., and Campbell, I., 1989, A serological comparison of bovine coronavirus strains, *Arch. Virol.* **104**:241.

Enjuanes, L., Gebauer, F., Correa, I., Bullido, M. J., Sune, C., Smerdon, C., Sanchez, C., Lenstra, J. A., Posthumus, W. P. A., and Meloen, R. H., 1990, Location of antigenic sites of the S-glycoprotein of transmissible gastroenteritis virus and their conservation in coronaviruses, in: *Coronaviruses and Their Diseases* (D. Cavanagh and T. D. K. Brown, eds.), pp. 159–172, Plenum Press, New York.

Fazakerley, J. K., Parker, S. E., Bloom, F., and Buchmeier, M. J., 1992, The V5A 13.1 envelope glycoprotein deletion mutant of mouse hepatitis virus type-4 is neuroattenuated by its reduced rate of spread in the central nervous system, *Virology* **187**:178.

Fiscus, S. A., and Teramoto, Y. A., 1987a, Antigenic comparison of feline coronavirus isolates: Evidence for markedly different peplomer glycoproteins, *J. Virol.* **61**:2607.

Fiscus, S. A., and Teramoto, Y. A., 1987b, Functional differences in the peplomer glycoproteins of feline coronavirus isolates, *J. Virol.* **61**:2655.

Frana, M. F., Behnke, J. N., Sturman, L. S., and Holmes, K. V., 1985, Proteolytic cleavage of the E2 glycoprotein of murine coronavirus: Host-dependent differences in proteolytic cleavage and cell fusion, *J. Virol.* **56**:912.

Gallagher, T. M., and Buchmeier, M. J., 1990a, Monoclonal antibody-selected variants of MHV-4 contain substitutions and deletions in the E2 spike glycoprotein, in: *Coronaviruses and Their Diseases* (D. Cavanagh and T. D. K. Brown, eds.), pp. 385–393, Plenum Press, New York.

Gallagher, T. M., Parker, S. E., and Buchmeier, M. J., 1990, Neutralization-resistant variants of a neurotropic coronavirus are generated by deletions within the amino-terminal half of the spike glycoprotein, *J. Virol.* **64**:731.

Gallagher, T. M., Escarmis, C., and Buchmeier, M. J., 1991, Alteration of the pH dependence of coronavirus-induced cell fusion: Effect of mutations in the spike glycoprotein, *J. Virol.* **65**:1916.

Garwes, D. J., and Pocock, D. H., 1975, The polypeptide structure of transmissible gastroenteritis virus, *J. Gen. Virol.* **29**:25.

Garwes, D. J., and Reynolds, D. J., 1981, The polypeptide structure of canine coronavirus and its relationship to porcine transmissible gastroenteritis virus, *J. Gen. Virol.* **52**:153.

Garwes, D. J., Lucas, M. H., Higgins, D. A., Pike, B. V., and Cartwright, S. F., 1978/79, Antigenicity of structural components from porcine transmissible gastroenteritis virus, *Vet. Microbiol.* **3**:179.

Garwes, D. J., Stewart, F., Cartwright, S. F., and Brown, I., 1988, Differentiation of porcine coronavirus from transmissible gastroenteritis virus, *Vet. Rec.* **122**:86.

Gebauer, F., Posthumus, W. P. A., Correa, I., Suñé, C., Smerdou, C., Sánchez, C. M., Lenstra, J. A.,

Meloen, R. H., and Enjuanes, L., 1991. Residues involved in the antigenic sites of transmissible gastroenteritis coronavirus S glycoprotein, *Virology* **183:**225.
Godet, M., Rasschaert, D., and Laude, H., 1991, Processing and antigenicity of entire and anchor-free spike glycoprotein S of coronavirus TGEV expressed by recombinant baculovirus, *Virology* **185:**732.
Gombold, J. L., Hingley, S. T., and Weiss, S. R., 1993, Fusion-defective mutants of mouse hepatitis virus A59 contain a mutation in the spike protein cleavage signal, *J. Virol.* **67:**4504.
Grosse, B., and Siddell, S. G., 1994, Single amino acid changes in the S2 subunit of the MHV surface glycoprotein confer resistance to neutralization by S1 subunit-specific monoclonal antibody, *Virology* **202:**814.
Hasony, H. J., and Macnaughton, M. R., 1981, Antigenicity of mouse hepatitis virus strain 3 subcomponents in C57 strain mice, *Arch. Virol.* **69:**33.
Hogue, B. G., and Brian, D. A., 1986, Structural proteins of human respiratory coronavirus OC43, *Virus Res.* **5:**131.
Hogue, B. G., King, B., and Brian, D. A., 1984, Antigenic relationships among proteins of bovine coronavirus, human respiratory coronavirus OC43, and mouse hepatitis coronavirus A59, *J. Virol.* **51:**384.
Hohdatsu, T., Okada, S., and Koyama, H., 1991, Characterization of monoclonal antibodies against feline infectious peritonitis virus type II and antigenic relationship between feline, porcine, and canine coronaviruses, *Arch. Virol.* **117:**85.
Holmes, K. V., Doller, E. W., and Sturman, L. S., 1981, Tunicamycin resistant glycosylation of a coronavirus glycoprotein: demonstration of a novel type of virus glycoprotein, *Virology* **115:**334.
Hopkins, S. R., 1974, Serological comparisons of strains of infectious bronchitis virus using plaque-purified isolants, *Avian Dis.* **18:**231.
Horsburgh, B. C., Brierley, I., and Brown, T. D. K., 1992, Analysis of a 9.6 KB sequence from the 3′ end of a canine coronavirus genomic RNA, *J. Gen. Virol.* **75:**2849.
Hussain, K. A., Storz, J., and Kousoulas, K. G., 1991, Comparison of bovine coronavirus (BCV) antigens: Monoclonal antibodies to the spike glycoprotein distinguish between vaccine and wild-type strains, *Virology* **183:**442.
Jacobs, L., de Groote, R., van der Zeijst, B. A. M., Horzinek, M. C., and Spaan, W., 1987, The nucleotide sequence of the peplomer gene of porcine transmissible gastroenteritis virus (TGEV): Comparison with the sequence of the peplomer protein of feline infectious perotonitis virus (FIPV), *Virus Res.* **8:**363.
Jia, W., Karaca, K., Parrish, C. R., and Naqi, S. A., 1993a, Sequences of the spike protein gene of IBV strains Arkansas 99 (Genbank Accession Number L10384) and CV-T2 (Gen Bank Accession Number U04739).
Jia, W., Karaca, K., Naqi, S., Fabricant, J., Bauman, B., and Andriguetto, A., 1993b, Significance of genetic recombination in the evolution of variant IBV in the field, in: *Proceedings of the Xth International Congress of the World Veterinary Poultry Association*, p. 146, (J. York, ed.), Australian Veterinary Poultry Society, Sydney
Jimenez, G., Correa, I., Melgosa, M. P., Bullido, M. J., and Enjuanes, L., 1986, Critical epitopes in transmissible gastroenteritis virus neutralization, *J. Virol.* **60:**131.
Johnson, R. B., and Marquardt, W. W., 1975, The neutralizing characteristics of strains of infectious bronchitis virus as measured by the constant-virus variable-serum method in chicken tracheal cultures, *Avian Dis.* **19:**82.
Jordi, B. J. A. M., Kremers, D. A. W. M., Kusters, H. G., and van der Zeijst, B. A. M., 1989, Nucleotide sequence of the gene coding for the peplomer protein (= spike protein) of infectious bronchitis virus, strain D274, *Nucleic Acids Res.* **17:**6726.
Kant, A., Koch, G., Van Roozelaar, D. J., Kusters, J. G., Poelwijk, F. A. J., and van der Zeijst, B. A. M., 1992, Location of antigenic sites defined by neutralizing monoclonal antibodies on the S1 avian infectious bronchitis virus glycopeptide, *J. Gen. Virol.* **73:**591.
Karaca, K., Naqi, S., and Gelb, J., 1992, Production and characterization of monoclonal antibodies to three infectious bronchitis virus serotypes, *Avian Dis.* **36:**903.
Keck, J. G., Soe, L. H., Makino, S., Stohlman, S. S., and Lai, M. M. C., 1988, RNA recombination of

murine coronavirus: Recombination between fusion-positive mouse hepatitis virus A59 and fusion-negative mouse hepatitis virus 2, *J. Virol.* **62:**1989.

Kemp, M. C., Hierholzer, J. C., Harrison, A., and Burks, J. S., 1984, Characterisation of viral proteins synthesized in 299E infected cells and effect(s) of inhibition of glycosylation and glycoprotein transport, *Adv. Exp. Med. Biol.* **173:**65.

King, D. J., and Cavanagh, D., 1991, Infectious Bronchitis, in: *Diseases of Poultry* ( B. W. Calnek, H. J. Barnes, C. W. Beard, W. M. Reid, and H. W. Yoder, Jr., eds.), pp 471– 484, Iowa State University Press, Ames.

Koch, G., and Kant, A., 1990a, Binding of antibodies that strongly neutralize infectious bronchitis virus is dependent on the glycosylation of the viral peplomer protein, in: *Coronaviruses and Their Diseases* (D. Cavanagh and T. D. K. Brown, eds.), pp. 143–150, Plenum Press, New York.

Koch, G., and Kant, A., 1990b, Nucleotide and amino acid sequence of the S1 subunit of the spike glycoprotein of avian infectious bronchitis virus strain D3896, *Nucleic Acids Res.* **18:**3063.

Koch, G., Hartog, L., Kant, A., and Van Roozelaar, D. J., 1990, Antigenic domains on the peplomer protein of avian infectious bronchitis virus: Correlation with biological functions, *J. Gen. Virol.* **71:**1929.

Kooi, C., Mizzen, L, Alderson, C., Daya, M., and Anderson, R., 1988, Early events of importance in determining host cell permissiveness to mouse hepatitis virus infection, *J. Gen. Virol.* **69:**1125.

Kooi, C., Cervin, M., and Anderson, R., 1991, Differentiation of acid pH-dependent and nondependent entry pathways for mouse hepatitis virus, *Virology* **180:**108.

Koolen, M. J. M., Borst, M. A. J., Horzinek, M. C., and Spaan, W. J. M., 1990, Immunogenic peptide comprising a mouse hepatitis virus A59 B-cell epitope and an influenza virus T-cell epitope protects against lethal infection, *J. Virol.* **64:**6270.

Kryzstyniak, K., and Dupuy, J. M., 1984, Entry of mouse hepatitis virus 3 into cells, *J. Gen. Virol.* **65:**227.

Kubo, H., and Taguchi, F., 1993, Expression of the S1 and S2 subunits of murine coronavirus JHMV spike protein by a vaccinia virus transient expression system, *J. Gen. Virol.* **74:**2373.

Kubo, H., Takase-Yoden, S., and Taguchi, F., 1993, Neutralization and fusion inhibition activities of monoclonal antibodies specific for the S1 subunit of the spike protein of neurovirulent murine coronavirus JHMV cl-2 variant, *J. Gen. Virol.* **74:**1421.

Künkel, F., and Herrler, G., 1993, Structural and functional analysis of the surface protein of human coronavirus OC43, *Virology* **195:**195.

Kusters, J. G., Niesters, H. G. M., Lenstra, J. A., Horzinek, M. C., and van der Zeijst, B. A. M., 1989a, Phylogeny of antigenic variants of avian coronavirus IBV, *Virology* **169:**217.

Kusters, J. G., Jager, E. J., Lenstra, J. A., Koch, G., Posthumus, W. P. A., Meloen, R. H., and van der Zeijst, B. A. M., 1989b, Analysis of an immunodominant region of infectious bronchitis virus, *J. Immunol.* **143:**2692.

Kusters, J. G., Jager, E. J., Niesters, H. G. M., and van der Zeijst, B. A. M., 1990, Sequence evidence for RNA recombination in field isolates of avian coronavirus infectious bronchitis virus, *Vaccine* **8:**605.

Kwon, H., and Jackwood, M., 1993, Sequence of the spike S1 subunit gene of IBV strain JMK (Gen Bank Accession Number L14070) and Gray (Acc. No. L14069).

La Monica, N., Banner, L. R., Morris, V. L., and Lai, M. M. C., 1991, Localization of extensive deletions in the structural genes of two neurotropic variants of murine coronavirus JHM, *Virology* **182:**883.

Laviada, M. D., Videgain, S. P., Moreno, L., Alonso, F., Enjuanes, L., and Escribano, J. M., 1990, Expression of swine transmissible gastroenteritis virus envelope antigens on the surface of infected cells: Epitopes externally exposed, *Virus Res.* **16:**247.

Lenstra, J. A., Kusters, J. G., Koch, G., Van Der Jeijst, B. A. M., 1989, Antigenicity of the peplomer protein of infectious bronchitis virus, *Molecular Immunology* **1:**7.

Li, D., and Cavanagh, D., 1990, Role of pH in syncytium induction and genome uncoating of avian infectious bronchitis coronavirus (IBV), in: *Coronaviruses and Their Diseases* (D. Cavanagh and T. D. K. Brown, eds.), pp. 33–36, Plenum Press, New York.

Li, D., and Cavanagh, D., 1992, Coronavirus IBV-induced membrane fusion occurs at near neutral pH, *Arch. Virol.* **122:**307.

Luytjes, W., Sturman, L. S., Bredenbeek, P. J., Charite, J., van der Zeijst, B. A. M., Horzinek, M. C., and Spaan, W. J. M., 1987, Primary structure of the glycoprotein E2 of coronavirus MHV-A59 and identification of the trypsin cleavage site, *Virology* **161**:479.

Luytjes, W., Geerts, D., Posthumus, W., Meloen, E., and Spaan, W., 1989, Amino acid sequence of a conserved neutralising epitope of murine coronaviruses, *J. Virol.* **63**:1408.

Martin, J. P., Chen, W., Obert, G., and Koehren, F., 1990, Characterization of attenuated mutants of MHV3: Importance of the E2 protein in organ tropism and infection of isolated liver cells, in: *Coronaviruses and Their Diseases* (D. Cavanagh and T. D. K. Brown, eds.), pp. 403–410, Plenum Press, New York.

Matsubara, Y., Watanabe, R., and Taguchi, F., 1991, Neurovirulence of six different murine coronavirus JHMV variants for rats, *Virus Res.* **20**:45.

Michaud, L., and Dea, S., 1993, Characterization of monoclonal antibodies to bovine enteric coronavirus and antigenic variability among Quebec isolates, *Arch. Virol.* **131**:455.

Mizzen, L., Hilton, A., Cheley, S., and Anderson, R., 1985, Attenuation of murine coronavirus infection by ammonium chloride, *Virology* **142**:378.

Mockett, A. P. A., 1985, Envelope proteins of avian infectious bronchitis virus: Purification and biological properties, *J. Virol. Methods* **12**:271.

Mockett, A. P. A., Cavanagh, D., and Brown, T. D. K., 1984, Monoclonal antibodies to the S1 spike and membrane proteins of avian infectious bronchitis coronavirus strain Massachusetts M41, *J. Gen. Virol.* **65**:2281.

Morris, V. L., Tieszer, C., Mackinnon, J., and Percy, D., 1989, Characterisation of coronavirus JHM variants isolated from Wistar Furth rats with a viral-induced demyelinating disease, *Virology* **169**:127.

Morris, V. L., Wilson, G. A. R., Mckenzie, C. E., Tieszer, C., La Monica, M., Banner, L., Percy, D., Lai, M. M. C., and Dales, S., 1990, Murine hepatitis virus JHM variants isolated from Wistar Furth rats with viral-induced neurological disease, in: *Coronaviruses and Their Diseases* (D. Cavanagh and T. D. K. Brown, eds.), pp. 411–416, Plenum Press, New York.

Mounir, S., and Talbot, P. J., 1993, Molecular characterization of the S protein gene of human coronavirus OC43, *J. Gen. Virol.* **74**:1981.

Nakanaga, K., Yamanouchi, K., and Fujiwara, K., 1986, Protective effect of monoclonal antibodies on lethal mouse hepatitis virus infection in mice, *J. Virol.* **59**:165.

Niesters, H. G. M., Lenstra, J. A., Spaan, W. J. M., Zijderveld, A. J., Bleumink-Pluym, N. M. C., Hong, F., Van Scharrenburg, G. J. M., Horzinek, M. C., and van der Zeijst, B. A. M., 1986, The peplomer protein sequence of the M41 strain of coronavirus IBV and its comparison with Beaudette strains, *Virus Res.* **5**:253.

Noda, M., Yamashita, H., Koide, F., Kadoi, K., Omori, T., Asagi, M., and Inaba, Y., 1987, Hemagglutination with transmissible gastroenteritis virus, *Arch. Virol.* **96**:109.

Noda, M., Koide, F., Asagi, M., and Inaba, Y., 1988, Physicochemical properties of transmissible gastroenteritis virus hemagglutinin, *Arch. Virol.* **99**:163.

Oleszak, E. L., and Leibowitz, J. L., 1990, Fc receptor-like activity of mouse hepatitis virus E2 glycoprotein, in: *Coronaviruses and Their Diseases* (D. Cavanagh and T. D. K. Brown, eds.), pp. 51–58, Plenum Press, New York.

Oleszak, E. L., Perlman, S., and Leibowitz, J. L., 1992, MHV S peplomer protein expressed by a recombinant vaccinia virus vector exhibits IgG Fc-receptor activity, *Virology* **186**:122.

Opstelten, D-J E., de. Groote, P., Horzinek, M. C., Vennema, H., and Rottier, P. J. M., 1993, Disulfide bonds in folding and transport of mouse hepatitis coronavirus glycoproteins, *J. Virol.* **67**:7394.

Parker, S. E., Gallagher, T. M., and Buchmeier, M. J., 1989, Sequence analysis reveals extensive polymorphism and evidence of deletions within the E2 glycoprotein gene of several strains of murine hepatitis virus, *Virology* **173**:664.

Parker, M. D., Yoo, D., Cox, G. J., and Babiuk, L. A., 1990, Primary structure of the S peplomer gene of bovine coronavirus and surface expression in insect cells, *J. Gen. Virol.* **71**:263.

Parr, R. L., and Collisson, E. W., 1993, Epitopes on the spike protein of a nephropathogenic strain in infectious bronchitis virus, *Arch. Virol.* **133**:369.

Patterson, S., and Bingham, R. W., 1976, Electron microscope observations on the entry of avian infectious bronchitis virus into susceptible cells, *Arch. Virol.* **52**:191.

Payne, H. R., and Storz, J., 1988, Analysis of cell fusion induced by bovine coronavirus infection, *Arch. Virol.* **103**:27.

Payne, H. R., Storz, J., and Henk, W. G., 1990, Initial events in bovine coronavirus infection: analysis through immunogold probes and lysosomotropic inhibitors, *Arch. Virol.* **114:**175.

Pfleiderer, M., Routledge, E., and Siddell, S. G., 1990, Functional analysis of the coronavirus MHV-JHM surface glycoproteins in vaccinia virus recombinants, in: *Coronaviruses and Their Diseases* (D. Cavanagh and T. D. K. Brown, eds.), pp. 21–31, Plenum Press, New York.

Posthumus, W. P. A., Meloen, R. H., Enjuanes, L., Correa, I., Van Nieuwstadt, A. P., Kock, G., de Groot, R. J., Kusters, J. G., Luytjes, W., Spaan, W. J., van der Zeijst, B. A. M., and Lenstra, J. A., 1990a, Linear neutralizing epitopes on the peplomer protein of coronavirus, in: *Coronaviruses and Their Diseases* (D. Cavanagh and T. D. K. Brown, eds.), pp. 181–188, Plenum Press, New York.

Posthumus, W. P. A., Lenstra, J. A., Schaaper, W. M. M., Van Nieuwstadt, A. P., Enjuanes, L., and Meloen, R. H., 1990b, Analysis and simulation of a neutralising epitope of transmissible gastroenteritis virus, *J. Virol.* **64:**3304.

Posthumus, W. P. A., Lenstra, J. A., Van Nieuwstadt, A. P., Schaaper, W. M. M., van der Zeijst, B. A. M., and Meloen, R. H., 1991, Immunogenicity of peptides simulating a neutralization epitope of transmissible gastroenteritis virus, *Virology* **182:**371.

Pulford, D. J., and Britton, P., 1991, Intracellular processing of the porcine coronavirus transmissible gastroenteritis virus spike protein expressed by recombinant vaccinia virus, *Virology* **182:**765.

Raabe, T., Schelle-Prinz, B., and Siddell, S. G., 1990, Nucleotide sequence of the gene encoding the spike glycoprotein of human coronavirus HCV-229E, *J. Gen. Virol.* **71:**1065.

Rasschaert, D., and Laude, H., 1987, The predicted primary structure of the peplomer protein E2 of the porcine coronavirus transmissible gastroenteritis virus, *J. Gen. Virol.* **68:**1883.

Rasschaert, D., Duarte, M., and Laude, H., 1990, Porcine respiratory coronavirus differs from transmissible gastroenteritis virus by a few genomic deletions, *J. Gen. Virol.* **71:**2599.

Roos, D. S., Duchala, C. S., Stephensen, C. B., Holmes, K. V., and Choppin, P. W., 1990, Control of virus-induced cell fusion by host cell lipid composition, *Virology* **175:**345.

Rottier, P. J. M., Horzinek, M. C., and van der Zeijst, B. A. M., 1981, Viral protein synthesis in mouse hepatitis virus strain A59 infected cells: Effect of tunicamycin, *J. Virol.* **40:**350.

Routledge, E., Stauber, R., Pfleiderer, M., and Siddell, S. G., 1991, Analysis of murine coronavirus surface glycoprotein functions by using monoclonal antibodies, *J. Virol.* **65:**254.

Sanchez, C. M., Jimenez, G., Laviada, M. D., Correa, I., Sune, C., Bullido, M. J., Gebauer, F., Smerdou, C., Callebaut, P., Escribano, J. M., and Enjuanes, L., 1990, Antigenic homology among coronaviruses related to transmissible gastroenteritis virus, *Virology* **175:**410.

Sawicki, S. G., 1987, Characterisation of a small plaque mutant of the A59 strain of mouse hepatitis virus defective in cell fusion, *Adv. Exp. Med. Biol.* **218:**169.

Sawicki, S. G., and Sawicki, D. L., 1986, Coronavirus minus-strand RNA synthesis and effect of cycloheximide on coronavirus RNA synthesis, *J. Virol.* **57:**328.

Schmidt, I., Skinner, M., and Siddell, S., 1987, Nucleotide sequence of the gene encoding the surface projection glycoprotein of coronavirus MHV-JHM, *J. Gen. Virol.* **68:**47.

Schmidt, M. F. G., 1982, Acylation of viral spike glycoproteins: A feature of enveloped RNA viruses, *Virology* **116:**327.

Schmidt, O. W., and Kenny, G. E., 1982, Polypeptides and functions of antigens from human coronaviruses 229E and OC43, *Infect. Immun.* **35:**515.

Schultze, B., Hess, R. G., Rott, R., Klenk, H. D., and Herrler, G., 1990, Isolation and characterization of the acetylesterase of hemagglutinating encephalomyelitis virus (HEV), in: *Coronaviruses and Their Diseases* (D. Cavanagh and T. D. K. Brown, eds.), pp. 109–113, Plenum Press, New York.

Schultze, B., Gross, H-J., Brossmer, R., and Herrler, G., 1991, The S protein of bovine coronavirus is a hemagglutinin recognizing 9-*O*-acetylated sialic acid as a receptor determinant, *J. Virol.* **65:**6232.

Schultze, B., Cavanagh, D., and Herrler, G., 1992. Neuraminidase treatment of avian infectious bronchitis coronavirus reveals a hemagglutinating activity that is dependent on sialic acid-containing receptors on erythrocytes, *Virology* **189**:792.

Siddell, S. G., 1982, Coronavirus JHM: Tryptic peptide fingerprinting of virion proteins and intracellular polypeptides, *J. Gen. Virol.* **62:**259.

Siddell, S. G., Wege, H., and Ter Meulen, V. T., 1983, The biology of coronaviruses, *J. Gen. Virol.* **64:**761.

Simkins, R. A., Saif, L. J., and Weilnau, P. A., 1989, Epitope mapping and the detections of transmissible gastroenteritis viral proteins in cell culture using biotinylated monoclonal antibodies in a fixed-cell ELISA, *Arch. Virol.* **107:**179.

Simkins, R. A., Weilnau, P. A., Bias, J., and Saif, L. J., 1992, Antigenic variation among transmissible gastroenteritis virus (TGEV) and porcine respiratory coronavirus strains detected with monoclonal antibodies to the S protein of TGEV, *Am. J. Vet. Res.* **53:**1253.

Spaan, W., Cavanagh, D., and Horzinek, M. C., 1990, Coronaviruses, in: *Immunochemistry of Viruses II* (M. H. V. van Regenmortel and T. D. K. Brown, eds.), pp. 359–380, Elsevier, Amsterdam.

Stauber, R., Pfleiderer, M., and Siddell, S., 1993, Proteolytic cleavage of the murine coronavirus surface glycoprotein is not required for fusion activity, *J. Gen. Virol.* **74:**183.

Stern, D. F., and Sefton, B. M., 1982a, Coronavirus proteins: Biogenesis of avian infectious bronchitis virus virion proteins, *J. Virol.* **44:**794.

Stern, D. F., and Sefton, B. M., 1982b, Coronavirus proteins: Structure and function of the oligosaccharides of the avian infectious bronchitis virus glycoproteins, *J. Virol.* **44:**804.

Storz, J., Rott, R., and Kaluza, G., 1981, Enhancement of plaque formation and cell fusion of an enteropathogenic coronavirus by trypsin treatment, *Infect. Immun.* **31:**1214.

Stuhler, A., Wege, H., and Siddell, S. G., 1991, Localization of antigenic sites on the surface glycoprotein of mouse hepatitis virus, *J. Gen. Virol.* **72:**1655.

Sturman, L. S., Holmes, K. V., and Behnke, J., 1980, Isolation of coronavirus envelope glycoproteins and interaction with the viral nucleocapsid, *J. Virol.* **33:**449.

Sturman, L. S., Ricard, C. S., and Holmes, K. V., 1985, Proteolytic cleavage of the E2 glycoprotein of murine coronavirus: Activation of cell-fusing activity of virions by trypsin and separation of two different 90K cleavage fragments, *J. Virol.* **56:**904.

Sturman, L. S., Ricard, C. S., and Holmes, K. V., 1990, Conformational change of the coronavirus peplomer glycoprotein at pH 8.0 and 37°C correlates with virus aggregation and virus-induced cell fusion, *J. Virol.* **64:**3042.

Sutou, S., Sato, S., Okabe, T., Nakai, M., and Sasaki, N., 1988, Cloning and sequencing of genes encoding structural proteins of avian infectious bronchitis virus, *Virology* **65:**589.

Taguchi, F., 1993, Fusion formation by the uncleaved spike protein of murine coronavirus JHMV variant cl-2, *J. Virol.* **67:**1195.

Taguchi, F., and Fleming, J. O., 1989, Comparison of six different murine coronavirus JHM variants by monoclonal antibodies against the E2 glycoprotein, *Virology.* **169:**233.

Taguchi, F., Siddell, S. G., Wege, H., and Ter Meulen, V., 1985, Characterisation of a variant virus selected in rat brains after infection by coronavirus mouse hepatitis virus JHM, *J. Virol.* **54:**429.

Taguchi, F., Yoden, S., Siddell, S., and Kikuchi, T., 1990, Expression of spike protein of murine coronavirus JHM using baculovirus vector, in: *Coronaviruses and Their Diseases* (D. Cavanagh and T. D. K. Brown, eds.), pp. 211–216, Plenum Press, New York.

Taguchi, F., Ikeda, T., and Shida, H., 1992, Molecular cloning and expression of a spike protein of neurovirulent murine coronavirus JHMV variant cl-2, *J. Gen. Virol.* **73:**1065.

Takase-Yoden, S., Kikuchi, T., Siddell, S. G., and Taguchi, F., 1990, Localization of major neutralizing epitopes on the S1 polypeptide of the murine coronavirus peplomer glycoprotein, *Virus Res.* **18:**99.

Talbot, P. J., and Buchmeier, M., 1985, Antigenic variation among murine coronaviruses: Evidence for polymorphism on the peplomer glycoprotein E2, *Virus Res.* **2:**317.

Talbot, P. J., Salmi, A. A., Knobler, R. L., and Buchmeier, M. J., 1984, Topographical mapping of epitopes on the glycoproteins of murine hepatitis virus-4 (strain JHM): Correlation with biological activities, *Virology* **132:**250.

Talbot, P. J., Dionne, G., and Lacroix, M., 1988, Vaccination against lethal coronavirus-induced encephalitis with a synthetic decapeptide homologous to a domain in the predicted peplomer stalk, *J. Virol.* **62:**3032.

Tijssen, P., Verbeek, A. J., and Dea, S., 1990, Evidence of close relatedness between turkey and bovine coronaviruses, in: *Coronaviruses and Their Diseases* (D. Cavanagh and T. D. K. Brown, eds.), pp. 457–460, Plenum Press, New York.

Tomley, F. M., Mockett, A. P. A., Boursnell, M. E. G., Binns, M. M., Cook, J. K. A., Brown, T. D. K., and Smith, G. L., 1987, Expression of the infectious bronchitis virus spike protein by recombinant vaccinia virus and induction of neutralizing antibodies in vaccinated mice, *J. Gen. Virol.* **68:**2291.

van Berlo, M. F., van den Brink, W. J., Horzinek, M. C., and van der Zeijst, B. A. M., 1987, Fatty acid acylation of viral proteins in murine hepatitis virus-infected cells, *Arch. Virol.* **95:**123.

Van Dinter, S., and Flintoff, W. F., 1987, Rat glial C6 cells are defective in murine coronavirus internalization, *J. Gen. Virol.* **68:**1677.

Vautherot, J. F., Madelaine, M. F., and Laporte, J., 1990, Topological and functional analysis of epitopes on the S(E2) and the HE(E3) glycoproteins of bovine enteric coronavirus, in: *Coronaviruses and Their Diseases* (D. Cavanagh and T. D. K. Brown, eds.), pp. 173–180, Plenum Press, New York.

Vautherot, J-F., Madelaine, M-F., Boireau, P., and Laporte, J., 1992a, Bovine coronavirus peplomer glycoproteins: Detailed antigenic analyses of S1, S2 and HE, *J. Gen. Virol.* **73:**1725.

Vautherot, J-F., Laporte, J., and Boireau, P., 1992b, Bovine coronavirus S glycoprotein: Localisation of an immunodominant region at the amino-terminal end of S2, *J. Gen. Virol.* **73:**3289.

Vennema, H., de Groot, R. J., Harbour, D. A., Dalderup, M., Gruffydd-Jones, T., Horzinek, M. C., and Spaan, W. J. M., 1990a, Immunogenicity of recombinant feline infectious peritonitis virus spike protein in mice and kittens, in: *Coronaviruses and Their Diseases* (D. Cavanagh and T. D. K. Brown, eds.), pp. 217–222, Plenum Press, New York.

Vennema, H., Rottier, P. J. M., Heijnen, L., Godeke, G. J., Horzinek, M. C., and Spaan, W. J. M., 1990b, Biosynthesis and function of the coronavirus spike protein, in: *Coronaviruses and Their Diseases* (D. Cavanagh and T. D. K. Brown, eds.), pp. 9–19, Plenum Press, New York.

Vennema, H., Heijnen, L., Zijderfeld, A., Horzinek, M. C., and Spaan, W. J. M., 1990c, Intracellular transport of recombinant coronavirus spike proteins: Implications for virus assembly, *J. Virol.* **64:**339.

Vennema, H., de Groot, R. J., Harbour, D. A., Dalderup, M., Gruffydd-Jones, T., Horzinek, M. C., and Spaan, W. J. M., 1990d, Early death after feline infectious peritonitis virus challenge due to a recombinant vaccinia virus immunization, *J. Virol.* **64:**1407.

Wang, L., Junker, D., Hock, L., Ebiary, E., and Collisson, W. W., 1994, Evolutionary implications of genetic variations in the S1 gene of infectious bronchitis virus, *Virus Res.* **34:**327.

Wang, F-I., Fleming, J. O., and Lai, M. M. C., 1992, Sequence analysis of the spike protein gene of murine coronavirus variants: Study of genetic sites affecting neuropathogenicity, *Virology* **186:**742.

Wang, L., Junker, D., and Collisson, E. W., 1993, Evidence of natural recombination within the S1 gene of infectious bronchitis virus, *Virology* **192:**710.

Wege, H., Dorries, R., and Wege, H., 1984, Hybridoma antibodies to the murine coronavirus JHM: Characterisation of epitopes on the peplomer protein, *J. Gen. Virol.* **65:**1931.

Weismuller, D. G., Sturman, L. S., Buchmeier, M. J., Fleming, J. O., and Holmes, K. V., 1990, Monoclonal antibodies to the peplomer glycoprotein of coronavirus mouse hepatitis virus identify two subunits and detect a conformational change in the subunit released under mild alkaline conditions, *J. Virol.* **64:**3051.

Welch, S-K. W., and Saif, L. J., 1988, Monoclonal antibodies to a virulent strain of transmissible gastroenteritis virus: Comparison of reactivity with virulent and attenuated virus, *Arch. Virol.* **101:**221.

Wesley, R. D., 1990, Nucleotide sequence of the E2-peplomer protein gene and partial nucleotide sequence of the upstream polymerase gene of transmissible gastroenteritis virus (Miller strain), in: *Coronaviruses and Their Diseases* (D. Cavanagh and T. D. K. Brown, eds.), pp. 301–306, Plenum Press, New York.

Wesley, R. D., Woods, R. D., and Cheung, A. K., 1991, Genetic analysis of porcine respiratory coronavirus, an attenuated variant of transmissible gastroenteritis virus, *J. Virol.* **65:**3369.

Wesseling, J. G., Godeke, G-J., Schijns, V. E. C. J., Prevec, L., Graham, F. L., Horzinek, M. C., and Rottier, P. J. M., 1993, Mouse hepatitis virus spike and nucleocapsid proteins expressed by adenovirus vectors protect mice against a lethal infection, *J. Gen. Virol.* **74:**2061.

Wesseling, J. G., Vennema, H., Godeke, G-J., Horzinek, M. C., Spaan, W. J. M., and Rottier, P. J. M., 1994, Nucleotide sequence and expression of the spike (S) gene of canine coronavirus and comparison with the S proteins of feline and porcine coronaviruses, *J. Gen. Virol.* **75:**1789.

Wilson, I. A., Skehel, J. J., and Wiley, D. C., 1981, Structure of the haemagglutinin membrane glycoprotein of influenza virus at 3 A resolution, *Nature* **289:**366.

Wysocka, M., Korngold, R., Yewdell, J., and Bennink, J., 1989, Target and effector cell fusion accounts for B lymphocyte-mediated lysis of mouse hepatitis virus-infected cells, *J. Gen. Virol.* **70:**1465.

Yoo, D., Parker, M. D., and Babiuk, L. A., 1990, Analysis of the S spike (peplomer) glycoprotein of bovine coronavirus synthesized in insect cells, *Virology* **179:**121.

Yoo, D., Parker, M. D., and Babiuk, L. A., 1991a, the S2 subunit of the spike glycoprotein of bovine coronavirus mediates membrane fusion in insect cells, *Virology* **180:**395.

Yoo, D., Parker, M. D., Song, J., Cox, G. J., Deregt, D., and Babiuk, L. A., 1991b, Structural analysis of the conformational domains involved in neutralization of bovine coronavirus using deletion mutants of the spike glycoprotein S1 subunit expressed by recombinant baculoviruses, *Virology* **183:**91.

Zhang, X., Kousoulas, K. G., and Storz, J., 1991, Comparison of the nucleotide and deduced amino acid sequences of the S genes specified by virulent and avirulent strains of bovine coronaviruses, *Virology* **183:**397.

Zhang, X., Herbst, W., Kousoulas, K. G., and Storz, J., 1994, Comparison of the S genes and the biological properties of respiratory and enteropathogenic bovine coronaviruses, *Arch. Virol.* **134:**421.

CHAPTER 6

# The Coronavirus Membrane Glycoprotein

PETER J. M. ROTTIER

## I. INTRODUCTION

Coronaviruses have a simple protein composition. While there is some variation among different members, a basic set of four protein species universally occurs: the nucleocapsid protein (N), the spike protein (S), a small membrane protein (SM), and the membrane glycoprotein (M). Some coronaviruses have an additional membrane glycoprotein (HE). The M protein, previously also called E1, is the subject of this chapter. As will become clear, M is a peculiar glycoprotein, different from all other viral glycoproteins in its structural and biochemical features. These unique features may be responsible for important biological properties of coronaviruses, in particular for their intracellular budding.

M is the most abundant virion protein. In murine hepatitis virus (MHV) it was estimated by isotopic labeling to comprise some 40% of the particle's protein mass, exceeding the nucleocapsid protein on a molar basis at a ratio of 2 : 1 (Sturman *et al.*, 1980). A similar ratio was obtained from incorporation studies with avian infectious bronchitis virus (IBV) (Stern *et al.*, 1982), while equimolar ratios were determined for bovine coronavirus (BCV) (King and Brian, 1982) and human coronavirus (HCV) OC43 (Hogue and Brian, 1986). In the latter case, M and N were calculated to be present at a rate of 726 molecules per virion.

The M gene, together with the genes for the other structural proteins, is located in the 3′ one third of the coronaviral genome, downstream from the spike protein gene and upstream from the nucleoprotein gene. As a conse-

PETER J. M. ROTTIER • Institute of Virology, Department of Infectious Diseases and Immunology, University of Utrecht, 3584 CL Utrecht, The Netherlands.

*The Coronaviridae*, edited by Stuart G. Siddell, Plenum Press, New York, 1995.

quence of the specific mode of transcription of coronaviruses, the M protein is expressed from a mRNA that, in addition to the M sequence, carries extra genetic information including the N sequence. This additional information is located 3′ from the M gene and is functionally redundant; the M protein is equally well translated from a mRNA derived from a cloned copy of the M gene (see Machamer and Rose, 1987; Mayer *et al.*, 1988; Rottier and Rose, 1987).

## II. PHYSICOCHEMICAL PROPERTIES

### A. Covalent Modifications

The M protein is usually found as a family of differentially glycosylated proteins, including the unglycosylated precursor. These proteins span a $M_r$ range of 20 to 38kDa. The M protein of a particular coronavirus carries either N- or O-linked oligosaccharides (Table I). The only other modification identified so far is the addition of sulfate (Garwes *et al.*, 1976), but it is unknown whether this constituent is attached to the oligosaccharide side chains or bound directly to the polypeptide through tyrosine. M is not acylated (Niemann and Klenk, 1981; Schmidt, 1982), nor does it contain phosphate (e.g., Stohlman and Lai, 1979; Rottier *et al.*, 1981a; King and Brian, 1982; Hogue and Brian, 1986).

The structures of the N-linked oligosaccharides carried by coronavirus M proteins have not been studied in detail, but are probably not different from those found in other viral and cellular N-glycosylated proteins. Carbohydrates bound to a polypeptide through O-linkage to serine or threonine residues are quite uncommon among viral proteins. Of the O-glycosylated coronaviral M proteins, the side chains of the M protein of MHV strain A59 grown in 17C11 cells, a spontaneously transformed BALBC/3T3 line, have been analyzed (Niemann and Klenk, 1981; Niemann *et al.*, 1984). Two size classes of oligosaccharides were released from the protein by β-elimination. Their structures, as

TABLE I. Type of Glycosylation of M Proteins of Coronaviruses

| Common name | Designation |
|---|---|
| *N-glycosylation* | |
| Canine coronavirus | CCV |
| Feline infectious peritonitis virus | FIPV |
| Feline enteric coronavirus | FECV |
| Human coronavirus strain 229E | HCV-229E |
| Infectious bronchitis virus | IBV |
| Turkey coronavirus | TCV |
| Transmissible gastroenteritis virus | TGEV |
| *O-glycosylation* | |
| Bovine coronavirus | BCV |
| Diarrhoea virus of infant mice | DVIM |
| Human coronavirus strain OC43 | HCV-OC43 |
| Mouse hepatitis virus | MHV |

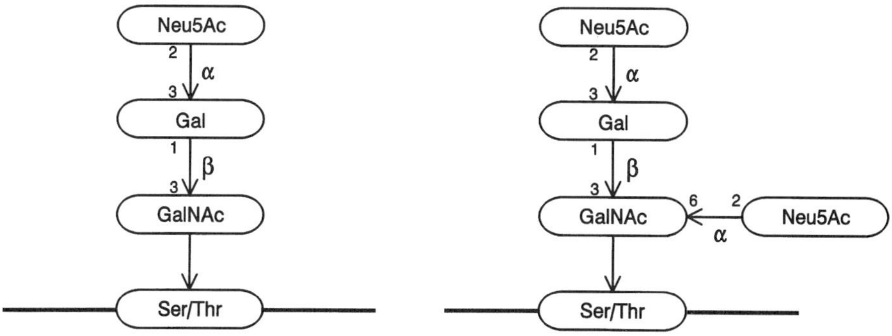


FIGURE 1. Structures of the O-linked oligosaccharide side chains of the MHV-A59 M glycoprotein.

determined by a combined gas chromatographic and mass spectrometric analysis, are shown in Fig. 1. The branched form was most abundant, comprising about 65% of the carbohydrate structures. Both forms were attached to the M polypeptide via *N*-acetylgalactosamine. Of the sialic acids (*N*-acetylneuraminic acid, Neu5Ac), some 20% were identified as the O-acetylated derivative Neu5,9$Ac_2$ (Niemann *et al.*, 1984).

The oligosaccharides are attached to the N-terminal region of the M molecule and exposed at the virion surface. Treatment of MHV (Sturman, 1977; Sturman and Holmes, 1977; Rottier *et al.*, 1984) and BCV (King and Brian, 1982) with various proteases removed some 5kDa from the polypeptide, including all the O-linked sugars. Similarly, the N-linked oligosaccharides of the IBV M protein were all detached by hydrolysis of the exposed domain of the molecule (Cavanagh, 1981; Cavanagh *et al.*, 1986a). Using [$^{35}$S]formyl-methionine to terminally label the polypeptide, this ectodomain was identified as the N-terminus (Stern *et al.*, 1982; Rottier *et al.*, 1984).

## B. Solubility

The M protein has unusual solubility properties. This was demonstrated for MHV M in the pioneering studies of Sturman and Holmes (Sturman, 1977, 1981; Sturman and Holmes, 1977; Sturman *et al.*, 1980). When isolated at 4 °C from virions after nonionic detergent solubilization of the viral membrane, the protein formed globular, irregular aggregates of various sizes. Raising the temperature to 37 °C induced a conformational change in the molecules that led the aggregates to associate with the nucleocapsid through interaction with the viral RNA (Sturman *et al.*, 1980). Thermosensitivity of the M protein was also prominent in sodium dodecyl sulfate (SDS) solution: heating from 25 °C or 37 °C, in which range the protein is in a monomeric state, to 100 °C resulted in the formation of various self-aggregates, an effect that was even more pronounced in the presence of reducing agents (Sturman, 1977; Sturman and Holmes, 1977). Unlike the M protein of the MHV-A59 strain, the M protein of MHV-JHM additionally appeared to form heterogeneous complexes with the S glycoprotein under these conditions (Wege *et al.*, 1979; Siddell *et al.*, 1981). The tendency of the MHV M protein to aggregate was also observed after various other denaturing treatments (Sturman, 1977). Apparently, any condition that promotes the unfolding of the M protein can expose a (probably hydrophobic) domain of the molecule which then engages in interactions.

Aggregation of M in SDS has been described for a number of other coronaviruses such as hemagglutinating encephalitis virus (HEV) (Callebaut and Pensaert, 1980), HCV-OC43 (Schmidt and Kenny, 1982; Hogue and Brian, 1986), and BCV (Niemann and Klenk, 1981; Deregt *et al.*, 1987). On the other hand, it is not a universal feature as the effect was not observed after heat denaturation of the M polypeptide of IBV (Cavanagh, 1981; Stern and Sefton, 1982b) or HCV-229E (Schmidt and Kenny, 1982), not even in the presence of 2-mercaptoethanol. The reason for this variable behavior is not understood. The effect seems to be independent of the degree and type of glycosylation. Even complete removal of the N-terminal ectodomain did not abolish the aggregation properties of the polypeptide (Sturman, 1977).

## III. PROTEIN STRUCTURE

### A. Primary Structure

During the past years, the M genes of several coronaviruses have been sequenced. The deduced amino acid sequences of a number of M proteins are compiled in the alignment presented in Fig. 2. The primary structures are 225–230 amino acids long, with the exception of the transmissible gastroenteritis virus (TGEV) and feline infectious peritonitis virus (FIPV) sequences which amount to 262 residues, due to an extension at the $NH_2$-terminus. The polypeptides are slightly basic with net charges at neutral pH ranging from +4 (HCV-229E) to +9 (BCV). Their cysteine content is quite variable, with only two such residues in the HCV-229E sequence and nine residues in IBV M. These numbers do not explain the effects of reducing agents on the solubility of the coronavirus M proteins.

Pairwise sequence comparisons support the long-standing classification of coronaviruses made on the basis of antigenic relationships (McIntosh *et al.*, 1969; Bradburne, 1970; Pedersen *et al.*, 1978; Horzinek *et al.*, 1982). The M proteins of MHV and BCV, viruses of the same antigenic subgroup, are very closely related (86% identity), but differ largely from all the others (e.g., MHV/IBV 29%, MHV/HCV-229E 32%, MHV/FIPV 36%). Similarly, the M sequences of TGEV and FIPV, also antigenically related viruses, show a strong homology (84% identity), but are only distantly related to the others (e.g., TGEV/IBV 17%). The avian IBV M polypeptide has only low homologies to the mammalian proteins (e.g., IBV/FIPV 20% identity), in agreement with it being classified separately. HCV-229E has previously been placed into the TGEV group on the basis of weak serological cross-reactivities (Pedersen *et al.*, 1978; Macnaughton, 1981). The M protein of this virus has only little sequence similarity with the TGEV or FIPV protein (e.g., HCV-229E/TGEV 44% identity) or with any of the other M proteins. The same holds true for the S proteins (Wesseling *et al.*, 1994). The data therefore support the recent proposition by Sanchez *et al.* (1990) to

```
           1                                                                       70
MHV        .......... .......... .......... .MSSTTQAPE PVYQWTADEA VQFLKEWNFS LGIILLFITI
BCV        .......... .......... .......... ..MSSVTTPA PVYTWTADEA IKFLKEWNFS LGIILLFITI
HCV        .......... .......... .......... .......... MSNDNCTGDI VTHLKNWNFG WNVILTIFIV
IBV        .......... .......... .......... ..MPNETNCT LDFE....QS VQLFKEYNLF ITAFLLFLTI
FIPV       MKYILLILAC IIACVYGERY CAMQ.DSGLQ CINGTNSRCQ TCFE..RGDL IWHLANWNFS WSVILIVFIT
TGEV       MK.ILLILAC VIACACGERY CAMKSDTDLS CRNSTASDCE SCFN..GGDL IWHLANWNFS WSIILIVFIT
Consensus  ---------- ---------- ---------- ---------- --F-----D- ---L--WNF- --IIL-----

           71                                                                     140
MHV        ILQFGYTSRS MFIYVVKMII LWLMWPLTIV LCIFN..CVY ALNN.VYLGF SIVFTIVSIV IWIMYFVNSI
BCV        ILQFGYTSRS MFVYVIKMII LWLMWPLTII LTIFN..CVY ALNN.VYLGF SIVFTIVAII MWIVYFVNSI
HCV        ILQFGHYKYS RLFYGLKMLV LWLLWPLVLA LSIFDTWANW D.SNWAFVAF SFFMAVSTLV MWVMYFANSF
IBV        ILQYGYATRS KVIYTLKMIV LWCFWPLNIA VGVIS..CTY PPNTGGLVA. AIILTVFACL SFVGYWIQSI
FIPV       VLQYGRPQFS WLVYGIKMLI MWLLWPIVLA LTIFNAYSEY QVSRYVMFGF SVAGAVVTFA LWMMYFVRSV
TGEV       VLQYGRPQFS WFVYGIKMLI MWLLWPVVLA LTIFNAYSEY QVSRYVMFGF SIAGAIVTFV LWIMYFVRSI
Consensus  ILQ-G----S -FVY--KM-I LWLLWP---- L-IF-----Y ---------F SI-------- -WI-YFV-SI

           141                                                                    210
MHV        RLFIRTGSWW SFNPETNNLM CIDMKGTVYV RPIIEDYHTL TATIIRGHLY MQGVKLGTGF SLSDLPAYVT
BCV        RLFIRTGSWW SFNPETNNLM CIDMKGRMYV RPIIEDYHTL TVTIIRGHLY MQGIKLGTGY SLSDLPAYVT
HCV        RLFRRARTFW AWNPEVNAIT VTTVLGQTYY QPIQQAPTGI TVTLLSGVLY VDGHRLASGV QVHNLPEYMT
IBV        RLFKRCRSWW SFNPESNAVG SILLTNGQQC NFAIESVPMV LSPIIKNGVL YCEGQWLAKC EPDHLPKDIF
FIPV       QLYRRTKSWW SFNPETNAIL CVNALGRSYV LPLDGTPTGV TLTLLSGNLY AEGFKMAGGL TIEHLPKYVM
TGEV       QLYRRTNSWW SFNPETKAIL CVSALGRSYV LPLEGVPTGV TLTLLSGNLY AEGFKIADGM NIDNLPKYVM
Consensus  -LF-R--SWW SFNPE-N--- -I---G--Y- -P-------- T-T---G-LY --G-----G- ----LP-YV-

           211                                                          272
MHV        VAKVSHLCTY K...RAFLDK VDGVSGFAVY VKSK....VG NYRLPSNKPS G..ADTALLR I.        228
BCV        VAKVSHLLTY KR.GF..LDK IGDTSGFAVY VKSKV....G NYRLPSTQKG SGMDTALLRN NI        230
HCV        VAVPSTTIIY SRVGR..SVN SQNCTGWVFY VRVKH....G DFSAVSSPMS NMTENERLLH FF        225
IBV        VCTPDRRNIY RMVQKYTGDQ SGNKKRFATF VYAKQSVDTG ELESVATGGS SLYT...... ..        225
FIPV       IATPSRTIVY TLVGK..QLK ATTATGWAYY VKSKA....G DYST.EARTD NLSEHEKLLH MV        262
TGEV       VALPSRTIVY TLVGK..KLK ASSATGWAYY VKSKA....G DYST.EARTD NLSEQEKLLH MV        262
Consensus  VA--S----Y ---------- -----G-A-Y V--K-----G -Y-------- -------L-- --
```

FIGURE 2. Sequence alignment of the M proteins of MHV-A59 (Armstrong *et al.*, 1984), BCV (Lapps *et al.*, 1987), HCV-229E (Raabe and Siddell, 1989), IBV (Baudette strain; Boursnell *et al.*, 1984), FIPV (Vennema *et al.*, 1991a), and TGEV (Laude *et al.*, 1987). Not included are the recently determined sequences of TCV (Verbeek and Tijssen, 1991), HCV-OC43 (Mounir and Talbot, 1992), CCV (Horsburgh *et al.*, 1992), and FECV (A. Herrewegh, H. Vennema, R. de Groot, and P. Rottier, unpublished data). Comparison shows the former two to be very similar to the sequences of MHV and BCV, while the latter have a high similarity to the sequences of FIPV and TGEV.

classify HCV-229E in a distinct taxonomic cluster; this suggestion was based on an extensive antigenic comparison of coronaviruses using monoclonal antibodies (see Chapter 1).

As indicated by its solubility, M is a very hydrophobic protein. It contains 44–51% of hydrophobic amino acids which are concentrated in the $NH_2$-terminal half of the molecule (Fig. 3). Despite the high degree of sequence variation between the coronavirus M polypeptides, the hydropathicity profiles are remarkably similar. The dominant common feature is the occurrence of three hydrophobic domains alternating with short hydrophilic regions. In the TGEV and FIPV sequence a fourth hydrophobic domain is present at the $NH_2$-terminus, which functions as a cleavable membrane insertion signal (Kapke *et al.*, 1988; Vennema *et al.*, 1991b). In the other M proteins, no such signal appears to be operative; rather, these proteins have a hydrophilic amino terminus. The carboxy terminal half of all M proteins is amphiphilic, with a hydrophilic domain at the carboxy end.

In view of the large differences in primary sequences, the surprising conser-

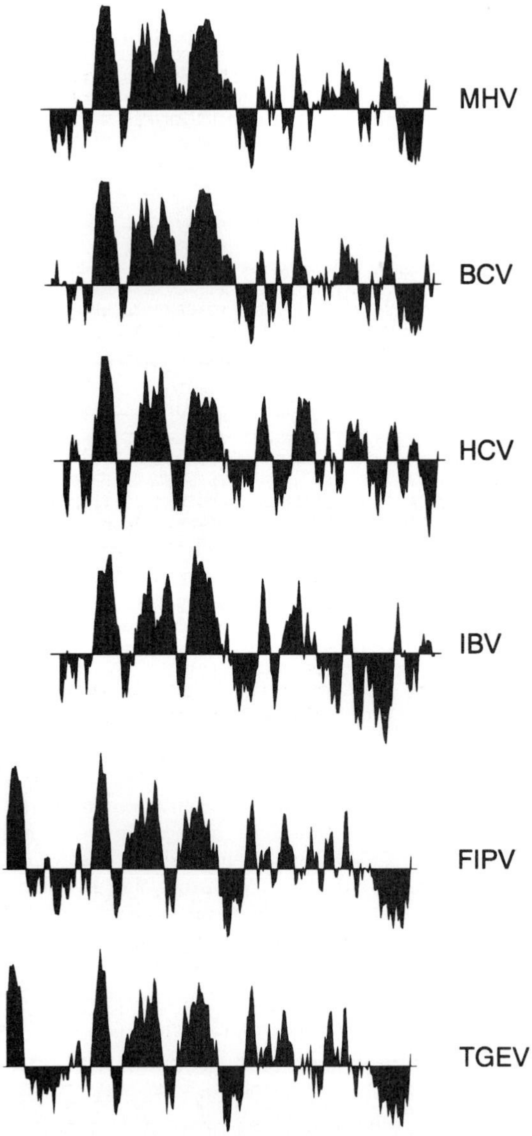


FIGURE 3. Hydropathicity profiles of M proteins from MHV-A59, BCV, HCV-229E, IBV, FIPV, and TGEV determined according to Kyte and Doolittle (1982) using a seven residue moving window. Peaks extending upward indicate hydrophobic regions, downward peaks correspond to hydrophilic regions.

vation of the overall chemical features indicates that there are rigid structural constraints on M as a result of functional requirements. The highest levels of sequence conservation appear in the hydrophobic regions and in the center of the polypeptide directly adjacent to the third internal hydrophobic region, where a stretch of 8 amino acids is extremely well conserved (see consensus in Fig. 2). This suggests a selective pressure on these domains for the maintenance of important structural characteristics.

As mentioned above, coronavirus M proteins are glycosylated in their $NH_2$-terminal ectodomain, i.e., the hydrophilic domain preceding the first internal hydrophobic region (see Section III.B). In HCV-229E, IBV, FIPV, and TGEV, the protein is N-glycosylated. Potential oligosaccharide attachment sites occur once in the relevant part of the HCV-229E M sequence (position 45 in Fig. 2), twice in the case of IBV M (positions 35 and 38), and once in the cases of FIPV and TGEV M (position 33). These sites are indeed used as the numbers of side chains are in agreement with those experimentally observed (Stern and Sefton, 1982b; Kapke *et al.*, 1988; Vennema *et al.*, 1991b). Another N-glycosylation consensus sequence (positions 58–60 in Fig. 2) occurs at the start of the first internal hydrophobic region of some M proteins but appears not to be used. Apparently, this sequence is not exposed to the modifying enzymes in the endoplasmic reticulum and resides within the membrane.

It is not known which of the serine and threonine residues are substituted in the O-glycosylated coronavirus M proteins. The structural features that determine a functional O-glycosylation site have not yet been established. In MHV, a cluster of four hydroxy amino acids is located next to the initiating methionine which presumably is removed posttranslationally as has been shown for IBV M (Cavanagh *et al.*, 1986b). The resulting $NH_2$-terminal tetrapeptide sequence (Ser-Ser-Thr-Thr) is identical to the O-glycosylated aminoterminus of glycophorin A, the major glycoprotein of the human erythrocyte membrane. It was shown by Niemann *et al.* (1984) that the O-linked sugar structures of MHV M are identical to those found in glycophorin. As the coronavirus protein also appeared to exhibit blood group M activity, as does glycophorin, the authors inferred that the hydroxy amino acid cluster contains the functional oligosaccharide acceptor sites in MHV M. On the basis of the heterogeneity in the glycosylation of the M protein, they concluded that up to three of the four residues in the cluster are modified by oligosaccharide side chains. In BCV a slightly different cluster of hydroxy amino acids occurs at the $NH_2$-terminus due to the presence of a valine residue (Fig. 2). Assuming that the sequence requirements for glycosylation allow for this difference, Lapps *et al.* (1987) suggested, by analogy, that the additions of the glycans take place in this terminal segment. They argued that, in the case of BCV, up to two O-linked oligosaccharide side chains per M molecule are attached.

Recent sequence information shows that considerable genetic diversity exists among different strains or isolates of coronaviruses in their spike protein (e.g., for MHV, see Luytjes *et al.*, 1987; Schmidt *et al.*, 1987; Parker *et al.*, 1989; Gallagher *et al.*, 1990; Wang *et al.*, 1992). The limited data available for the M gene suggest that variations are less extensive. Comparison, for instance, of the A59 and JHM strains of MHV (Pfleiderer *et al.*, 1986) revealed 21 nucleotide changes scattered over the entire molecule, which result in only seven conservative amino acid changes (3.5%). Cavanagh and Davis (1988) analyzed the $NH_2$-terminal domain, including the first hydrophobic region, of the M sequence of 23 strains of IBV. Both base substitutions and small deletions and insertions were detected. A fourfold greater extent of amino acid variation was found in the ectodomain of the protein as compared to the membrane-embedded segment. Notably, one of the N-glycosylation sequences (positions 38–40 in Fig. 2) was highly conserved while the other (positions 35–37) was not. Also, based on a complete comparison of M sequences of two IBV strains (Binns *et al.*,

1986), it was concluded that the exposed $NH_2$-terminal domain is the most variable part of the molecule. The M proteins of TGEV and PCRV (Rasschaert *et al.*, 1990), two closely related porcine coronaviruses, also differ in their amino acid sequences at 13 positions, 8 of which occur in the amino terminal part preceding the first hydrophobic region.

## B. Membrane Topology

The disposition of the M molecule in the lipid bilayer was studied through protease protection analyses. Digestion of *in vitro* assembled MHV-A59 M protein showed the bulk of the polypeptide to be resistant to proteolysis: only a small (1.5-kDa) portion was removed when the membranes were intact while another 2.5-kDa fragment was digested after detergent permeabilization (Rottier *et al.*, 1984). The latter fragment is located luminally and represents the $NH_2$-terminus of the molecule, as was shown by selective labeling. In virions, this 2.5-kDa fragment is exposed on the outside and is glycosylated to a variable extent, while the 1.5kDa COOH-terminal end protrudes from the inner face of the membrane. Experiments with IBV (Cavanagh *et al.*, 1986a) suggest that the tertiary structure of the M protein of this virus, and probably of other coronaviruses as well, is very similar.

The results of the biochemical studies, combined with a theoretical analysis of the primary structure of the M polypeptide, have led to a general topological model of the assembled protein as shown in Fig. 4A (Armstrong *et al.*, 1984; Rottier *et al.*, 1986). The structure is characterized by the presence of three membrane-spanning helices in the $NH_2$-terminal half that anchor the protein in the lipid bilayer. This segment is flanked on the one side by the hydrophilic $NH_2$-terminus and on the other side by a region that contains the extremely well-conserved 8 amino acids domain (Fig. 2) and in which a surface helix is predicted for some (but not all) M proteins (Rottier *et al.*, 1986). The bulk of the carboxy-terminal half of the M molecule is supposedly embedded in the polar surface of the membrane. In line with this, a mutant M protein lacking all the

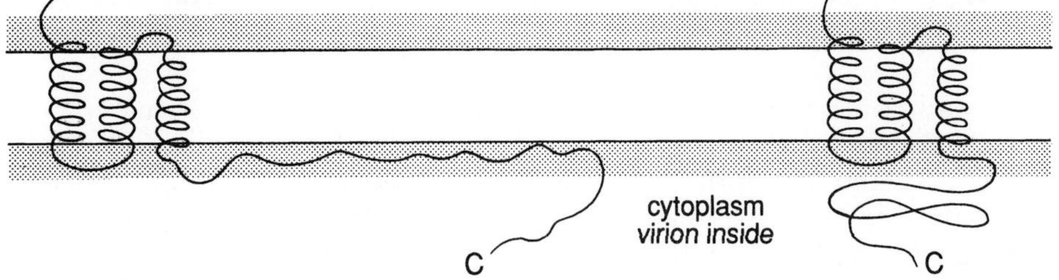


FIGURE 4. Topological models of the membrane-assembled M protein. Note that the two models differ only in the disposition of the protease-resistant region in the carboxy-terminal half. The hexagon symbol attached to the $NH_2$-terminal region indicates the potential glycosylation sites.

membrane-spanning domains was found to associate with membranes *in vitro* (Mayer *et al.*, 1988). However, a location of this extremely protease resistant region adjacent to the membrane, as in Fig. 4B, cannot be excluded.

## IV. ASSEMBLY IN THE ENDOPLASMIC RETICULUM

Viral proteins use mechanisms and intracellular pathways as do normal cellular proteins. The process by which a nascent polypeptide is directed to and assembled in the endoplasmic reticulum occurs quite rapidly. Therefore, these processes are most conveniently studied in *in vitro* systems, which are easier to manipulate and in which the reaction kinetics are inherently slower. Though these systems have given us insight into the processes involved in the insertion and translocation of simple membrane and secretory proteins, little is known about the events that generate the more complex, multispanning membrane proteins. The coronavirus M protein, relatively simple as it is with only three transmembrane domains, thus represents an attractive model.

The MHV-A59 M protein is translated on membrane-associated polysomes (Niemann *et al.*, 1982). Its membrane insertion involves the action of the ribonucleoprotein complex called the signal-recognition particle (SRP). In a wheat germ translation system devoid of membranes, the synthesis of the protein could be specifically and stably blocked by the addition of SRP. Subsequent addition of salt-washed microsomes fully released the translation arrest and resulted in the correct membrane integration of the polypeptide (Rottier *et al.*, 1985).

In agreement with the absence of a hydrophobic $NH_2$-terminal peptide, the protein is assembled without a cleavable signal sequence (Rottier *et al.*, 1984). The polypeptide chain is inserted into the lipid bilayer in a cotranslational manner. Time course experiments in a synchronized *in vitro* translation of the M mRNA showed that the nascent polypeptide chain was able to integrate when membranes were added before 140–150 of the 228 residues had been polymerized. Later additions no longer allowed membrane insertion. Once beyond a critical point in synthesis, the domain containing the insertion signal(s) is apparently no longer accessible to the insertion machinery. Additional time course experiments revealed that such signal(s) may be located anywhere within the hydrophobic $NH_2$-terminal half of the molecule. During a synchronized translation in a wheat germ extract, SRP was able to induce an arrest until the most C-terminal hydrophobic domain had emerged from the ribosome (Rottier *et al.*, 1985).

Direct evidence that the topogenic signals in the M protein reside within the hydrophobic part of the polypeptide came from expression studies with cloned cDNA copies of M genes. *In vitro* mutagenesis followed by transcription and translation in the presence of microsomal membranes showed that each hydrophobic domain can individually insert and anchor the polypeptide in the membrane (Mayer *et al.*, 1988; Rottier *et al.*, 1990; Krijnse Locker *et al.*, 1992b). Similar results were obtained by expression of mutant M genes of MHV-A59 M (Mayer *et al.*, 1988; Armstrong *et al.*, 1990; Krijnse Locker *et al.*, 1992b) and IBV

M (Machamer and Rose, 1987). The mutant proteins with only the first or the third transmembrane domain integrate in membranes such that their $NH_2$-terminus is translocated into the lumen, while their COOH-terminus remains on the cytoplasmic side. Large deletions in the hydrophilic $NH_2$-terminal region did not affect this ability, nor did the mutations alter the topology of the assembled protein (Mayer *et al.*, 1988; Krijnse Locker *et al.*, 1992b). This part of the protein consequently plays no role in the membrane integration process. As expected, similar conclusions could be drawn from studies of mutations in the carboxy-terminal domain (Armstrong *et al.*, 1990; Rottier *et al.*, 1990; Krijnse Locker *et al.*, 1992b).

These observations lead to the model shown in Fig. 5. SRP interacts with the first hydrophobic domain as soon as it appears from the ribosome; elongation halts until the complex has attached to the membrane where the hydrophobic domain interacts with the signal sequence receptor and is inserted into the membrane, probably as a hairpin, while SRP is released. Presumably, the hydrophilic $NH_2$-terminus is translocated and the two following hydrophobic domains are then sequentially inserted. Completion of the polypeptide chain accompanied by further cotranslational folding finally leads to the fully assembled protein.

Interestingly, the deduced amino acid sequences of the TGEV (Laude *et al.*, 1987; Kapke *et al.*, 1988), FIPV (Vennema *et al.*, 1991a), and canine coronavirus (CCV) M protein (Horsburgh *et al.*, 1992) have a hydrophobic amino terminal extension (see Section III.A). Sequencing of the mature M polypeptide from purified TGEV showed that the first 17 residues were absent, indicating a cleavable signal sequence (Laude *et al.*, 1987). Translation studies with mRNA specifying TGEV M showed the signal-directed membrane insertion of the protein *in vitro*, although cleavage did not occur under these conditions (Kapke *et al.*, 1988). These authors also tested the expression of a truncated version of

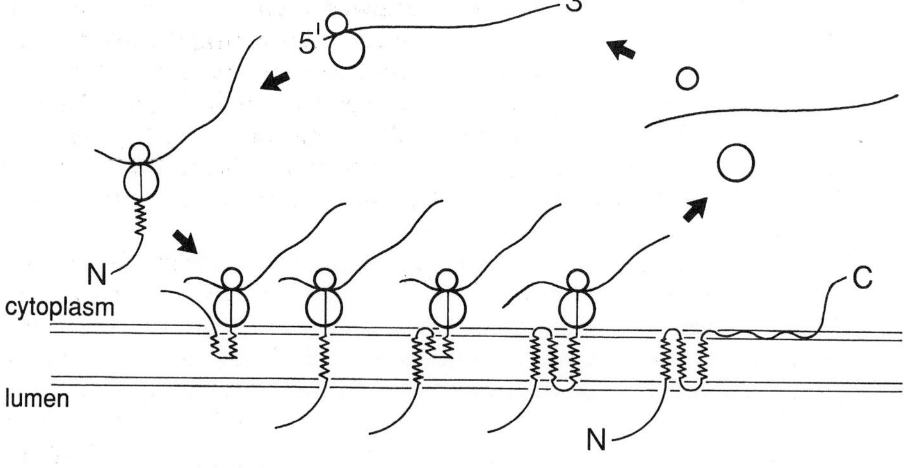


FIGURE 5. Model of the membrane assembly process of the coronavirus M protein.

the gene which lacked the information for the signal-containing first 21 residues. Though the polypeptide appeared to be integrated efficiently into microsomal membranes, translocation of the $NH_2$-terminal domain was poor, as judged from the small extent of N-glycosylation of the protein. In contrast, deletion of the signal sequence had no effect when the FIPV M protein was expressed in a vaccinia virus/cell system (Vennema *et al.*, 1991b). The mutant protein was glycosylated as efficiently as the wild-type protein. Why, in these viruses, the M protein contains a cleavable amino-terminal signal sequence thus remains unclear.

## V. INTRACELLULAR TRANSPORT AND MATURATION

### A. Transport and Processing in Coronavirus-Infected Cells

Intracellular transport of the coronavirus M protein differs from that of most other viral glycoproteins, including the coronaviral spike protein. Whereas these proteins are usually targeted to the cell surface, migration of the M protein is limited to the perinuclear region of the cell. This restricted mobility correlates with the intracellular location of coronavirus assembly in the infected cell.

#### 1. Intracellular Budding of Coronaviruses

Already from the early studies of the coronavirus infection process, it became clear that the entire replication cycle takes place in the cytoplasm. Ultrastructural observations with different coronaviruses demonstrated that morphogenesis occurs at intracellular membrane-bound compartments as virions were seen in the lumina of the rough endoplasmic reticulum, smooth-walled vesicles and Golgi apparatus (e.g., David-Ferreira and Manaker, 1965; Becker *et al.*, 1967; Chasey and Alexander, 1976; Holmes and Behnke, 1981; Dubois-Dalcq *et al.*, 1982, 1984). Virus budding never takes place at the plasma membrane. A thorough investigation by J. and S. A. Tooze *et al.* (J. Tooze *et al.*, 1984, 1985, 1987; S. A. Tooze *et al.*, 1988), using a combination of biochemical and morphological techniques, has elucidated the temporal sequence of events in the maturation of MHV-A59 in Sac(−) cells, a line of transformed murine fibroblasts. Early in infection, shortly after the appearance of M and S, the first progeny virions are seen by electron microscopy in the perinuclear region. Budding of these early particles occurs into small, smooth vesicles or tubules lying between the rough endoplasmic reticulum and the *cis* side of the Golgi stack. This smooth membrane compartment, termed *budding compartment* by the authors (S. A. Tooze *et al.*, 1988), is distinct from and does not form part of the Golgi complex but is connected to the rough endoplasmic reticulum (Krijnse Locker *et al.*, 1994a). At early times, budding does not occur in either of these two organelles. Later, however, budding also starts to occur into the rough membranes of the endoplasmic reticulum, and this compartment becomes the major site of virus assembly late in infection. Budding into the Golgi cisternae is a rare event. Virions are transported from their site of synthesis to and through the Golgi complex by vesicular carriers. In the Golgi complex they are usually seen only at the rims of the stacks. At the *trans* side of the Golgi system, particles are collected into vesicles of the constitutive exocytic pathway (J. Tooze *et al.*, 1987) and released from the cell.

The budding compartment also exists in uninfected Sac(−) cells, where it is associated with transitional elements and vesicles of the rough endoplasmic reticulum and the *cis* side of the Golgi complex. Presumably, it is equivalent to the intermediate or salvage compartment which plays a crucial role in the sorting of resident proteins of the endoplasmic reticulum (for review, see Hauri and Schweizer, 1992). As virions accumulate in the budding compartment during MHV-A59 infection, it becomes strongly dilated.

Budding into the intermediate compartment seems to be a general feature of coronaviruses. An ultrastructural analysis of cells infected with IBV, TGEV, or FIPV revealed that these viruses also use the smooth perinuclear membranes (Klumperman *et al.*, 1994). Budding into such tubulovesicular membrane structures has also been described in neural cells infected with either MHV-A59 or the JHM strain of MHV (Dubois-Dalcq *et al.*, 1982).

### 2. Localization and Transport of the M Protein

After its synthesis on membrane-bound polyribosomes, the M protein is transported from the endoplasmic reticulum and through the Golgi complex to finally appcar outside the cell as part of virions. The rate and extent of this process differs depending on the particular virus–cell system and on the time of infection. Pulse-chase experiments early in MHV-A59 infection showed the M protein to be chased almost quantitatively from 17C11 cells within about 2 hr (Holmes *et al.*, 1981a). In contrast, very inefficient clearance of M was observed, both from 17C11 cells (Holmes *et al.*, 1981b) and from Sac(−) cells (Rottier *et al.*, 1981b) during a similar period somewhat later in the infection cycle. It appears that in early stages the synthesis of the protein is balanced with its release in virions.

To relate the localization of the MHV-A59 M protein to the site of budding, J. and S. A. Tooze *et al.* (J. Tooze *et al.*, 1984; S. A. Tooze *et al.*, 1988) labeled infected Sac(−) cells for indirect immunofluorescence with a monoclonal antibody to the protein. Early in infection, a perinuclear fluorescence pattern was observed that was similar to but more extensive than that obtained with antibodies specific for the Golgi apparatus. As infection proceeded, a more reticular staining throughout the cell appeared. No labeling of the plasma membrane occurred until late in infection when patches of fluorescence indicated association of released virions with the cell surface. Immunoperoxidase staining confirmed that the M protein had accumulated in the budding compartment and not in the endoplasmic reticulum early in infection. There was, however, also some labeling of the stacked cisternal membranes of the Golgi complex, indicating that some M protein reaches the Golgi apparatus as free integral membrane protein. As no virus assembly is observed in these compartments, it was inferred that the density of the protein in this organelle apparently does not reach the critical threshold level required for budding.

Similar data were obtained by Klumperman *et al.* (1994) in a study of the subcellular localization of M proteins in coronavirus-infected cells. Using immunogold labeling and electron microscopy, they demonstrated that in MHV and IBV infection, both in Sac(−) cells, free M protein is present in the membranes of the budding compartment as well as in the Golgi complex. Within the Golgi complex, the distribution patterns of MHV-M and IBV-M were found to be different. MHV-M was localized more toward the *trans* side, while IBV-M was concentrated on the *cis* side, similar to the patterns found when these proteins are expressed independently (Machamer *et al.*, 1990; Krijnse Locker *et al.*, 1992a; Klumperman *et al.*, 1994).

The combined data indicate that at early times the M protein is transported as an integral membrane protein to the smooth membranes of the budding compartment where it accumulates and is incorporated into virions and from where it is exported through the Golgi complex out of the cell. At later times, when the rate of M synthesis in the rough endoplasmic reticulum exceeds the rate of its exit to the budding compartment, the protein allows virion assembly also in these reticular membranes. A fraction of M apparently escapes from being incorporated into virions and accumulates in the Golgi complex.

Interestingly, when the temperature of Sac(−) cells infected with MHV-A59 was lowered to 31 °C, release of virions was strongly inhibited (S. A. Tooze *et al.*, 1988). Electron microscopic analysis showed that virion formation at this temperature occurred normally, but that the particles accumulated in the budding compartment and in the rough endoplasmic reticulum. Entry of virions into the Golgi complex appears to be inhibited at 31 °C, as few virions were observed in or beyond this compartment and the M protein did not undergo the oligosaccharide modifications known to occur there.

MHV M protein is not transported as free integral membrane protein to the plasma membrane of infected cells (Dubois-Dalcq *et al.*, 1982; J. Tooze *et al.*, 1984, 1987; S. A. Tooze *et al.*, 1988; S. A. Tooze and Stanley, 1986). The protein is detected there only in extracellular virions that have readsorbed in clusters to the cell surface (Sugiyama and Amano, 1981; Dubois-Dalcq *et al.*, 1982; S. A. Tooze *et al.*, 1988). This is in keeping with the Golgi localization of MHV-M when expressed from its cloned gene (see Section VB). It was surprising, therefore, that free M protein was observed at the surface of cells infected with TGEV. This was shown both biochemically (surface iodination) and serologically using M-specific monoclonal antibodies by Laviada *et al.* (1990). The protein was detected at the plasma membrane as early as 4 hr after infection, i.e., before any infectious virus had been released from the cells. In another study, however, Pulford and Britton (1990) could not confirm these findings.

### 3. Oligosaccharide Modifications during Transport

Transport of glycoproteins through the biosynthetic pathway of the cell is accompanied by modifications of their glycan moieties as they encounter the modifying enzymes. Due to the specific locations of these enzymes, the modifications occur sequentially in time and place. As a consequence, the state of glycosylation is an indicator of a glycoprotein's progress on the exocytic route. Many of the oligosaccharide-processing enzymes occur in the Golgi apparatus. The formation of coronavirions in a pre-Golgi compartment thus implies that these enzymes act on glycoprotein molecules which present themselves as parts of huge macromolecular structures.

The M protein of a number of coronaviruses is glycosylated only by O-linked carbohydrates (see Table I). This type of linkage is rare among viral glycoproteins, but is found quite frequently in various cellular glycoproteins. Usually, O-linked oligosaccharides occur on a polypeptide in combination with N-glycosidically linked side chains. Sometimes the O-glycosylation pattern is very extensive and complex such as in mucin-type molecules. For these and other reasons, O-glycosylation has been difficult to study, and our knowledge lags far behind that of N-glycosylation. Because of its relative simplicity, the coronavirus M protein seems an attractive tool to catch up.

The M protein of MHV-A59 has been resolved into a number of differentially O-glysosylated forms. As mentioned previously, two types of oligosaccharide side chains are bound to M in the virions produced by 17C11 cells (see Fig. 1) (Niemann *et al.*, 1984). The same structures were found to predominate when the virus was grown in Sac(−) cells (S. A. Tooze *et al.*, 1988), but at least two additional glycosylated species were detected in these cells. One was identified as carrying only *N*-acetyl-galactosamine (GalNAc); the other was not identified. Two glycosylated forms were observed in AtT20 cells, a pituitary tumor cell line, but their structures were not clearly resolved (J. Tooze *et al.*, 1987). In all cases, a significant proportion of the M molecules remained unglycosylated. These observations indicate that the extent and possibly also the nature of O-glycosylation varies and is determined by the host cell.

Early studies on the biosynthesis of MHV-A59 M, both in infected Sac(−) cells (Rottier *et al.*, 1981b) and in 17C11 cells (Holmes *et al.*, 1981b; Niemann *et al.*, 1982), suggested that the addition of O-linked oligosaccharides is a posttranslational event. This idea is now well established (S. A. Tooze *et al.*, 1988; Krijnse Locker *et al.*, 1992a, 1994a). The M protein is synthesized in a nonglycosylated form in the rough endoplasmic reticulum and is transported to the intermediate compartment where the first sugar, GalNAc, is added. Due to the membrane-continuities between these compartments, no vesicular transport step is required for this addition (Krijnse Locker *et al.*, 1994a). All subsequent modifications occur in the Golgi apparatus. First, a galactose unit is added, followed immediately by sialic acid. The transferase enzymes responsible for these additions are both located in Golgi cisternae, proximal to the *trans*-Golgi network. Finally, further modifications can lead to the appearance of two more forms of the M protein, the oligosaccharides of which have not been characterized. These modifications occur in the *trans*-Golgi network as they can be inhibited specifically by treatment of the cells with the drug brefeldin A (Krijnse Locker *et al.*, 1992a).

Coronavirus M proteins that are modified by N-linked oligosaccharides acquire their sugars cotranslationally in the endoplasmic reticulum. Core-glycosylation is, however, not a very efficient process with these proteins. Unglycosylated and partially glycosylated molecules occur in infected cells as well as in the virions released from them (Stern and Sefton, 1982a, b; Stern *et al.*, 1982; Cavanagh, 1981, 1983; Vennema *et al.*, 1990b; Jacobs *et al.*, 1986). Incomplete glycosylation does not correlate with the involvement of a cleavable $NH_2$-terminal signal sequence during biosynthesis; it was observed both with IBV and with FIPV and TGEV M proteins which have an internal and an amino terminal insertion signal, respectively. During transport to the budding compartment, and on passing through the Golgi complex en route to the plasma membrane, the high-mannose oligosaccharides are trimmed and processed, having been converted to the complex type as they are released from the cell. Again, these processes occur very inefficiently, since a significant proportion of the M molecules in extracellular virions remains sensitive to the enzyme endoglycosydase H which only recognizes immature glycans (Stern and Sefton, 1982b; Cavanagh, 1983; Vennema *et al.*, 1990b). The incomplete maturation of coronavirus glycoproteins is not specific for the M protein. The same applies to the S protein (Stern and Sefton, 1982b; Cavanagh, 1983; Vennema *et al.*,1990b). The reasons for these findings are unclear, but it is plausible that they are caused by steric effects. Since the proteins move through the Golgi compartments as part of and protruding from the viral envelope, access of the modifying enzymes to the oligosaccharides might well be severely hindered. Incomplete maturation of N-linked carbohydrates is not unprecedented and has occasionally been observed with other glycoproteins (Doyle *et al.*, 1986; Geyer *et al.*, 1988; Earl *et al.*, 1991).

## B. Transport of the Expressed M Protein

The interesting membrane structure and the intracellular restriction in coronavirus-infected cells prompted studies of the biogenesis and transport of the M protein in the absence of other coronaviral proteins. The M proteins of MHV-A59, IBV, TGEV, and FIPV have been expressed in cells from cloned cDNA (e.g., Machamer and Rose, 1987; Rottier and Rose, 1987; Machamer *et al.*, 1990; Klumperman *et al.*, 1994) as well as by microinjection of an *in vitro* transcribed mRNA (Armstrong *et al.*, 1987, 1990; Mayer *et al.*, 1988). These studies unequivocally demonstrate that the protein accumulates in the Golgi apparatus and does not reach the plasma membrane. Analysis of the expressed MHV-A59 M protein by immunofluorescence in various cell types revealed the perinuclear appearance typical for the Golgi complex (Armstrong *et al.*, 1987; Rottier and Rose, 1987; Mayer *et al.*, 1988; Krijnse Locker *et al.*, 1992a; Klumperman *et al.*, 1994 (see also Fig. 6). This localization was confirmed by electron microscopy using immunogold labeling and Golgi-specific markers (Krijnse Locker *et al.*, 1992a; Klumperman *et al.*, 1994), showing that within the Golgi complex the MHV M protein is concentrated in the *trans*-most compartments. Consistently, biochemical labeling revealed that, although a proportion of the molecules usually remained unglycosylated, as in coronavirus-infected cells, the large majority acquired the O-linked oligosaccharides added in the Golgi

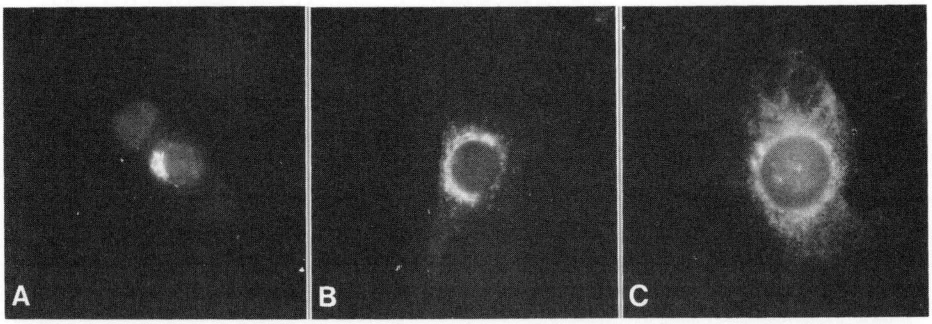


FIGURE 6. Localization of the MHV-A59 M and S protein by indirect immunofluorescence. Sac(−) cells expressing the M protein from (A) a vaccinia virus vector or, (B, C) during infection with MHV-A59 were stained with a M-specific antipeptide serum (A, B) or an S-specific monoclonal antibody after permeabilization.

apparatus (Armstrong *et al.*, 1987; Rottier and Rose, 1987; Krijnse Locker *et al.*, 1992a; Klumperman *et al.*, 1994). In contrast, the IBV M protein expressed in AtT20 cells (Machamer *et al.*, 1990) or in HepG2 cells (Klumperman *et al.*, 1994) was localized by immunoelectron microscopy to the *cis* side of the Golgi complex. Accordingly, its N-linked sugars remain largely immature. Taken together, these studies indicate that the intracellular restriction of coronavirus M is an intrinsic property of the protein and is not dependent on other coronaviral factors.

Its accumulation in the Golgi apparatus, midway on the exocytic pathway, has made the M protein an attractive tool for studying the principles of targeting to and retention in this organelle. For the IBV protein, Machamer and coworkers (Machamer and Rose, 1987; Machamer *et al.*, 1990, 1993; Swift and Machamer, 1991) have shown that the first transmembrane domain carries the signal necessary and sufficient for localization in the *cis*-Golgi region. This was demonstrated most convincingly by transferring the hydrophobic domain into reporter molecules normally transported to the plasma membrane and showing their Golgi retention (Swift and Machamer, 1991). Several uncharged polar residues were found to be critical for the functioning of the domain. These residues line one face of a predicted α-helix formed by the transmembrane domain (Swift and Machamer, 1991; Machamer *et al.*, 1993). Surprisingly, no such role is played by the first transmembrane domain of the MHV M protein. A mutant M protein with only this transmembrane domain did not leave the endoplasmic reticulum (Armstrong *et al.*, 1990; Rottier *et al.*, 1990), although inserting this domain in place of the membrane-spanning domain of a plasma membrane protein still allowed this chimeric molecule to reach the cell surface (Machamer *et al.*, 1993). Mutation studies with the MHV M protein indicate that, in this case, the carboxy-terminal domain, probably in combination with an internal domain, determines the Golgi localization (Armstrong and Patel, 1991; Krijnse Locker *et al.*, 1994b). It was suggested that different principles may act in the differential localization of the IBV and MHV M proteins in the Golgi complex (Weisz *et al.*, 1993).

# VI. BIOLOGICAL FUNCTIONS

Glycoproteins of enveloped viruses are involved in a number of biological activities. Envelope proteins function in the binding of virions to cell receptors. They mediate the introduction of the nucleocapsid into the cytoplasm by triggering the fusion of viral and cellular membranes. Viral membrane proteins are also essential at the end of the infection cycle during the process of virion assembly. In some viruses, envelope glycoproteins exhibit receptor-destroying activities. Finally, the proteins generally induce immunological responses in the host by eliciting neutralizing antibodies and cytotoxicity. In coronaviruses, several of these properties have been attributed to the S and HE proteins. The M protein has been demonstrated to play a key role in coronavirus budding and to induce immunological reactions during infection in the host.

## A. Role of M in Coronavirus Budding

There are two types of findings pointing to a predominant function of the M protein in the intracellular formation of progeny virus particles. First, growth of coronaviruses in the presence of tunicamycin gave rise to the production of spikeless, noninfectious virions (Holmes *et al.*, 1981b; Rottier *et al.*, 1981b; Stern and Sefton, 1982b; Mounir and Talbot, 1992). These particles were indeed devoid of S and HE protein but contained M, suggesting that M is the only envelope glycoprotein required for virus budding. Second, throughout the infection there is a correlation between the intracellular sites at which progeny virions bud and the perinuclear location of the M protein (see Section V). Concentration and location of M apparently are decisive in determining where and when budding occurs.

Molecular details of the budding process have not yet been elucidated. Conceivably, the M protein is transported as an integral membrane protein to the budding compartment or, later in infection, backs up in the endoplasmic reticulum. The protein accumulates locally to a concentration high enough to be recognized by nucleocapsids formed in the cytosol. The nucleocapsids associate with the M protein by interacting with its cytoplasmic domain. These interactions initiate budding into the lumen of the membrane compartment, whereby M is sequestered into progeny virions. An affinity of M for nucleocapsids has been observed *in vitro* with several coronaviruses. Subviral particles prepared by Nonidet P-40 disruption of purified MHV (Wege *et al.*, 1979; Sturman *et al.*, 1980), HEV (Callebaut and Pensaert, 1980), IBV (Lanser and Howard, 1980), or BCV (King and Brian, 1982) still contained M protein associated with the nucleocapsids. The significance of these interactions remains to be assessed. Binding of MHV-M to nucleocapsids did occur through an interaction with the RNA but was not specific for viral RNA and was dependent on a temperature-induced conformational change in the M protein (Sturman *et al.*, 1980). The distribution of positive charges over the M molecule certainly seems to favor an association with the viral RNA. For instance, all but one of the 18 arginines and lysines present in the MHV M protein are located at the cytoplasmic face of the intracellular membrane; of these, 5 occur in the carboxy-terminal 24 amino acids.

Little is still known about where and how the other viral membrane proteins are assembled into coronavirions. Electron microscopic observations have shown that the envelopes of budding virions were covered with peplomers (Chasey and Alexander, 1976; Dubois-Dalcq *et al.*, 1982). Apparently, budding involves the collective incorporation of all viral envelope proteins. This implies that the spike glycoproteins convene with the M proteins at the sites of budding and, by consequence, that the different viral envelope proteins recognize each other and interact. Association of the M and S protein in MHV-A59 infected cells has indeed been demonstrated (Opstelten *et al.*, 1993a, 1994). After their synthesis, M molecules rapidly engage in formation of noncovalently linked heteromultimeric structures with S proteins, probably already in the endoplasmic reticulum. In contrast, newly synthesized S protein first has to undergo a number of folding reactions before it reaches a conformation competent to interact with M. If proper folding of S is prevented, e.g., by inhibiting the formation of disulfide bonds, no complexes between the two proteins are formed and the M protein is transported to the Golgi complex (Opstelten *et al.*, 1993b).

These data suggest a model of virus assembly in which the viral envelope proteins form complexes that accumulate in the budding compartment to generate a microenvironment where nucleocapsids can bind and assemble into virions. Such a process would explain the exclusion of host cellular membrane proteins from virions as the molecular selection for viral membrane proteins would preclude their incorporation into the heteromultimeric complexes. It might also explain why the M and S proteins are efficiently assembled into viral particles in the budding compartment, while their intrinsically preferred destinations are the Golgi complex and the plasma membrane, respectively. Consistent with such a model, the MHV M and S proteins when coexpressed in cells were found to associate and form complexes (D.-J. Opstelten and P. Rottier, unpublished observations). Surprisingly, however, these complexes were not retained in the budding compartment, but accumulated in the Golgi complex. Apparently, other viral factors are required to localize budding in pre-Golgi membranes. One possibility is that the nucleocapsid plays an organizing role. Alternatively, the small nonglycosylated virion membrane protein (SM) might be important. So far, this protein has largely been neglected. Further studies are warranted, however, because the protein was recently shown to play a key role in the formation of viruslike particles (H. Vennema, G.-J. Godeke, and P. Rottier, unpublished observations).

While the M protein is required for budding, its glycosylation is not. Nonglycosylated coronavirus M proteins are able to efficiently direct the formation of virions as was shown using tunicamycin (Stern and Sefton, 1982b) and monensin (Niemann *et al.*, 1982), inhibitors of N- and O-glycosylation, respectively. Moreover, in the absence of drugs no particular form of the differentially glycosylated M species is preferentially incorporated into virions. Their relative abundance in extracellular viral particles correlates with the ratio of their appearance in the infected cell (Stern and Sefton, 1982b; Holmes *et al.*, 1981a). Additional evidence that glycosylation of M is not essential for virus assembly comes from the demonstration by Laude *et al.* (1992) that disruption of the consensus sequence of the sole N-glycosylation site in the M protein of TGEV results in a viable mutant virus. The authors mention that the mutation did not affect the specific infectivity of the virus nor its multiplication rate. Apparently, the oligosaccharides attached to the coronaviral M proteins are not important for the infection process at the level of the cell. They probably play a role in the interaction with the host at the level of the organism.

## B. Induction of Immunological Responses

With the exception of FIPV (Vennema *et al.*, 1990a), the S glycoprotein of coronaviruses is believed to be the prime inducer of protective immunity. Nevertheless, several observations indicate that immune responses to other viral proteins, including the M protein, may also play a role. Though largely buried within the membrane, the M protein does elicit specific antibodies during infection in the host and also when expressed separately through a live carrier virus (e.g., Pulford and Britton, 1990; Vennema *et al.*, 1991a; Wesseling *et al.*, 1993).

Monoclonal antibodies to the M protein can neutralize infectivity *in vitro*, but only in the presence of complement, as was shown for some monoclonal antibodies to MHV-M (Collins *et al.*, 1982; Fleming *et al.*, 1989) and TGEV-M (Woods *et al.*, 1988). In most cases, however, anti-M antibodies appear non-neutralizing *in vitro* (Buchmeier *et al.*, 1984; Laude *et al.*, 1986; Deregt and Babiuk, 1987; Fiscus and Teramoto, 1987; Fleming *et al.*, 1989).

Little is known about the protective effects of antibodies to the M protein in animals. Monovalent antibodies elicited by immunization of mice with purified MHV M protein failed to protect against a virus challenge (Hasony and Macnaughton, 1981). In contrast, two of four monoclonal antibodies to MHV-JHM M did protect mice from a normally lethal challenge after passive transfer (Fleming *et al.*, 1989). This protection was not associated with a particular antigenic determinant in the M protein nor was it mediated by complement. Interestingly, one of the two antibodies was not neutralizing *in vitro*, not even in the presence of complement. Some level of protection against another coronavirus was also observed in cats after immunization with a recombinant vaccinia virus expressing the FIPV M protein (Vennema *et al.*, 1991a). Although all animals seroconverted after a challenge with a lethal dose of FIPV and developed clinical signs, three of eight kittens survived.

The M protein of TGEV can induce the production of α-interferon in lymphocytes as was shown by the inhibiting effect of anti-M monoclonal antibodies on interferon induction by fixed TGEV-infected cells (Charley and Laude, 1988). By analysis of mutant viruses with reduced interferogenic potential, this biological activity was mapped to the exposed amino-terminal region of the M molecule (Laude *et al.*, 1992). The significance of α-interferon action for immunity to viral infection needs to be established.

Little is known about the cellular immune response to coronavirus infections. A cytotoxic T-cell response to the M protein has not been described. It has been suggested, however, that cell-mediated recognition of the M protein might be an important part of an effective immune response to MHV-JHM (Mobley *et al.*, 1992).

## VII. SUMMARY AND PERSPECTIVES

The coronavirus M protein has a unique molecular structure. The different domains in its structure may explain the specific properties of the protein and reflect its biological features in infection.

The N-terminal hydrophilic virion ectodomain carries the N- or O-linked oligosaccharides. It is the most variable part of the molecule which presumably contains the major antigenic determinants. It is responsible for immunological reactions such as antibody and interferon induction.

The hydrophobic region containing the three transmembrane helixes constitutes the core of the protein and is probably responsible for its peculiar physical properties. This region dictates the protein's topology in the membrane and its intracellular transport. It is the part of the protein most likely to be engaged in the intermolecular interactions that must occur at the membrane during the virion assembly process, both mutually between M molecules and with other viral membrane proteins.

The carboxy-terminal half of the molecule consists of an amphiphilic portion and an exposed tail facing the cytoplasm in infected cells. Virion budding is accomplished by the association of nucleocapsids with this part of the M protein.

The data reviewed in this chapter demonstrate that the coronavirus M protein has a number of interesting features both as a virion protein and as a model membrane protein. Though much has already been learned, many fundamental questions remain to be answered. Clearly, more needs to be known about the precise mechanism by which the protein is assembled in the membrane, about the disposition of the amphiphilic region of the molecule, and about the signals governing its intracellular transport. Little is known about the molecular details of the assembly of coronaviruses, the interactions of the M proteins with the nucleocapsid, or the interactions with the other membrane components. Nothing is known about the importance of the differential glycosylation among coronaviral M proteins. Some of these questions may be approached through mutagenesis and expression of the M gene. The answer to most questions, however, will require the generation of mutant viral genomes, either by the manipulation of infectious cDNA clones or by RNA recombination. This holds true not only with respect to the role of the M protein but for the molecular analysis of coronavirus infection in general. Evidently, this will be the major challenge for coronavirologists in the years to come.

## VIII. REFERENCES


Armstrong, J., and Patel, S., 1991, The Golgi sorting domain of coronavirus E1 protein, *J. Cell Sci.* **98:**567.

Armstrong, J., Niemann, H., Smeekens, S., Rottier, P., and Warren, G., 1984, Sequence and topology of a model intracellular membrane protein, E1 glycoprotein, from a coronavirus, *Nature* **308:**751.

Armstrong, J., McCrae, M., and Colman, A., 1987, Expression of coronavirus E1 and rotavirus VP10 membrane proteins from synthetic RNA, *J. Cell. Biochem.* **35:**129.

Armstrong, J., Patel, S., and Riddle, P., 1990, Lysosomal sorting mutants of coronavirus E1 protein, a Golgi membrane protein, *J. Cell Sci.* **95:**191.

Becker, W. B., McIntosh, K., Dees, J. H., and Chanock, R. M., 1967, Morphogenesis of avian infectious bronchitis virus and a related human virus (strain 229E), *J. Virol..* **1:**1019.

Binns, M. M., Boursnell, M. E. G., Tomley, F. M., and Brown, T. D. K., 1986, Nucleotide sequence encoding the membrane protein of the IBV strain 6/82, *Nucleic Acids Res.* **14:**5558.

Boursnell, M. E. G., Brown, T. D. K., and Binns, M. M., 1984, Sequence of the membrane protein gene from avian coronavirus IBV, *Virus Res.* **1:**303.

Bradburne, A. F., 1970, Antigenic relationships amongst coronaviruses, *Arch. Gesamte Virusforsch.* **31:**352.

Buchmeier, M. J., Lewicki, H. A., Talbot, P. J., and Knobler, R. L., 1984, Murine hepatitis virus-4 (strain JHM)-induced neurologic disease is modulated *in vivo* by monoclonal antibody, *Virology* **132:**261.

Callebaut, P. E., and Pensaert, M. B., 1980, Characterization and isolation of structural polypeptides in haemagglutinating encephalomyelitis virus, *J. Gen. Virol.* **48:**193.

Cavanagh, D., 1981, Structural polypeptides of coronavirus IBV, *J. Gen. Virol.* **53:**93.

Cavanagh, D., 1983, Coronavirus IBV glycopolypeptides: Size of their polypeptide moieties and nature of their oligosaccharides, *J. Gen. Virol.* **64:**1187.

Cavanagh, D., and Davis, P. J., 1988. Evolution of avian coronavirus IBV: Sequence of the matrix glycoprotein gene and intergenic region of several serotypes, *J. Gen. Virol.* **69:**621.

Cavanagh, D., Davis, P. J., and Pappin, D. J. C., 1986a, Coronavirus IBV glycopolypeptides: Locational studies using proteases and saponin, a membrane permeabilizer, *Virus Res.* **4:**145.

Cavanagh, D., Davis, P. J., Pappin, D. J. C., Binns, M. M., Boursnell, M. E. G., and Brown, T. D. K., 1986b, Coronavirus IBV: Partial amino terminal sequencing of spike polypeptide S2 identifies the sequence Arg-Arg-Phe-Arg-Arg at the cleavage site of the spike precursor propolypeptide of IBV strains Beaudette and M41, *Virus Res.* **4:**133.

Charley, B., and Laude, H., 1988, Induction of alpha interferon by transmissible gastroenteritis coronavirus: Role of transmembrane glycoprotein E1, *J. Virol.* **62:**8.

Chasey, D., and Alexander, D. J., 1976. Morphogenesis of avian infectious bronchitis virus in primary chick kidney cells, *Arch. Virol.* **52:**101.

Collins, A. R., Knobler, R. L., Powell, H., and Buchmeier, M. J., 1982, Monoclonal antibodies to murine hepatitis virus 4 (strain JHM) define the viral glycoprotein responsible for attachment and cell–cell fusion, *Virology* **119:**358.

David-Ferreira, J. F., and Manaker, R. A., 1965, An electron microscope study of the development of a mouse hepatitis virus in tissue culture cells, *J. Cell Biol.* **24:**57.

Deregt, D., and Babiuk, L. A., 1987, Monoclonal antibodies to bovine coronavirus: Characteristics and topographical mapping of neutralizing epitopes on the E2 and E3 glycoproteins, *Virology* **161:**410.

Deregt, D., Sabara, M., and Babiuk, L. A., 1987, Structural proteins of bovine coronavirus and their intracellular processing, *J. Gen. Virol.* **68:**2863.

Doyle, C., Sambrook, J., and Gething, M.-J., 1986, Analysis of progressive deletions of the transmembrane and cytoplasmic domains of influenza hemagglutinin, *J. Cell Biol.* **103:**1193.

Dubois-Dalcq, M. E., Doller, E. W., Haspel, M. V., and Holmes, K. V., 1982, Cell tropism and expression of mouse hepatitis viruses (MHV) in mouse spinal cord cultures, *Virology* **119:**317.

Dubois-Dalcq, M. E., Holmes, K. V., and Rentier, B., 1984, *Assembly of RNA Viruses*, Springer Verlag, New York.

Earl, P. L., Moss, B., and Doms, R. W., 1991, Folding, interaction with GRP78-BiP, assembly, and transport of the human immunodeficiency virus type 1 envelope protein, *J. Virol.* **65**:2047.

Fiscus, S. A., and Teramoto, Y. A., 1987, Antigenic comparison of feline coronavirus isolates: Evidence for markedly different peplomer glycoproteins, *J. Virol.* **61**:2607.

Fleming, J. O., Shubin, R. A., Sussman, M. A., Casteel, N., and Stohlman, S. A., 1989, Monoclonal antibodies to the matrix (E1) glycoprotein of mouse hepatitis virus protect mice from encephalitis, *Virology* **168**:162.

Gallagher, T. M., Parker, S. E., and Buchmeier, M. J., 1990, Neutralization-resistant variants of the neurotropic coronavirus are generated by deletions within the amino-terminal half of the spike glycoprotein, *J. Virol.* **64**:731.

Garwes, D. J., Pocock, D. H., and Pike, B. V., 1976, Isolation of subviral components from transmissible gastroenteritis virus, *J. Gen. Virol.* **32**:283.

Geyer, H., Holschbach, C., Hunsmann, G., and Schneider, J., 1988, Carbohydrates of human immunodeficiency virus. Structures of olisosaccharides linked to the envelope glycoprotein 120, *J. Biol. Chem.* **263**:11760.

Hasony, H. J., and Macnaughton, M. R., 1981, Antigenicity of mouse hepatitis virus strain 3 subcomponents in C57 strain mice, *Arch. Virol.* **69**:33.

Hauri, H.-P., and Schweizer, A., 1992, The endoplasmic reticulum-Golgi intermediate compartment, *Curr. Opin. Cell Biol.* **4**:600.

Hogue, B. G., and Brian, D. A., 1986, Structural proteins of human respiratory coronavirus OC43, *Virus Res.* **5**:131.

Holmes, K. V., and Behnke, J. N., 1981, Evolution of a coronavirus during persistent infection *in vitro, Adv. Exp. Med. Biol.* **142**:287.

Holmes, K. V., Doller, E. W., and Behnke, J. N., 1981a, Analysis of the functions of coronavirus glycoproteins by differential inhibition of synthesis with tunicamycin, *Adv. Exp. Med. Biol.* **142**:133.

Holmes, K. V., Doller, E. W., and Sturman, L. S., 1981b, Tunicamycin resistant glycosylation of a coronavirus glycoprotein: Demonstration of a novel type of viral glycoprotein, *Virology* **115**:334.

Horsburgh, B. C., Brierley, I., and Brown, T. D. K., 1992, Analysis of a 9.6 kb sequence from the 3′ end of canine coronavirus genomic RNA, *J. Gen. Virol.* **73**:2849.

Horzinek, M. C., Lutz, H., and Pedersen, N. C., 1982, Antigenic relationships among homologous structural polypeptides of porcine, feline, and canine coronaviruses. *Infect, Immunology* **37**:1148.

Jacobs, L., Van der Zeijst, B. A. M., and Horzinek, M. C., 1986, Characterization and translation of transmissible gastroenteritis virus mRNAs, *J. Virol.* **57**:1010.

Kapke, P. A., Tung, F. Y. T., Hogue, B. G., Brian, D. A., Woods, R. D., and Wesley, R., 1988, The amino-terminal signal peptide on the porcine transmissible gastroenteritis coronavirus matrix protein is not an absolute requirement for membrane translocation and glycosylation, *Virology* **165**:367.

King, B., and Brian, D. A., 1982, Bovine coronavirus structural proteins, *J. Virol.* **42**:700.

Klumperman, J., Krijnse Locker, J., Meijer, A., Horzinek, M. C., Geuze, H. J., and Rottier, P. J. M., 1994, Coronavirus M proteins accumulate in the Golgi complex beyond the site of virion budding, *J. Virol.* **68**:6523.

Krijnse Locker, J., Griffiths, G., Horzinek, M. C., and Rottier, P. J. M., 1992a, O-glycosylation of the coronavirus M protein, *J. Biol. Chem.* **267**:14094.

Krijnse Locker, J., Rose, J. K., Horzinek, M. C., and Rottier, P. J. M., 1992b, Membrane assembly of the triple-spanning coronavirus M protein, *J. Biol. Chem.* **267**:21911.

Krijnse Locker, J., Ericsson, M., Rottier, P. J. M., and Griffiths, G., 1994, Characterization of the budding compartment of mouse hepatitis virus, *J. Cell Biol.* **124**:55.

Krijnse Locker, J., Klumperman, J., Oorschot, V., Geuze, H. J., Horzinek, M. C., and Rottier, P. J. M., 1995, The cytoplasmic tail of mouse hepatitis virus M protein is essential but not sufficient for its retention in the Golgi complex, *J. Biol. Chem.* **269**:28263.

Kyte, J., and Doolittle, R. F., 1982, A simple method for displaying the hydropathic character of a protein, *J. Mol. Biol.* **157**:105.

Lanser, J. A., and Howard, C. R., 1980, The polypeptides of infectious bronchitis virus (IBV-41 strain). *J. Gen. Virol.* **46**:349.

Lapps, W., Hogue, B. G., and Brian, D. A., 1987, Sequence analysis of the bovine coronavirus nucleocapsid and matrix protein genes, *Virology* **157:**47.

Laude, H., Chapsal, J.-M., Gelfi, J., Labiau, S., and Grosclaude, J., 1986, Antigenic structure of transmissible gastroenteritis virus. I. Properties of monoclonal antibodies directed against virion proteins, *J. Gen. Virol.* **67:**119.

Laude, H., Rasschaert, D., and Huet, J.-C., 1987, Sequence and N-terminal processing of the transmissible protein El of the coronavirus transmissible gastroenteritis virus, *J. Gen. Virol.* **68:**1687.

Laude, H., Gelfi, J., Lavenant, L., and Charley, B., 1992, Single amino acid changes in the viral glycoprotein M affect induction of alpha interferon by the coronavirus transmissible gastroenteritis virus, *J. Virol.* **66:**743.

Laviada, M. D., Videgain, S. P., Moreno, L., Alonso, F., Enjuanes, L., and Escribano, J. M., 1990, Expression of swine transmissible gastroenteritis virus envelope antigens on the surface of infected cells: Epitopes externally exposed, *Virus Res.* **16:**247.

Luytjes, W., Sturman, L. S., Bredenbeek, P. J., Charité, J., Van der Zeijst, B. A. M., Horzinek, M. C., and Spaan, W. J. M., 1987, Primary structure of the glycoprotein E2 of coronavirus MHV-A59 and identification of the trypsin cleavage site, *Virology* **161:**479.

Machamer, C. E., and Rose, J. K., 1987, A specific transmembrane domain of a coronavirus El glycoprotein is required for its retention in the Golgi region, *J. Cell Biol.* **105:**1205.

Machamer, C. E., Mentone, S. A., Rose, J. K., and Farquhar, M. G., 1990, The El glycoprotein of an avian coronavirus is targeted to the *cis* Golgi complex, *Proc. Natl. Acad. Sci. USA* **87:**6944.

Machamer, C. E., Grim, M. G., Esquela, A., Chung, S. W., Rolls, M., Ryan, K., and Swift, A. M., 1993, Retention of a *cis* Golgi protein requires polar residues on one face of a predicted α-helix in the transmembrane domain, *Mol. Biol. Cell* **4:**695.

Macnaughton, M. R., 1981, Structural and antigenic relationships between human, murine and avian coronaviruses, *Adv. Exp. Med. Biol.* **142:**19.

Mayer, T., Tamura, T., Falk, M., and Niemann, H., 1988, Membrane integration and intracellular transport of the coronavirus glycoprotein El, a class III membrane glycoprotein, *J. Biol. Chem.* **263:**14956.

McIntosh, K., Kapikian, A. Z., Hardison, K. A., Hartley, J. W., and Chanock, R. M., 1969, Antigenic relationships among the coronaviruses of man and between human and animal coronaviruses, *J. Immunol.* **102:**1109.

Mobley, J., Evans, G., Dailey, M. O., and Perlman, S., 1992, Immune response to a murine coronavirus: Identification of a homing receptor-negative $CD4^+$ T cell subset that responds to viral glycoproteins, *Virology* **187:**443.

Mounir, S., and Talbot, P. J., 1992, Sequence analysis of the membrane protein gene of human coronavirus OC43 and evidence for O-glycosylation, *J. Gen. Virol.* **73:**2731.

Niemann, H., and Klenk, H.-D., 1981, Coronavirus glycoprotein El, a new type of viral glycoprotein, *J. Mol. Biol.* **153:**993.

Niemann, H., Boschek, B., Evans, D., Rosing, M., Tamura, T., and Klenk, H.-D., 1982, Post-translational glycosylation of coronavirus glycoprotein El: Inhibition by monensin, *EMBO J.* **1:**1499.

Niemann, H., Geyer, R., Klenk, H.-D., Linder, D., Stirm, S., and Wirth, M., 1984, The carbohydrates of mouse hepatitis virus (MHV) A59: Structures of the O-glycosidically linked oligosaccharides of glycoprotein El, *EMBO J.* **3:**665.

Opstelten, D.-J. E., Horzinek, M. C., and Rottier, P. J. M., 1993a, Complex formation between the spike protein and the membrane protein during mouse hepatitis virus assembly, *Adv. Exp. Med. Biol.* **342:**189.

Opstelten, D.-J. E., De Groote, P., Horzinek, M. C., Vennema, H., and Rottier, P. J. M., 1993b, Disulfide bonds in folding and transport of mouse hepatitis coronavirus glycoproteins, *J. Virol.* **67:**7394.

Opstelten, D.-J. E., De Groote, P., Horzinek, M. C., and Rottier, P. J. M., 1994, Folding of the mouse hepatitis virus spike protein and its association with the membrane protein, *Arch. Virol.* (Suppl) **9:**319.

Parker, S. E., Gallagher, T. M., and Buchmeier, M. J., 1989, Sequence analysis reveals extensive polymorphism and evidence of deletions within the E2 glycoprotein gene of several strains of murine hepatitis virus, *Virology* **173:**664.

Pedersen, N. C., Ward, I., and Mengeling, W. L., 1978, Antigenic relationships of the feline infectious peritonitis virus to coronaviruses of other species, *Arch. Virol.* **58:**45.

Pfleiderer, M., Skinner, M. A., and Siddell, S. G., 1986, Coronavirus MHV-JHM: Nucleotide sequence of the mRNA that encodes the membrane protein, *Nucleic Acids Res.* **14:**6338.

Pulford, D. J., and Britton, P., 1990, Expression and cellular localisation of porcine transmissible gastroenteritis virus N and M proteins by recombinant vaccinia viruses, *Virus Res.* **18:**203.

Raabe, T., and Siddell, S. G., 1989, Nucleotide sequence of the gene encoding the membrane protein of human coronavirus 229 E, *Arch. Virol.* **107:**323.

Rasschaert, D., Duarte, M., and Laude, H., 1990, Porcine respiratory coronavirus differs from transmissible gastroenteritis virus by a few genomic deletions, *J. Gen. Virol.* **71:**2599.

Rottier, P. J. M., and Rose, J. K., 1987, Coronavirus El glycoprotein expressed from cloned cDNA localizes in the Golgi region, *J. Virol.* **61:**2042.

Rottier, P. J. M., Spaan, W. J. M., Horzinek, M. C., and Van der Zeijst, B. A. M., 1981a, Translation of three mouse hepatitis virus strain A59 subgenomic RNAs in *Xenopus laevis* oocytes, *J. Virol.* **38:**20.

Rottier, P. J. M., Horzinek, M. C., and Van der Zeijst, B. A. M., 1981b, Viral protein synthesis in mouse hepatitis virus strain A59-infected cells: Effect of tunicamycin, *J. Virol.* **40:**350.

Rottier, P., Brandenburg, D., Armstrong, J., Van der Zeijst, B., and Warren, G., 1984, Assembly *in vitro* of a spanning membrane protein of the endoplasmic reticulum: The El glycoprotein of coronavirus mouse hepatitis virus A59, *Proc. Natl. Acad. Sci. USA* **81:**1421.

Rottier, P., Armstrong, J., and Meyer, D. I., 1985, Signal recognition particle-dependent insertion of coronavirus El, an intracellular membrane glycoprotein, *J. Biol. Chem.* **260:**4648.

Rottier, P. J. M., Welling, G. W., Welling-Wester, S., Niesters, H. G. M., Lenstra, J. A., and Van der Zeijst, B. A. M., 1986, Predicted membrane topology of the coronavirus protein El, *Biochemistry* **25:**1335.

Rottier, P. J. M., Krijnse Locker, J., Horzinek, M. C., and Spaan, W. J. M., 1990, Expression of MHV-A59 M glycoprotein: Effects of deletions on membrane integration and intracellular transport, *Adv. Exp. Med. Biol.* **276:**127.

Sanchez, C. M., Jimenez, G., Laviada, M. D., Correa, I., Sune, C., Bullido, M. J., Gebauer, F., Smerdou, C., Callebaut, P., Escribano, J. M., and Enjuanes, L., 1990, Antigenic homology among coronaviruses related to transmissible gastroenteritis virus, *Virology* **174:**410.

Schmidt, I., Skinner, M., and Siddell, S., 1987, Nucleotide sequence of the gene encoding the surface projection glycoprotein of coronavirus MHV-JHM, *J. Gen. Virol.* **68:**47.

Schmidt, M. F. G., 1982, Acylation of viral spike glycoproteins, a feature of enveloped RNA viruses, *Virology* **116:**327.

Schmidt, O. W., and Kenny, G. E., 1982, Polypeptides and functions of antigens from human coronaviruses 229E and OC43, *Infect. Immun.* **35:**515.

Siddell, S. G., Wege, H., Barthel, A., and Ter Meulen, V., 1981, Coronavirus JHM. Intracellular protein synthesis, *J. Gen. Virol.* **53:**145.

Stern, D. F., and Sefton, B. M., 1982a, Coronavirus proteins: Biogenesis of avian infectious bronchitis virus virion proteins, *J. Virol.* **44:**794.

Stern, D. F., and Sefton, B. M., 1982b, Coronavirus proteins: Structure and function of the oligosaccharides of the avian infectious bronchitis virus glycoproteins, *J. Virol.* **44:**804.

Stern, D. F., Burgess, L., and Sefton, B. M., 1982, Structural analysis of virion proteins of the avian coronavirus infectious bronchitis virus, *J. Virol.* **42:**208.

Stohlman, S. A., and Lai, M. M. C., 1979, Phosphoproteins of murine hepatitis viruses, *J. Virol.* **32:**672.

Sturman, L. S., 1977, Characterization of a coronavirus. I. Structural proteins: Effects of preparative conditions on the migration of protein in polyacrylamide gels, *Virology* **77:**637.

Sturman, L. S., 1981, The structure and behaviour of coronavirus A59 glycoproteins, *Adv. Exp. Med. Biol.* **142:**1.

Sturman, L. S., and Holmes, K. V., 1977, Characterization of a coronavirus. II. Glycoproteins of the viral envelope: Tryptic peptide analysis, *Virology* **77:**650.

Sturman, L. S., Holmes, K. V., and Behnke, J., 1980, Isolation of coronavirus envelope glycoproteins and interaction with the viral nucleocapsid, *J. Virol.* **33:**449.

Sugiyama, K., and Amano, Y., 1981, Morphological and biological properties of a new coronavirus associated with diarrhea in infant mice, *Arch. Virol.* **67:**241.

Swift, A. M., and Machamer, C. E., 1991, A Golgi retention signal in a membrane-spanning domain of coronavirus E1 protein, *J. Cell Biol.* **115**:19.

Tooze, J., Tooze, S., and Warren, G., 1984, Replication of coronavirus MHV-A59 in sac$^-$ cells: Determination of the first site of budding of progeny virions, *Eur. J. Cell Biol.* **33**:281.

Tooze, J., Tooze, S. A., and Warren, G., 1985, Laminated cisternae of the rough endoplasmic reticulum induced by coronavirus MHV-A59 infection, *Eur. J. Cell Biol.* **36**:108.

Tooze, J., Tooze, S. A., and Fuller, S. D., 1987, Sorting of progeny coronavirus from condensed secretory proteins at the exit from the *trans*-Golgi network of AtT20 cells, *J. Cell Biol.* **105**:1215.

Tooze, S. A., and Stanley, K. K., 1986, Identification of two epitopes in the carboxyterminal 15 amino acids of the E1 glycoprotein of mouse hepatitis virus A59 by using hybrid proteins, *J. Virol.* **60**:928.

Tooze, S. A., Tooze, J., and Warren, G., 1988, Site of addition of *N*-acetyl-galactosamine to the E1 glycoprotein of mouse hepatitis virus-A59, *J. Cell. Biol.* **106**:1475.

Vennema, H., De Groot, R. J., Harbour, D. A., Dalderup, M., Gruffydd-Jones, T., Horzinek, M. C., and Spaan, W. J. M., 1990a, Early death after feline infectious peritonitis virus challenge due to recombinant vaccinia virus immunization, *J. Virol.* **64**:1407.

Vennema, H., Heijnen, L., Zijderveld, A., Horzinek, M. C., and Spaan, W. J. M., 1990b, Intracellular transport of recombinant coronavirus spike proteins: Implications for virus assembly, *J. Virol.* **64**:339.

Vennema, H., De Groot, R. J., Harbour, D. A., Horzinek, M. C., and Spaan, W. J. M., 1991a, Primary structure of the membrane and nucleocapsid protein genes of feline infectious peritonitis virus and immunogenicity of recombinant vaccinia viruses in kittens, *Virology* **181**:327.

Vennema, H., Rijnbrand, R., Heijnen, L., Horzinek, M. C., and Spaan, W. J. M., 1991b, Enhancement of the vaccinia virus/phage T7 RNA polymerase expression system using encephalomyocarditis virus 5′-untranslated region sequences, *Gene* **108**:201.

Verbeek, A., and Tijssen, P., 1991, Sequence analysis of the turkey enteric coronavirus nucleocapsid and membrane protein genes: A close genomic relationship with bovine coronavirus, *J. Gen. Virol.* **72**:1659.

Wang, F.-I., Fleming, J. O., and Lai, M. M. C., 1992, Sequence analysis of the spike protein gene of murine coronavirus variants: Study of genetic sites affecting neuropathogenicity, *Virology* **186**:742.

Wege, H., Wege, H., Nagashima, K., and Ter Meulen, V., 1979, Structural polypeptides of the murine coronavirus JHM, *J. Gen. Virol.* **42**:37.

Weisz, O. A., Swift, A. M., and Machamer, C. E., 1993, Oligomerization of a membrane protein correlates with its retention in the Golgi complex, *J. Cell Biol.* **122**:1185.

Wesseling, J. G., Godeke, G.-J., Schijns, V. E. C. J., Prevec, L., Graham, F. L., Horzinek, M. C., and Rottier, P. J. M., 1993, Mouse hepatitis virus spike and nucleocapsid proteins expressed by adenovirus vectors protect mice against a lethal infection, *J. Gen. Virol.* **74**:2061.

Wesseling, J. G., Vennema, H., Godeke, G.-J., Horzinek, M. C., and Rottier, P. J. M., 1994, Nucleotide sequence and expression of the spike (S) gene of canine coronavirus and comparison with the S proteins of feline and porcine coronaviruses, *J. Gen. Virol.* **75**:1789.

Woods, R. D., Wesley, R. D., and Kapke, P. A., 1988, Neutralization of porcine transmissible gastroenteritis virus by complement-dependent monoclonal antibodies, *Am. J. Vet. Res.* **49**:300.

CHAPTER 7

# The Coronavirus Nucleocapsid Protein

HUBERT LAUDE AND PAUL S. MASTERS

## I. INTRODUCTION

The nucleocapsid (N) proteins of coronaviruses, as with most enveloped viruses, have received less attention than the surface glycoproteins and generally have been perceived to be of lesser concern. Interest in this class of proteins, however, has been stimulated in recent years by such diverse developments as the recognition of the importance of RNA–protein interactions, most notably in the study of spliceosomes and related ribonucleoproteins, as well as the finding in numerous viral systems that internal virion proteins can be major determinants of the cellular immune response.

Coronaviruses represent, together with the recently characterized toroviruses, the only positive-strand enveloped viruses with helically symmetric nucleocapsids. Thus, coronavirus N proteins present a number of intriguing molecular biological problems: What are the geometric parameters of these nucleocapsid structures and how do they compare to their negative strand RNA virus counterparts? What is the nature of the N–RNA and N–N interactions that stabilize the helix? Where are the loci of the multiple phosphorylations in the N molecule and what regulatory role (if any) do these play? How does N interact with the membrane protein (M) during virion assembly and uncoating? How does N participate in the unique mechanism of coronavirus RNA synthesis? Finally, what is the antigenicity of the N molecule? This chapter re-

HUBERT LAUDE • Laboratoire de Virologie et Immunologie Moléculaires, INRA, Centre de Recherches de Jouy-en-Josas, 78350 Jouy-en-Josas, France. PAUL S. MASTERS • Wadsworth Center for Laboratories and Research, New York State Department of Health, Albany, New York 12201-0509.

*The Coronaviridae*, edited by Stuart G. Siddell, Plenum Press, New York, 1995.

views our present state of knowledge about these largely unanswered questions.

## II. CORONAVIRUS RIBONUCLEOPROTEIN

The structure of the nucleocapsid (or ribonucleoprotein, RNP) of coronaviruses has not yet been completely elucidated, although electron microscopic studies are in general agreement about the nature of this internal viral component. Ideally, a complete description of the coronaviral RNP would include: the diameter of the helical nucleocapsid as well as the diameter of the central cylindrical hole around which the helix is wound; the number of helical turns per unit length; the stoichiometry of nucleotides of RNA per monomer of N protein; the overall length of the encapsidated genome; and, finally, the superstructure of the helix packed within the virion. The existing studies, carried out by thin sectioning of virion particles (Apostolov *et al.*, 1970) or by negative staining of spontaneously or detergent-disrupted virions (Kennedy and Johnson-Lussenburg, 1975/76; Macnaughton *et al.*, 1978; Caul *et al.*, 1979; Davies *et al.*, 1981), provide some of these parameters for some coronaviruses.

The earlier morphological work described the coronavirus RNP as a thread-like strand 7–8 nm in diameter [avian infectious bronchitis virus (IBV): Apostolov *et al.*, 1970)] or 8–9 nm in diameter [human coronavirus-229E (HCV-229E): Kennedy and Johnson-Lussenburg, 1975/76)]. Subsequent work yielded images of clearly helical structures from HCV-229E (Macnaughton *et al.*, 1978; Caul *et al.*, 1979), murine hepatitis virus-3 (MHV-3) (Macnaughton *et al.*, 1978), and IBV (Davies *et al.*, 1981). These had reported diameters of 9–11 nm, 11–13 nm (Caul *et al.*, 1979), or 14–16 nm (Macnaughton *et al.*, 1978), with 3- to 4-nm diameter hollow cores. The variations in observed dimensions possibly reflect differences in the methods of sample preparation. These helix diameters fall at the most narrow extreme for those observed with nucleocapsids in the paramyxovirus family (i.e., in the pneumovirus genus). RNP structures of IBV were the most difficult to prepare among the viruses studied (Macnaughton *et al.*, 1978; Davies *et al.*, 1981) and fully elongated unwound helices were often observed, the longest of these being 6.7 $\mu$m in length.

The long, helical coronavirus RNP apparently must assume a higher-order structure to be packaged into the viral particle. For HCV-229E, a tightly coiled superstructure (approximately 60-nm diameter) was sometimes observed in material from disrupted virions, although more often this appeared as a tangled or knotted mass (Kennedy and Johnson-Lussenburg, 1975/76). Similarly, a coiled or helical superstructure was seen for the RNP of MHV-A59 virions budding into the endoplasmic reticulum of infected cells (Holmes and Behnke, 1981). This coiling suggests that the coronavirus helical nucleocapsid is fairly flexible, another structural similarity it shares with paramyxovirus nucleocapsids.

Several biochemical features of virion-associated N protein, established in early studies, were consistent with its postulated role as the structural protein of the coronavirus RNP. N was found to be the sole virus polypeptide not to be glycosylated and to be phosphorylated (Pocock and Garwes, 1977; Wege *et al.*, 1979; Stohlman and Lai, 1979; Callebaut and Pensaert, 1980; Macnaughton, 1980; Cavanagh, 1981; Lomniczi and Morser, 1981; Stern *et al.*, 1981; Rottier *et al.*,1981a; Siddell *et al.*, 1981b; King and Brian, 1982). A relatively high content in arginine and glutamine residues was established, consistent with the basic character anticipated for a nucleic acid binding protein (Garwes *et al.*, 1976; Sturman, 1977). Also, protease digestion of virions left the N polypeptide completely unaffected, arguing for its internal location (Sturman, 1977; Callebaut and Pensaert, 1980; Obert *et al.*, 1981; King and Brian, 1982; Schmidt and Kenny, 1982).

Direct experimental support for the assignment of the N polypeptide as the nucleocapsid protein came from the analysis of the composition of subviral components from detergent-solubilized virions separated on density gradients. In several reports, treatment of the viral envelope with a nonionic detergent in low salt concentration failed to liberate true RNP structures. Thus, gel electrophoretic analysis of subviral particles released by NP40 treatment at 20 °C from transmissible gastroenteritis virus (TGEV) virions (buoyant density 1.295 g/ml in cesium sulfate: Garwes *et al.*, 1976), MHV-JHM virions (1.26 g/ml in sucrose: Wege *et al.*, 1979), hemagglutinating encephalitis virus (HEV) virions (Pocock and Garwes, 1977), or from IBV virions treated with Triton X-100 at 37 °C (1.32 g/ml in sucrose: Lancer and Howard, 1980) revealed the presence of two polypeptides, corresponding to the nucleocapsid protein N and the membrane glycoprotein M. The recovery of RNP structures containing essentially only the N polypeptide species was reported following NP40 treatment at 0–4 °C of MHV-A59 virions (Sturman *et al.*, 1980) and of IBV virions (1.27 g/ml in sucrose: Davies *et al.*, 1981). The MHV study recognized a subtle influence of temperature on the interaction of M protein with the RNP: virions disrupted by NP40 at 4 °C released a structure banding at 1.28 g/ml in sucrose and containing only N protein, while subsequent incubation at 37 °C produced an M protein–nucleocapsid complex banding at 1.22 g/ml.

## III. N PROTEIN STRUCTURE

Almost all of our present structural knowledge of the N proteins of coronaviruses derives from amino acid sequences deduced from nucleotide sequences of cloned N gene cDNAs. To date, these have been reported for 11 coronaviruses: MHV (Skinner and Siddell, 1983, 1984; Armstrong *et al.*, 1983, 1984; Parker and Masters, 1990; Kunita *et al.*, 1992; Decimo *et al.*, 1993), IBV (Boursnell *et al.*, 1985; Sutou *et al.*, 1988; Williams *et al.*, 1992), TGEV (Kapke and Brian, 1986; Rasschaert *et al.*, 1987; Britton *et al.*, 1988), bovine coronavirus (BCV) (Lapps *et al.*, 1987; Cruciere and Laporte, 1988), HCV-OC43 (Kamahora *et al.*, 1989), HCV-229E (Schreiber *et al.*, 1989; Myint *et al.*, 1990), feline infectious peritonitis virus (FIPV) (Vennema *et al.*, 1991), turkey coronavirus (TCV) (Verbeek and Tijssen, 1991), canine coronavirus (CCV) (Horsburgh *et al.*, 1992; Vennema *et al.*, 1992), porcine epidemic diarrhea virus (PEDV) (Bridgen *et al.*, 1993), and sialodacryoadenitis virus (SDAV) (Kunita *et al.*, 1993).

The general physical properties of the N proteins are markedly similar (Table I). The encoded polypeptides range from 377 to 455 amino acids in length. Notably, MHV, SDAV, BCV, TCV, and HCV-OC43, which belong to the same antigenic cluster, have N proteins roughly 10% larger than those of the remainder of the family. Also striking is the presence in PEDV N protein of an approximately 40-residue-long stretch that has no counterpart in the other members (Bridgen *et al.*, 1993). This unique sequence, particularly rich in serine, arginine, and asparagine residues, is located in the central portion of the molecule (see Fig. 1) and might reflect a recombinational event or a stuttering of the viral polymerase since it exhibits some periodicity. The virion-associated N protein of coronaviruses resolves by sodium dodecyl sulfate–polyacrylamide gel electrophoresis (SDS-PAGE) as a single species with molecular weight ranging from 45 to 63 kDa, depending on the virus and on the strain (for references, see Siddell *et al.*, 1982). These apparent sizes are significantly larger than the molecular weights calculated from sequence information, possibly due both to anomalous distributions of amino acid compositions and to phosphorylation of the proteins. Even among five closely related strains of MHV, molecular weight differences of up to 3 to 5 kDa have been observed (Cheley *et al.*, 1981), although the calculated molecular weights for these species are almost identical (Parker and Masters, 1990).

Coronavirus N proteins are highly basic, having an excess of at least 12 lysine and arginine residues over aspartate and glutamate residues. One measure of this is seen in the calculated overall isoelectric points (pIs) of these polypeptides, which fall in the range 10.3–10.7 (Table I). Basic amino acids are not clustered in strings as in proteins such as simian virus 40 (SV40) T antigen or the capsid proteins of alphaviruses, but local densities of positive charge can be found, particularly in two loci in the middle of the N molecule (Fig. 1). In marked contrast to the bulk of the molecule, the carboxy-terminus of each coronavirus N protein is quite acidic, as indicated by pIs ranging from 4.3 to 5.5 calculated for the 45 carboxy-terminal amino acids (Table I). Another noteworthy feature of the N proteins is that they all have a relatively high serine content (7–11%) (Table I). These potential targets for phosphorylation (see Section V) are distributed throughout the molecule, but many are interspersed in the first of the two basic regions mentioned above. It is interesting that most of the above-cited characteristics of coronavirus N proteins are also shared by the nucleocapsid (NP) proteins of influenzaviruses (Table I), perhaps due to an ancestral relationship between these two families.

In spite of the general similarities indicated for the coronavirus N proteins, there exists only a low degree of sequence homology among them. For example, the N protein of BCV shows an overall sequence homology of 29% with the N protein of TGEV or of IBV; the sequence homology of the N proteins of MHV and BCV, which belong to the same antigenic cluster, is 70%, i.e., notably less than between the M proteins (Lapps *et al.*, 1987). An optimal alignment of the 11 sequences reveals about 30 residues common to all, including 8 glycines and 3 prolines; in contrast, the few cysteine residues present in the sequences are not at conserved positions. An exception to this overall lack of N protein homology is a region of roughly 50 amino acids, falling within the amino-terminal third of

TABLE I. Amino Acid Compositions of the Coronavirus N Proteins[a]

| | Amino acids | Lys + arg | Asp + glu | Ser | pI total[b] | pI C-terminus[c] | Predicted mol. wt. (kDa) | Reported $M_r$ (kDa)[d] |
|---|---|---|---|---|---|---|---|---|
| MHV-A59 | 454 | 63 | 46 | 41 | 10.5 | 5.1 | 49.7 | 50–55 |
| SDAV | 454 | 60 | 42 | 40 | 10.6 | 5.3 | 49.4 | 50 |
| BCV | 448 | 59 | 46 | 42 | 10.6 | 5.3 | 49.4 | 52 |
| TCV | 448 | 59 | 46 | 43 | 10.6 | 5.3 | 49.4 | 52 |
| HCV-OC43 | 448 | 58 | 46 | 42 | 10.5 | 5.3 | 49.3 | 52 |
| TGEV | 382 | 69 | 46 | 42 | 10.7 | 5.3 | 43.4 | 47 |
| CCV | 381 | 69 | 47 | 36 | 10.6 | 5.1 | 43.4 | 50 |
| FIPV | 377 | 65 | 45 | 35 | 10.7 | 5.5 | 42.7 | 45 |
| HCV-229E | 389 | 60 | 46 | 39 | 10.3 | 4.5 | 43.4 | 47–50 |
| PEDV | 441 | 72 | 50 | 34 | 10.7 | 4.3 | 48.9 | 55–58 |
| IBV | 409 | 72 | 53 | 28 | 10.5 | 4.9 | 45.0 | 50–54 |
| Influenza (A/NT/60/68) | 498 | 70 | 59 | 39 | 10.4 | 4.5 | 55.9 | 50–54 |

[a]Sources of coronavirus sequences are given in the text. Influenza N protein data was deduced from the sequence data of Huddleston and Brownlee (1982).
[b]Computed isoelectric point of the entire N molecule.
[c]Computed isoelectric point of the C-terminal 45 amino acid residues.
[d]Relative mass, from more recent reports, as determined by SDS-PAGE.

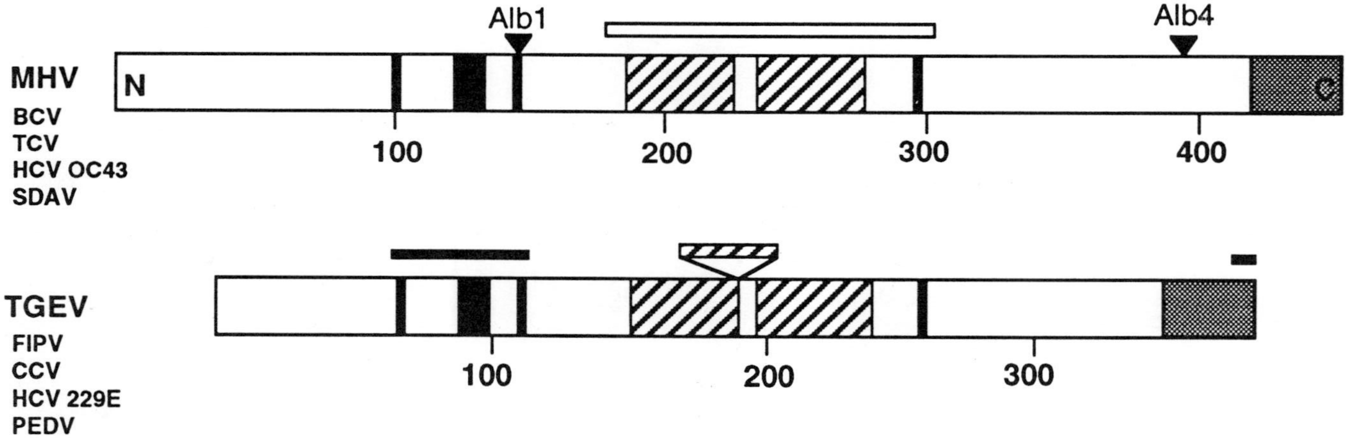

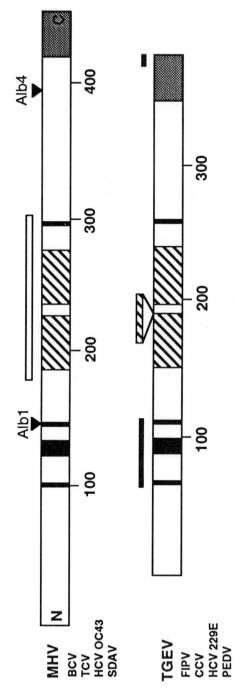

FIGURE 1. Structural organization of the coronavirus N protein. Common structural features and their distribution along the polypeptide chain are shown for two coronaviruses, MHV and TGEV, each being representative of a coronavirus subgroup, as indicated. Regions homologous to all the viruses are indicated by closed boxes (at least 3 consecutive amino acids conserved; the sequence stretch encompassing the amino-most 3 boxes is shown in Fig. 2). The striped boxes indicate two major hydrophylic, basic domains: the first, rich in Arg and Ser residues; the second, rich in Lys residues. The shaded box indicates a markedly acidic domain. The bars above the frames indicate particular domains mapped in the N protein of the relevant virus: open bar, RNA binding domain; striped bar, relative position of the 40 residue insertion in PEDV N; closed bars, antigenic sites. The arrowheads indicate the position of sequence alterations in two *ts* mutants (see text).

each molecule, which contains a relatively high degree of sequence identity (Fig. 2). Such a marked degree of conservation may signal an important function common to the N proteins. The strong proportion of aromatic residues in this segment possibly participate in stabilizing N binding to RNA via ring stacking interactions (Ollis and White, 1987); alternatively, the conserved segment may be involved in other hydrophobic interactions not directly related to RNA binding.

To date, two coronavirus N protein mutants have been described, both in MHV-A59. Each is thermolabile as well as temperature-sensitive (*ts*), forming small plaques at the nonpermissive temperature. The first of these, Albl (Masters *et al.*, 1994), has a single critical amino acid change within the highly conserved region mentioned above, as well as a closely linked amino acid change that is phenotypically silent. The second, Alb4 (Koetzner *et al.*, 1992), contains a 29-amino-acid deletion of a putative spacer region that separates the carboxy-terminal domain of the N protein from the remainder of the molecule. In addition to these, the *ts*209 mutant of MHV-A59 (Koolen *et al.*, 1983) is potentially an N protein mutant (see Section IV) but has not yet been mapped. Further study of these mutants may provide new insights into N protein structure and function.

```
MHV-A59    102  G Y W Y R H N R R S F K T P D G Q Q K Q L L P
SDAV       102  G Y W Y R H N R R S F K T P D G Q Q K Q L L P
BCV         99  G Y W Y R H N R R S F K T A D G N Q R Q L L P
TCV         99  G Y W Y R H N R R S F K T A D G N Q R Q L L P
HCV-OC43    99  G Y W Y R H N R G S F K T A D G N Q R Q L L P
TGEV        68  G Y W N R Q T R - - Y R M V K G Q R K E L P E
CCV         68  G Y W N R Q T R - - Y R M V K G R R K E L P E
FIPV        68  G Y W N R Q I R - - Y R I V K G Q R K E L A E
HCV-229E    56  G Y W N V Q K R - - F R T R K G K R V D L S P
PEDV        51  G Y W N E Q I R - - W R M R R G E R I E Q P S
IBV         69  G Y W R R Q A R - - F K P G K G G R K P V P D

R W Y F Y Y L G T G P H A G A S Y G D S I E G V F W V A  152
R W Y F Y Y L G T G P H A G A S F G D S I E G V F W V A  152
R W Y F Y Y L G T G P H A K D Q Y G T D I D G V F W V A  149
R W Y F Y Y L G T G P H A K D Q Y G T D I D G V F W V A  149
R W Y F Y Y L G T G P H A K D Q Y G T D I D G V Y W V A  149
R W F F Y Y L G T G P H A D A K F K D K L D G V V W V A  116
R W F F Y Y L G T G P H A D A K F K Q K L D G V V W V A  116
K W F F Y F L G T G P H A D A K F K D K I D G V F W V A  116
K L H F Y Y L G T G P H K D A K F R E R V E G V V W V A  105
N W H F Y Y L G T G P H G D L R Y R T R T E G V F W V A  99
A W Y F Y Y T G T G P A A D L N W G D T Q D G I V W V A  117
```

FIGURE 2. Region of greatest homology among coronavirus N proteins. Numbers indicate the amino acid residues of each N protein shown in the alignment. Boxed residues are those which are identical in at least seven of the eleven sequences. The sources of the sequence data are given in the text.

## IV. SYNTHESIS OF N PROTEIN AND NUCLEOCAPSID FORMATION

The nucleocapsid protein appears to be the most abundant viral polypeptide in coronavirus-infected cells during all stages of infection. N protein is readily detected from 3 to 5 hr after infection onward, at the same time or earlier than the other structural proteins, and its synthesis is maintained throughout the virus cycle (Anderson *et al.*, 1979; Bond *et al.*, 1979; Holmes *et al.*, 1981; Siddell *et al.*, 1981b; Stern *et al.*, 1981; Rottier *et al.*, 1981b; Stohlman *et al.*, 1983; Garwes *et al.*, 1984; Talbot *et al.*, 1984; Keck *et al.*, 1988; Dea and Tijssen, 1989; Simkins *et al.*, 1989). In latently infected Schwannoma cells, MHV-JHM N protein was shown to be produced for as long as 7 days and beyond in the absence of infectious virus production (Coulter-Mackie *et al.*, 1985). In cells infected with MHV-JHM defective-interfering RNAs, the level of N synthesis was found to be not significantly altered throughout multiple passages, in contrast to the other viral polypeptides (Makino *et al.*, 1985). Thus, it seems that a strong and permanent expression of N gene product accompanies both productive and nonproductive coronavirus infection.

Coronavirus N protein is synthesized on free polysomes (Niemann *et al.*, 1982), but appears to become rapidly associated to intracellular membranes (Stohlman *et al.*, 1983). Its subcellular location has been examined by immunofluorescence labeling of MHV-, TGEV-, and TCV-infected cells. Anti-N monoclonal antibodies produce a bright particulate fluorescence which tends to distribute evenly in the cytoplasm (Robb and Bond, 1979; Laude *et al.*, 1986; Holmes *et al.*, 1987; Welch and Saif, 1988; Dea and Tijssen, 1989). Intriguingly, two groups of investigators have reported an important concentration of N-antigen within the nuclei of MHV-infected cells (Robb and Bond, 1979; Holmes *et al.*, 1987). The MHV-A59 *ts* mutant Alb4, shown to have a short deletion in its N gene (Koetzner *et al.*, 1992), exhibited an altered distribution of N antigen in the cytoplasm and the nucleus (Holmes *et al.*, 1987). Translocation of N into the nucleus might reflect the ability of the protein to bind to nucleic acids in a nonspecific manner (see Section VI), or it may indicate the fortuitous presence of a nuclear translocation signal in this highly basic protein. However, nuclear translocation cannot be essential for productive infection since MHV is capable of replicating in enucleated cells (Brayton *et al.*, 1981; Wilhelmsen *et al.*, 1981).

Studies on the biosynthesis of coronavirus nucleocapsid proteins in infected cells have revealed the presence of one major intracellular form, designated N, and of several N-related polypeptides. The major N species has been shown to comigrate with virion-associated N protein in denaturing gels (Siddell *et al.*, 1981b; Rottier *et al.*, 1981b; Laude *et al.*, 1986; Dea and Tijssen, 1989) and to be phosphorylated (MHV: Siddell *et al.*, 1981a; Stohlman *et al.*, 1983; TGEV: Garwes *et al.*, 1984), basic (Siddell *et al.*, 1981b), and heterogeneously charged (Bond *et al.*, 1979). The identity between intracellular and virion N has been further established by tryptic peptide fingerprinting in the case of MHV-JHM and MHV-A59 (Siddell *et al.*, 1980; Bond *et al.*, 1984). N polypeptide is a primary gene product, as demonstrated by pulse-chase analysis of cells infected by MHV-JHM and MHV-A59 (Siddell *et al.*, 1981b; Rottier *et al.*, 1981b; Holmes *et al.*, 1981; Perlman *et al.*, 1986/87), MHV-3 (Cheley and Anderson, 1981), TGEV (Laude *et al.*, 1986), and BCV (Deregt *et al.*, 1987). In agreement with this are the results of cell-free translation experiments, where N-assigned mRNA isolated from cells infected by MHV (Rottier *et al.*,1981a; Siddell *et al.*, 1980), TGEV (Jacobs *et al.*, 1986), or FIPV (De Groot *et al.*, 1987) programmed synthesis of a polypeptide species of the same size as the virion-associated protein. The same appears to hold true for recombinant N proteins expressed independently of other viral components in mammalian cells using vaccinia or adenovirus vectors (Pulford and Britton, 1990; Vennema *et al.*, 1991; Stohlman *et al.*, 1992; Wesseling *et al.*, 1993) or in yeast cells (Britton *et al.*, 1988).

Several groups of investigators have mentioned the existence of one or more intracellular polypeptides with a slightly lower molecular weight (2–5 kDa less) than the major N species, sometimes designated N′, N″, etc. These latter have been observed in immunoprecipitated or immunoblotted material, but also in direct lysates from MHV-JHM and MHV-A59, MHV-3, TGEV, BCV, IBV, and TCV, i.e., most of the coronavirus–host cell systems studied (Cheley and Anderson, 1981; Stern *et al.*, 1981; Rottier *et al.*, 1981b; Bond *et al.*, 1984; Garwes *et al.*, 1984; Coulter-Mackie *et al.*, 1985; Robbins *et al.*, 1986; Keck *et al.*, 1988; Welch and Saif, 1988; Dea and Tijssen, 1989). The relationship between N and these faster migrating species has been confirmed by peptide fingerprinting in the case of MHV (Cheley and Anderson, 1981; Bond *et al.*, 1984) and IBV (Stern *et al.*, 1981). Both the N and N′ species found in TGEV grown in LLC-PK1 cells were phosphorylated (Garwes *et al.*, 1984).

While there is general agreement that only the full-length N species is incorporated into virions, it is as yet unclear whether the N subspecies result from a different level of phosphorylation, proteolytic degradation at preferential cleavage sites, premature termination, or translation of deleted transcripts arising from defective particles. It has been proposed that the 57-kDa species of MHV-JHM corresponds to a nonphosphorylated precursor; a phosphorylated species of 60kDa became apparent only after a 10-min chase and the ratio of 57 : 60 kDa decreased as infection proceeded (Stohlman *et al.*, 1983). In contrast, several groups of investigators reported that the N subspecies are made in increasing amounts during the course of infection (Anderson *et al.*, 1979; Holmes *et al.*, 1987; Talbot *et al.*, 1984). Furthermore, pulse-chase experiments with both MHV-JHM and MHV-3 provided strong evidence that the lower molecular weight species are derived from the major N species (Anderson *et al.*, 1979; Cheley and Anderson, 1981). Comparison of immunoblots of MHV-A59- or MHV-JHM-infected cell extracts done with antibody to full-length N or with antibody to a synthetic peptide of the carboxy-terminus led to the conclusion that the intracellular faster-migrating species lacked the carboxy-terminal domain (Holmes *et al.*, 1987). Similarly, the subspecies observed following vaccinia-vectored expression of TGEV N protein in two different cell system were both unreactive toward an anti-N antibody directed to the C-terminal end (Pulford and Britton, 1990).

The virus strain and/or the host cell might determine the number, size, and amount of N subspecies. Thus, cells infected by the MHV-A59 mutant *ts*209 have been found to contain an extra N-related band (Koolen *et al.*, 1983). On the other hand, an extra band of N was detected in LLC-PK1 but not in thyroid TGEV-infected cells (Garwes *et al.*, 1984), and the N subspecies found in LLC-PK1 and CV1 cells expressing recombinant TGEV N protein differed in size (Pulford and Britton, 1990). In several virus–cell systems, N subspecies were not observed (MHV-JHM: Siddell *et al.*, 1980; Niemann *et al.*, 1982; TGEV: Laude *et al.*, 1986; BCV: Deregt *et al.*, 1987). From the above observations it would appear that the N subspecies most likely represent molecules partially degraded by cellular or serum proteases, although this was not formally proved in any of the published data. Such a phenomenon, which may be of no biological significance, is reminiscent of the limited proteolytic cleavage of the nucleocapsid protein of paramyxoviruses (Mountcastle *et al.*, 1974) and toroviruses (Horzinek *et al.*, 1985). Faster-migrating N species have also been observed during *in vitro* mRNA translation and were assumed to be the result of a proteolytic degradation during immunoprecipitation (MHV: Rottier *et al.*, 1981a; FIPV: De Groot *et al.*, 1987) or of premature termination (MHV: Leibowitz *et al.*, 1982). In addition, various truncated species of N protein have been characterized in the brain tissue from MHV-JHM-infected mice (Talbot *et al.*, 1984) and in immune complexes present in the body fluids of FIPV-infected cats (Horzinek *et al.*, 1986).

There is limited information about the fate of N protein in infected cells. Studies performed with MHV indicated that only a very small fraction of N is chased out of the cell into mature virions even after a 6-hr chase period (Holmes *et al.*, 1981), suggesting that a large pool of free N protein is built up early in the infection. MHV-A59- or MHV-JHM-infected cells appeared to contain a substantial amount of free nucleocapsids; intracellular EDTA-resistant structures of 200–230S, which contained N protein and genomic RNA, were identified (Robb and Bond, 1979; Spaan *et al.*, 1981). Perlman *et al.* (1986/87) determined the rate at which N protein was transferred into such nucleocapsid structures in the absence of *de novo* protein synthesis. The pool of free N protein in cycloheximide-treated cells was found to decay with a half-time of approximately 1 hr, which is, for example, quite slow compared with rhabdoviruses. Nearly all the genomic RNA was found to be incorporated in nucleocapsids (Spaan *et al.*, 1981; Perlman *et al.*, 1986/87).

## V. N PHOSPHORYLATION

Phosphorylation is the only known posttranslational modification of coronavirus N proteins (Stohlman and Lai, 1979; Lomniczi and Morser, 1981; King and Brian, 1982; Garwes *et al.*, 1984), and it is also another point of resemblance between the nucleoproteins of coronaviruses and influenza viruses (Kistner *et al.*, 1989). MHV N protein is phosphorylated in the cytoplasm within about 10 min of its synthesis and it concomitantly becomes tightly associated with a cell membrane fraction (Stohlman *et al.*, 1983). N protein represents the major phosphopolypeptide in both MHV- and TGEV-infected cells (Siddell *et al.*, 1981a; Garwes *et al.*, 1984). For MHV, the phosphate linkage has been shown to be exclusively to serine residues (Stohlman and Lai, 1979;

Siddell *et al.*, 1981a). Even though the exact number and location of the phosphoserines are still undefined, the process appears to be highly specific. An high-pressure liquid chromatography (HPLC) analysis of the tryptic phosphopeptides of MHV N revealed that phosphorylation may occur at only two or three sites (Wilbur *et al.*, 1986), a number relatively small in comparison to the 30 to 40 potential target residues in the N proteins. Moreover, the two plaque morphology variants of MHV-JHM examined in this study were found to differ in their phosphorylation patterns. It remains unclear whether the process involves autophosphorylation or a protein kinase activity of either viral or host origin. For MHV, a virion-associated protein kinase activity has been described that can transfer additional phosphate from ATP to the N protein as well as to other substrates *in vitro*. This enzyme appeared to be cyclic AMP-independent and to require divalent cations (Siddell *et al.*, 1981a), features common to those of protein kinases identified in a variety of other enveloped viruses (Tan, 1975). Whether the same enzyme is responsible for the kinase activity *in vivo*, however, has not yet been shown.

The finding that some anti-MHV N monoclonal antibodies recognized the phosphorylated form but not a nonphosphorylated, faster-migrating species assumed to be its precursor, led to the proposal that phosphorylation may induce a drastic conformational change of the protein (Stohlman *et al.*, 1983), possibly relevant to its biological function. However, the question of the role of N protein phosphorylation is largely unanswered. It has been speculated by a number of workers that it may govern the tightness of the association between N and RNA, but there is as yet no direct evidence that bears on this possibility. MHV N proteins derived from purified virions, cellular membranes, and the cytosol of infected cells were reported to have similar phosphopeptide maps (Wilbur *et al.*, 1986), a finding in agreement with the observed rapidity of the phosphorylation process. On this basis, phosphorylation might not be expected to play a critical role in the regulation of virus assembly. Finally, it has been hypothesized that phosphatase-catalyzed nucleocapsid dephosphorylation might control the processes of coronavirus penetration and uncoating, and potentially relevant host enzyme activities have been sought. An acid phosphatase activity associated with the particulate fraction of oligodendrocytes has been shown to convert the MHV N protein to a lower molecular weight product (Beushausen *et al.*, 1987). Also, a neutral, serine-threonine phosphatase activity was detected in the endosomal fraction of L-2 murine fibroblasts and in other susceptible cells, which recognized MHV N protein as a more specific substrate than acid or alkaline phosphatases (Mohandas and Dales, 1991). The relevance of such data would certainly be greater if the site of fusion of entering coronavirus particles were known more precisely.

## VI. N PROTEIN BINDING TO RNA

The most noteworthy function of coronavirus N proteins is that they bind to the RNA genomes of their respective viruses. Neither the nature nor the mechanism of this N–RNA complex formation has been well elucidated. Analogies with other RNA-binding proteins are limited. Coronavirus N proteins do not contain zinc finger motifs (Klug and Rhodes, 1987) or sequences similar to the RNA consensus and RNA-binding motifs found in many cellular RNA-binding proteins such as nucleolin, poly(A)-binding protein, and various hnRNP and snRNP proteins (Dreyfuss *et al.*, 1988). With respect to the properties of other RNA viruses containing helical nucleocapsids, coronaviruses appear to most closely resemble the orthomyxoviruses. The N (or NP) proteins of both of these families form complexes with RNA that tend to dissociate in high concentrations of salt and that provide only limited protection against the action of ribonucleases (Bukrinskraya, 1973; Macnaughton *et al.*, 1978; Robbins *et al.*, 1986; Kingsbury *et al.*, 1987). This contrasts markedly with the N–RNA complexes of the rhabdoviruses and paramyxoviruses, which are stable in high salt and are markedly ribonuclease-resistant (Bukrinskraya, 1973; Leppart *et al.*, 1979).

Efforts to characterize coronavirus N–RNA binding have sought to address two fundamental questions. First, what RNA sequences or structures are specifically bound by N? Second, what parts of the N molecule participate in RNA recognition? The ability of the MHV N protein to bind to nucleic acids has been examined by an RNA overlay protein blot assay (ROPBA) in which electrophoretically separated proteins from purified virions or from MHV-infected cells were transferred to nitrocellulose and probed with labeled RNA (Robbins *et al.*, 1986). Monomeric N as well as a minor multimeric N component, possibly an N trimer, were the only virus-specific RNA-binding proteins detected in this manner. Nucleic acid sequence binding specificity was not demonstrated, however, since N bound equally well to MHV RNA or to single- or double-stranded RNA or DNA of heterologous origin.

Subsequent adaptation of the ROPBA procedure allowed the demonstration of sequence-specific binding of RNA by MHV N protein (Stohlman *et al.*, 1988). The inclusion of a large molar excess of unlabeled (uninfected) cellular RNA with the labeled probe of MHV-infected cellular RNA was found to compete out low-affinity binding of RNA by both N and cellular proteins. Synthetic RNA probes containing defined extents of the 5′ leader RNA adjacent to either mRNA 1 or mRNA 6 were then used to localize a sequence-specific, high-affinity binding site for N to the region of nucleotides 56 to 67 of the MHV genome. This conclusion was reinforced by the observation that an anti-N monoclonal antibody could selectively immunoprecipitate from MHV-infected cells all leader-containing RNA species greater than or equal to 65 nucleotides in length (Baric *et al.*, 1988). Since the leader sequence is contained in all subgenomic positive-strand RNAs, as well as in the genome, this may suggest that some recognition event other than binding by N protein mediates the selective incorporation of genome RNA into assembled virions.

The assignment of the RNA-binding property of N to one or more particular portions of the N molecule is at present incomplete. The detection by ROPBA of the species N′ and N″, presumed to be products of small proteolytic truncations, suggests that at least limited removal of one of the two termini of N does not alter RNA binding (Robbins *et al.*, 1986). Recently, this point has been investigated using a nondenaturing gel assay of MHV N protein translated in a rabbit reticulocyte lysate programmed with engineered synthetic mRNAs (Masters, 1992). In this system, full-length N protein was shown to bind to an endogenous RNA species in the reticulocyte lysate. Construction of a set of N gene deletions allowed the mapping of this RNA-binding activity to the central one of three domains of N protein proposed previously on the basis of sequence comparisons of different strains of MHV (Parker and Masters, 1990). Deletions entering the central domain abolished this non-sequence-specific RNA-binding activity. Sequence-specific RNA binding, as measured by ROPBA, was also mapped to the central portion of purified MHV N protein following its fragmentation by treatment with formic acid (Nelson and Stohlman, 1993). The isolated RNA-binding domain, comprising amino acids 169 to 308, retained the ability to interact specifically with its target RNA sequence. Considerable work remains to be done to delineate the minimal RNA-binding domain of N and to define the exact nature of the association between this molecule and RNA.

Evidence recently has been presented that MHV N protein binds specifically to cell membranes (Anderson and Wong, 1993). The binding was inhibited by nonviral RNA and DNA, as well as by various membrane phospholipids such as cardiolipin, a property shared by certain DNA binding proteins. This observation led the authors to speculate that membrane lipid association of the N protein may compete for RNA binding sites on the N protein, a mechanism possibly relevant to the processes of nucleocapsid uncoating and assembly.

## VII. N–N AND N–M PROTEIN INTERACTIONS

Very little information is presently available about the protein–protein contacts N makes in assembled virions or in infected cells. The structure of the coronavirus nucleocapsid suggests that, at the very least, there must be some form of interactions between N monomers that neighbor each other upon each successive turn of the helix. There may also be an interaction between adjacent N monomers bound along the RNA strand, although these would not necessarily be required for helical encapsidation.

It has been observed for some coronaviruses that SDS-PAGE, under nonreducing conditions, detected an N-related protein of 140–160kDa, present at a level of a few percent compared to N. This species was identified by immunoblotting of BCV, HCV-OC43, and MHV virions (Hogue *et al.*, 1984; Robbins *et al.*, 1986; Deregt *et al.*, 1987), and for MHV it was demonstrated to bind to RNA (Robbins *et al.*, 1986). The N multimer appeared to be the size of a trimer, and it was shown to be held together by disulfide linkages, since it disappeared in the presence of mercaptoethanol. It is possible that this species is a fundamental unit of the nucleocapsid structure. However, since the N protein of each of these three viruses contains just two cysteine residues separated by only four amino acids, it seems likely that other types of N–N binding must also contribute to stabilizing the helix.

The coronavirus M protein is thought to be the principal determinant of virion assembly and budding. Consequently, it might be expected that there exist N–M interactions that play a major role in viral structure, although little experimental evidence bears on this issue. During the development of techniques to fractionate and purify viral components, it was found that a temperature-dependent binding of M and the nucleocapsid occurred in NP40-disrupted MHV virions (Sturman *et al.*, 1980). At 4 °C, M protein was solubilized by the non-ionic detergent, but at ambient temperature or 37 °C, M remained associated with the nucleocapsid, cosedimenting in sucrose gradients. However, the same study demonstrated that NP40-solubilized M was able to bind to RNA in the absence of N protein; thus, it is not clear whether the M–nucleocapsid binding being examined was due to an N–M interaction. Clearly, it would be desirable to examine the N-binding properties of the cytosolic carboxy-terminal segment of M in isolation from the remainder of the M molecule.

## VIII. POTENTIAL ROLE OF N PROTEIN IN RNA SYNTHESIS

Various lines of evidence indicate a possible role for N protein in coronavirus RNA synthesis. An earlier, perplexing observation in this regard was the finding that, after multiple passages, MHV defective-interfering (DI) particles inhibited the synthesis of all normal viral RNA species except RNA7, which was translated abundantly into N protein (Makino *et al.*, 1985). Although the mechanism of this selective resistance is unknown, it may imply that N is required for MHV RNA synthesis, or at least for the synthesis of DI RNA.

More directly, it has been shown that antibodies to N protein almost totally inhibited viral RNA synthesis in an *in vitro* system prepared from MHV-infected cells, whereas, under the same conditions, no significant inhibition was seen with anti-M or anti-S antibodies (Compton *et al.*, 1987). Moreover, the product RNA synthesized in this system, which was mostly of genome length, was found to be encapsidated by N protein. This might suggest that coronavirus RNA synthesis is coupled to the encapsidation of nascent RNA, analogous to the replication of viruses with helical negative-strand RNA nucleocapsids. For coronaviruses, however, the requirement for N may also apply to the synthesis of subgenomic RNAs. The immunoprecipitation of all leader-containing MHV RNAs, as well as replicative intermediate RNA complexes by an anti-N monoclonal antibody, has been taken to mean that the N–leader RNA interaction must be important to the mechanism of leader-primed transcription (Baric *et al.*, 1988).

The analogy with some families of negative-strand RNA viruses may be inexact in another respect. The protein synthesis inhibitor cycloheximide has been shown to prevent or inhibit genomic and subgenomic RNA synthesis in MHV-infected cells (Sawicki and Sawicki, 1986), and this effect was mirrored by puromycin or cycloheximide in the *in vitro* RNA-synthesizing system (Compton *et al.*, 1987). For vesicular stomatitis virus (VSV), a rhabdovirus, the necessity for continuous protein synthesis in order to sustain RNA replication is due solely to a requirement for VSV N protein (Patton *et al.*, 1984). However, for MHV-infected cells it was shown that, even in the presence of cycloheximide, substantial pools of soluble free N protein were available for the encapsidation of nascent RNA (Perlman *et al.*, 1986/87). Thus, synthesis of N may not be rate-limiting for coronavirus RNA synthesis. Given the complexity and enormity of the putative coronavirus polymerase and the uniqueness of the mechanism of coronavirus RNA synthesis, much needs to be done to clarify how N protein participates in this process.

## IX. ANTIGENIC PROPERTIES

The nature of antigenic relatedness between the nucleocapsid proteins of different coronaviruses was first investigated using polyclonal sera raised against complete or subviral particles. The presence of common antigenic determinants on the N protein as well as on the other structural polypeptides of closely related viruses such as TGEV, FIPV, and CCV has been demonstrated. For instance, anti-TGEV sera have been shown to react with FIPV N and CCV N to about the same extent as with TGEV N in immunoblotting or enzyme-linked immunosorbent assay (ELISA) tests (Horzinek *et al.*, 1982; Have *et al.*, 1992). Similarly, antibodies directed against BCV N protein recognize HCV-OC43 N and MHV-A59 N (Hogue *et al.*, 1984). However, it would appear that cross-reactivity at the level of the N protein is not restricted to members belonging to the same antigenic cluster. Thus, HCV-229E RNP could be detected by immunodiffusion using antisera to MHV, TGEV, or HEV (Yassen and Johnson-Lussenburg, 1978). An antigenic relationship was also observed between HCV-229E and MHV-3 using heterologous anti-RNP antisera but not antisera against virion external components (Hasony and Macnaughton, 1982). A two-way cross-reaction between the N proteins of FIPV and PEDV has been evidenced using both blotting and immunoprecipitation tests, whereas the viral envelope proteins exhibited no heterologous reactivity (Yaling *et al.*, 1988). These authors also obtained preliminary evidence of a cross-reactivity between the N proteins of IBV and MHV and of FIPV and HEV. Finally, cross-reactivity has been reported for the N proteins of TCV and IBV (Dea *et al.*, 1990) and of TGEV and PEDV (Have *et al.*, 1992). These observations suggest that a few common determinants may exist on the nucleocapsid antigen that are suitable for the identification of many or all members of the family.

The N protein appears to be a substantially conserved antigen according to studies examining the reactivity of monoclonal antibodies (MAbs) toward different strains of the same virus. The N protein of MHV exhibits a moderate degree of antigenic variation; several determinants common to up to 11 strains or isolates have been identified (Fleming *et al.*, 1983; Talbot and Buchmeier, 1985). In both of these studies, the antigenic polymorphism of N protein was found to be less extensive than that of S protein. The first authors divided the MHV strains into two groups according to the relative binding of MAbs to N and proposed that such variation was correlated with their pathogenesis, whereas the second found that antigenic variation of S protein was a better correlate. The IBV N protein may also be subjected to antigenic variation since no fully conserved epitope could be detected within a panel of Australian field strains (Ignjatovic and McWaters, 1991). By contrast, N protein from different strains of TGEV exhibited little or no significant qualitative or quantitative antigenic divergence (Laude *et al.*, 1986; Welch and Saif, 1988; Sanchez *et al.*, 1990). Similarly, FIPV N epitopes were generally highly conserved, although a few type-specific epitopes were discriminated (Fiscus and Teramoto, 1987). Thus, the interstrain relatedness is closer for TGEV and FIPV than in the case of MHV, as is also reflected by a lesser variation of S antigen. Finally, a proportion of common epitopes have been identified in viruses showing a close antigenic relationship, like TGEV and FIPV or BCV and TCV (Fiscus and Teramoto, 1987; Sanchez *et al.*, 1990; Dea *et al.*, 1990).

The identification of B cell epitopes as well as their localization on the N primary structure is less documented than for the S protein. Epitope mapping by competitive assays using a small panel of anti-MHV N or anti-TGEV N MAbs delineated at least two nonoverlapping determinants (Talbot *et al.*, 1985). A translation product obtained from an intracellular DI RNA of MHV-JHM and encoding the 89 C-terminal amino acids of N protein was found to be immunoprecipitated by an anti-N MAb (Makino *et al.*, 1988). Four subfragments altogether covering the whole MHV-JHM N protein sequence were individually able to elicit an antibody response and reacted with sera from diseased Lewis rats, thus indicating that B cell epitopes are distributed throughout the entire length of the molecule (Wege *et al.*, 1993). Western blotting immunoscreening of a series of bacterial fusion products has allowed a more detailed antigenic analysis of the TGEV N protein (Martin Alonso *et al.*, 1992). Seven of eleven MAbs recognized the amino-terminal half of the polypeptide chain. Interestingly, one epitope was localized within a 60 amino acid stretch that essentially overlaps the most conserved region of the coronavirus N protein (see Fig. 1). A second antigenic domain was delineated within the carboxy-terminal half, to which the heterogeneity reported for the N protein of PRCV relative to TGEV appears to be restricted. Additionally, an epitope 11 amino acids long has been localized very near the carboxy-terminus; this well-defined epitope has been used as a portable marker allowing detection of various fusion proteins (Parra *et al.*, 1989). Taken together, the above observations suggest that a major epitope situated in the carboxy end might be a common feature of the coronavirus N proteins. Detergent-resistant epitopes have also been detected but not mapped in the case of MHV-JHM, BCV, and IBV N proteins (Talbot *et al.*, 1984; Deregt and Babiuk, 1987; Ignjatovic and McWaters, 1991).

Due to its internal position in the virus particle, the N protein might not be expected to bear neutralization-mediating determinants. Indeed, none of the coronavirus anti-N MAbs isolated thus far exhibited significant neutralizing activity. On the other hand, no B cell epitopes externally exposed at the cell membrane could be demonstrated in the case of MHV-JHM and TGEV N proteins (Collins *et al.*, 1982; Laude *et al.*, 1986; Laviada *et al.*, 1990; Pulford and Britton, 1990). It has been claimed that one MAb specific for MHV-3 N protein reacted with the surface of infected cultured cells and neutralized viral cytopathic effect in the presence of complement, suggesting that at least one epitope of N protein is accessible to antibodies on the cell membrane (Lecomte *et al.*, 1987). Of possible relevance to this point is the apparent protective activity conferred by certain anti-N MAbs when transferred to mice subsequently lethally infected by MHV-2 or MHV-3 (Nakanaga *et al.*, 1986; Lecomte *et al.*,

1987). This paradoxical observation parallels the well-documented protection by anti-N Ab to rabies (Lodmell *et al.*, 1993).

Attention has been drawn only recently to the T-cell determinants potentially expressed on coronavirus antigens. Two MHC class II (I-$E^d$)-restricted murine T-cell hybridomas generated after immunization with IBV have been shown to be responsive to the N protein (Boots *et al.*, 1991a). This response was strain-specific, thus confirming an antigenic variation of the protein among the different serotypes. Both the antigenic determinants were mapped within the region spanning amino acids 71 to 78 by using recombinant expression products and synthetic peptides. Furthermore, the epitope was shown to prime cellular immune response to IBV in the chicken (Boots *et al.*, 1991b). Studies on MHV-JHM-infected Lewis rats allowed the demonstration of an early and strong T-helper cell response specific for the N protein. N-specific $CD4^+$ T-cell lines were established and shown to confer protection against acute encephalitis upon adoptive transfer to otherwise lethally infected animals (Körner *et al.*, 1991). The carboxy-terminally located, 95-residue-long fragment of N expressed as a bacterial fusion protein was shown to induce the most pronounced response and to mediate protection (Wege *et al.*, 1993). Finally, the MHV-JHM strain has been reported to elicit an immunodominant anti-N protein cytotoxic T lymphocyte (CTL) response in BALB/c mice (Stohlmann *et al.*, 1992). Indeed, among 21 $CD8^+$ T-cell lines derived from animals undergoing an acute demyelinating encephalitis, 17 were found to recognize the N protein as a target (Stohlman *et al.*, 1993). By using a combination of truncated forms of N expressed by vaccinia virus recombinants and a series of overlapping peptides, the response has been mapped to an epitope comprising residues 316 to 330, which contains a described $L^d$ binding motif (Bergmann *et al.*, 1993). Despite a natural sequence variation affecting two residues of the motif, a cross-reactivity of the JHM N-specific CTL with six other MHV strains was observed. Altogether, the above studies lend support to the view that the N protein may be an important antigen with respect to both helper and cytotoxic immune response.

## X. CONCLUSION

The N proteins have occupied a more peripheral position thus far in the study of coronaviruses. A substantial fraction of the papers cited in this chapter have been concerned with N as an entity tangential to the principal focus of the study in hand. Thus, our information about this protein is incomplete, and the answers to many important questions remain obscure. An understanding of the organization of the coronaviral nucleocapsid is of utmost importance for a complete elucidation of how the virus expresses its genetic information once it reaches the cytoplasm. The persistent association of N protein with the genome RNA, and possibly with the subgenomic RNAs, suggests that exploring the roles of N in coronavirus transcription and translation will be a fertile area of research for years to come. Techniques recently developed have taken advantage of the high rate of RNA recombination that occurs during coronavirus replication in order to engineer site-specific N gene mutations into the genome of MHV. It is expected that this approach will greatly facilitate the elucidation of N protein structure function relationships in the future. It should also be possible to extend this methodology to the study of other genes and other members of the coronavirus family.

## XI. REFERENCES


Anderson, R., and Wong, F., 1993, Membrane and phospholipid binding by murine coronaviral nucleocapsid N protein, *Virology* **194**:224.

Anderson, R., Cheley, S., and Haworth-Hatherell, E., 1979, Comparison of polypeptides of two strains of murine hepatitis virus, *Virology* **97**:492.

Apostolov, K., Flewett, T. H., and Kendal, A. P., 1970, Morphology of influenza A, B, C and infectious bronchitis virus (IBV) virions and their replication, in: *The Biology of Large RNA viruses* (R. D. Barry and B. W. J. Mahy, eds.), pp. 3–26, Academic Press, London.

Armstrong, J., Smeekens, S., and Rottier, P., 1983, Sequence of the nucleocapsid gene from murine coronavirus MHV-A59, *Nucl. Acids Res.* **11**:883.

Armstrong, J., Smeekens, S., Spaan, W., Rottier, P., and Van der Zeijst, B., 1984, Cloning and sequencing the nucleocapsid and E1 genes of coronavirus, *Adv. Exp. Med. Biol.* **173**:155.

Baric, R. S., Nelson, G. W., Fleming, J. O., Deans, R. J., Keck, J. G., Casteel, N., and Stohlman, S. A., 1988, Interactions between coronavirus nucleocapsid protein and viral RNAs: Implications for viral transcription, *J. Virol.* **62**:4280.

Bergmann, C., McMillan, M., and Stohlman, S., 1993, Characterization of the $L^d$-restricted cytotoxic T-lymphocyte epitope in the mouse hepatitis virus nucleocapsid protein, *J. Virol.* **67**:7041.

Beushausen, S., Narindrasorasak, S., Sanwal, B. D., and Dales, S., 1987, *In vivo* and *in vitro* models of demyelinating disease: Activation of the adenylate cyclase system influences JHM virus expression in explanted rat oligodendrocytes, *J. Virol.* **61**:3795.

Bond, C. W., Leibowitz, J. L., and Robb, J. A., 1979, Pathogenic murine coronaviruses, II. Characterization of virus-specific proteins of murine coronaviruses JHMV and A59V, *Virology* **94**:371.

Bond, C. W., Anderson, K., and Leibowitz, J. L., 1984, Protein synthesis in cells infected by murine hepatitis viruses JHM and A59: Tryptic peptide analysis, *Arch. Virol.* **80**:333.

Boots, A. M. H., Kusters, J. G., Van Noort, J. M., Zwaagstra, K. A., Rijke, E., Van der Zeijst, B. A. M., and Hensen, E. J., 1991a, Localization of a T-cell epitope within the nucleocapsid protein of avian coronavirus, *Immunology* **74**:8.

Boots, A. M. H., Van Lierop, M. J., Kusters, J. G., Van Kooten, P. J. S., Van der Zeijst, B. A. M., and Hensen, E. J., 1991b, MHC class II-restricted T-cell hybridomas recognizing the nucleocapsid protein of avian coronavirus IBV, *Immunology* **72**:10.

Boursnell, M. E. G., Binns, M. M., Foulds, I. J., and Brown, T. D. K., 1985, Sequences of the nucleocapsid genes from two strains of avian infectious bronchitis virus, *J. Gen. Virol.* **66**:573.

Brayton, P. R., Ganges, R. G., and Stohlman, S. A., 1981, Host cell nuclear function and murine hepatitis virus replication, *J. Gen. Virol.* **56**:457.

Bridgen, A., Duarte, M., Tobler, K., Laude, H., and Ackermann, M., 1993, Sequence determination of the nucleocapsid protein gene of the porcine epidemic diarrhoea virus confirms that this virus is a coronavirus related to human coronavirus 229E and porcine transmissible gastroenteritis virus, *J. of Gen. Virol.* **74**:1795.

Britton, P., Carmenes, R. S., Page, K. W., Garwes, D. J., and Parra, F., 1988, Sequence of the nucleoprotein gene from a virulent British field isolate of transmissible gastroenteritis virus and its expression in *Saccharomyces cerevisiae*, *Molec. Microbiol.* **2**:89.

Bukrinskaya, A. G., 1973, Nucleocapsids of large RNA viruses as functionally active units in transcription, *Adv. Virus Res.* **18**:195.

Callebaut, P. E., and Pensaert, M. B., 1980, Characterization and isolation of structural polypeptides in haemagglutinating encephalomyelitis virus, *J. Gen. Virol.* **48**:193.

Caul, E. O., Ashley, C. R., Ferguson, M., and Egglestone, S. I., 1979, Preliminary studies on the isolation of coronavirus 229E nucleocapsids, *FEMS Microbiol. Lett.* **5**:101.

Cavanagh, D., 1981, Structural polypeptides of coronavirus IBV, *J. Gen. Virol.* **53**:93.

Cheley, S., and Anderson, R., 1981, Cellular synthesis and modification of murine hepatitis virus polypeptides, *J. Gen. Virol.* **54**:301.

Cheley, S., Morris, V. L., Cupples, M. J., and Anderson, R., 1981, RNA and polypeptide homology among murine coronaviruses, *Virology* **115**:310.

Collins, A. R., Knobler, R. L., Powell, H., and Buchmeier, M. J., 1982, Monoclonal antibodies to murine hepatitis virus-4 (strain JHM) define the viral glycoprotein responsible for attachment and cell–cell fusion, *Virology* **119**:358.

Compton, S. R., Rogers, D. B., Holmes, K. V., Fertsch, D., Remenick, J., and McGowan, J. J., 1987, *In vitro* replication of mouse hepatitis virus strain A59, *J. Virol.* **61**:1814.

Coulter-Mackie, M., Adler, R., Wilson, G., and Dales, S., 1985, *In vivo* and *in vitro* models of demyelinating diseases; XII. Persistence and expression of corona JHM virus functions in RN2-2 Schwannoma cells during latency, *Virus Res.* **3**:245.

Cruciere, C., and Laporte, J., 1988, Sequence analysis of bovine enteric coronavirus (F15) genome. 1. Sequence of the gene coding for the nucleocapsid protein; analysis of the predicted protein. *Ann. Inst. Past.* **139**:123.

Davies, H. A., Dourmashkin, R. R., and Macnaughton, M. R., 1981, Ribonucleoprotein of avian infectious bronchitis virus, *J. Gen. Virol.* **53**:67.

Dea, S., and Tijssen, P., 1989, Antigenic and polypeptide structure of turkey enteric coronaviruses as defined by monoclonal antibodies, *J. Gen. Virol.* **70**:1725.

Dea, S., Verbeek, A. J., and Tijssen, P., 1990, Antigenic and genomic relationships among turkey and bovine enteric coronaviruses, *J. Virol.* **64**:3112.

Decimo, D., Philippe, H., Hadchouel, M., Tardieu, M., and Meunier-Rotival, M., 1993, The gene encoding the nucleocapsid protein: Sequence analysis in murine hepatitis virus type 3 and evolution in *Coronaviridae, Arch. Virol.* **130**:279.

De Groot, R. J., Ter Haar, R., Horzinek, M. C., and Van der Zeijst, B. A. M., 1987, Intracellular RNAs of the feline infectious peritonitis coronavirus strain 79-1146, *J. Gen. Virol.* **68**:995.

Deregt, D., and Babiuk, L. A., 1987, Monoclonal antibodies to bovine coronavirus: Characteristics and topographical mapping of neutralizing epitopes on the E2 and E3 glycoproteins, *Virology* **161**:410.

Deregt, D., Sabara, M., and Babiuk, L. A., 1987, Structural proteins of bovine coronavirus and their intracellular processing, *J. Gen. Virol.* **68**:2863.

Dreyfuss, G., Swanson, M. S., and Pinol-Roma, S., 1988, Heterogeneous nuclear ribonucleoprotein particles and the pathway of mRNA formation, *Trends Biochem. Sci.* **13**:86.

Fiscus, S. A., and Teramoto, Y. A., 1987, Antigenic comparison of feline coronavirus isolates: Evidence for markedly different peplomer glycoproteins, *J. Virol.* **61**:2607.

Fleming, J. O., Stohlman, S. A., Harmon, R. C., Lai, M. M. C., Frelinger, J. A., and Weiner, L. P., 1983, Antigenic relationships of murine coronaviruses: Analysis using monoclonal antibodies to JHM (MHV-4) virus, *Virology* **131**:296.

Garwes, D. J., Pocock, D. H., and Pike, B. V., 1976, Isolation of subviral components from transmissible gastroenteritis virus, *J. Gen. Virol.* **32**:283.

Garwes, D. J., Bountiff, L., Millson, G. C., and Elleman, C. J., 1984, Defective replication of porcine transmissible gastroenteritis virus in a continuous cell line, *Adv. Exp. Med. Biol.* **178**:79.

Hasony, H. J., and Macnaughton, M. R., 1982, Serological relationships of the subcomponents of human coronavirus strain 229E and mouse hepatitis virus strain 3, *J. Gen. Virol.* **58**:449.

Have, P., Moving, V., Svansson, V., Uttenthal, A., and Bloch, B., 1992, Coronavirus infection in mink (*Mustela vison*). Serological evidence of infection with a coronavirus related to transmissible gastroenteritis virus and porcine epidemic diarrhea virus, *Vet. Microbiol.* **31**:1.

Hogue, B. G., King, B., and Brian, D. A., 1984, Antigenic relationships among proteins of bovine coronavirus, human respiratory coronavirus OC43, and mouse hepatitis coronavirus A59, *J. Virol.* **51**:384.

Holmes, K. V., and Behnke, J. N., 1981, Evolution of a coronavirus during persistent infection *in vitro, Adv. Exp. Med. Biol.* **142**:287.

Holmes, K. V., Doller, E. W., and Behnke, J. N., 1981, Analysis of the functions of coronavirus glycoproteins by differential synthesis with tunicamycin, *Adv. Exp. Med. Biol.* **142**:133.

Holmes, K. V., Boyle, J. F., Williams, R. K., Stephensen, C. B., Robbins, S. G., Bauer, E. C., Duchala, C. S., Frana, M. F., Weismiller, D. G., Compton, S., McGowan, J. J., and Sturman, L. S., 1987,

Processing of coronavirus proteins and assembly of virions, in: *Positive Strand RNA Viruses* (M. A. Brinton and R. R. Rueckert, eds.), pp. 339–349, Alan R. Liss, New York.

Horsburgh, B. C., Brierley, I., and Brown, T. D. K., 1992, Analysis of a 9.6 kb sequence from the 3′ end of canine coronavirus genomic RNA, *J. Gen. Virol.* **73:**2849.

Horzinek, M. C., Lutz, H., and Pedersen, N. C., 1982, Antigenic relationships among homologous structural polypeptides of porcine, feline, and canine coronaviruses, *Infect. Immun.* **37:**1148

Horzinek, M. C., Ederveen, J., and Weiss, M., 1985, The nucleocapsid of Berne virus, *J. Gen. Virol.* **66:**1287.

Horzinek, M. C., Ederveen, J., Egberink, H., Jacobse-Geels, H. E. L., Niewold, T., and Prins, J., 1986, Virion polypeptide specificity of immune complexes and antibodies in cats inoculated with feline infectious peritonitis virus, *Am. J. Vet. Res.* **47:**754.

Huddleston, J. A., and Brownlee, G. G., 1982, The sequence of the nucleoprotein gene of human influenza A virus, strain A/NT/60/68, *Nucl. Acids Res.* **10:**1029.

Ignjatovic, J., and McWaters, P. G., 1991, Monoclonal antibodies to three structural proteins of avian infectious bronchitis virus: Characterization of epitopes and antigenic differentiation of Australian strains, *J. Gen. Virol.* **72:**2915.

Jacobs, L., Van der Zeijst, B. A. M., and Horzinek, M. C., 1986, Characterization and translation of transmissible gastroenteritis virus mRNAs, *J. Virol.* **57:**1010.

Kamahora, T., Soe, L. H., and Lai, M. M. C., 1989, Sequence analysis of nucleocapsid gene and leader RNA of human coronavirus OC43, *Virus Res.* **12:**1.

Kapke, P. A., and Brian, D. A., 1986, Sequence analysis of the porcine transmissible gastroenteritis coronavirus nucleocapsid protein gene, *Virology* **151:**41.

Keck, J. G., Hogue, B. G., Brian, D. A., and Lai, M. M. C., 1988, Temporal regulation of bovine coronavirus RNA synthesis, *Virus Res.* **9:**343.

Kennedy, D. A., and Johnson-Lussenburg, C. M., 1975/76, Isolation and morphology of the internal component of human coronavirus, strain 229E, *Intervirology* **6:**197.

King, B., and Brian, D. A., 1982, Bovine coronavirus structural proteins, *J. Virol.* **42:**700.

Kingsbury, D. W., Jones, I. M., and Murti, K. G., 1987, Assembly of influenza ribonucleoprotein *in vitro* using recombinant nucleoprotein, *Virology* **156:**396.

Kistner, O., Muller, K., and Scholtissek, C., 1989, Differential phosphorylation of influenza A viruses, *J. Gen. Virol.* **70:**2421.

Klug, A., and Rhodes, D., 1987, "Zinc fingers": A novel protein motif for nucleic acid recognition, *Trends Biochem. Sci.* **12:**464.

Koetzner, C. A., Parker, M. M., Ricard, C. S., Sturman, L. S., and Masters, P. S., 1992, Repair and mutagenesis of the genome of a deletion mutant of the coronavirus mouse hepatitis virus by targeted RNA recombination, *J. Virol.* **66:**1841.

Koolen, M. J. M., Osterhaus, A. D. M. E., Van Steenis, G., Horzinek, M. C., and Van der Zeijst, B. A. M., 1983, Temperature-sensitive mutants of mouse hepatitis virus strain A59: Isolation, characterization, and neuropathogenic properties, *Virology* **125:**393.

Körner, H., Schliephake, A., Winter, J., Zimprich, F., Lassmann, H., Sedgwick, J., Siddell, S., and Wege, H., 1991, Nucleocapsid or spike protein-specific $CD4^+$ T lymphocytes protect against coronavirus-induced encephalomyelitis in the absence of $CD8^+$ T cells. *J. Immunol.* **147:**2317.

Kunita, S., Terada, E., Goto, K., and Kagiyama, N., 1992, Sequence analysis and molecular detection of mouse hepatitis virus using the polymerase chain reaction, *Lab. Anim. Sci.* **42:**593.

Kunita, S., Mori, M., and Terada, E., 1993, Sequence analysis of the nucleocapsid protein gene of rat coronavirus SDAV-681, *Virology* **193:**520.

Lancer, J. A., and Howard, C. R., 1980, The disruption of infectious bronchitis virus (IBV-41 strain) with triton X-100 detergent, *J. Virol. Meth.* **1:**121.

Lapps, W., Hogue, B. G., and Brian, D. A., 1987, Sequence analysis of the bovine coronavirus nucleocapsid and matrix protein genes, *Virology* **157:**47.

Laude, H., Chapsal, J. M., Gelfi, J., Labiau, S., and Grosclaude, J., 1986, Antigenic structure of transmissible gastroenteritis virus. I. Properties of monoclonal antibodies directed against virion proteins, *J. Gen. Virol.* **67:**119.

Laviada, M. D., Videgain, S. P., Moreno, L., Alonso, F., Enjuanes, L., and Escribano, J. M., 1990, Expression of swine transmissible gastroenteritis virus envelope antigens on the surface of infected cells: Epitopes externally exposed, *Virus Res.* **16:**247.

Lecomte, J., Cainelli-Gebara, V., Mercier, G., Mansour, S., Talbot, P. J., Lussier, G., and Oth, D., 1987, Protection from mouse hepatitis virus type 3-induced acute disease by an anti-nucleoprotein monoclonal antibody, *Arch. Virol.* **97**:123.

Leibowitz, J. L., Weiss, S. R., Paavola, E., and Bond, C. W., 1982, Cell-free translation of murine coronavirus RNA, *J. Virol.* **43**:905.

Leppart, M., Rittenhouse, L., Perrault, J., Summers, D. F., and Kolakofsky, D., 1979, Plus and minus strand leader RNAs in negative strand virus-infected cells, *Cell* **18**:735.

Lodmell, D. L., Esposito, J. J., and Ewalt, L. C., 1993, Rabies virus antinucleoprotein antibody protects against rabies virus challenge *in vivo* and inhibits rabies virus replication *in vitro*, *J. Virol.* **67**:6080.

Lomniczi, B., and Morser, J., 1981, Polypeptides of infectious bronchitis virus. I. polypeptides of the virion, *J. Gen. Virol.* **55**:155.

Macnaughton, M. R., 1980, The polypeptides of human and mouse coronaviruses, *Arch. Virol.* **63**:75.

Macnaughton, M. R., Davies, H. A., and Nermut, M. V., 1978, Ribonucleoprotein-like structures from coronavirus particles, *J. Gen. Virol.* **39**:545.

Makino, S., Fujioka, N., and Fujiwara, K., 1985, Structure of the intracellular defective viral RNAs of defective interfering particles of mouse hepatitis virus, *J. Virol.* **54**:329.

Makino, S., Shieh, C. K., Soe, L. H., Baker, S. C., and Lai, M. M. C., 1988, Primary structure and translation of a defective interfering RNA of murine coronavirus, *Virology* **166**:550.

Martin Alonso, J. M., Balbin, M., Garwes, D. J., Enjuanes, L., Gascon, S., and Parra, F., 1992, Antigenic structure of transmissible gastroenteritis virus nucleoprotein, *Virology* **188**:168.

Masters, P. S., 1992, Localization of an RNA-binding domain in the nucleocapsid protein of the coronavirus mouse hepatitis virus, *Arch. Virol.* **125**:141.

Masters, P. S., Koetzner, C. A., Kerr, C. A., and Heo, Y., 1994, Optimization of targeted recombination and mapping of a novel nucleocapsid gene mutation in the coronavirus mouse hepatitis virus, *J. Virol.* **68**:328.

Mohandas, D. V., and Dales, S., 1991, Endosomal association of a protein phosphatase with high dephosphorylating activity against a coronavirus nucleocapsid protein, *FEBS Lett.* **282**:419.

Mountcastle, W. E., Compans, R. W., Lackland, H., and Choppin, P. W., 1974, Proteolytic cleavage of subunits of the nucleocapsid of the paramyxovirus simian virus 5, *J. Virol.* **14**:1253.

Myint, S., Harmsen, D., Raabe, T., and Siddell, S. G., 1990, Characterization of a nucleic acid probe for the diagnosis of human coronavirus 229E infections, *J. Med. Virol.* **31**:165.

Nakanaga, K., Yamanouchi, K., and Fujiwara, K., 1986, Protective effect of monoclonal antibodies on lethal mouse hepatitis virus infection in mice, *J. Virol.* **59**:168.

Nelson, G., and Stohlman, S. A., 1993, Localization of the RNA-binding domain ou mouse hepatitis virus nucleocapsid protein, *J. Gen. Virol.* **74**:1975.

Niemann, H., Boschek, B., Evans, D., Rosing, M., Tamura, T., and Klenk, H. D., 1982, Post-translational glycosylation of coronavirus glycoprotein E1: Inhibition by monensin, *EMBO J.* **12**:1499.

Obert, G., Grollemund, E., Nonnenmacher, H., and Kirn, A., 1981, Analysis and localisation of mouse hepatitis virus 3 (MHV3) polypeptides, *Ann. Virol. (Inst. Pasteur)* **132E**:109.

Ollis, D. L., and White, S. W., 1987, Structural basis of protein–nucleic acid interactions, *Chem. Rev.* **87**:981.

Parker, M. M., and Masters, P. S., 1990, Sequence comparison of the n genes of five strains of the coronavirus mouse hepatitis virus suggests a three domain structure for the nucleocapsid protein, *Virology* **179**:463.

Parra, F., Balbin, M, Carmenes, R. S., Alonso, J. M. M., and Gascon, S., 1989, Production of glycopolypeptide antigens of porcine transmissible gastroenteritis virus (TGEV), in: *Animal Cell Technology, Biotechnology Action Program Report* (D. De Nettancourt, H. Bazin, and A. Klepsch, eds.), pp. 24–25, The Commission of European Community,

Patton, J. J., Davis, N. L., and Wertz, G. W., 1984, N protein alone satisfies the requirement for protein synthesis during RNA replication of vesicular stomatitis virus, *J. Virol.* **49**:303.

Perlman, S., Ries, D., Bolger, E., Chang, L. J., and Stoltzfus, C. M., 1986/87, MHV nucleocapsid synthesis in the presence of cycloheximide and accumulation of negative strand MHV RNA, *Virus Res.* **6**:261.

Pocock, D. H., and Garwes, D. J., 1977, The polypeptides of haemagglutinating encephalomyelitis virus and isolated subviral particles, *J. Gen. Virol.* **37**:487.

Pulford, D. J., and Britton, P., 1990, Expression and cellular localization of porcine transmissible gastroenteritis virus N and M proteins by recombinant vaccinia viruses, *Virus Research* **18**:203.

Rasschaert, D., Gelfi, J., and Laude, H., 1987, Enteric coronavirus TGEV: Partial sequence of the genomic RNA, its organization and expression, *Biochimie* **69**:591.

Robb, J. A., and Bond, C. W., 1979, Pathogenic murine coronaviruses. I. Characterization of biological behavior *in vitro* and virus-specific intracellular RNA of strongly neurotropic JHMV and weakly neurotropic A59V viruses, *Virology* **94**:352.

Robbins, S. G., Frana, M. F., McGowan, J. J., Boyle, J. F., and Holmes, K. V., 1986, RNA-binding proteins of coronavirus MHV: Detection of monomeric and multimeric N protein with an RNA overlay-protein blot assay, *Virology* **150**:402.

Rottier, P. J. M., Spaan, W. J. M., Horzinek, M. C., and Van der Zeijst, B. A. M., 1981a, Translation of three mouse hepatitis virus strain A59 subgenomic RNAs in Xenopus laevis oocytes, *J. Virol.* **38**:20.

Rottier, P. J. M., Horzinek, M. C., and Van der Zeijst, B. A. M., 1981b, Viral protein synthesis in mouse hepatitis virus strain A59-infected cells: Effect of tunicamycin, *J. Virol.* **40**:350.

Sanchez, C. M., Jimenez, G., Laviada, M. D., Correa, I., Sune, C., Bullido, M. J., Gebauer, F., Smerdou, C., Callebaut, P., Escribano, J. M., and Enjuanes, L., 1990, Antigenic homology among coronaviruses related to transmissible gastroenteritis virus, *Virology* **174**:410.

Sawicki, S. G., and Sawicki, D. L., 1986, Coronavirus minus-strand RNA synthesis and effect of cycloheximide on coronavirus RNA synthesis, *J. Virol.* **57**:328.

Schmidt, O. W., and Kenny, G. E., 1982, Polypeptides and functions of antigens from human coronaviruses 229E and OC43, *Infect. Immun.* **35**:515.

Schreiber, S. S., Kamahora, T., and Lai, M. M. C., 1989, Sequence analysis of the nucleocapsid protein gene of human coronavirus 229E, *Virology* **169**:142.

Siddell, S. G., Wege, H., Barthel, A., and Ter Meulen, V., 1980, Coronavirus JHM: Cell-free synthesis of structural protein p60, *J. Virol.* **33**:10.

Siddell, S. G., Barthel, A., and Ter Meulen, V., 1981a, Coronavirus JHM: A virion-associated protein kinase, *J. Gen. Virol.* **52**:235.

Siddell, S., Wege, H., Barthel, A., and Ter Meulen, V., 1981b, Coronavirus JHM: Intracellular protein synthesis, *J. Gen. Virol.* **53**:145.

Siddell, S., Wege, H., and Ter Meulen, V., 1982, The structure and replication of coronavirus, *Curr. Topics Microbiol. Immunol.* **99**:131.

Simkins, R. A., Saif, L. J., and Weilnau, P. A., 1989, Epitope mapping and the detection of transmissible gastroenteritis viral proteins in cell culture using biotinylated monoclonal antibodies in a fixed-cell ELISA, *Arch. Virol.* **107**:179.

Skinner, M. A., and Siddell, S. G., 1983, Coronavirus JHM: Nucleotide sequence of the mRNA that encodes the nucleocapsid protein, *Nucl. Acids Res.* **11**:5045.

Skinner, M., and Siddell, S., 1984, Nucleotide sequencing of mouse hepatitis virus strain JHM messenger RNA7, *Adv. Exp. Med. Biol.* **173**:163.

Spaan, W. J. M., Rottier, P. J. M., Horzinek, M. C., and Van der Zeijst, B. A. M., 1981, Isolation and identification of virus-specific mRNAs in cells infected with mouse hepatitis virus (MHV-A59), *Virology* **108**:424.

Stern, D., Burgess, L., Linesh, S., Kennedy, I., 1981, The avian coronavirus multiplication strategy, *Adv. Exp. Biol. Med.* **142**:185.

Stohlman, S. A., and Lai, M. M. C., 1979, Phosphoproteins of murine hepatitis viruses, *J. Virol.* **32**:672.

Stohlman, S. A., Fleming, J. O., Patton, C. D., and Lai, M. M. C., 1983, Synthesis and subcellular localization of the murine coronavirus nucleocapsid protein, *Virology* **130**:527.

Stohlman, S. A., Baric, R. S., Nelson, G. N., Soe, L. H., Welter, L. M., and Deans, R. J., 1988, Specific interaction between coronavirus leader RNA and nucleocapsid protein, *J. Virol.* **62**:4288.

Stohlman, S. A., Kyuwa, S., Cohen, M., Bergmann, C., Polo, J. M., Yeh, J., Anthony, R., and Keck, J. G., 1992, Mouse hepatitis virus nucleocapsid protein-specific cytotoxic T lymphocytes are $L^d$ restricted and specific for the carboxy terminus, *Virology* **189**:217.

Stohlman, S. A., Kyuwa, S., Polo, J. M., Brady, D., Lai, M. M. C., and Bergmann, C., 1993, Characterization of mouse hepatitis virus-specific cytotoxic T cells derived from the central nervous system of mice infected with the JHM strain, *J. Virol.* **67**:7050.

Sturman, L. S., 1977, Characterization of a coronavirus. I. Structural proteins: Effects of preparative conditions on the migration of protein in polyacrylamide gels, *Virology* **77**:637.

Sturman, L. S., Holmes, K. V., and Behnke, J., 1980, Isolation of coronavirus envelope glycoproteins and interaction with the viral nucleocapsid, *J. Virol.* **33**:449.

Sutou, S., Sato, S., Okabe, T., Nakai, M., and Sasaki, N., 1988, Cloning and sequencing of genes encoding structural proteins of avian infectious bronchitis virus, *Virology* **165**:589.

Talbot, P. J., and Buchmeier, M. J., 1985, Antigenic variation among murine coronaviruses: Evidence for polymorphism on the peplomer glycoprotein, E2, *Virus Res.* **2**:317.

Talbot, P. J., Knobler, R. L., and Buchmeier, M. J., 1984, Western and dot immunoblotting analysis of viral antigens and antibodies: Application to murine hepatitis virus, *J. Immunol. Methods* **73**:177.

Talbot, P. J., Salmi, A. A., Knobler, R. L., and Buchmeier, M. J., 1985, Epitope-specific antibody response to murine hepatitis virus-4 (strain JHM), *J. Immunol.* **134**:1217.

Tan, K. B., 1975, Comparative study of the protein kinase associated with animal viruses, *Virology* **64**:566.

Van der Most, R. G., Heijnen, L., Spaan, W. J. M., and de Groot, R. J., 1992, Homologous RNA recombination allows efficient introduction of site-specific mutations into the genome of coronavirus MHV-A59 via synthetic co-replicating RNAs, *Nucl. Acids Res.* **20**:3375.

Vennema, H., de Groot, R. J., Harbour, D. A., Horzinek, M. C., and Spaan, W. J. M., 1991, Primary structure of the membrane and nucleocapsid protein genes of feline infectious peritonitis virus and immunogenicity of recombinant vaccinia viruses in kittens, *Virology* **181**:327.

Vennema, H., Rossen, J. W. A., Wesseling, J., Horzinek, M. C., and Rottier, P. J. M., 1992, Genomic organization and expression of the 3′ end of the canine and feline enteric coronaviruses, *Virology* **191**:134.

Verbeek, A., and Tijssen, P., 1991, Sequence analysis of the turkey enteric coronavirus nucleocapsid and membrane protein genes: A close genomic relationship with bovine coronavirus, *J. Gen. Virol.* **72**:1659.

Wege, H., Wege, H., Nagashima, K., and Ter Meulen, V., 1979, Structural polypeptides of the murine coronavirus JHM, *J. Gen. Virol.* **42**:37.

Wege, H., Schliephake, A., Körner, H., Flory, E., and Wege, H., 1993, An immunodominant CD4+ T cell site on the nucleocapsid protein of murine coronavirus contributes to protection against encephalomyelitis, *J. Gen. Virol.* **74**:1287.

Welch, S. K. W., and Saif, L. J., 1988, Monoclonal antibodies to a virulent strain of transmissible gastroenteritis virus: Comparison of reactivity with virulent and attenuated virus, *Arch. Virol.* **101**:221.

Wesseling, J. G., Godeke, G. J., Schijns, V. E. C. J., Prevec, L., Graham, F. L., Horzinek, M. C., and Rottier, P. J. M., 1993, Mouse hepatitis virus spike and nucleocapsid proteins expressed by adenovirus vectors protect mice against a lethal infection, *J. Gen. Virol.* **74**:2061.

Wilbur, S. M., Nelson, G. W., Lai, M. M. C., McMillan, M., and Stohlman, S. A., 1986, Phosphorylation of the mouse hepatitis virus nucleocapsid protein, *Biochem. Biophys. Res. Commun.* **141**:7.

Wilhelmsen, K. C., Leibowitz, J. L., Bond, C. W., and Robb, J. A., 1981, The replication of murine coronaviruses in enucleated cells, *Virology* **110**:225.

Williams, A. K., Wang, L., Sneed, L. W., and Collisson, E. W., 1992, Comparative analyses of the nucleocapsid genes of several strains of infectious bronchitis virus and other coronaviruses, *Virus Research* **25**:213.

Yaling, Z., Ederveen, J., Egberink, H., Pensaert, M., and Horzinek, M. C., 1988, Porcine epidemic diarrhea virus (CV777) and feline infectious peritonitis virus (FIPV) are antigenically related, *Arch. Virol.* **102**:63.

Yassen, S. A., and Johnson-Lussenburg, C. M., 1978, Comparative antigenic studies on coronaviruses, *Int. Virol.* **4**:451.

CHAPTER 8

# The Coronavirus Hemagglutinin Esterase Glycoprotein

David A. Brian, Brenda G. Hogue, and Thomas E. Kienzle

## I. DISCOVERY OF THE HEMAGGLUTININ ESTERASE PROTEIN

Early reviews on coronavirus structure (Siddell *et al.*, 1983a,b; Sturman and Holmes, 1985) described coronaviruses as having three major structural proteins: a large-surface (or peplomer) glycoprotein of around 200 kDa, a phosphorylated nucleocapsid protein of around 50 kDa, and a glycosylated, multispanning membrane protein of around 30 kDa. This description was based primarily on studies of the prototypic avian infectious bronchitis virus (IBV) and the highly studied mouse hepatitis virus (MHV), strain A59. Although IBV was shown early on to have a weak hemagglutinating property, detection of the hemagglutinating activity required that the virus first be treated with phospholipase C or concentrated by centrifugation in sucrose gradients (Bingham *et al.*, 1975). Not all strains of IBV demonstrated hemagglutination, however, and hemagglutinating activity by IBV, as well as by the porcine transmissible gastroenteritis virus (TGEV) (Noda *et al.*, 1987, 1988) was probably a cryptic property

DAVID A. BRIAN • Department of Microbiology, University of Tennessee, Knoxville, Tennessee 37996-0845. BRENDA G. HOGUE • Department of Microbiology and Immunology, Division of Molecular Virology, Baylor College of Medicine, Houston, Texas 77030. THOMAS E. KIENZLE • John L. McClellan Memorial Veterans Administration Medical Research Service, Little Rock, Arkansas 72205; *present address*: Department of Ophthalmology, Baylor College of Medicine, Houston, Texas 77030.

*The Coronaviridae*, edited by Stuart G. Siddell, Plenum Press, New York, 1995.

of the spike protein (Cavanagh and Davis, 1986). The nature of this hemagglutinating activity is not well understood.

Several coronaviruses, unlike those described above, have been shown to have a strong hemagglutinating property. The activity requires no pretreatment of virus. These are the bovine coronavirus (BCV) (Sharpee *et al.*, 1976), the hemagglutinating encephalomyelitis virus of swine (HEV) (Callebaut and Pensaert, 1980), human coronavirus OC43 (HCV OC43) (and OC38 and other close relatives) (Kaye and Dowdle, 1969), the turkey coronavirus (TCV) (Dea and Tijssen, 1988), and the MHV-related diarrhea virus of infant mice (DVIM) (Sugiyama and Amano, 1980, 1981; Sugiyama *et al.*, 1986). Interestingly, in all cases where it has been examined, these viruses show a second, shorter fringe of peplomers by electronmicroscopy (Bridger *et al.*, 1978; Sugiyama and Amano, 1981).

The first report showing the possibility of a fourth major structural protein for the strongly hemagglutinating coronaviruses was that by Callebaut and Pensaert for HEV (1980). A glycoprotein of 140 kDa was described that reduced to subunits of 76 kDa after treatment with 2-mercaptoethanol. The gradient-purified 140-kDa glycoprotein appeared to have hemagglutinating properties. The character of the fourth major structural protein was more clearly shown by King *et al.* for the bovine coronavirus (King and Brian, 1982; King *et al.*, 1985). The 140-kDa glycoprotein was not removed from BCV by bromelain (whereas the surface protein was) and BCV retained its ability to hemagglutinate mouse erythrocytes. The property of hemagglutination was therefore associated with the 140-kDa glycoprotein. This association was further supported by four other observations. (1) Monospecific, polyclonal rabbit antiserum against the BCV hemagglutinin failed to identify a homologous protein on the nonhemagglutinating MHV A59 virion, an otherwise close antigenic relative of BCV (Hogue *et al.*, 1984). (2) Monoclonal antibodies to the BCV 140-kDa protein inhibited hemagglutination by BCV (in addition to showing neutralizing activity) (Deregt and Babiuk, 1987; Deregt *et al.*, 1989). (3) The presence of the 140-kDa protein on HCV OC43 correlated with hemagglutinating activity during protease digestion experiments (in a pattern similar to that for BCV) (Hogue and Brian, 1986). (4) The dimeric 140-kDa protein has been found on all of the strongly hemagglutinating coronaviruses studied to date.

## II. DEDUCED AMINO ACID SEQUENCE AND PROPERTIES OF THE BOVINE CORONAVIRUS HEMAGGLUTININ ESTERASE PROTEIN

Extensive studies demonstrating the pattern of synthesis, assembly, and glycosylation of the hemagglutinin protein (Deregt *et al.*, 1987; Hogue *et al.*, 1989) were done prior to knowledge of the hemagglutinin gene sequence or knowledge of its acetyl esterase activity. These are discussed here, however, in the context of the deduced amino acid sequence for the protein.

The gene encoding the hemagglutinin esterase (HE) protein maps on the 5′ side of the spike protein gene. This was predicted from the size of a mRNA (species 2b) that was present in BCV, but not in MHV A59 or some strains of

MHV-JHM, and from the 3′ nested-set pattern of coronavirus mRNAs (Abraham *et al.*, 1990; Keck *et al.*, 1988; Siddell, 1983). Interestingly, the (defective) gene encoding the HE protein was first sequenced for the nonhemagglutinating MHV A59 (Luytjes *et al.*, 1988). Examination of the gene sequence revealed three important features. (1) The MHV-A59 genome contains a nonfunctional, apparently vestigial, hemagglutinin gene with no initiation codon for translation of the HE open reading frame (ORF). This protein, as demonstrated by Hogue *et al.* (1984), is not made in MHV-A59-infected cells. (2) There is no UCUAAAC consensus intergenic sequence of the type thought to play a role in the initiation of transcription immediately upstream of the HE ORF. (3) Within the deduced HE amino acid sequence there is an F-G-D-S sequence identical to the acetyl esterase active site on the HE molecule of influenza C (Luytjes *et al.*, 1988).

Acetyl esterase activity, which functions to destroy a virus receptor (discussed in Section VI), was first demonstrated for the BCV hemagglutinin (Vlasak *et al.*, 1988a). The hemagglutinin was therefore renamed the hemagglutinin esterase protein.

Some strains of MHV-JHM express a homologous 65-kDa protein (Makino *et al.*, 1983; Siddell, 1981; Taguchi *et al.*, 1986). However, MHV-JHM and also, for example, MHV-S isolates, hemagglutinate, if at all, only very poorly (Shieh *et al.*, 1989; Sugiyama *et al.*, 1986; Talbot, 1989; Yokomori *et al.*, 1989). The reasons for this are unclear, as the cloned MHV-JHM HE gene, expressed using a vaccinia virus/T7 RNA polymerase system, encodes a protein that dimerizes, has acetyl-esterase activity, and is able to hemadsorb rat erythrocytes (Pfleiderer *et al.*, 1991).

Studies on the synthesis and glycosylation of the BCV HE protein established that it is synthesized as a monomeric apoprotein of approximately 42 kDa that becomes cotranslationally glycosylated and rapidly disulfide-linked (Deregt *et al.*, 1987; Hogue *et al.*, 1989). Only asparagine-linked glycosylation is found, with six to seven sugar chains of either the high mannose or hybrid type and three to four chains of the complex type attached per subunit, as deduced from a final subunit size of 65 kDa. Since coronaviruses assemble by budding into transitional elements occurring between the endoplasmic reticulum and pre-Golgi (Tooze *et al.*, 1988), the maturation of HE carbohydrates to complex forms must take place on the virion during migration through the exocytic pathway to the cell surface.

The deduced amino acid sequence of the HE protein, as determined from cDNA clones of BCV genomic RNA (Fig. 1), shows it to be a 47,700-Da apoprotein of 424 amino acids (Kienzle *et al.*, 1990; Parker *et al.*, 1989; Zhang *et al.*, 1991). There are two very hydrophobic regions. The first encompasses the first 18 amino acids and has the properties of a signal peptide. Amino-terminal amino acid sequencing of the processed virion HE protein identifies a phenylalanine, indicating that the first 18 amino acids indeed function as a signal peptide for membrane translocation (Hogue *et al.*, 1989). The deduced molecular weight of the nonglycosylated protein without its signal peptide is 45 kDa. The second hydrophobic region extends from amino acid 390 through 415 and is known to serve as the membrane anchor for HE (Kienzle *et al.*, 1990; Parker *et al.*, 1990). Therefore, the resulting HE is a type I glycoprotein with an amino-

```
                                                                   60
CTAAACTCAGTGAAAATGTTTTTGCTTCTTAGATTTGTTCTAGTTAGCTGCATAATTGGT
                 M  F  L  L  L  R  F  V  L  V  S  C  I  I  G
                 HE →                                             120
AGCCTAGGTTTTGATAACCCTCCTACCAATGTTGTTTCGCATTTAAATGGAGATTGGTTT
S  L  G  F  D  N  P  P  T  N  V  V  S  H  L  N  G  D  W  F
                                                 M  E  I  G  F
                                            IORF1 →               180
TTATTTGGTGACAGTCGTTCAGATTGTAATCATGTTGTTAATACCAACCCCCGTAATTAT
L  F  G  D  S  R  S  D  C  N  H  V  V  N  T  N  P  R  N  Y
 Y  L  V  T  V  V  Q  I  V  I  M  L  L  I  P  T  P  V  I  I
                                                                  240
TCTTATATGGACCTTAATCCTGCCCTGTGTGATTCTGGTAAAATATCATCTAAAGCTGGC
S  Y  M  D  L  N  P  A  L  C  D  S  G  K  I  S  S  K  A  G
 L  I  W  T  L  I  L  P  C  V  I  L  V  K  Y  H  L  K  L  A
                                                                  300
AACTCCATTTTTAGGAGTTTTCACTTTACCGATTTTTATAATTACACAGGCGAAGGTCAA
N  S  I  F  R  S  F  H  F  T  D  F  Y  N  Y  T  G  E  G  Q
 T  P  F  L  G  V  F  T  L  P  I  F  I  I  T  Q  A  K  V  N
                                                                  360
CAAATTATTTTTTATGAGGGTGTTAATTTTACGCCTTATCATGCCTTTAAATGCACCACT
Q  I  I  F  Y  E  G  V  N  F  T  P  Y  H  A  F  K  C  T  T
 K  L  F  F  M  R  V  L  I  L  R  L  I  M  P  L  N  A  P  L
                                                                  420
TCTGGTAGTAATGATATTTGGATGCAGAATAAAGGCTTGTTTTACACTCAGGTTTATAAG
S  G  S  N  D  I  W  M  Q  N  K  G  L  F  Y  T  Q  V  Y  K
 L  V  V  M  I  F  G  C  R  I  K  A  C  F  T  L  R  F  I  R
                                                                  480
AATATGGCTGTGTATCGCAGCCTTACTTTTGTTAATGTACCATATGTTTATAATGGCTCT
N  M  A  V  Y  R  S  L  T  F  V  N  V  P  Y  V  Y  N  G  S
 I  W  L  C  I  A  A  L  L  L  L  M  Y  H  M  F  I  M  A  L
                                                                  540
GCACAATCTACAGCTCTTTGTAAATCTGGTAGTTTAGTTCTTAATAACCCTGCATATATA
A  Q  S  T  A  L  C  K  S  G  S  L  V  L  N  N  P  A  Y  I
 H  N  L  Q  L  F  V  N  L  V  V  *
                                                                  600
GCTCGTGAAGCTAATTTTGGGGATTATTATTATAAGGTTGAAGCTGACTTTTATTTGTCA
A  R  E  A  N  F  G  D  Y  Y  Y  K  V  E  A  D  F  Y  L  S
                                                                  660
GGTTGTGACGAGTATATCGTACCACTTTGTATTTTTAACGGCAAGTTTTTGTCGAATACA
G  C  D  E  Y  I  V  P  L  C  I  F  N  G  K  F  L  S  N  T
                                                                  720
AAGTATTATGATGATAGTCAATATTATTTTAATAAAGACACTGGTGTTATTTATGGTCTC
K  Y  Y  D  D  S  Q  Y  Y  F  N  K  D  T  G  V  I  Y  G  L
                                                                  780
AATTCTACTGAAACCATTACCACTGGTTTTGATTTTAATTGTCATTATTTAGTTTTACCC
N  S  T  E  T  I  T  T  G  F  D  F  N  C  H  Y  L  V  L  P
                                                                  840
TCTGGTAATTATTTAGCCATTTCAAATGAGCTATTGTTAACTGTTCCTACGAAAGCAATC
S  G  N  Y  L  A  I  S  N  E  L  L  L  T  V  P  T  K  A  I
                                                                  900
TGTCTTAACAAGCGTAAGGATTTTACGCCTGTACAGGTTGTTGATTCACGGTGGAACAAT
C  L  N  K  R  K  D  F  T  P  V  Q  V  V  D  S  R  W  N  N
                                                                  960
GCCAGGCAGTCTGATAACATGACGGCGGTTGCTTGTCAACCCCCGTACTGTTATTTTCGT
A  R  Q  S  D  N  M  T  A  V  A  C  Q  P  P  Y  C  Y  F  R
                                                                 1020
AATTCTACTACCAACTATGTTGGTGTTTATGATATCAATCATGGGGATGCTGGTTTTACT
N  S  T  T  N  Y  V  G  V  Y  D  I  N  H  G  D  A  G  F  T
                 M  L  V  F  M  I  S  I  M  G  M  L  V  L  L
                 IORF2 →
```

FIGURE 1.

```
         .         .         .         .         .          1080
AGCATACTCAGTGGTTTGTTATATGATTCACCTTGTTTTTCGCAGCAAGGTGTTTTTAGG
S  I  L  S  G  L  L  Y  D  S  P  C  F  S  Q  Q  G  V  F  R
 A  Y  S  V  V  C  Y  M  I  H  L  V  F  R  S  K  V  F  L  G
         .         .         .         .         .          1140
TATGATAATGTTAGCAGTGTCTGGCCTCTCTATTCCTATGGCAGATGCCCTACTGCTGCT
Y  D [N] V  S  S  V  W  P  L  Y  S  Y  G  R  C  P  T  A  A
 M  I  M  L  A  V  S  G  L  S  I  P  M  A  D  A  L  L  L  L
         .         .         .         .         .          1200
GATATTAATACCCCTGATGTACCTATTTGTGTGTATGATCCGCTACCACTTATTTTGCTT
D  I  N  T  P  D  V  P  I  C  V  Y  D  P  L  P  L  I  L  L
 I  L  I  P  L  M  Y  L  F  V  C  M  I  R  Y  H  L  F  C  L
         .         .         .         .         .          1260
GGCATCCTTTTGGGTGTTGCGGTCATAATTATTGTAGTTTTGTTGTTATATTTTATGGTG
G  I  L  L  G  V  A  V  I  I  I  V  V  L  L  L  Y  F  M  V
         .         .         .         .      2007
GATAATGGTACTAGGCTGCATGATGCTTAGACCATAATCTAAACATG
D [N] G  T  R  L  H  D  A  *                 M
 M  V  L  G  C  M  M  L  R  P  *              S →
```

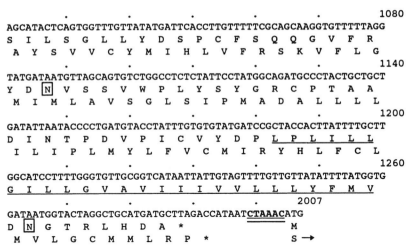

FIGURE 1. Sequence of the BCV Mebus strain HE gene and deduced amino acid sequences of the HE, IORF 1, and IORF 2 proteins. The nucleotide sequence begins 15 bases upstream from the presumed start codon of the HE gene [8,694 bases from the 3′ end of the genome, excluding the poly(A) tail] at the CTAAAC consensus sequence (doubly underlined) and ends with the ATG initiation codon of the spike protein gene. The amino-terminal signal peptide and presumed carboxy-terminal anchor sequences are underlined. The amino terminal signal cleavage site is indicated by an arrow (↑). The putative esterase active site (F-G-D-S) and potential N-linked glycosylation sites (NXS or NXT, where X ≠ P) are boxed. The CTAAAC consensus sequence preceding the spike protein gene is also doubly underlined. The GenBank accession number of the nucleotide sequence is M31052.

terminal ectodomain and a carboxy-terminal anchor with nine amino acids residing on the cytoplasmic (intravirion) side of the membrane. There are eight potential N-glycosylation sites on the ectodomain of the protein. The F-G-D-S putative esterase-active site is at amino acid positions 19 through 22 of the processed protein. There are 13 cysteine residues in the processed subunit, but it is not known which of these are used in interchain disulfide bonding to form the dimer or intrachain folding.

The BCV HE gene contains within it two nonoverlapping ORFs, one (IORF 1) that could encode a 136 amino acid protein of 15,741 Da, and another (IORF 2) that could encode an 83 amino acid protein of 9,514 Da (Fig. 1). It is not known whether these are expressed during infection. The existence of these proteins is questionable since both are very hydrophobic throughout their length and there are no large internal ORFs within the HE gene of MHV-JHM or the HE pseudogene of MHV A59 [from our analysis of data in Luytjes *et al.* (1988) and Shieh *et al.* (1989)].

Expression of the cloned BCV HE gene, both *in vitro* and *in vivo*, has established that the HE protein alone contains all the necessary information for disulfide-linked homodimerization, membrane translocation, anchorage, transport to the cell surface as an integral membrane protein, erythrocyte-binding activity, and esterase activity (Kienzle *et al.*, 1990; Parker *et al.*, 1990; Yoo *et al.*, 1992). Cell surface expression of HE may explain hemadsorption by infected cells, although other explanations such as binding by nonreleased surface-adherent viruses have been given (Bucknall *et al.*, 1972; Kapikian *et al.*, 1972). Cells infected with an adenovirus recombinant expressing the BCV HE exhibited both hemadsorption and esterase activities, both of which could be blocked by monoclonal antibodies having neutralizing activity (Parker *et al.*, 1990, Yoo *et al.*, 1992).

## III. COMPARISON OF THE HEs OF BCV, HCV-OC43, MHV, AND INFLUENZA C VIRUS

A remarkable conservation of amino acid sequence between the deduced HE proteins of MHV A59 (Luytjes *et al.*, 1988) and influenza C virus (Nakada *et al.*, 1984) suggested a common evolutionary origin for these proteins (Luytjes *et al.*, 1988). A similarly high degree of homology among the HEs of influenza C virus, BCV, and HCV-OC43 (Parker *et al.*, 1989; Kienzle *et al.*, 1990; Zhang *et al.*, 1992) and a striking conservation in the position of the putative acetylesterase active site and several of the cysteine residues (Fig. 2) further supported the idea of a common origin for the HE proteins. Although the HEs of BCV and HCV OC43 are quite similar to each other, they as a group appear to be no closer to influenza C virus HE in sequence similarity than does MHV-JHM HE (Fig. 2), suggesting they are equidistant in evolutionary time. Because of the ability of coronaviruses to undergo recombination at a high frequency (Lai *et al.*, 1987), it is conceivable that during coinfection with influenza C virus, a coronavirus (e.g., HCV OC43) could have undergone a similar recombination to give rise to the HE gene in the coronavirus genome (Luytjes *et al.*, 1988). One major interesting difference between the HEs of BCV and influenza C is that the influenza C HE is formed by the proteolytic cleavage of a large precursor (Herrler *et al.*, 1981), whereas the BCV HE is not.

Interestingly, the HE of influenza C virus (Nakada *et al.*, 1984) with which BCV HE shows 29% amino acid sequence homology does form disulfide-linked dimeric structures (Herrler *et al.*, 1981). Comparative structural analyses might also suggest the site of the hemagglutinating (cell binding) domain, assuming it is common among the various HEs.

## IV. MODEL OF THE BCV HE ON THE VIRION

From evidence of the membrane orientation and dimeric structure of BCV HE, we have drawn a tentative model for the dimeric protein (Fig. 3). Placement

FIGURE 2. Alignment of the deduced amino acid sequences for the HE genes of MHV (JHM), BCV (Mebus), HCV-OC43, MHV-JHM, and influenza C virus (C/Cal/78). The MHV (Shieh *et al.*, 1989) and BCV (Kienzle *et al.*, 1990), HCV-OC43 (Zhang *et al.*, 1992), and influenza C (Nakada *et al.*, 1984) sequences were each aligned using the PileUp program in the Wisconsin GCG Sequence Analysis Package. The sequences begin with the signal peptide of each protein. The influenza C sequence contains all of the HA1 sunbunit and the first 26 amino acids of the HA2 sunbunit. Gaps (identified by dots) were introduced to maximize alignment. Aligned cysteine residues in all four sequences are boxed. The esterase site (FGDS) is boldfaced.

```
      1                                                50
BCV   .......... ..MFLL.... ...LRFVLVS CIIGSLGFDN PPTNVVSHLN
OC43  .......... ..MFLL.... ...PRFILVS CIIGSLGFYN PPTNVVSHVN
JHM   MGSTCIAMAP RTLLLL.... ...IGCQLV. .....FGF.N EPLNIVSHLN
FluC  .......MFF SLLLMLGLTE AEKIKICLQK QVNSSFSLHN GFGGNLYATE

      51                                              100
BCV   GDWFLFGDSR SDCNHVVNTN PRNYSYMDLN PALCDSG... .KISSKAGNS
OC43  GDWFLFGDSR SDCNHIVNIN PHNYSYMDLN PVLCDSG... .KISSKAGNS
JHM   DDWFLFGDSR SDCTYVENNG HPKLDWLDLD PKLCNSG... .KISAKSGNS
FluC  EKR.MFELVK PKAGASVLNQ STWIGFGDSR TDQSNSAFPR SLMSAKTADK

      101                                             150
BCV   IFRSFHFTDF .......... ..YNYTGEGQ QIIFYEGVNF TPYHAFKCTT
OC43  IFRSFHFTDF .......... ..YNYTGEGQ QIIFYEGVNF TPYHAFKCNR
JHM   LFRSFHFTDF .......... ..YNYTGEGD QIVFYEGVNF SPNHGFKCLA
FluC  .FRSLSGGSL MLSMFGPPGK VDYLYQGCGK HKVFYEGVNW SPHAAIDCYR

      151                                             200
BCV   SGSNDIWMQN KGLFYTQVYK NMAVYRSLTF VNVPYVYNGS AQSTALCK.S
OC43  SGSNDIWMQN KGLFYTQVYK NMAVYRSLTF VNVPYVYNGS AQSTALCK.S
JHM   YGDNKRWMGN KARFYARVYE KMAQYRSLSF VNVPYAYGGK AKPTSICK.H
FluC  KNWTDIKLN. ...FQKSIYE LASQSHCMSL VNALDKTIPL QVTKGVAKNC

      201                                             250
BCV   GSLVLNNPAY IAREANFGD. .......YYY KVEADFYLSG CDEYIVPLC.
OC43  GSLVLNNPAY IAPQANSGD. .......YYY KVEADFYLSG CDEYIVPLC.
JHM   KTLTLNNPTF ISKESNYVD. .......YYY ESEANFTLAG CDEFIVPLC.
FluC  NNSFLKNPAL YTQEVKPLEQ ICGEENLAFF TLPTQFGTYE CKLHLVASCY

      251                                             300
BCV   .IFNGKFLSN T.....KYYD DSQYYFNKDT GVIYGLNSTE TIT.....TG
OC43  .IFNGKFLSN T.....KYYD DSQYYFNKDT GVIYGLNSTE TIT.....TG
JHM   .VFNGHSKGS SSDPANKYYM DSQSYYNMDT GVLYGFNSTL DVGNTAKDPG
FluC  FIYDSKEVYN KRGCGNYF.. ..QVIYDSSG KVVGGLDNRV SPYTGNSGDT

      301                                             350
BCV   FDFNCHYLVL PSGNYLAISN ELLLTVPTKA ICLNKRKDFT PVQVVDSRWN
OC43  FDLNCYYLVL PSGNYLAISN ELLLTVPTKA ICLNKRKDFT PVQVVDSRWN
JHM   LDLTCRYLAL TPGNYKAVSL EYLLSLPSKA ICLRKPKRFM PVQVVDSRWN
FluC  PTMQCDMLQL KPGRYSVRSS PRFLLMPERS YCFDM.KEKG PVTAVQSIWG

      351                                             400
BCV   NARQSDNMTA VACQPPYCYF RNSTTNY.VG VYDINHGDAG FTSILSGLLY
OC43  NARQSDNMTA VACQPPYCYF RNSTTNY.VG VYDINHGDAG FTSILSGLLY
JHM   STRQSDNMTA VACQLPYCFF RNTSADYSGG THDVHHGDFH FRQLLSGLLL
FluC  KGRKSDYAVD QACLSTPGCM LIQKQKPYIG EADDHHGDQE MRELLSGLDY

      401                                             450
BCV   DSPCFSQQGV FRYDNVSSVW PLY....SYG RCPTAA.DIN TPDVPICVYD
OC43  NSPCFSQQGV FRYDNVSSVW PLY....PYG RCPTAA.DIN NPDLPICVYD
JHM   NVSCIAQQGA FLYNNVSSSW PAY....GYG QCPTAA.NIG YMA.PVCIYD
FluC  EARCISQSG. .WVNETSPFT EEYLLPPKFG RCPLAAKEES IPKIPDGLLI

      451                                             500
BCV   PLPLILLGI. .....LLGVA VIIIVVLLLY FMVDN..... .GTRLHDA
OC43  PLPVILLGI. .....LLGVA VIIIVVLLLY FMVDN..... .GTRLHDA
JHM   PLPVVLLGV. .....LLGIA VLIIVFLILY FMTDS..... .GVRLHEA
FluC  PTSGTDTTVT KPKSRIFGID DLIIGLLFVA IVEAGIGGYL LGSRKESGGG
```

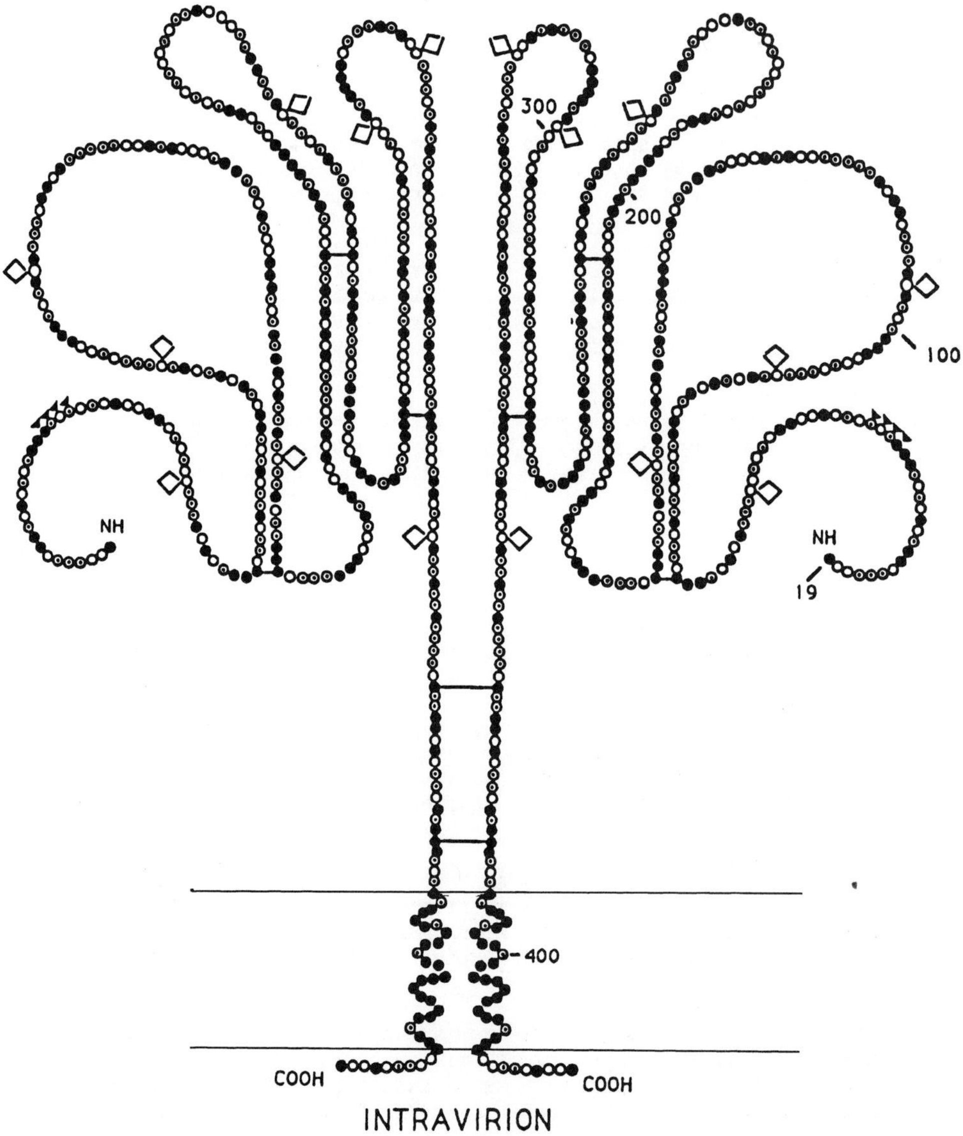


FIGURE 3. Model of the BCV HE protein. Identical chains representing the 65-kDa subunits (without N-terminal signal peptides) are shown as disulfide-linked monomers. Open circles represent hydrophilic amino acid residues (Asp, Glu, Lys, Arg, His, Asn, Gln), dotted circles represent intermediate amino acid residues (Gly, Ser, Thr, Tyr, Try, Pro), and solid circles represent hydrophobic amino acid residues (Ala, Val, Leu, Ile, Phe, Cys, Met). Amino acids were classified by the scale of Kyte and Doolittle (1982). Amino acid position numbers are shown (beginning with the N terminus of the signal peptide, not shown). Disulfide bridges between cysteine residues are represented by connected solid circles. The potentially glycosylated Asn residues in the ectodomain are identified by open diamonds. Triangles identify the putative active site of the esterase (F-G-D-S).

of disulfide bonds and protein loops is purely arbitrary but done in a way to maximize exposure of the hydrophilic, sugar-containing regions and the F-G-D-S esterase active site. It remains to be determined whether higher-order noncovalent oligomeric forms of the dimeric HE exist.

## V. HEMAGGLUTINATING ACTIVITY AND ITS POSSIBLE ROLE *IN VIVO*

It has been proposed that HE may function as a second attachment protein that, in addition to the spike protein, serves to initiate infection (Parker *et al.*, 1989; Vlasak *et al.*, 1988a). This proposal was based primarily on the following observations: (1) The process of hemagglutination itself is a cell-binding process that may mimic cell binding during infection. (2) The HE homologue on influenza C virus was known to be an attachment protein that recognizes 9-*O*-acetylated sialic acid-containing cell receptors (Herrler *et al.*, 1985; Herrler and Klenk, 1987; Rogers *et al.*, 1986; Vlasak *et al.*, 1987, 1989). (3) A panel of monoclonal antibodies to the BCV HE protein exhibited neutralization of infection in cell culture and in the calf (Deregt *et al.*, 1989). With these antibodies, four neutralizing epitopes could be identified on the HE molecule (Deregt and Babiuk, 1987).

The idea that HE may function as a second attachment protein seemed to be further strengthened by the striking correlation between its presence and the neuronotropism expressed by the virus (Table I). Targeting of the HE-containing coronaviruses to the brain of experimentally infected mice was demonstrated for BCV (Akashi *et al.*, 1981; Kubo *et al.*, 1982; Barthold *et al.*, 1990), HCV OC43 (Pearson and Mims, 1983; Barthold *et al.*, 1990), and HEV (Yagami *et al.*, 1986, 1993). Except for the restricted growth of HE-less MHV A59 in neurons of the subthalamic nucleus and substantia nigra of mice (Fishman *et al.*, 1985) and the neuronotropism demonstrated by the HE-less MHV3 (Talbot, 1989; Tardieu *et al.*, 1986), it is particularly striking that coronaviruses possessing HE are neu-

TABLE I. Neuronotropism of HE-Containing Coronaviruses Cell Tropism in Central Nervous System

| Coronavirus | Hemagglutinin form | Astrocytes | Oligodendrocytes | Neurons | References |
|---|---|---|---|---|---|
| MHV A59 | No HE | + | + | –[a] | b |
| MHV-JHM | Dimer | + | + | + | b–e |
| DVIM | Dimer | ? | ? | ? | f–h |
| BCV | Dimer | ? | ? | + | i |
| HCV OC43 | Dimer | – | + | + | j–l |
| HEV | Dimer | ? | ? | + | m–o |
| TCV | Dimer | ? | ? | ? | p |

[a]There is restricted growth in neurons of the subthalamic nucleus and substantia nigra in mice (Fishman *et al.*, 1985).
References: b, Dubois-Dalcq *et al.* (1982); c, Knobler *et al.* (1981); d, Parham *et al.* (1986); e, Taguchi *et al.* (1986); f, Sugiyama and Amano (1980); g, Sugiyama and Amano (1981); h, Sugiyama *et al.* (1986); i, Kubo *et al.* (1982); j, Pearson and Mims (1983); k, Pearson and Mims (1985); l, Barthold *et al.* (1990); m, Andries and Pensaert (1981); n, Yagami *et al.* (1986); o, Yagami *et al.* (1993); p, Dea and Tijssen (1988).

ronotropic in either the natural infection (e.g., HEV) or during experimental infection in mice (Table I). One possible explanation for this neuronotropism might be the high concentration of 9-*O*-acetylated sialic acids on brain gangliosides (Haverkamp *et al.*, 1977; Herrler and Klenk, 1987). It will be important to determine whether the neuronotropism exhibited by some, but not all, strains of MHV-JMH (Dubois-Dalcq *et al.*, 1982; Fleming *et al.*, 1987; Goto *et al.*, 1986; Knobler *et al.*, 1981; Massa *et al.*, 1988; Morris *et al.*, 1989; Parham *et al.*, 1986; Taguchi *et al.*, 1986; Wege *et al.*, 1982) is a function of the variability in expression of the HE protein (Shieh *et al.*, 1989). One recent study showed that JHM-infected mice were protected when passively immunized with HE-specific nonneutralizing monoclonal antibodies, suggesting that this protein does play a role in infection by MHV-JHM (Yokomori *et al.*, 1992).

Recently the potential role of HE as an attachment protein for initiating infection has required reevaluation since the peplomeric spike proteins on BCV have also been shown to bind 9-*O*-acetylated sialic acid, the receptor determinant for BCV infection, and to cause hemagglutination (Kinkel and Herrler, 1993; Schultze *et al.*, 1991a; Schultze and Herrler, 1992; Storz *et al.*, 1991). Whether the S protein of BCV also recognizes proteins similar to the carcinoembryonic antigen as does MHV or the aminopeptidase N protein as does HCV 229E or TGEV (see Chapter 4, this volume) remain to be determined. But the fact that S binds to the same receptor determinant as HE would suggest that they both use the same receptor protein to initiate infection, if in fact they both do initiate infection. In a study where separately purified S and HE proteins from BCV were used to measure hemagglutination, a difference in the agglutinability of erythrocytes from different animal species was found (Schultze *et al.*, 1991a,b). Both S and HE were able to agglutinate mouse and rat erythrocytes; however, only S could agglutinate chicken erythrocytes. This was thought to reflect a more powerful binding by S since the chicken erythrocytes have fewer surface sialic acid molecules. These observations have led Herrler and coworkers to suggest that S and HE recognize the same receptor determinant, but that S is the important molecule for initiating infection (Schultze, 1991a,b). HE, it was proposed, may function primarily as a receptor-destroying enzyme to remove glycoconjugates that act as false receptors.

## VI. ESTERASE ACTIVITY AND ITS POSSIBLE ROLE *IN VIVO*

The enzymatic activity associated with the HE is neuraminate-*O*-acetylesterase, which destroys the receptor determinants on erythrocytes bound by BCV, HCV OC43, and HEV (Vlasak *et al.*, 1988a,b). The receptors therefore contain 9-*O*-acetylated sialic acid and the activity is a serine esterase indistinguishable from the esterase of influenza C virus (Herrler *et al.*, 1985, 1988; Vlasak *et al.*, 1987, 1988a,b). The only other viruses possessing receptor-destroying activity are the orthomyxovirus types A and B and the paramyxoviruses, for which the receptor-destroying enzyme is neuraminidase.

Enzymatic association with the hemagglutinin molecule was shown in studies using diisopropyl fluorophosphate (Vlasak *et al.*, 1988a), which binds to the serine in the active site and acts as a serine esterase inhibitor (Muchmore and Varki, 1987). Diisopropyl fluorophosphate completely inhibited the receptor-destroying activity of BCV, and 3H-diisopropyl fluorophosphate treatment of BCV with subsequent sodium dodecyl sulfate–polyacrylamide gel electrophoresis of the protein revealed that the 140-kDa hemagglutinin molecule was specifically labeled. Presence of the F-G-D-S esterase-active site in the amino acid sequence of the hemagglutinin protein (Kienzle *et al.*, 1990; Parker *et al.*, 1989) and the existence of esterase activity on HE expressed from the cloned gene (Parker *et al.*, 1990) confirmed this association. The studies by Vlasak *et al.* (1988a) suggested that the active esterase may be required for either endocytosis of the virus and/or uncoating with subsequent release of the virus into the cytoplasm during the initial stages of infection. This notion comes from the fact that diisopropyl fluorophosphate treatment of BCV destroyed the esterase activity and at the same time reduced the infectious titer of the virus 100- to 400-fold. This treatment, however, did not prevent binding of the virus to sialic acid-containing receptors. The acetyl esterase, however, may act at the end of the replication cycle by releasing progeny virus from infected cells. In this way it would behave analogously to the neuraminidase of influenza A virus (Palese *et al.*, 1974).

## VII. CONCLUDING REMARKS

Many interesting features of HE expression, structure, and function remain to be characterized. It remains to be established, for example, what genetic mechanisms cause variable expression of HE in some strains of MHV. One hypothesis still needing direct proof is that it is the number of UCUAA repeat units at the 5′ end of the HE gene that regulates transcription (Shieh *et al.*, 1989). It also remains to be shown whether the two ORFs within the BCV HE gene (IORF 1 and IORF 2) are expressed during virus replication, and, if so, what the functions of their products might be. Is their function related to those of HE?

Regarding HE structure, it remains to be determined where the intramolecular and intermolecular disulfide bridges occur in the dimeric molecule, which asparagine residues become glycosylated, and the exact nature of the carbohydrate modifications. Detailed mapping will be required to identify the sites on the HE molecule involved in hemagglutination, cell binding (if different from hemagglutination), recognition by neutralizing monoclonal antibodies (Deregt and Babiuk, 1987), and interaction (if any) with other structural and nonstructural proteins. Do higher-level noncovalent multimeric forms of HE exist? Ultimately it will be important to determine the three-dimensional structure of HE and establish the antigenic epitopes on this structure.

The full range of functions of HE in virus replication remains to be determined. Further studies are needed to test whether HE is, in fact, required for replication in those viruses that express it constituitively. Is it responsible for expanded tissue tropism or host range in those viruses that possess it? If so, by what mechanism? Is it responsible for the strong neuronotropism exhibited by the HE-containing coronaviruses? If so, then special importance should be placed on understanding its role since the HE-containing HCV OC43 (and its close relatives) is a common human respiratory pathogen and has the potential for involving the CNS in human disease. The precise role(s) of the esterase in the infection process remains to be determined.

Finally, the usefulness of HE as a vaccine immunogen for the control of BCV-induced gastrointestinal disease should be explored. The existence of HE as an integral membrane protein on the surface of infected cells (Kienzle *et al.*, 1989), as well as a viral structural protein that can induce neutralizing antibodies (Deregt *et al.*, 1989), suggests that both cellular and humoral immune responses are important in protection against BCV. BCV-neutralizing antibodies can be induced in mice inoculated with an adenovirus recombinant expressing HE (Yoo *et al.*, 1992), raising the possibility that protective mucosal immunity can be induced by HE. The usefulness of HE in vaccines given by the mucosal route should be assessed.

## VIII. REFERENCES


Abraham, S., Kienzle, T. E., Lapps, W., and Brian, D. A. 1990, Deduced sequence of the bovine coronavirus spike protein and identification of the internal proteolytic cleavage site, *Virology* **176:**296.

Akashi, H., Inaba, Y., Minra, Y., Sato, K., Tokuhisa, S., Asagi, M., and Hayashi, Y., 1981, Propagation of the Kakagawa strain of bovine coronavirus in suckling mice, rats, and hamsters, *Arch. Virol.* **67:**367.

Andries, K., and Pensaert, M., 1981, Vomiting and wasting disease, a coronavirus infection of pigs, *Adv. Exp. Med. Biol.* **142:**399.

Barthold, S. W., de Souza, M. S., and Smith, A. L., 1990, Susceptibility of laboratory mice to intranasal and contact infection with coronaviruses of other species, *Lab. Anim. Sci.* **40:**481.

Bingham, R. W., Madge, M. H., and Tyrrell, D. A. J., 1975, Hemagglutination by avian infectious bronchitis virus—A coronavirus, *J. Gen. Virol.* **28:**381.

Bridger, J. C., Caul, E. O., and Egglestone, S. I., 1978, Replication of an enteric bovine coronavirus in intestinal organ cultures, *Arch. Virol.* **57:**43.

Bucknall, R. A., Kalica, A. R., and Chanock, R. M., 1972, Intracellular development and mechanism of hemadsorption of human coronavirus, OC43, *Proc. Soc. Exp. Biol. Med.* **139:**811.

Callebaut, P. E., and Pensaert, M. B., 1980, Characterization and isolation of structural polypeptides in hemagglutinating encephalomyelitis virus, *J. Gen. Virol.* **48:**193.

Cavanagh, D., and Davis, P. J., 1986, Coronavirus IBV: Removal of spike glycopolypeptide S1 by urea abolishes infectivity and hemagglutination but not attachment to cells, *J. Gen. Virol.* **67:**1443.

Dea, S., and Tijssen, P., 1988, Identification of the structural proteins of the turkey enteric coronavirus, *Arch. Virol.* **99:**173.

Deregt, D., and Babiuk, L. A., 1987, Monoclonal antibodies to bovine coronavirus: Characteristics and topographical mapping of neutralizing epitopes on the E2 and E3 glycoproteins, *Virology* **161:**410.

Deregt, D., Sabara, M., and Babiuk, L. A., 1987, Structural proteins of bovine coronavirus and their intracellular processing, *J. Gen. Virol.* **68:**2863.

Deregt, D., Gifford, G. A., Ijaz, M. K., Watts, T. C., Gilchrist, J. E., Haines, D. M., and Babiuk, L. A., 1989, Monoclonal antibodies to bovine coroanvirus glycoproteins E2 and E3: Demonstration of *in vivo* virus-neutralizing activity, *J. Gen. Virol.* **70:**993.

Dubois-Dalcq, M. E., Doller, E. W., Haspel, M. V., and Holmes, K. V., 1982, Cell tropism and expression of mouse hepatitis viruses (MHV) in mouse spinal cord cultures, *Virology* **119:**317.

Fishman, P. S., Gass, J. S., Swoveland, P. T., Lavi, E., Highkin, M. K., and Weiss, S. R., 1985, Infection of the basal ganglia by a murine coronavirus, *Science* **229:**877.

Fleming, J. O., Trousdale, M. D., Bradbury, J., Stohlman, S. A., and Weiner, L. P., 1987, Experimental demyelination induced by coronavirus JHM (MHV-4): Molecular identification of a virus determinant of paralytic disease, *Microbial Pathogen.* **3**:9.

Goto, N., Makino, S., and Fujiwara, K., 1986, Neuropathogenicity of mutant strains of mouse hepatitis virus, 1a and 2c, from DBT cells persistently infected with JHM strain, *Adv. Exp. Med. Biol.* **218**:439.

Haverkamp, J., Veh, R. W., Sander, M., Shauer, R., Kamerling, J. P., and Vliegenthart, J. F. G., 1977, Demonstration of 9-*O*-acetyl-*N*-acetylneuraminic acid in brain gangliosides from various vertebrates including man, *Hoppe-Seyler Physiol. Chem.* **358**:1609.

Herrler, G., and Klenk, H.-D., 1987, The surface receptor is a major determinant of the cell tropism of influenza C virus, *Virology* **159**:102.

Herrler, G., Nagele, A., Meier-Ewert, H., Bhown, A. S., and Compans, R. W., 1981, Isolation and structural analysis of influenza C virion glycoproteins, *Virology* **113**:439.

Herrler, G., Rott, R., Klenk, H. D., Muller, H. P., Shukla, A. K., and Schauer, R., 1985, The receptor-destroying enzyme of influenza C virus is neuraminate-*O*-acetylesterase, *EMBO J.* **4**:1503.

Herrler, G., Durkop, I., Becht, H., and Klenk, H.-D., 1988, The glycoprotein of influenza C virus is the hemagglutinin, esterase, and fusion factor, *J. Gen. Virol.* **69**:839.

Hogue, B. G., and Brian, D. A., 1986, Structural proteins of human respiratory coronavirus OC43, *Virus Res.* **5**:131.

Hogue, B. G., King, B., and Brian, D. A., 1984, Antigenic relationships among proteins of bovine coronavirus, human respiratory coronavirus OC43, and mouse hepatitis coronavirus A59, *J. Virol.* **51**:384.

Hogue, B. G., Kienzle, T. E., and Brian, D. A., 1989, Synthesis and processing of the bovine enteric coronavirus hemagglutinin protein, *J. Gen. Virol.* **70**:345.

Kapikian, A. Z., James, H. D., Kelly, S. J., King, L. M., Vaughn, A. L., and Chanock, R. M., 1972, Hemadsorption by coronavirus strain OC43, *Proc. Soc. Exp. Biol. Med.* **139**:179.

Kaye, H. S., and Dowdle, W. R., 1969, Some characteristics of hemagglutination of certain strains of "IBV-like" virus, *J. Infect. Dis.* **120**:576.

Keck, J., Hogue, B. G., Brian, D. A., and Lai, M. M. C., 1988, Temporal regulation of bovine coronavirus RNA synthesis, *Virus Res.* **9**:343.

Kienzle, T. E., Abraham, S., Hogue, B. G., and Brian, D. A., 1990, Structure and orientation of expressed bovine coronavirus hemagglutinin esterase protein, *J. Virol.* **64**:1834.

King, B., and Brian, D. A., 1982, Bovine coronavirus structural proteins, *J. Virol.* **42**:700.

King, B., Potts, B. J., and Brian, D. A., 1985, Bovine coronavirus hemagglutinin protein, *Virus Res.* **2**:53.

Knobler, R. L., Dubois-Dalcq, M., Haspel, M. V., Claysmith, A. P., Lampert, P. W., and Oldstone, M. B. A., 1981, Selective localization of wild type mutant mouse hepatitis (JHM strain) antigens in CNS tissue by fluorescence, light and electron microscopy, *J. Neuroimmunol.* **1**:81.

Kubo, M., Akashi, H., Inaba, Y., Osada, M., and Konno, S., 1982, Pathological studies on encephalitis in mice experimentally inoculated with bovine coronavirus, *Natl. Inst. Anim. Health Q.* **22**:45.

Kunkel, F., and Herrler, G., 1993, Structural and functional analysis of the surface protein of human coronavirus OC43, *Virology* **195**:195.

Kyte, J., and Doolittle, R. F., 1982, A simple method for displaying the hydrophatic character of a protein, *J. Mol. Biol.* **157**:105.

Lai, M. M. C., Makino, S., Soe, L. H., Shieh, C.-K., Keck, J. G., and Fleming, J. O., 1987, Coronavirus: A jumping RNA transcription, *Cold Spring Harbor Symp. Quant. Biol.* **52**:359.

Luytjes, W., Bredenbeek, P., Noten, A., Horzinek, M., and Spaan, W., 1988, Sequence of mouse hepatitis virus A59 mRNA2: Indications for RNA recombination between coronaviruses and influenza C virus, *Virology* **166**:415.

Makino, S., Taguchi, F., Hayami, M., and Fujiwara, K., 1983, Characterization of small plaque mutants of mouse hepatitis virus, JHM strain, *Microbiol. Immunol.* **27**:445.

Massa, P. T., Wege, H., and ter Meulen, V., 1988, Growth pattern of various JHM coronavirus isolates in primary rat glial cell cultures correlates with differing neurotropism *in vivo*, *Virus Res.* **9**:133.

Morris, V. L., Tieszer, C., Mackinnon, J., and Percy, D., 1989, Characterization of coronavirus JHM variants isolated from Wistar Furth rats with a viral-induced demyelinating disease, *Virology* **169**:127.

Muchmore, E. A., and Varki, A., 1987, Selective inhibition of influenza C esterase: A probe for detecting 9-*O*-acetylated sialic acids, *Science* **236:**1293.

Nakada, S., Creager, R. S., Krystal, M., Aaronson, R. P., and Palese, P., 1984, Influenza C virus hemagglutinin: Comparison with influenza A and B virus hemagglutinins, *J. Virol.* **50:**118.

Noda, M., Yamashita, H., Koide, F., Kadoi, K., Omori, T., Asagi, M., and Inaba, Y., 1987, Hemagglutination with transmissible gastroenteritis virus, *Arch. Virol.* **96:**109.

Noda, M., Koide, F., Asagi, M., and Inaba, Y., 1988, Physicochemical properties of transmissible gastroenteritis virus hemagglutinin, *Arch. Virol.* **99:**163.

Palese, P., Tobita, K., Ueda, M., and Compans, R. W., 1974, Characterization of temperature sensitive influenza A virus mutants defective in neuraminidase, *Virology* **61:**397.

Parham, D., Tereba, A., Talbot, P., Jackson, D., and Morris, V., 1986, Analysis of JHM central nervous system infections in rats, *Arch. Neurol.* **43:**702.

Parker, M. D., Cox, G. J., Deregt, D., Fitzpatrick, D. R., and Babiuk, L. A., 1989, Cloning and *in vitro* expression of the gene for the E3 hemagglutinin glycoprotein of bovine coronavirus, *J. Gen. Virol.* **70:**155.

Parker, M. D., Yoo, D., and Babiuk, L. A., 1990, Expression and secretion of the bovine coronavirus hemagglutinin-esterase glycoprotein by insect cells infected with recombinant baculovirus, *J. Virol.* **64:**1625.

Pearson, J., and Mims, C. A., 1983, Selective vulnerability of neural cells and age-related susceptibility to OC43 virus in mice, *Arch. Virol.* **77:**109.

Pearson, J., and Mims, C. A., 1985, Differential susceptibility of cultured neural cells to the human coronavirus OC43, *J. Virol.* **5:**1016.

Pfleiderer, M., Routledge., E., Herrler, G., and Siddell, S. G., 1991, High-level transient expression of the murine coronavirus haemagglutinin-esterase, *J. Gen. Virol.* **72:**1309.

Rogers, G., Herrler, G., Paulson, J., and Klenk, H.-D., 1986, Influenza C virus uses 9-*O*-acetyl-*N*-acetylneuraminic acid as high affinity receptor determinant for attachment to cells, *J. Biol. Chem.* **261:**5947.

Schultze, B., and Herrler, G., 1992, Bovine coronavirus uses *N*-acetyl-9-*O*-acetylneuraminiac acid as a receptor determinant to initiate the infection of cultured cells, *J. Gen. Virol.* **73:**901.

Schultze, B., Gross, H.-J., Brossmer, R., Klenk, H.-D. and Herrler, G., 1990, Hemagglutinating encephalomyelitis virus attaches to *N*-acetyl-9-*O*-acetylneuraminic acid-containing receptors on erythrocytes: Comparison with bovine coronavirus and influenza C virus, *Virus Res.* **16:**185.

Schultze, B., Gross, H. J., Brossmer, R., and Herrler, G., 1991a, The S protein of bovine coronavirus is a hemagglutinin recognizing 9-*O*-acetylated sialic acid as a receptor determinant, *J. Virol.* **65:**6232.

Schultze, B., Wahn, K., Klenk, K., and Herrler, G., 1991b, Isolated HE-protein from hemagglutinating encephalomyelitis virus and bovine coronavirus has receptor-destroying and receptor-binding activity, *Virology* **180:**221.

Sharpee, R. L., Mebus, C. A., and Bass, E. P., 1976, Characterization of a calf diarrheal coronavirus, *Am. J. Vet. Res.* **37:**1031.

Shieh, C.-K., Lee, H.-J., Yokomori, K., Monica, N. L., Makino, S., and Lai, M. M. C., 1989, Identification of new transcriptional initiation site and the corresponding functional gene 2b in the murine coronavirus RNA genome, *J. Virol.* **63:**3729.

Siddell, S., 1983, Coronavirus JHM: Coding assignments of subgenomic mRNAs, *J. Gen. Virol.* **64:**113.

Siddell, S. G., Wege, H., Barthel, A., and ter Meulen, V., 1981, Coronavirus JHM: Intracellular protein synthesis. *J. Gen. Virol.* **53:**145.

Siddell, S. G., Anderson, R., Cavanagh, D., Fujiwara, K., Klenk, H. D., Macnaughton, M. R., Pensaert, M., Stohlman, S. A., Sturman, L., and van der Zeijst, B. A. M., 1983a, Coronaviridae, *Intervirology* **20:**181.

Siddell, S., Wege, H., and ter Meulen, V., 1983b, The biology of coronaviruses, *J. Gen. Virol.* **64:**761.

Storz, J., Herrler, G., Snodgrass, D. R., Hussain, K. A., Zhang, X. M., Clark, M. A., and Rott, R., 1991, Monoclonal antibodies differentiate between the haemagglutinating and the receptor-destroying activities of bovine coronavirus, *J. Gen. Virol.* **72:**2817.

Sturman, L., and Holmes, K., 1985, The novel glycoproteins of coronaviruses, Trends in Biological Science **10:**17.

Sugiyama, K., and Amano, Y., 1980, Hemagglutination and structural polypeptides of a new coronavirus associated with diarrhea in infant mice, *Arch. Virol.* **66:**95.

Sugiyama, K., and Amano Y., 1981, Morphological and biological properties of a new coronavirus associated with diarrhea in infant mice, *Arch. Virol.* **67:**241.

Sugiyama, K., Ishikawa, R., and Fukuhara, N., 1986, Structural polypeptides of the murine coronavirus DVIM, *Arch. Virol.* **89:**245.

Taguchi, F., Massa, P. T., and ter Meulen, V., 1986, Characterization of a variant virus isolated from neural cell culture after infection of mouse coronavirus JHMV, *Virology* **155:**267.

Talbot, P., 1989, Hemagglutination by murine hepatitis viruses: Absence of detectable activity in strains 3, A59, and S grown on DBT cells, *Intervirology* **30:**117.

Tardieu, M., Boespflug, O., and Barbe, T., 1986, Selective tropism of a neurotropic coronavirus for ependymal cells, neurons, and meningeal cells, *J. Virol.* **60:**574.

Tooze, S. A., Tooze, J., and Warren, G., 1988, Site of addition of *N*-acetyl-galactosamine to the E1 glycoprotein of mouse hepatitis virus-A59, *J. Cell Biol.* **106:**1475.

Vlasak, R., Krystal, M., Nacht, M., and Palese, P., 1987, The influenza C virus glycoprotein (HE) exhibits receptor-binding (hemagglutinin) and receptor-destroying (esterase) activities, *Virology* **160:**419.

Vlasak, R., Luytjes, W., Leider, J., Spaan, W., and Palese, P., 1988a, The E3 protein of bovine coronavirus is a receptor-destroying enzyme with acetyl esterase activity, *J. Virol.* **62:**4686.

Vlasak, R., Luytjes, W., Spaan, W., and Palese, P., 1988b, Human and bovine coronaviruses recognize sialic acid-containing receptors similar to those of influenza C viruses, *Proc. Natl. Acad. Sci. USA* **85:**4526.

Vlasak, R., Muster, T., Lauro, A. M., Powers, J. C., and Palese, P., 1989, Influenza C virus esterase: Analysis of catalytic site, inhibition, and possible function, *J. Virol.* **63:**2056.

Wege, H., Siddell, S., and ter Meulen, V., 1982, The biology and pathogenesis of coronaviruses, *Curr. Top. Microbiol. Immunol.* **99:**165.

Yagami, K., Hirai, K., and Hirano, N., 1986, Pathogenesis of hemagglutinating encephalomyelitis virus (HEV) in mice experimentally infected by different routes, *J. Comp. Pathol.* **96:**645.

Yagami, K., Isumi Y., Kajiwara, N., Sugiyama, F., and Sugiyama, Y., 1993, Neurotropism of mouse-adapted haemagglutinating encephalomyelitis virus, *J. Comp. Pathol.* **108:**21.

Yokomori, K., Monica, N. L., Makino, S., Shieh, S.-K., and Lai, M. M. C., 1989, Biosynthesis, structure, and biological activities of envelope protein gp65 of murine coronavirus, *Virology* **173:**683.

Yokomori, K., Baker, S. C., Stohlman, S. A., and Lai, M. M. C., 1992, Hemagglutinin-esterase-specific monoclonal antibodies alter the neuropathogenicity of mouse hepatitis virus, *J. Virol.* **66:**2865.

Yoo, D., Graham, F. L., Prevec, L., Parker, M. D., Benko, M., and Babiuk, L. A., 1992, Synthesis and processing of the haemagglutinin-esterase glycoprotein of bovine coronavirus encoded in the E3 region of adenovirus, *J. Gen. Virol.* **73:**2591.

Zhang, X., Kousoulas, K. G., and Storz, J., 1991, The hemagglutinin/esterase glycoprotein of bovine coronaviruses: Sequence and functional comparisons between virulent and avirulent strains, *Virology* **185:**847.

Zhang, X., Kousoulas, K. G., and Storz, J., 1992, The hemagglutinin/esterase gene of human coronavirus strain OC43: Phylogenetic relationships to bovine and murine coronaviruses and influenza C virus, *Virology* **186:**318.

CHAPTER 9

# The Small-Membrane Protein

Stuart G. Siddell

## I. INTRODUCTION

Recently, it has become evident that, in addition to the S, M, and N proteins, the coronavirus genome encodes a further structural protein, the small-membrane or sM protein (Liu and Inglis, 1991; Godet *et al.*, 1992; Yu *et al.*, 1994). In this chapter, I shall describe the structure and expression of the sM protein and consider its possible function. Although the available information is modest, it already seems clear that the sM protein may have a more important role in the biology of coronaviruses than has been recognized previously.

## II. STRUCTURE

In the coronavirus genomes that have been analyzed todate, the sM protein gene is invariably located upstream and adjacent to the M protein gene. The sM protein genes encode polypeptides with predicted molecular weights between 9,100 and 12,400. An optimal alignment of the amino acid sequences of the sM proteins of bovine coronavirus (BCV), murine hepatitis virus (MHV), porcine transmissible gastroenteritis virus (TGEV), canine coronavirus (CCV), human coronavirus 229E (HCV 229E), porcine epidemic diarrhea virus (PEDV), and avian infectious bronchitis virus (IBV) reveals a number of conserved features that are illustrated in Fig. 1.

- A hydrophobic region located in the amino-terminal half of the polypeptide. This region is flanked by negatively charged residues (amino-terminus) and positively charged residues (carboxy-terminus).

STUART G. SIDDELL • Institute of Virology and Immunobiology, University of Würzburg, 97078 Würzburg, Germany.

*The Coronaviridae*, edited by Stuart G. Siddell, Plenum Press, New York, 1995.

```
               (-)                     (+)            *
1  M-FMADAYFADTVWYVGQIIFIVAICLLVIIVVVAFLATFKLCIQLCGMCNTLVLSPS--   BCV  (F15)
1  M-F--NLFLTDTVWYVGQIIFIVAVCLMVTIIVVAFLASIKRCIQLCGLCNTLLLSPS--   MHV  (JHM)
1  MTFPRALTVIDDNGMVINIIFWFLLIIILILLSIALLNIIKLCMVCCNLGRTVIIVPA--   TGEV (Purdue)
1  MTFPRALTVIDDKGMVISIIFWFLLIIILILFSIALLNIIKLCMVCCNLGRTVIIVPA--   CCV  (Insavc/1)
1  M-F---LKLVDDHALVVNVLLWCVVLIVILLVCITIIKLIKLCFTCHMFCNRTVYGPI--   HCV  (229E)
1  M-----LQLVNDNGLVVNVILWLFVLFFLLIISITFVQLVNLCFTCHRLCNSAVYTPI--   PEDV (CV777)
1  MMNLLNKSLEENGSFLTA-LYIIVGFLALYLLGRALQAFVQAADACCLFWYTWVVIPGAK   IBV  (Beaudette)
              <------hydrophobic region------>  <-cysteine->

                   *          *
58 ----IYVFNRGRQFYE------FYND-VK-------PPVLDVDD-----V    84 a.a.   BCV  (F15)
56 ----IYLYNRSKQLYK------YYNEEVR-------PPPLEVDDNIIQTL    88 a.a.   MHV  (JHM)
59 -----------QHAYD------AYKNFMR-------IKAYNPDGALL--A    82 a.a.   TGEV (Purdue)
59 -----------RHAYD------AYKNFMQ-------IRAYNPDEALL--V    82 a.a.   CCV  (Insavc/1)
55 -----------KNVYH------IYQSYMH-------IDPF-PKRVID--F    77 a.a.   HCV  (229E)
54 -----------GRLYR------VYKSYMR-------IDPL-PSTVID--V    76 a.a.   PEDV (CV777)
60 GTAFVYKYTYGRKLNNPELEAVIVNEFPKNGWNNKNPANFQDAQRDKLYS   109 a.a.   IBV  (Beaudette)
```

FIGURE 1. Alignment of the amino acid sequences of coronavirus sM proteins.

- A cysteine-rich region adjacent to the hydrophobic domain.
- A conserved proline residue near the middle of the polypeptide.
- With the exception of IBV, conserved tyrosine residues in the carboxy-half of the polypeptide.
- An abundance of charged residues in the carboxy-half of the polypeptide.

These features are indicative of an integral membrane protein and a membrane location for the sM protein, both in the infected cell and in the virion, has been demonstrated (see Section V). The distribution of charged residues on either side of the hydrophobic region predict a Nexo-Cendo orientation (Hartmann *et al.*, 1989), but the available experimental data are consistent with a Cexo-Nendo configuration. Thus, monoclonal antibodies, specific for epitopes located in the C-terminus of the TGEV sM protein, were able to induce cell-surface fluorescence in paraformaldehyde-fixed, TGEV-infected swine testicle (ST) cells (Godet *et al.*, 1992). Further experiments are needed to assign a definitive transmembrane orientation.

At present, there is no evidence to suggest that the coronavirus sM protein is glycoslylated or phosphorylated. However, Yu *et al.*, (1994) have suggested that the MHV sM protein is posttranslationally acylated. These authors noted a small difference in the electrophoretic mobility of the MHV-A59 sM protein synthesized either in infected L2 cells or by *in vitro* translation of a synthetic sM mRNA. This difference was abolished by treatment of the *in vivo* synthesized protein with 1 M hydroxlamine at pH 8.0. These data are consistent with the linkage of palmitic acid to the sM protein, and it seems reasonable to suggest that the cysteine residues located adjacent to the hydrophobic domain of the protein may be the sites of esterification. It should be noted, however, that Godet *et al.* (1992) were unable to detect the incorporation of palmitic acid chains in a recombinant TGEV sM protein expressed in insect cells. An sM protein gene has not been identified in the Berne virus (BEV) genome.

## III. EXPRESSION

The mRNAs that encode the sM proteins of coronaviruses are listed in Table I. In the cases of HCV 229E, PEDV, TGEV, and CCV, these assignments are based mainly on colinearity of the genomic sM open reading frame (ORF)

TABLE I. sM Protein mRNAs

| Virus | mRNA | ORFs | Expression | Reference |
|---|---|---|---|---|
| IBV | 3 | 3a, 3b, 3c (sM) | Tricistronic | Liu *et al.* (1991) |
| MHV | 5 | 5a, 5b (sM) | Bicistronic | Budzilowicz and Weiss (1987) |
| HCV 229E | 5 | sM | Monocistronic | Raabe *et al.* (1990) |
| BCV | 5.1 | sM | Monocistronic | Abraham *et al.* (1990) |
| PEDV | 4 | sM | Monocistronic | Duarte *et al.* (1994) |
| TGEV | 4 | sM | Monocistronic | Rasschaert *et al.* (1987) |
| CCV | 4 | sM | Monocistronic | Horsburgh *et al.* (1992) |

and the 5′ unique region of the mRNA, and by extrapolation of the coronavirus translational "model," which, in fact, was deduced by the analysis of other mRNAs (Rottier *et al.*, 1981; Leibowitz *et al.*, 1982; Siddell, 1983; Stern and Sefton, 1984; Jacobs *et al.*, 1986; de Groot *et al.*, 1987). In general, the sM mRNAs contain only one ORF in their 5′ unique region, and, in these cases, the mRNAs almost certainly function monocistronically. Accordingly, they may now be called sM mRNAs. There are, however, two exceptions. First, the mRNA 5 of MHV contains two ORFs (5a and 5b) in its unique region, and second, the mRNA 3 of IBV contains three ORFs (3a, 3b, and 3c) in its unique region. In both cases, the sM gene is the 3′ proximal ORF. In these two special cases, there is experimental evidence that the mRNAs function bi- or tricistronically, respectively.

## IV. *IN VITRO* STUDIES

### A. MHV mRNA 5

The MHV mRNA 5 contains two ORFs in its unique region. These ORFs overlap by five nucleotides. The upstream ORF, ORF 5a, is predicted to encode a basic protein of 12,500–13,000 molecular weight. The function of this protein is unknown, and it has not yet been detected in MHV-infected cells. With the exception of the AUG codon initiating the downstream ORF, no internal AUG codons are found in the sequence covered by ORF 5a. The downstream ORF, previously designated as ORF 5b, encodes the sM protein (Skinner *et al.*, 1985; Budzilowicz and Weiss, 1987). The *in vitro* translation of synthetic RNAs transcribed from cDNAs containing both ORF 5a and the sM ORF led Budzilowicz and Weiss (1987) to conclude that the MHV-A59 mRNA 5 is functionally bicistronic.

If the MHV mRNA 5 is functionally bicistronic, then expression of the sM ORF has to involve the internal initiation of protein synthesis. Two models can be proposed. First, as originally suggested on the basis of the structural data alone (Skinner *et al.*, 1985), a mechanism involving leaky ribosome scanning (Kozak, 1989) is possible. Alternatively, and especially in the light of recent data on the translation of picornavirus RNAs (Meerovitch and Sonenberg, 1993), a cap-independent mechanism involving ribosome entry at an internal position on the MHV mRNA 5 has to be considered.

Recently, Thiel and Siddell (1994) have analyzed the *in vitro* translation of synthetic mRNAs that contain the unique region of MHV (strain JHM) mRNA 5 preceded by an ORF derived from the β-galactosidase gene of *Escherichia coli*. These experiments show that the β-galactosidase ORF is an effective barrier to the movement of ribosomes from the 5′ end of the mRNA but, nevertheless, the sM ORF is efficiently translated. The authors conclude that the translation of the sM ORF is mediated by the internal entry of ribosomes. *In vivo* studies using reporter gene constructs and the naturally occurring mRNA 5 are required to strengthen this conclusion. However, if correct, it will be great interest to examine interactions between the putative MHV mRNA 5 "internal ribosome entry site" (IRES) and *trans*-acting factors that, directly or indirectly, mediate the initiation of protein synthesis at the sM ORF.

## B. IVB mRNA 3

The unique region of the IBV mRNA 3 contains three ORFs. The upstream ORFs, ORF 3a and ORF 3b, encode proteins of 6700 and 7400 molecular weight, respectively. The function of these polypeptides is unknown, although it has been shown that they are expressed in IBV-infected chick kidney (CK) cells (Liu *et al.*, 1991). Again, it seems noteworthy that, with the exception of the AUG that initiates the 3′ proximal ORF, there are no internal AUG codons in the sequences covered by ORFs 3a and 3b. The 3′ proximal ORF, previously designated ORF 3c, encodes the sM protein (Boursnell *et al.*, 1985). *In vitro* translation studies using synthetic mRNAs containing the 3a, 3b, and sM ORFs strongly suggest that all three proteins can be translated from a single molecular species, indicating that the IBV mRNA 3 is functionally tricistronic (Liu *et al.*, 1991).

As with the MHV mRNA 5, there is evidence that the IBV sM ORF of mRNA 3 is expressed by a mechanism involving the internal entry of ribosomes. Thus, Liu and Inglis (1992), again using synthetic mRNAs and *in vitro* translation, were able to show that a tricistronic mRNA whose peculiar 5′ end structure prevents the translation of the 5′ proximal ORFs (3a and 3b) directs the synthesis of sM normally. Moreover, the translation of sM, unlike that of 3a and 3b, was insensitive to the presence of the 5′ cap analogue, 7-methyl GTP, and it was unaffected by alteration of the sequence contexts for initiation on the 3a and 3b ORFs.

The same authors (Liu and Inglis, 1992) were able to show that an mRNA, in which the IBV 3a/3b/sM ORFs were placed downstream of the influenza A virus nucleocapsid protein gene, directed the efficient synthesis of sM, as well as nucleocapsid protein, whereas initiation at ORFs 3a and 3b could not be detected. Furthermore, expression of the sM ORF from this construct was abolished when the 3a and 3b regions were deleted. This suggests that this region contains a putative IRES element responsible for sM expression.

In a theoretical paper, Le *et al.* (1994) performed Monte Carlo simulations of RNA folding in the unique region of the IBV mRNA 3 and determined that a region of significant folding occurred prior to the initiation codon of the sM ORF. This region encompassed 265 nucleotides within the ORF 3a and ORF 3b coding regions. By computer modeling, the authors were then able to predict, within this region, a tertiary structure containing five highly significant RNA stem-loops modeled into a compact superstructure by the interaction of two pseudoknots. A further interesting feature of this model is the presence of putative base complimentarity between unpaired nucleotides in the mRNA together with 18S ribosomal RNA. This interaction would be located immediately upstream of the sM ORF initiation codon. The authors point out that these are features also found in their predictions of tertiary structure for picornavirus IRES elements (Le *et al.*, 1992, 1993)

The translation mechanism of the monocistronic sM mRNAs of coronaviruses has not been studied in any detail (see, however, Abraham *et al.*, 1990). In some cases, for example, the HCV 229E sM mRNA, there is a lengthy 5′ NTR that could, at least theoretically, encompass an IRES element. In other cases, for example, the PEDV sM mRNA, the 5′ NTR is predicted to be relatively short.

## V. *IN VIVO* STUDIES

Only a small number of studies have convincingly demonstrated the expression of the sM protein in coronavirus-infected cells and its incorporation into virion particles. The standard approach has been to construct bacterial fusion proteins containing sM protein sequences and to generate monoclonal antibodies or polyclonal sera which are then used in immunoprecipitation (IP), immunocytochemistry, or immunofluorescence (IF).

First, Smith *et al.* (1990) used an sM-fusion protein specific polyclonal serum to immunoprecipitate an approximately 12,000 molecular weight protein (the expected size for the IBV sM polypeptide is 12,400) from infected CK or Vero cells. As judged by cell fractionation and IP, the sM protein was associated with a large membrane fraction of the cell. IF studies demonstrated a juxtanuclear location with the transport of at least some sM protein to the cell surface. These conclusions were essentially confirmed by Liu *et al.* (1991).

Liu and Inglis (1991) subsequently demonstrated, for the first time, the association of the IBV sM protein with the virion envelope. Using the same polyclonal antiserum, these authors were able to immunoprecipitate the sM protein from highly purified radiolabeled preparations of IBV. Moreover, the sM protein was found to cofractionate with the other virion envelope proteins (S and M) upon detergent disruption of virus particles. The molar ratios of the S:N:M:sM proteins in purified IBV virions was calculated to be approximately 1:11:10:2.

Second, Godet *et al.* (1992), using monoclonal antibodies generated by immunization with a recombinant sM protein expressed in insect cells, were able to detect a 10,000 molecular weight protein (the predicted size of the TGEV sM polypeptide is 9200) in TGEV-infected porcine (PD5) cells. Indirect IF showed that the subcellular location of the sM protein was juxtanuclear and, again, a proportion of sM molecules were detected at the cell surface. Furthermore, incorporation of the sM protein into TGEV virions was detected by the analysis of purified, radiolabeled virus and the sM protein could be immunoprecipitated from the virion membrane fraction using monoclonal antibodies. The molar ratios of the TGEV sM:S:M proteins were calculated to be 1:20:300.

Third, Leibowitz *et al.* (1988), using an MHV sM-fusion protein specific polyclonal antiserum were able to detect expression of the sM protein in MHV-infected L2 cells by immunocytochemistry, or IP followed by one- or two-dimensional polyacrylamide gel electrophoresis. Subsequently, Yu *et al.* (1994) were able to show that the MHV sM protein is located in the virion envelope. In these studies, a radiolabeled protein, with the same molecular weight as the sM protein immunoprecipitated from MHV infected L2 cells, could be detected in highly purified MHV virions. Additionally, these authors were able to show that the rabbit antiserum was able to neutralize viral infectivity in the presence of complement. The IF staining of MHV-infected cells with two goat antipeptide antisera also revealed that the sM protein is membrane associated and, at least partially, transported to the cell surface. Similar IF data have been obtained for BCV-infected cells, using the MHV sM specific antiserum (Abraham *et al.*, 1990).

## VI. FUNCTION

There are no published experimental data on the function of the coronavirus sM protein. However, it appears to be essential because, in contrast to some other coronavirus genes [e.g., ns2 of MHV; Schwarz *et al.* (1990)], there have been no reports of viable coronaviruses with a defective sM gene. Nevertheless, almost all authors have speculated on a possible role for the protein, namely, in the assembly of virions. This is because small integral membrane proteins have been described in several enveloped viruses, including alphaviruses, influenza viruses, and paramyxoviruses (Garoff *et al.*, 1980; Welch and Sefton, 1980; Gaedigk-Nitschko and Schlesinger, 1990; Lamb *et al.*, 1985; Hiebert *et al.*, 1985; Olmsted and Collins, 1989) and they have been implicated in the assembly process. Thus, for example, site-directed mutagenesis studies on Semliki Forest virus have shown that the sM (6 kDa) protein probably has a role in virus assembly and the budding process (Liljström *et al.*, 1991)

Recently, a system has been developed that may provide important insights into the coronavirus assembly process. This system is based on the cellular coexpression of coronavirus structural proteins, using a T7/vaccinia virus-based format, and the detection of viruslike particles in the culture supernatant (H. Venemma, Utrecht and E. Bos, Leiden, personal communication). Preliminary results suggest that the sM protein may, indeed, have an important role in particle assembly and/or budding.

## VII. REFERENCES


Abraham, S., Kienzle, T. E., Lapps, W. E., and Brian, D. A., 1990, Sequence and expression analysis of potential non-structural proteins of 4.9, 4.8, 12.7 and 9.5 kDa encoded betwen the spike and membrane protein genes of the bovine coronavirus, *Virology* **177**:488.

Boursnell, M. E. G., Binns, M. M., and Brown, T. D. K., 1985, Sequence of coronavirus IBV genomic RNA: Three open reading frames in the unique region of mRNA D, *J. Gen. Virol.* **66**:2253.

Budzilowicz, C. J., and Weiss, S. R., 1987, *In vitro* synthesis of two polypeptides from a non-structural gene of coronavirus mouse hepatitis virus strain A59, *Virology* **157**:509.

De Groot, R. J., ter Haar, R. J., Horzinek, M. C., and van der Zeijst, B. A. M., 1987, Intracellular RNAs of the feline infectious peritonitis coronavirus strain 79-1146, *J. Gen. Virol.* **68**:995.

Duarte, M., Tobler, K., Brigden, A., Rasschsert, D., Ackermann, M., and Laude, H., 1994, Sequence analysis of the porcine epidemic diarrhea virus genome between the nucleocapsid and spike protein genes reveals a polymorphic ORF, *Virology* **198**:466.

Gaedigk-Nitschko, K., and Schlesinger, M. J., 1990, The Sindbis virus 6K protein can be detected in virions and is acylated with fatty acids, *Virology* **175**:274.

Garoff, H., Frischauf, A-M., Simons, K., Lehrach, H., and Delius, H., 1980, Nucleotide sequence of cDNA coding for Semliki Forest virus membrane glycoproteins, *Nature* **288:**236.

Godet, M., L'Haridon, R., Vautherot, J-F., and Laude, H., 1992, TGEV corona virus ORF 4 encodes a membrane protein that is incorporated into virions, *Virology* **188:**666.

Hartmann, E., Rapoport, T. A., and Lodish, H. F., 1989, Predicting the orientation of eucaryoutic membrane-spanning proteins, *Proc. Natl. Acad. Sci. USA* **86:**5786.

Hiebert, S. W., Paterson, R. G., and Lamb, R. A., 1985, Identification and predicted sequence of a previously unrecognized small hydrophobic protein, SH, of the paramyxovirus simian virus 5, *J. Virol.* **55:**744.

Horsburgh, B. C., Brierley, I., and Brown, T. D. K., 1992, Analysis of 9.6 kb sequence from the 3′ end of canine coronavirus genomic RNA, *J. Gen. Virol.* **73:**2849.

Jacobs, L., van der Zeijst, B. A. M., and Horzinek, M. C., 1986, Characterization and translation of transmissible gastroenteritis virus mRNAs, *J. Virol.* **57:**1010.

Kozak, M., 1989, The scanning model for translation: An update, *J. Cell Biol.* **108:**229.

Lamb, R. A., Zebedee, S. L., and Richardson, C. D., 1985, Influenza virus M2 protein is an integral membrane protein expressed on the infected cell surface, *Cell* **40:**627.

Le, S-Y., Chen, J-H., Sonenberg, N., and Maizel, J. V., 1992, Conserved tertiary structure elements in the 5′ untranslated region of human enteroviruses and rhinoviruses, *Virology* **191:**858.

Le, S-Y., Chen, J-H., Sonenberg, N., and Maizel, J. V., Jr., 1993, Conserved tertiary structural elements in the 5′ nontranslated region of cardiovirus, apthovirus and hepatitis A virus RNAs, *Nucleic Acids Res.* **21:**2445.

Le, S., Sonenberg, N., and Maizel Jr., J. V., 1994, Distinct structural elements and internal entry of ribosomes in mRNA 3 encoded by infectious bronchitis virus, *Virology* **198:**405.

Leibowitz, J. L., Weiss, S. R. Paavola, E., and Bond, C. W., 1982, Cell-free translation of murine coronavirus RNA, *J. Virol.* **43:**905.

Leibowitz, J. L., Perlman, S., Weinstock, G., De Vries, J. R., Budzilowicz, C., Weissemann, J. M., and Weiss, S. R., 1988, Detection of a murine coronavirus nonstructural protein encoded in a downstream open reading frame, *Virology* **164:**156.

Liljeström, P., Lusa, S., Huylebroeck, D., and Garoff, H., 1991, *In vitro* mutagenesis of a full length cDNA clone of Semliki Forest virus: The small 6,000 molecular weight membrane protein modulates virus release, *J. Virol.* **65:**4107.

Liu, D. X., and Inglis, S. C., 1991, Association of the infectious bronchitis virus 3c protein with the virion envelope, *Virology* **185:**911.

Liu, D. X., and Inglis, S. C., 1992, Internal entry of ribosomes on a tricistronic mRNA encoded by infectious bronchitis virus, *J. Virol.* **66:**6143.

Liu, D. X., Cavanagh, D., Green, P., and Inglis, S. C., 1991, A polycistronic mRNA specified by the coronavirus infectious bronchitis virus, *Virology* **184:**531.

Meerovitch, K., and Sonenberg, N., 1993, Internal initiation of picornavirus RNA translation, *Semin. Virol.* **4:**217.

Olmsted, R. A., and Collins, P. L., 1989, The 1A protein of respiratory syncytial virus is an integral membrane protein present as multiple, structurally distinct species, *J. Virol.* **63:**2019.

Raabe, T., Schelle-Prinz, B., and Siddell, S. G., 1990, Nucleotide sequence of the gene encoding the spike glycoprotein of human coronavirus HCV 229E, *J. Gen. Virol.* **71:**1065.

Rasschaert, D., Gelfi, J., and Laude, H., 1987, Enteric coronavirus TGEV: Partial sequence of the genomic RNA, its organization and expression, *Biochemie* **69:**591.

Rottier, P. J. M., Spann, W. J. M., Horzinek, M. C., and van der Zeijst, B. A. M., 1981, Translation of three mouse hepatitis virus strain A59 subgenomic RNAs in *Xenopus laevis* oocytes, *J. Virol.* **38:**20.

Schwarz, B., Routledge, E., and Siddell, S. G., 1990, Murine coronavirus nonstructural protein ns2 is not essential for virus replication in transformed cells, *J. Virol.* **64:**4784.

Siddell, S., 1983, Coronavirus JHM: Coding assignments of subgenomic mRNAs, *J. Gen. Virol.* **64:**113.

Skinner, M. A., Ebner, D., and Siddell, S. G., 1985, Coronavirus MHV-JHM mRNA 5 has a sequence arrangement which potentially allows translation of a second, downstream open reading frame, *J. Gen. Virol.* **66:**581.

Smith, A. R., Binns, M. M., Boursnell, M. E. G., Brown, T. D. K., and Inglis, S. C., 1990, Identification of a new membrane-associated polypeptide specified by the coronavirus infectious bronchitis virus, *J. Gen. Virol.* **71**:3.

Stern, D. F., and Sefton, B. M., 1984, Coronavirus multiplication: Location of genes for virion proteins on the avian infectious bronchitis virus genome, *J. Virol.* **50**:22.

Thiel, V., and Siddell, S. G., 1994, Internal ribosome entry in the coding region of murine hepatitis virus mRNA 5, *J. Gen. Virol.* **75**:3041.

Welch, W. J., and Sefton, B. M., 1980, Characterization of a small nonstructural viral polypeptide present late during infection of BHK cells by Semliki Forest virus, *J. Virol.* **33**:230.

Yu, X., Bi, W., Weiss, S. R., and Leibiwitz, J. L., 1994, Mouse hepatitis virus gene 5b is a new virion envelope protein, *Virology* **202**:1018.

CHAPTER 10

# The Coronavirus Nonstructural Proteins

T. D. K. Brown and I. Brierley

## I. INTRODUCTION

The definition of a group of virus-coded proteins/polypeptides, the nonstructurals, as those not found in virions is crude, but has some merit in that it immediately focuses attention on the potential difficulties encountered in defining them. It leaves aside the practical difficulties implicit in attempting to detect small amounts of virus-coded polypeptides, particularly if they are of low molecular weight, which may be present in virions, and, on the other hand, in deciding whether or not such polypeptides, if detected, are present adventitiously. This chapter will thus consider currently available information relating to coronavirus polypeptides (or predicted polypeptides) other than those routinely present in the virions of all coronaviruses subjected to detailed analysis, i.e., the spike glycoprotein (S), the integral membrane glycoprotein (M), the small-membrane protein (sM), and the nucleocapsid protein (N), or present in only some coronaviruses, i.e., the hemagglutinin esterase (HE) glycoprotein.

A brief consideration of the range of methods that have been employed to detect nonstructural proteins in virus-infected cells and the limitations of such methods is relevant to an analysis of available data on coronavirus nonstructural proteins. Prior to the introduction of techniques for molecular cloning, only a limited number of approaches were available for detecting nonstructural polypeptides. The most obvious of these was the detection, using radiolabeling techniques, of products absent from virions but present in infected cells. The success of this type of approach is dependent in part on the efficiency of host

T. D. K. BROWN and I. BRIERLEY • Virology Division, Department of Pathology, Cambridge University, Cambridge CB2 1QP, England.

*The Coronaviridae*, edited by Stuart G. Siddell, Plenum Press, New York, 1995.

cell shutoff; this is frequently inefficient in coronavirus-infected cells. The *in vitro* translation of purified virus mRNAs provided a potential alternative approach to identification of virus-coded polypeptides not present in virions. Assay of enzymatic activity predicted to be associated with a nonstructural protein, e.g., virus-coded RNA-dependent RNA polymerase activity, is also possible. These approaches have been applied with varying success to coronaviruses.

The prospects for analyzing nonstructural proteins were dramatically improved by the advent of molecular cloning techniques and consequent ease of sequence determination. The cloning and sequencing of coronavirus cDNAs has revealed substantial numbers of open reading frames (ORFs) in the unique regions of virus mRNAs known not to be involved in the synthesis of virion proteins. In most cases, no candidate product of these ORFs had previously been detected in infected cells. Cloned sequences can be employed to develop reagents for the detection of putative nonstructural proteins coded by the ORFs; production of monospecific antisera is particularly valuable in this context.

Detection is, however, only an initial, albeit essential, step in the characterization of nonstructural proteins. A range of functional studies are ultimately required for an understanding of the roles of these proteins; such studies have scarcely been started for coronaviruses. We are thus, at the present time, aware of the existence of a substantial number of ORFs potentially coding for nonstructural polypeptides without, in most cases, significant insights into their functions other than those which can be gained by more-or-less sophisticated computer-based sequence analyses.

The minimum function (RNA-dependent RNA synthesis) of the nonstructural proteins encoded by mRNA 1 (genomic/virion RNA) is defined by its infectivity (Lomniczi and Kennedy, 1977; Schochetman *et al.*, 1977); this makes it unique among coronavirus nonstructural proteins. For this reason the products encoded by mRNA 1s will be considered together for all coronaviruses; other ORFs for which no such unifying knowledge of function exists will be considered on the basis of individual viruses or groups of related coronaviruses.

## II. PRODUCTS OF mRNA 1

### A. Sequence Analysis of the Unique Regions of mRNA 1s

#### 1. General Analysis

Four complete sequences of RNA 1 unique regions are currently available: those of avian infectious bronchitis virus (IBV) strain Beaudette, murine hepatitis virus (MHV) strains JHM and A59, and human coronavirus (HCV) strain 229E. Sequencing of the putative unique region of mRNA 1 of IBV represented in 17 overlapping cDNA clones (Boursnell *et al.*, 1987) provided the first convincing evidence that this region of the coronavirus genome had an extremely large coding capacity (approximately 20 kb in the case of IBV) and that it contained two overlapping ORFs (1a and 1b) capable, in the case of IBV, of coding for polypeptides of 441,000 and 300,000, Da respectively. The downstream 1b ORF was found to be in the −1 frame with respect to the upstream 1a ORF. There was a 42-nucleotide overlap between the ORFs. The implication was that coronavirus polymerases were likely to be complex and that their components might be expressed by an unusual mechanism.

The unique region of RNA 1 of MHV strains JHM and A59 were subsequently cloned and sequenced; they are approximately 22-kb long and again were found to contain two overlapping ORFs (1a and 1b) (Baker *et al.*, 1989; Bonilla *et al.*, 1994; Bredenbeek *et al.*, 1990b; Lee *et al.*, 1991; Pachuk *et al.*, 1989; Soe *et al.*, 1987). The relationship between the 1a and 1b ORFs is similar to that seen in IBV; the 1b ORF is in the −1 frame with respect to 1a and there is a 75-nucleotide overlap between them. Attention has been drawn to apparent minor errors in the published sequence of MHV-JHM and also to a region of genuine heterogeneity detected also in MHV-2 (Bonilla *et al.*, 1994).

The cloning and sequencing of the HCV 229E mRNA 1 unique region has also been reported (Herold *et al.*, 1993). Again it was found to be large (approximately 20 kb) and to contain two overlapping reading frames with the 1b ORF being in the −1 frame with respect to the 1a ORF and overlapping it by 43 nucleotides. A summary of the major features of the available mRNA 1 unique region sequences are presented in Table I.

The overall relationships between the amino acid sequences of the predicted products of the 1a and 1b ORFs have been determined by computer analysis (summarized in Table I). It is clear that the 1b ORF is much more highly conserved than the 1a ORF, and that within 1a the greatest divergence is seen at the amino-terminus. The substantial additional sequences present in MHV 1a, but not in IBV 1a, are found at the 5′ end (Bredenbeek *et al.*, 1990b; Herold *et al.*, 1993; Lee *et al.*, 1991; Pachuk *et al.*, 1989; Soe *et al.*, 1987). The high degree of amino acid sequence conservation in 1b is also indicated by the limited data available for porcine transmissible gastroenteritis virus (TGEV), feline infectious peritonitis virus (FIPV), and canine coronavirus (CCV). This high degree of conservation is in marked contrast to that observed for other coronavirus nonstructural proteins (see Sections III to X).

TABLE I. Summary of the Features of Sequenced mRNA 1 Unique Regions[a]

| | IBV Beaudette | MHV-JHM | MHV A59 | HCV 229E |
|---|---|---|---|---|
| Length of 1a ORF (aa) | 3951 | 4488 | 4468 | 4085 |
| Length of 1b ORF (aa) | 2691 | 2731 | 2733 | 2686 |
| Length of 1a/1b fusion (aa)[b] | 6629 | 7203 | 7176 | 6758 |
| 1a/1b Overlap (nt) | 42 | 75 | 75 | 43 |
| 1a Amino acid identity to IBV (%) | (100) | 29 | 29 | 27 |
| 1a Amino acid similarity to IBV (%) | (100) | 52 | 52 | 51 |
| 1b Amino acid identity to IBV (%) | (100) | 56 | 57 | 54 |
| 1b Amino acid similarity to IBV (%) | (100) | 72 | 72 | 70 |

[a]Data obtained from published sequences using the GCG GAP program with default settings.
[b]based on simultaneous slippage model of Brierley *et al.* (1987, 1989, 1992).

## 2. Motif Analysis

Computer-based motif analyses, in the absence of substantial efforts to obtain experimental data, have been extensively applied to the nucleotide sequences of the mRNA 1 unique regions and to the amino acid sequences derived from them. Initial analyses of the IBV sequence demonstrated amino acid homologies in the 1b ORF with RNA-dependent RNA polymerase-related polypeptides of Sindbis virus, an alphavirus, and brome mosaic virus, a tricornavirus; these lay between amino acids 1200 and 1500 (Boursnell *et al.*, 1987). More sophisticated analyses of the 1a and 1b ORFs were subsequently carried out (Gorbalenya *et al.*, 1988a,b, 1989b; Hodgman, 1988). The homologies detected by Boursnell *et al.* (1987) were confirmed and their relationships extended using sequences from a wide range of RNA viruses. It was suggested that they included an nucleoside triphosphate (NTP) binding domain characteristic of helicases.

Other motifs characteristic of positive-strand virus RNA-dependent RNA polymerases were also detected. The most "obvious" of these was the cryptic GDD motif present in the IBV sequence as SDD. A region with extremely high homology to potyvirus polymerases is noteworthy. Various other features were identified in the 1a and 1b ORFs. These included a potential zinc finger nucleic acid binding domain in 1b, a homologue of the 3C protease of picornaviruses in 1a, and 13 putative Q/G,S target sites distributed in both 1a and 1b. Homology to a fragment of the catalytic center of another class of cysteine protease from *Streptococcus pneumoniae* (SPL protease) was detected in 1a. Attention was also drawn to a possible growth factor-related domain and two potential membrane spanning regions in 1a.

The availability of the IBV sequence and the analytical abilities of the Russian group laid the foundations for the dissection of other coronavirus mRNA 1 sequences as they became available. The possibilities for both narrow and broad phylogenetic comparisons are now more extensive and it is possible to consider the features of a core set of coronavirus mRNA 1 motifs, to which putative functions can be ascribed, and their relationships to both viral and cellular homologues. Those currently identified are: in 1a (1) a papainlike cysteine protease domain(s), (2) a chymotrypsin/picornaviral 3C proteaselike domain, and (3) a cysteine-rich growth factor-related domain; and in 1b (4) a classical positive-strand RNA-dependent RNA polymerase domain, (5) a zinc finger nucleic acid binding domain, and (6) an NTP binding/helicase domain. To this list it is perhaps reasonable to add the conserved membrane spanning domains. The locations of these motifs/domains are presented in Fig. 1 and Table II.

### *a. Papainlike Cysteine Protease Domains*

Viral homologues of the classical papainlike proteases (PLP) were first identified in poty- (Oh and Carrington, 1989) and alphaviruses (Hardy and Strauss, 1989; Strauss *et al.*, 1992). The identities of their putative active site cysteine and histidine residues were confirmed by site-directed mutagenesis.

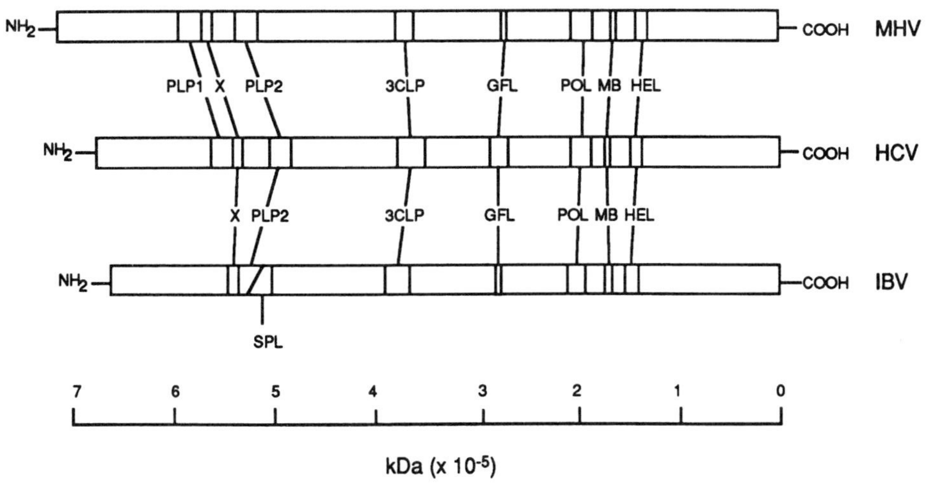


FIGURE 1. The positions of motifs identified in the predicted polypeptide sequence of the 1a/b polyprotein encoded by the mRNA 1 unique regions of IBV, MHV, and HCV are shown. PLP = papain-like protease, 3CLP = 3C-like protease, GFL = growth factor-like, POL = RNA-dependent RNA polymerase, MB = metal binding, HEL = helicase. The data are from computer assisted analyses (Gorbalenya *et al.*, 1989; Lee *et al.*, 1991; Herold *et al.*, 1993).

Homologues were subsequently identified in coronaviruses (Gorbalenya *et al.*, 1989b; Herold *et al.*, 1993; Lee *et al.*, 1991). Two PLP domains are present in both MHV and HCV, but only one is present in IBV. The IBV domain lies upstream of and overlaps the SPL protease domain. The role of the upstream PLP (PLP1) domain in the cleavage of p28 from the amino-terminus of MHV 1a has been demonstrated (Baker *et al.*, 1989, 1993). It has been suggested that this PLP

TABLE II. Positions of Motifs within the Predicted Amino Acid Sequences of mRNA 1a and b ORFs[a]

| | IBV | MHV-JHM | HCV |
|---|---|---|---|
| ORF 1a | | | |
| PLP1 | — | 1100–1349 | 1041–1234 |
| PLP2 | 1236–1497 | 1696–1953 | 1688–1886 |
| SPL | 1385–1677 | — | — |
| 3CLP | 2811–2943 | 3382–3517 | 2965–3268[b] |
| GFL | 3894–3917 | 4446–4468 | 3933–4069[b] |
| ORF 1b | | | |
| POL | 588–818 | 560–852 | 534–836 |
| MB | 931–1005 | 942–1015 | 924–999 |
| HEL | 1209–1500 | 1217–1506 | 1202–1330 |

[a]Based on data from Gorbalenya *et al.* (1988a,b, 1989a, b, 1991), Herold *et al.* (1993), Koonin (1991), and Lee *et al.* (1991).
[b]Boundaries of predicted processed polypeptide-containing motif rather than those of motif itself (based on assumptions about targets for the 3CLP).

domain is particularly closely related to the leader protease of foot-and-mouth disease virus (FMDV) (Gorbalenya *et al.*, 1991). There are no data available on the role of the downstream PLP domain (PLP2) of MHV. There are similarly no published data on the role of either of the PLP domains of HCV 229E or of the single PLP domain of IBV. The IBV PLP is homologous to the downstream PLP2 domains of MHV and HCV.

It is interesting that the arteriviruses, which contain a subset of the coronavirus polymerase sequences in their mRNA 1 unique regions, encode a PLP domain in a vestigial 1a ORF and that it cleaves 1a to generate a 30-kDA polypeptide. The cleavage is, however, downstream of the PLP domain (Snijder *et al.*, 1992). Attention has also been drawn to a region highly conserved between IBV and MHV, the so-called X domain, which is present immediately upstream of the IBV PLP domain but immediately downstream of the MHV PLP1 domain. A related domain has been detected close to the thiol protease domains of alpha- and rubiviruses (Gorbalenya *et al.*, 1991).

### *b. Chymotrypsin/Picornaviral 3C-like Protease Domains*

The chymotrypsin/picornaviral 3C-like protease (3CLP) domains of approximately 300 amino acids with homologies to proteases identified in picorna-, como-, nepo-, poty-, sobemo-, and luteoviruses are present in the 3′ region of the 1a ORFs of IBV, MHV, and HCV. The proposed catalytic histidine and cysteine residues initally identified in IBV are present in MHV and HCV (Herold *et al.*, 1993; Lee *et al.*, 1991). Unusual features of the coronavirus 3CLPs include, first, the lack of conservation in MHV and HCV of what appeared to be a conserved, catalytically active acidic residue in IBV and other viral 3CLPs (Gorbalenya *et al.*, 1989b), and, second, the substitution of coronavirus sequences of tyrosine for the conserved glycine normally seen in the putative 3CLP active site substrate binding region (Herold *et al.*, 1993; Lee *et al.*, 1991). Attempts have been made to predict cleavage sites for the putative 3CLP, based largely on studies of picornaviral enzymes; the cleavages occur primarily at Q,E/G,S,A but are selective, and the basis for the selectivity is unclear. Gorbalenya *et al.* (1989) derived a consensus sequence for IBV including Q/S or G using approaches based on comparison of sequences surrounding the Q/S and Q/G sites flanking putative functional domains. Further potential cleavage sites conforming to the consensus were then identified; a total of 13 sites was reported. It was suggested that a significant feature of the consensus was the presence of a hydrophobic residue at −1, which is a general feature of 3CLP target sequences. The sites flanking the 3CLP domain both had a positively charged residue at −3. Conservation of at least some of these potential cleavage sites has been observed in MHV and HCV (Herold *et al.*, 1993; Lee *et al.*, 1991). The membrane domains flank the 3CLP domain.

### *c. Cysteine-Rich Growth Factor-Related Domains*

This type of sequence is found in the predicted 1a polypeptide sequences of IBV, MHV, and HCV (Gorbalenya *et al.*, 1989b; Herold *et al.*, 1993; Lee *et al.*,

1991). It is clear that it is exceptionally cysteine-rich, but its relationship to growth factors may be distant and evidence of functional relationships to growth factors is lacking.

#### *d. RNA-Dependent RNA Polymerase Domains*

The amino acid sequence relationships between positive-strand virus RNA-dependent RNA polymerases are well established (Argos, 1988; Koonin, 1991). Notable features include the high degree of conservation of this region between IBV, MHV, and HCV, in particular the finding that the coronavirus sequences all contain an SDD motif rather than the classical GDD motif. No evidence for a second polymerase domain has been obtained.

#### *e. Zinc Finger Nucleic Acid-Binding Domain*

This domain, characterized by the presence of a defined series of histidine and cysteine residues, is again clearly conserved in the three coronavirus 1b sequences; but some of the structural features proposed on the basis of the IBV sequence cannot be present in the MHV or HCV polypeptides. Evidence for metal binding has yet to be obtained.

#### *f. NTP Binding/Helicase Domain*

The homology between this region of the coronavirus genome and sequences present in other positive-strand RNA viruses was detected in initial analyses of the IBV sequence, but its significance was not appreciated until the work of Gorbalenya and Koonin (Gorbalenya *et al.*, 1988a,b, 1989a,b). Not surprisingly, the domain is well conserved in MHV and HCV.

### B. Expression of the 1a and 1b ORFs *in Vitro*

Cell-free approaches have been used to analyze the translation of genomic RNA/mRNA 1. The translation of both genomic RNA preparations and *in vitro* transcripts prepared from cloned cDNAs has been studied. Virion RNA of MHV was translated in rabbit reticulocyte lysate (RRL) with the production of three structurally related polypeptides of molecular weight greater than 200 kDa (Leibowitz *et al.*, 1982). A series of minor products of lower molecular weight could also be discerned. These observations have been extended in a number of ways (Denison and Perlman, 1986). A series of products were detected in RRL translations with two quantitatively significant products with molecular weights greater than 200 kDa being emphasized (p220 and p250). A well-defined product of low molecular weight was characterized. It was demonstrated to be N-terminal and to be generated by proteolytic cleavage from a relatively large precursor. The precursor–product relationships of p28 and the discrete high-molecular-weight products was not however completely clarified. The inhibitory effects of leupeptin, a reversible inhibitor of trypsinlike serine/cysteine proteases, and $ZnCl_2$ on p28 cleavage were demonstrated. Both these studies were carried out prior to the publication of the IBV polymerase sequence (Boursnell *et al.*, 1987), which revealed for the first time the potential complexity of the coronavirus polymerase "problem."

The observations of Denison and Perlman (1986) were extended by Soe *et al.* (1987) using *in vitro* translation of *in vitro* transcripts prepared from the 5′-most 2 kb of the MHV genome. They confirmed the N-terminal location of p28. They also demonstrated that deletion of sequences upstream of the potential initation codon at position 215 increased the efficiency of translation in RRL of *in vitro* transcripts derived from the 2 kb fragment. It was suggested that the low levels of translation observed when the 5′ untranslated region (UTR) is intact result from the presence of highly stable RNA secondary structures. The work on MHV was extended (Baker *et al.*, 1989) with the demonstration that sequences involved in the cleavage of p28 were located between 3.9 and 5.3 kb from the 5′ end of the MHV genome and the observation that the protease was active only in *cis*. Following the identification of two PLP domains in the MHV 1a sequence (Lee *et al.*, 1991), a more detailed examination of the sequences responsible for the cleavage of p28 was carried out (Baker *et al.*, 1993). The catalytic functions of the putative active site cysteine and histidine residues of PLP2 identified by computer-based analysis were confirmed by site-directed mutagenesis. Its activity was shown to be sensitive to the deletion of at least some of the polypeptide sequences between it and the N-terminal target site.

*In vitro* translation of MHV genomic RNA has been used to study processing of the polypeptide sequences encoded by the 1b ORF (Denison *et al.*, 1991). A series of polypeptides recognized by antisera raised against the carboxy-terminus of ORF 1b were identified. Leupeptin, in contrast to its effect on p28 cleavage, did not alter the pattern of cleavage observed for the 1b products, thus providing preliminary evidence for a role for at least two proteolytic activities in the cleavage of the MHV gene 1 polypeptides. It proved impossible, however, to align convincingly the cleavage products (apparent molecular weights, 90, 74, 53, 44, and 32 kDa) with predicted 3CLP cleavage sites in 1b.

An important question that arose following completion of the sequence of IBV was the mechanism by which ORF 1b is expressed, since it overlaps 1a by 42 nucleotides and is in the −1 reading frame with respect to 1a. A number of observations raised the possibility that 1b may be expressed as a fusion protein with the upstream 1a ORF following a −1 ribosomal frameshift at the overlap region of the two ORFs (Boursnell *et al.*, 1987). First, the 70 or so bases preceding the first AUG codon of 1b were found to have strong codon bias, similar to the bias found in other IBV genes, suggesting they had coding function. Second, no subgenomic mRNA with 1b as its 5′-proximal ORF had been detected in IBV-infected cells. Third, nucleotide sequence comparisons of the 1a/1b overlap region with the *gag–pol* overlap region of Rous sarcoma virus (RSV) revealed a short but significant homology (8 out of 9 bases conserved) in the region where frameshifting was suspected to occur in RSV (Jacks and Varmus, 1985).

In order to test the possibility of ribosomal frameshifting in IBV, a region of cDNA corresponding to the 1a/1b overlap region was cloned within a reporter gene and tested for frameshifting by *in vitro* transcription and translation (Brierley *et al.*, 1987). It was confirmed that the 1a/1b region indeed specified a highly efficient −1 frameshift, with one in three ribosomes changing frame within the overlap region. This was the first nonretroviral example of the phenomenon. The IBV frameshift signal was also functional *in vivo*; synthetic transcripts containing the appropriate signals when injected into *Xenopus* oocytes produced the expected ratio of nonframeshifted to frameshifted products (Brierley *et al.*, 1990).

In further experiments, the signals for frameshifting were investigated using site-directed mutagenesis (Brierley *et al.*, 1991, 1992). These signals were shown to be located within an 86-nucleotide stretch encompassing the 1a/1b overlap and were comprised of two main elements (see Fig. 2). The first was the sequence UUUAAAC, which was shown to be the actual site of the frame change; of the ribosomes that enter in the 1a frame (U-UUA-AAC), some 30% leave in the 1b frame (UUU-AAA). This "slippery sequence" in itself was insufficient to evoke the change of frame, and additional information downstream is required. This information is in the form of an RNA pseudoknot, an unusual kind of RNA structure composed of two base-paired regions stacked coaxially in a quasi-continuous manner and connected by two single-stranded loop regions (Pleij and Bosch, 1989). The results indicated that no primary nucleotide sequence elements in the pseudoknot were required for the frameshift process; as long as the overall shape and predicted stability of the structure was maintained, frameshifting was highly efficient.

Frameshifting is thought to occur by simultaneous slippage into the −1 reading frame of two ribosome-bound tRNAs present in the aminoacyl and peptidyl sites of the ribosome during decoding of the slippery sequence. The role of the pseudoknot in this process is uncertain; the most plausible explanation advanced so far is that the presence of the pseudoknot may slow or stall the ribosome as it translates the slippery sequence, allowing realignment of the decoding tRNAs on the mRNA in a new frame (Jacks *et al.*, 1988). Consistent with this idea is that the RNA pseudoknot must be within 5 to 7 nucleotides of the slip site for efficient frameshifting to occur in the IBV signal (Brierley *et al.*, 1992), and additionally, that pausing at RNA pseudoknots can be demonstrated *in vitro* (Somogyi *et al.*, 1993; Tu *et al.*, 1992).

It is clear from recent work that the frameshift expression strategy appears to be conserved among coronaviruses (see Fig. 2). Frameshift signals have been described for MHV strains A59 (Bredenbeek *et al.*, 1990b) and JHM (Lee *et al.*, 1991) and also for HCV strain 229E (Herold *et al.*, 1993; Herold and Siddell, 1993). In each case, the slippery sequence UUUAAAC is used and is followed by an RNA pseudoknot. For IBV and MHV, the pseudoknots are similar in terms of the size of the stems and the loop lengths, although the stem 2 of MHV is predicted to be longer. The pseudoknot of HCV 229E, however, forms a more complex structure. Loop 2 of the HCV pseudoknot is 164 nucleotides in length, by far the longest loop 2 seen in those pseudoknots involved in frameshifting. Furthermore, the nucleotides that form the second arm of stem 2 of the pseudoknot are themselves part of a hairpin loop that is predicted to form at the end of loop 2 (Herold and Siddell, 1993). This generates an additional stem, stem 3. It is

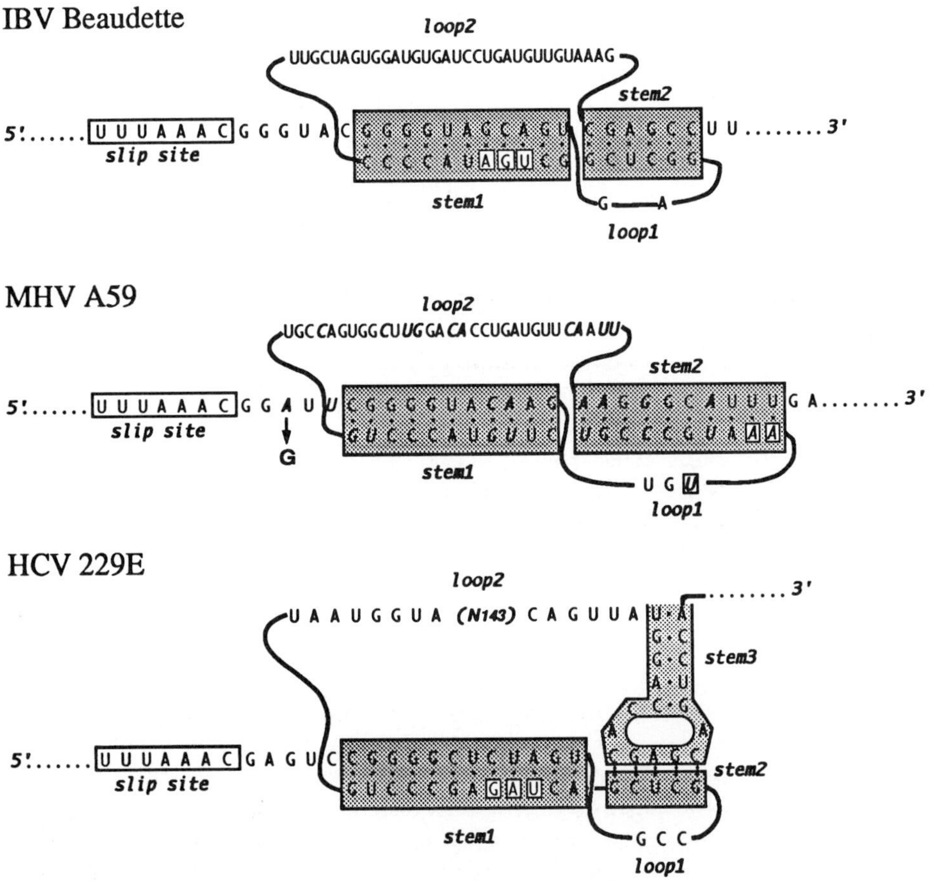


FIGURE 2. RNA pseudoknot structure associated with coronavirus frameshift sites. The italicized bases in the MHV A59 sequence are those that differ from those in the homologous IBV sequence (Boursnell *et al.*, 1987; Bredenbeek *et al.*, 1990). The frameshift signal of MHV JHM is identical to that of MHV A59 with exception of the indicated A to G change (Bredenbeek *et al.*, 1990; Lee *et al.*, 1991). In the case of each frameshift signal, the 1a termination codon is indicated by white, individually boxed characters. The data for HCV 229E are from Herold *et al.* (1993).

possible that the formation of this hairpin allows stem 2 to form more readily, compensating both for the extra length of loop 2 and the relative instability of the HCV stem 2, which has only five base pairs instead of the six base pairs seen for the other coronaviruses.

The discovery of the pseudoknot at the IBV frameshift site prompted a re-evaluation of the components of retrovirus frameshift signals. It became clear that many retroviral frameshift sites also had the potential to form RNA pseudoknots (Brierley *et al.*, 1989; ten Damm *et al.*, 1990). Over the past few years, several of these signals have been investigated and the frameshift efficiency of the sites measured *in vitro*. It seems that the coronaviruses have the most efficient frameshift signals, with 30–40% of ribosomes changing frame. For retroviruses, the efficiency varies from a few percent up to 20%. Clearly, the combination of the UUUAAAC sequence and a long, stable RNA pseudoknot in coronaviruses provides a very efficient frameshift signal. Why coronaviruses employ a frameshift strategy remains to be determined, but there are a number of possibilities. The most obvious is the likely requirement for production of a defined ratio of 1a:1a/1b products. In retroviruses, frameshifting generates the gag–pro, gag–pol, or gag–pro–pol polyproteins from which essential replication enzymes are derived. Disruption of the stoichiometry of the gag:gag–pol ratio is likely to have drastic consequences. The effect of modulation of the frameshift efficiency on virus viability has been studied for a double-stranded RNA virus, L-A, of *Saccharomyces cerevisae*, which employs a frameshift strategy to produce a gag–pol fusion protein. Experiments in which the ratio of gag:gag–pol proteins in cells was varied have indicated that virus viability would be greatly reduced if the ratio is even moderately distorted (Dinman and Wickner, 1992).

Frameshifting may also be a strategy to avoid packaging defective RNAs. In retroviruses, the packaging signal is located 3′-ward of the splice donor such that only unspliced RNAs are packaged. If the gag–pol fusion protein were produced from an RNA in which the *gag* termination codon was removed by a small splicing event, then this RNA would be packaged and would be nonviable (since it would not be able to produce the gag and pol products in the correct stoichiometric amounts). Retroviruses avoid this problem by employing a frameshift strategy. The discovery that the MHV 1b ORF contains the genomic packaging signal (van der Most *et al.*, 1991) provides a possible explanation for the absence of a subgenomic mRNA for 1b, since this RNA would compete with the full-length genome for packaging into virions. Coronaviruses may well produce 1b by frameshifting in order to avoid this predicament.

## C. Detection of the Products of mRNA 1 *in Vivo*

Only limited success has been achieved in directly detecting coronavirus polymerase components *in vivo*. A single report has been presented in which a component was detected following radiolabeling of infected cells (Denison and Perlman, 1987). The p28 product, first detected in *in vitro* translation experiments with genomic RNA, was found in infected BALB/c 17CL-1 cells, but not in purified virions using two-dimensional gel electrophoresis. This small basic protein accumulated in the cells at late time.

More recently, monospecific antisera raised against synthetic peptides or bacterial fusion proteins containing coronavirus polymerase sequences have been used to detect polymerase components in infected cells. Products from both the IBV and MHV mRNA 1 ORFs have been detected using Western blotting (Brierley *et al.*, 1990; Denison *et al.*, 1991; Zoltick *et al.*, 1990). A series of specific bands was observed, but it has not been possible to study their precursor–product relationships using this approach. Denison *et al.* (1991) used immunoprecipitation of radiolabeled MHV-A59-infected cell lysates with antipeptide or antifusion protein antisera to study the processing of the aminoterminus of the 1a polyprotein. The rapid appearance of p28 was demonstrated, but a complex spectrum of extremely high-molecular-weight products was detected by antisera directed against sequences downstream of the p28 coding region. Pulse-chase experiments suggested that polypeptides with apparent molecular weights of 240 and 50 kDa might represent mature cleavage products from the polyprotein region encoded within the 5′-most 10 kb of MHV 1a.

More success has been achieved in detecting and characterizing discrete products from the IBV ORF 1b. A 100-kDa polypeptide has been detected in infected cells by immunoprecipitation using an antibody raised against a region of the 1b ORF containing the SDD polymerase motif. Mapping and processing mechanisms have been studied using T7 vaccinia/transfection approaches. The approximate position of the carboxy-terminus has been mapped using deletion analysis and is consistent with generation of the 100-kDa polypeptide by cleavage at a Q/S site thought to be a target for the viral 3CLP (Gorbalenya *et al.*, 1989). The position of the upstream cleavage site has not been determined experimentally, but the size of the processed product is consistent with a further 3CLP-mediated Q/S cleavage. The role of the 3CLP in processing has been confirmed by deletion of the 3CLP domain from constructs containing 1a/1b sequences; this resulted in appearance of a full-length translation product and disappearance of the processed 100-kDa product (Liu *et al.*, 1994).

The detection of RNA-dependent RNA polymerase activities in TGEV- and MHV-infected cells has been reported (Brayton *et al.*, 1982, 1984; Dennis and Brian, 1982; Mahy *et al.*,1983) There is no direct evidence linking these activities to the expression of the unique region of mRNA 1, but it is highly likely that they are a reflection of complexes formed between the range of polypeptides generated by processing of the 1a and 1ab polyproteins. The activity is associated with membranes and evidence has been presented for several distinct activities; an early activity involved in synthesizing full-length negative strands and a late activity which can be further subdivided into a replication fraction synthesizing full-length positive strands and a transcription fraction involved in production of both full-length and subgenomic mRNAs have been identified (Brayton *et al.*, 1984).

## III. PUTATIVE NONSTRUCTURAL ORFs PRESENT IN THE SUBGENOMIC RNAs OF MHV

### A. mRNA 2

#### 1. The 2a ORF

The MHV strains JHM and A59 differ in the expression of ORFs present in the genome between the polymerase and the spike genes. In the case of MHV-JHM two mRNAs 2 and 2-1 have been identified; RNA 2 contains the 2a ORF in its unique region, while RNA 2-1 contains the HE-encoding 2b ORF in its unique region. MHV A59, on the other hand, appears to produce only a single mRNA 2 that contains the 2a ORF and a defective 2b ORF in its unique region. The 2a ORF is considered to encode a nonstructural polypeptide. The 2a ORF sequences are highly related in both MHV-JHM and A59 (Luytjes *et al.*, 1988; Shieh *et al.*, 1989), potentially coding for, respectively, a 265 and a 261 amino acid polypeptide. Computer analysis of the ORF has suggested the possible presence of a nucleotide binding site with the possible implication of a nucleic acid binding function (Luytjes *et al.*, 1988). No homologue of the 2a ORF is found in IBV, TGEV, FIPV, or CCV.

### 2. The 2a Product

A 30- to 35-kDa product expressed from purified MHV RNA 2 was detected following *in vitro* translation of purified RNA (Leibowitz *et al.*, 1982; Siddell, 1983). Products of this size had previously been identified in MHV-infected cells (Bond *et al.*, 1979; Siddell *et al.*, 1981). The expression of the 2a product in MHV-infected cells has been confirmed using antisera raised against bacterial fusion proteins containing parts of the putative 2a sequence (Bredenbeek *et al.*, 1990a; Zoltick *et al.*, 1990). A 30-kDa product was identified by both groups. It had a cytosolic localization (Zoltick *et al.*, 1990). The pattern of its expression was similar to that of virus structural proteins, but it had a more rapid turnover than the nucleocapsid protein. It was not detected in purified virions (Bredenbeek *et al.*, 1990a). The function of the 30-kDa product remains unknown, but it has been shown that strains of MHV-JHM deleted for 2a show normal growth properties *in vitro*. It is therefore possible that the 2a gene product is only of significance when the virus is growing *in vivo* (Schwarz *et al.*, 1990).

## B. mRNA 4

### 1. The 4 ORF

The unique region of mRNA 4 contains an ORF potentially capable of translation to produce a 139-residue polypeptide (Skinner and Siddell, 1985). Analysis of the predicted amino acid sequence revealed a number of interesting features. It has a large hydrophobic amino-terminus and a basic carboxy-terminus. It has a strikingly high content of threonine residues outside the hydrophobic region.

### 2. The 4 Product

Early *in vitro* translation studies failed to conclusively assign a polypeptide product of mRNA 4. A product of the size predicted to result from the translation of the RNA has been detected in infected cells (Siddell *et al.*, 1981) and in *in vitro* translation products of RNA fractions enriched for mRNAs 4 and 5 (Siddell, 1983). The similarity in size of the mRNA 4 ORF and ORFs found in the unique region of mRNA 5 made a coding assignment unsatisfactory on the basis of these data. The difficulty was resolved by the production of a specific antiserum against a bacterial fusion protein consisting of β-galactosidase fused to the predicted carboxy-terminus of the mRNA 4 polypeptide (Ebner *et al.*, 1988). The use of this antiserum in immunoprecipitation studies confirmed the expression of the predicted polypeptide in infected cells that was detected as a 15-kDa product. Its correlation with the previously observed *in vitro* translation product of approximately 15 kDa was also confirmed by immunoprecipitation. Immunofluorescence studies demonstrated a cytoplasmic location. The possibility that it was as major virion component was excluded; the antiserum possessed no neutralizing activity. The function of the 15-kDa product remains unknown. The mutation of the gene 4 transcriptional initiation signal in the MHV S strain, which prevents synthesis of the gene 4 mRNA, has no effect on growth in DBT cells (Yokomori and Lai, 1991).

## C. mRNA 5

Sequencing of the unique region of mRNA 5 of MHV-JHM and MHV-A59 revealed the presence of two ORFs, 5a and b (Budzilowicz and Weiss, 1987; Skinner *et al.*, 1985). No mRNA that might allow expression of the downstream ORF by a conventional scanning mechanism has been observed; no transcriptional signals were obvious in the sequence immediately upstream of the 5b ORF.

### 1. The 5a ORF

This is 321 bases in length in MHV-JHM and 336 bases in length in MHV-A59; it potentially encodes polypeptides of molecular weights 12.4 kDa (107 residues) and 13 kDa (112 residues), respectively. The predicted polypeptide is basic.

### 2. The 5a Product

The putative product of the 5a ORF has only been detected by *in vitro* translation of synthetic mRNA (Budzilowicz and Weiss, 1987). Deletion of almost the whole of the 5a ORF in the MHV S strain has no effect on its growth in DBT cells (Yokomori and Lai, 1991).

### 3. The 5b ORF

This is 264 bases in length in MHV-JHM and 249 bases in length in MHV-A59; it potentially encodes polypeptides of molecular weights 10.2 kDa (88 residues) and 9.6 kDa (83 residues), respectively. The predicted polypeptide has a strongly hydrophobic region toward the amino-terminus.

### 4. The 5b Product

Recently, it has been shown that the ORF 5b product is, in fact, the sM polypeptide of MHV (Yu *et al.*, 1994). Previously, a 9- to 10-kDa product was detected in MHV-infected cells by Skinner *et al.* (1985) and was proposed as a candidate for the product of the 5b ORF. Its synthesis in *in vitro* translation reactions was correlated to the abundance of mRNA 5 in the RNA preparations. The potential for expression of this ORF was confirmed by *in vitro* translation of synthetic mRNAs produced by T7 polymerase transcription of cloned cDNA sequences (Budzilowicz and Weiss, 1987). It is interesting to note that expression of the downstream 5b ORF from RNAs containing both ORFs was substantially more efficient than that of the upstream 5a ORF. However, the RNA species analyzed did not have the structure of the natural mRNA 5. The mechanism of the selectivity was not clarified.

The finding of a suboptimal initiator AUG context for the 5a ORF and the absence of internal methionine residues in its predicted sequence coupled with codon usage data had suggested that both ORFs might be expressed, possibly by a leaky scanning mechanism, but recent work using synthetic mRNAs has demonstrated that the translation of 5b is likely to involve internal entry of ribosomes (Thiel and Siddell, 1994). The expression of the 5b ORF in infected cells has been demonstrated using a specific antiserum raised against a recA/5b/β-galactosidase fusion protein expressed in *E. coli* (Leibowitz *et al.*, 1988). The antiserum detected the 5b product in infected cells using both peroxidase labeling of fixed cells and immunoprecipitation of infected cell extracts. The observed size was consistent with that of the 9- to 10-kDa product previously detected in infected cells by Skinner *et al.* (1985).

## IV. PUTATIVE NONSTRUCTURAL POLYPEPTIDE ORFs PRESENT IN THE SUBGENOMIC RNAs OF BCV

### A. mRNA 2-1

The close sequence relationship of BCV to MHV is reflected in part by the pattern of subgenomic RNAs detected in BCV infected cells. mRNA 2 has a unique region potentially coding for a nonstructural polypeptide and a mRNA 2-1 coding for the HE protein as in MHV-JHM (Keck *et al.*, 1988).

#### 1. The 2a ORF

The sequence of the single ORF present in the unique region of mRNA 2 has been determined (Cox *et al.*, 1989). The predicted polypeptide has a size of 32 kDa. It shares nucleotide (65%) and amino acid (45%) homology with the predicted product of the MHV 2a ORF. An antiserum raised against a β-galactosidase fusion protein has been used to detect expression of the 2a polypeptide (ns2) in infected cells. It has been demonstrated to be a nonstructural phosphoprotein that accumulates in the cytoplasm during infection. The phosphorylation occurs on serine and threonine residues (Cox *et al.*, 1991).

### B. mRNA 4

The unique region of mRNA 4 as defined by consensus intergenic homology regions contains two small ORFs (Abraham *et al.*, 1990). An RNA species likely to correspond to this unique region has been detected (Keck *et al.*, 1988).

The MHV RNA 4 codes for a single larger polypeptide; the possibility of an artifactual origin for the two BCV ORFs has been eliminated by sequencing of independently cloned cDNAs. It is suggested that the two ORFs have arisen as a result of a one-base deletion in the middle of a single ancestral ORF (Abraham *et al.*, 1990). The addition of a single base gives a single ORF of 11 kDa with substantial homology to the 15-kDa product of MHV ORF 4.

### 1. The 4a ORF

This predicts a polypeptide of 4.9 kDa.

### 2. The 4b ORF

This predicts a polypeptide of 4.8 kDa.

## C. mRNAs 5 and 5-1

Initial analyses of the subgenomic RNAs found in BCV-infected cells suggested a single mRNA between mRNA 4 and mRNA 6 (Keck *et al.*, 1988). Sequence analysis of the putative unique region of this RNA suggested that an additional consensus signal for transcriptional initiation is present in this region and a reexamination of the subgenomic RNAs corresponding to the region revealed the presence of an additional species, 5-1 (Abraham *et al.*, 1990). Thus, again, there is a contrast with the situation found in the related virus MHV where a single RNA species contains both the 5a and 5b ORFs in its unique region and the downstream ORF appears to be efficiently expressed (Budzilowicz and Weiss, 1987; Leibowitz *et al.*, 1988; Skinner *et al.*, 1985). The likely role of the 5-1 RNA in the expression of the homologue of MHV 5b is suggested by the observation that translation of *in vitro* transcripts containing both the 5 and 5-1 ORFs does not result in efficient synthesis of the downstream 5-1 product (Abraham *et al.*, 1990). It is, however, efficiently synthesized when in the upstream position.

### 1. The 5 ORF

This predicts a 109-amino acid, 12.7-kDa polypeptide. It has approximately 50% amino acid homology with the 12.4- and 13-kDa polypeptides of MHV-JHM and MHV-A59, respectively (Budzilowicz and Weiss, 1987; Skinner *et al.*, 1985).

### 2. The 5-1 ORF

This predicts an 84-amino acid, 9.5-kDa polypeptide. This has greater than 60% amino acid homology with the 10.2- and 9.6-kDa polypeptides of MHV-JHM and MHV-A59, respectively (Budzilowicz and Weiss, 1987; Skinner *et al.*, 1985).

### 3. The 5-1 Product

The synthesis of this product in infected cells has been demonstrated using immunofluorescence. The specific antiserum used was that prepared against the MHV 5b product (Leibowitz *et al.*, 1988). It is expressed both internally and on the surface of infected cells. This would be consistent with the presence of a potentially membrane-spanning hydrophobic region. As in the case of MHV, the 5b polypeptide may, in fact, be the sM polypeptide of BCV (Godet *et al.*, 1992).

## V. PUTATIVE NONSTRUCTURAL POLYPEPTIDE ORFs PRESENT IN THE SUBGENOMIC RNAs OF HCV-OC43

HCV-OC43 and BCV show remarkable similarity in terms of nucleotide sequence and immunological cross-reactivity of their structural proteins, implying a close "evolutionary" relationship. The region between the S and M proteins predicted on the basis of this relationship to encode nonstructural proteins has been cloned and sequenced (Mounir and Talbot, 1993). As expected it contains ORFs with high homology to those encoding nonstructural polypeptides in the corresponding region of BCV, but only two ORFs were identified. They appear to be expressed from separate mRNAs (5 and 5-1) as are their counterparts in BCV. The ORFs expected to be expressed from mRNA 4 are absent, but, surprisingly, the mRNA is still expressed.

### A. The 5 ORF

This predicts a 109 amino acid 12.7-kDa polypeptide. It has 96.3% amino acid identity with its BCV homologue.

### B. The 5-1 ORF

This predicts an 84-amino acid, 9.5-kDa polypeptide. This has 96.4% identity with its BCV counterpart. This may be the sM polypeptide of HCV OC43.

## VI. PUTATIVE NONSTRUCTURAL POLYPEPTIDE ORFs PRESENT IN THE SUBGENOMIC RNAs OF TGEV AND PRCV

Two large ORFs encoding putative nonstructural polypeptides have consistently been identified between the spike and small membrane genes of TGEV, but there have been differing reports of size and putative mechanisms of expression (Britton *et al.*, 1989; Kapke *et al.*, 1989; Rasschaert *et al.*, 1987; Wesley *et al.*, 1989). It is possible that in some strains of virus they are expressed from two mRNAs, whereas in others only one involved. In one case the existence of the two RNA species has been demonstrated by S1 mapping (Wesley *et al.*, 1989). However, for simplicity, the ORFs and their products will be described within both frameworks; the putative mRNAs will be referred to as 3 (3-1) and the ORFs as 3 (3a) and 3-1 (3b). An extreme 3′ ORF and corresponding mRNA (7) has been also been identified (Britton *et al.*, 1989; de Groot *et al.*, 1988; Jacobs *et al.*, 1986).

## A. mRNA 3

### 1. The 3 (3a) ORF

Two groups have sequenced the region containing this ORF using cDNA clones derived from the tissue culture-adapted Purdue strain of TGEV (Kapke *et al.*, 1989; Rasschaert *et al.*, 1987). The predicted polypeptides are identical, being 7.7 kDa in size. Two further sequences from virulent isolates, the FS772/70 strain (Britton *et al.*, 1989) and the Miller strain (Wesley *et al.*, 1989) have been determined. The predicted polypeptides differ in these cases, being 6.6 kDa and 7.9 kDa, respectively, in size. All groups have identified a putative upstream transcription reinitiation site thought to be involved in synthesis of mRNA 3. Its function has been confirmed using S1 mapping in one case (Wesley *et al.*, 1989). The sequences of these predicted polypeptides have been analyzed using various computer-based techniques, but no characteristics suggestive of function have been identified.

The putative initiation site for transcription mRNA 3 and part of the 3a ORF is missing from the porcine respiratory coronavirus (PRCV) sequence. It is unlikely, therefore, that the 3a ORF is expressed in this "variant" of TGEV (Page *et al.*, 1991).

### 2. The 3-1 (3b) ORF

The region containing this ORF has again been sequenced by four groups using the Purdue strain and the two virulent strains, FS772/70 and Miller (Britton *et al.*, 1989; Kapke *et al.*, 1989; Rasschaert *et al.*, 1987; Wesley *et al.*, 1989). A striking discrepancy between the two Purdue sequences has been noted. The sequence of Rasschaert *et al.* (1987) predicted a polypeptide of 18.8 kDa; the subsequent work of Kapke *et al.* (1988) suggested that there may have been a sequencing error in the initial study leading to a truncation of the predicted 3-1 ORF. The data of Kapke *et al.* (1988) are consistent with the observation of a 24-kDa *in vitro* translation product from mRNA 3-enriched preparations (Jacobs *et al.*, 1986) and the subsequent virulent strain ORF size predictions of 27.6 kDa (FS772/70) (Britton *et al.*, 1989) and 27.7 kDa (Miller) (Wesley *et al.*, 1989). The most notable feature of this predicted ORF is the hydrophobicity of its N-terminus and the clustering of charged residues at the carboxy-terminus.

In PRCV, the 3 (3b) ORF is likely to expressed from a 5′ truncated version of mRNA 3 transcribed from a novel initiation site immediately upstream of ORF 3 (Page *et al.*, 1991).

### B. mRNA 7

#### 1. The 7 ORF

The sequencing of the extreme 3′ end of the TGEV genome has revealed the presence of a small ORF that may be expressed from a 0.7-kb mRNA (Britton *et al.*, 1989; Kapke and Brian, 1986; Rasschaert *et al.*, 1987). The 9-kDa polypeptide predicted by the ORF has the characteristics of a membrane polypeptide; it contains hydrophobic sequences at the N- and C-termini separated by a central hydrophilic region.

## VII. PUTATIVE NONSTRUCTURAL POLYPEPTIDE ORFs PRESENT IN THE SUBGENOMIC RNAs OF FIPV AND FECV

Two mRNAs (other than the genomic RNA, RNA 1) that may encode nonstructural polypeptides have been identified in FIPV-infected cells; these are a 5.2-kb (RNA 3) and a 1.6-kb (RNA 6) species (de Groot *et al.*, 1987). Sequence data are only available for the unique region of RNA 6 (de Groot *et al.*, 1988).

### A. mRNA 6

#### 1. The 6a ORF

In the genomic sequence, this ORF overlaps with the 3′ end of the nucleocapsid ORF. S1 nuclease analysis, however, has demonstrated that only sequences downstream of the nucleocapsid ORF are contained in mRNA 6 (de Groot *et al.*, 1988). This is consistent with the finding of a consensus transcription initiation signal at the mRNA body junction detected by S1 mapping. On this basis the 6a ORF predicts an 11-kDa polypeptide. The 11-kDa polypeptide predicted by ORF 6a of FIPV is strongly homologous to the 9-kDa polypeptide predicted by the 7 ORF of TGEV. ORF 6a of FIPV is 69 nucleotides longer than ORF 7 of TGEV. The insertion extends the hydrophilic central part of the predicted polypeptide. The feline enteric coronavirus (FECV) 6a polypeptide is 99% identical (one amino acid difference) with that of FIPV (Vennema *et al.*, 1992).

#### 2. The 6b ORF

This ORF predicts a 22-kDa polypeptide that is not homologous to any TGEV polypeptide. It is hydrophilic with a short hydrophobic section at the N-terminus. There is no consensus transcription initiation signal appropriate for expression of this ORF from its own mRNA and no RNA smaller than the 1.6-kb species containing both 6a and 6b has been detected in infected cells (de Groot *et al.*, 1988) The putative methionine initiation codon for the 6b ORF is in an optimal context. The 6a ORF contains two internal methionine residues in addition to the initiator methionine, but these are in poor contexts. This situation, however, does contrast with that found in other cases where leaky scanning expression of downstream coronavirus ORFs has been invoked; in these cases, internal methionines are absent from the upstream ORFs (Boursnell *et al.*, 1985; Skinner *et al.*, 1985). The FECV 6b ORF shows a carboxy-terminal deletion when compared to that of FIPV; the amino-terminal 123 residues show 89% identity to the equivalent FIPV sequence (Vennema *et al.*, 1992).

## VIII. PUTATIVE NONSTRUCTURAL POLYPEPTIDE ORFs PRESENT IN THE SUBGENOMIC RNAs OF CCV

Of a CCV sequence, 9.6 kb extending from the region encoding the carboxy-terminus of ORF 1b to the 3′ end of the genome is available for strain Insavc-1 (Horsburgh *et al.*, 1992). Sequences from the extreme 3′ end of the genomes of strains K378 and I-71 are also available (Vennema *et al.*, 1992). Two mRNAs (3 and 7) each encoding two putative nonstructural polypeptides have been identified.

### A. mRNA 3

#### 1. The 3a ORF

This encodes a 8.6-kDa polypeptide that shows 83.5% identity to the TGEV 3a polypeptide (Horsburgh *et al.*, 1992).

#### 2. The 3b ORF

This encodes a 4.0-kDa polypeptide. This modest coding potential is the consequence of a termination codon present within a "potentially larger" ORF (28.4 kDa) that has 92.4% identity with the equivalent TGEV ORF (Horsburgh *et al.*, 1992). In spite of the presence of a potential transcriptional initiator sequence immediately upstream of the 3b ORF, no evidence for the synthesis of a corresponding mRNA has been obtained (Horsburgh *et al.*, 1992).

## IX. PUTATIVE NONSTRUCTURAL POLYPEPTIDE ORFs PRESENT IN THE SUBGENOMIC RNAs OF HCV 229E

The unique regions of mRNA 4 and 5 of HVC 229E have been sequenced (Jouvenne *et al.*,1992; Raabe *et al.*, 1990; Raabe and Siddell, 1989).

### A. mRNA 4

#### 1. The 4a ORF

This ORF predicts a polypeptide of 15.3 kDa that lacks internal methionines (Raabe *et al.*, 1990; Raabe and Siddell, 1989).

### 2. The 4b ORF

This ORF predicts a polypeptide of 10.2 kDa that is characterized by a hydrophobic amino-terminus (Raabe *et al.*, 1990; Raabe and Siddell, 1989). cDNAs encoding deleted versions of ORFs 4a and 4b have been cloned and sequenced; the predicted polypeptide products are 4.7 kDa and 9.6 kDa, respectively. It seems likely that the cDNAs were derived from a minor mRNA species (Jouvenne *et al.*, 1992).

## B. mRNA 5

### 1. The 5 ORF

This ORF predicts a hydrophobic polypeptide of 9.1 kDa that is homologous to the corresponding sM polypeptide of TGEV (Godet *et al.*, 1992).

# X. PUTATIVE NONSTRUCTURAL POLYPEPTIDE ORFs PRESENT IN THE SUBGENOMIC RNAs OF IBV

## A. mRNA 3

Nucleotide sequences for the unique region region of IBV mRNA 3 have been reported for three virus strains: Beaudette, M41, and KB8523 (Boursnell *et al.*, 1985; Niesters *et al.*, 1986; Sutou *et al.*, 1988). The predicted amino sequences of the three ORFs detected are conserved in all strains. The absence of internal methionines and the poor contexts of the initiator methionines of the upstream ORFs initially focused attention on the possibility of downstream ORF expression by leaky scanning, but it has been subsequently demonstrated that expression of the 3c ORF involves internal ribosome entry (Liu and Inglis, 1992). There is no evidence for individual mRNAs for the three ORFs.

### 1. The 3a ORF

The polypeptide predicted to be translated from this ORF has a molecular weight of 6.7 kDa. It is neutral and hydrophobic. It has been detected in infected cells using antifusion protein antisera (Liu *et al.*, 1991).

### 2. The 3b ORF

The polypeptide predicted by this ORF has a molecular weight of 7.4 kDa. It is highly acidic. It has been detected in infected cells using antifusion protein antisera (Liu *et al.*, 1991).

### 3. The 3c ORF

The polypeptide predicted by this ORF has molecular weight of 12.4 kDa. It has a hydrophobic domain near the N-terminus, potentially capable of spanning cellular membranes, and a hydrophilic C-terminal domain. The hydrophobic domain is flanked by hydrophilic residues. Codon usage in the 12.4-kDa ORF conforms to that of the IBV structural proteins; the codon usage of the 6.7- and 7.4-kDa ORFs does not.

#### 4. The 3c Product

The expression of the 3c polypeptide product in IBV-infected cells has been demonstrated using a specific antiserum raised against a bacterial fusion protein consisting of β-galactosidase and the complete 3c coding region (Smith *et al.*, 1990). A 12.4-kDa product was detected in infected cells by immunoprecipitation. It was largely located in the large-membrane subcellular fraction of infected cells along with the spike and membrane glycoproteins; nucleocapsid was, on the other hand, found largely in the microsomal fraction. This is consistent with a membrane location for the 3c product. Immunofluorescence studies demonstrated that the polypeptide was located internally and on the cell surface; there was some indication of the internal fraction of 3c being located in the Golgi and that the 3c polypeptide is present in virions (Liu and Inglis, 1991). It is clear that it is homologous to the sM polypeptide of TGEV.

### B. mRNA 5

The nucleotide sequence of the unique region of IBV mRNA 5 has been determined for the Beaudette strain (Boursnell and Brown, 1984). Two ORFs were identified. No evidence for an mRNA capable of expressing the downstream ORF independently has been obtained. Neither predicted polypeptide contains an internal methionine.

#### 1. The 5a ORF

The predicted polypeptide has a molecular weight of 7.5 kDa. The polypeptide is hydrophobic, having a very high leucine content (17 of 65 residues).

#### 2. The 5b ORF

The predicted polypeptide has a molecular weight of 9.5 kDa.

## XI. CONCLUSION

All coronaviruses appear to encode a substantial number of nonstructural polypeptides; this is consistent with their genetic complexity. The ORFs present in the unique regions of the genomic RNA (mRNA 1) from IBV, MHV, and HCV show substantial homology in amino acid sequence and in their organization. The smaller putative nonstructural ORFs are, however, more heterogeneous in predicted sequence, genomic location, and putative mode of expression even within clusters of related viruses. This plasticity is particularly striking in the case of polypeptides encoded between S and M. Expression in infected cells of many of the ORFs detected by sequencing has not yet been demonstrated and there is little or no insight into their functions. It will difficult to determine these functions in the absence of systems, for example, for genetic analysis of *in vitro* replication/transcription and for the production of site-directed mutations in the viral genome.

## XII. REFERENCES


Abraham, S., Kienzle, T. E., Lapps, W. E., and Brian, D. A., 1990, Sequence and expression analysis of potential nonstructural proteins of 4.9, 4.8, 12.7, and 9.5 kDa encoded between the spike and membrane protein genes of the bovine coronavirus, *Virology* **177:**488.

Argos, P., 1988, A sequence motif present in many polymerases, *Nucleic Acids Res.* **16:**9909.

Baker, S. C., Shieh, C. K., Soe, L. H., Chang, M. F., Vannier, D. M., and Lai, M. M., 1989, Identification of a domain required for autoproteolytic cleavage of murine coronavirus gene A polyprotein, *J. Virol.* **63:**3693.

Baker, S. C., Yokomori, K., Dong, S., Carlisle, R., Gorbalenya, A. E., Koonin, E. V., and Lai, M. M., 1993, Identification of the catalytic sites of a papain-like cysteine proteinase of murine coronavirus, *J. Virol.* **67:**6056.

Bond, C. W., Leibowitz, J. L., and Robb, J. A., 1979, Pathogenic murine coronaviruses. II. Characterisation of virus-specific proteins of murine coronaviruses JHMV and A59V, *Virology* **94:**371.

Bonilla, P. J., Gorbalenya, A. E., and Weiss, S. R., 1994, Mouse hepatitis virus strain A59 polymerase gene ORF 1a: heterogeneity among MHV strains, *Virology* **198:**736.

Boursnell, M. E. G., and Brown, T. D. K., 1984, Sequencing of coronavirus IBV genomic RNA: A 195-base open reading frame encoded by mRNA B, *Gene* **29:**87.

Boursnell, M. E., Binns, M. M., and Brown, T. D. K., 1985, Sequencing of coronavirus IBV genomic RNA: three open reading frames in the 5′ "unique" region of mRNA D, *J. Gen. Virol.* **66:**2253.

Boursnell, M. E., Brown, T. D. K., Foulds, I. J., Green, P. F., Tomley, F. M., and Binns, M. M., 1987, Completion of the sequence of the genome of the coronavirus avian infectious bronchitis virus, *J. Gen. Virol.* **68:**57.

Brayton, P. R., Lai, M. M. C., Patton, C. D., and Stohlman, S. A., 1982, Characterization of two RNA polymerase activities induced by mouse hepatitis virus, *J. Virol.* **42:**847.

Brayton, P. R., Stohlman, S. A., and Lai, M. M. C., 1984, Further characterization of mouse hepatitis virus RNA-dependent RNA polymerases, *Virology* **133:**197.

Bredenbeek, P. J., Noten, A. F. H., Horzinek, M. C., and Spaan, W. J. M., 1990a, Identification and stability of a 30-kDa non-structural protein encoded by mRNA 2 of mouse hepatitis virus in infected cells, *Virology* **175:**303.

Bredenbeek, P. J., Pachuk, C. J., Noten, A. F., Charite, J., Luytjes, W., Weiss, S. R., and Spaan, W. J., 1990b, The primary structure and expression of the second open reading frame of the polymerase gene of the coronavirus MHV-A59; a highly conserved polymerase is expressed by an efficient ribosomal frameshifting mechanism, *Nucleic Acids Res.* **18:**1825.

Brierley, I., Boursnell, M. E., Binns, M. M., Bilimoria, B., Blok, V. C., Brown, T. D. K., and Inglis, S. C., 1987, An efficient ribosomal frame-shifting signal in the polymerase-encoding region of the coronavirus IBV, *EMBO J.* **6:**3779.

Brierley, I., Digard, P., and Inglis, S. C., 1989, Characterization of an efficient coronavirus ribosomal frameshifting signal: Requirement for an RNA pseudoknot, *Cell* **57:**537.

Brierley, I., Boursnell, M. E., Binns, M. M., Bilimoria, B., Rolley, N. J., Brown, T. D. K., and Inglis, S. C., 1990, Products of the polymerase-encoding region of the coronavirus IBV, in: *Coronaviruses and Their Diseases, Advances in Experimental Biology and Medicine,* 276 (D. Cavanagh and T. D. K. Brown, eds.), p. 275, Plenum Press, New York.

Brierley, I., Jenner, A. J., and Inglis, S. C., 1992, Mutational analysis of the "slippery-sequence" component of a coronavirus ribosomal frameshifting signal, *J. Mol. Biol.* **227:**463.

Britton, P., Lopez Otin, C., Martin Alonso, J., and Parra, F., 1989, Sequence of the coding regions from the 3.0 kb and 3.9 kb mRNA. Subgenomic species from a virulent isolate of transmissible gastroenteritis virus, *Arch. Virol.* **105:**165.

Budzilowicz, C. J., and Weiss, S. R., 1987, *In vitro* synthesis of two polypeptides from a nonstructural gene of coronavirus mouse hepatitis virus strain A59, *Virology* **157:**509.

Cox, G. J., Parker, M. D., and Babiuk, L. A., 1989, The sequence of cDNA of bovine coronavirus 32K nonstructural gene, *Nucleic Acids Res.* **17:**5847.

Cox, G. J., Parker, M. D., and Babiuk, L. A., 1991, Bovine coronavirus nonstructural protein ns2 is a phosphoprotein, *Virology* **185:**509.

de Groot, R. J., ter-Haar, R. J., Horzinek, M. C., and van-der-Zeijst, B. A., 1987, Intracellular RNAs of the feline infectious peritonitis coronavirus strain 79-1146, *J. Gen. Virol.* **68:**995.

de Groot, R. J., Andeweg, A. C., Horzinek, M. C., and Spaan, W. J., 1988, Sequence analysis of the 3′-end of the feline coronavirus FIPV 79-1146 genome: Comparison with the genome of porcine coronavirus TGEV reveals large insertions, *Virology* **167:**370.

Denison, M. R., and Perlman, S., 1986, Translation and processing of mouse hepatitis virus virion RNA in a cell-free system, *J. Virol.* **60:**12.

Denison, M., and Perlman, S., 1987, Identification of putative polymerase gene product in cells infected with murine coronavirus A59, *Virology* **157:**565.

Denison, M. R., Zoltick, P. W., Leibowitz, J. L., Pachuk, C. J., and Weiss, S. R., 1991, Identification of polypeptides encoded in open reading frame 1b of the putative polymerase gene of the murine coronavirus mouse hepatitis virus A59, *J. Virol.* **65:**3076.

Dennis, D. E., and Brian, D. A., 1982, RNA-dependent RNA polymerase activity in coronavirus-infected cells, *J. Virol.* **42:**153.

Dinman, J. D., and Wickner, R. B., 1992, Ribosomal frameshifting efficiency and gag/gag-pol ratio are critical for yeast M1 double-stranded RNA virus propagation, *J. Virol.* **66:**3369.

Ebner, D., Raabe, T., and Siddell, S. G., 1988, Identification of the coronavirus MHV-JHM mRNA 4 product, *J. Gen. Virol.* **69:**1041.

Godet, M., l'Haridon, R., Vautherot, J. F., and Laude, H., 1992, TGEV corona virus ORF4 encodes a membrane protein that is incorporated into virions, *Virology* **188:**666.

Gorbalenya, A. E., Koonin, E. V., Donchenko, A. P., and Blinov, V. M., 1988a, A conserved NTP-motif in putative helicases, *Nature* **333:**22.

Gorbalenya, A. E., Koonin, E. V., Donchenko, A. P., and Blinov, V. M., 1988b, A novel superfamily of nucleoside triphosphate-binding motif containing proteins which are probably involved in duplex unwinding in DNA and RNA replication and recombination, *FEBS Lett.* **235:**16.

Gorbalenya, A. E., Blinov, V. M., Donchenko, A. P., and Koonin, E. V., 1989a, An NTP-binding motif is the most conserved feature in a highly diverged monophyletic group of proteins involved in positive strand RNA viral replication, *J. Mol. Evol.* **28:**256.

Gorbalenya, A. E., Koonin, E. V., Donchenko, A. P., and Blinov, V. M., 1989b, Coronavirus genome: Prediction of putative functional domains in the non-structural polyprotein by comparative amino acid sequence analysis, *Nucleic Acids Res.* **17:**4847.

Gorbalenya, A. E., Koonin, E. V., and Lai, M. M., 1991, Putative papain-related thiol proteases of positive-strand RNA viruses. Identification of rubi- and aphthovirus proteases and delineation of a novel conserved domain associated with proteases of rubi-, alpha- and coronaviruses, *FEBS Lett.* **288:**201.

Hardy, W. R., and Strauss, J. H., 1989, Processing the non-structural polyproteins of Sindbis virus: Non-structural proteinase is in the C-terminal half of nsP2 and functions both in *cis* and in *trans*, *J. Virol.* **63:**4653.

Herold, J., and Siddell, S. G., 1993, An elaborated pseudoknot is required for high-frequency frameshifting during translation of HCV 229E polymerase messenger RNA, *Nucleic Acids Res.* **21:**5838.

Herold, J., Raabe, T., Schelle-Prinz, B., and Siddell, S. G., 1993, Nucleotide sequence of the human coronavirus 229E RNA polymerase locus, *Virology* **195:**680.

Hodgman, T. C., 1988, A new superfamily of replicative proteins, *Nature* **333:**22.

Horsburgh, B. C., Brierley, I., and Brown, T. D., 1992, Analysis of a 9.6 kb sequence from the 3′ end of canine coronavirus genomic RNA, *J. Gen. Virol.* **73:**2849.

Jacks, T., and Varmus, H. E., 1985, Expression of the Rous sarcoma virus pol gene by ribosomal frameshifting, *Science* **230:**1237.

Jacks, T., Madhani, H. D., Masiarz, F. R., and Varmus, H. E., 1988, Signals for ribosomal frameshifting in the Rous sarcoma virus gag-pol region, *Cell* **55:**447.

Jacobs, L., van der Zeijst, B. A. M., and Horzinek, M. C., 1986, Characterisation and translation of transmissible gastroenteritis virus mRNAs, *J. Gen. Virol.* **57:**1010.

Jouvenne, P., Mounir, S., Stewart, J. N., Richardson, C. D., and Talbot, P. J., 1992, Sequence analysis of human coronavirus 229E mRNAs 4 and 5: Evidence for polymorphism and homology with myelin basic protein, *Virus Res.* **22:**125.

Kapke, P. A., and Brian, D. A., 1986, Sequence analysis of the porcine transmissible gastroenteritis coronavirus nucleocapsid protein gene, *Virology* **151:**41.

Kapke, P. A., Tung, F. Y., and Brian, D. A., 1989, Nucleotide sequence between the peplomer and matrix protein genes of the porcine transmissible gastroenteritis coronavirus identifies three large open reading frames, *Virus Genes* **2:**293.

Keck, J. G., Hogue, B. G., Brian, D. A., and Lai, M. M., 1988, Temporal regulation of bovine coronavirus RNA synthesis, *Virus Res.* **9:**343.

Koonin, E. V., 1991, The phylogeny of RNA-dependent RNA polymerases of positive-strand RNA viruses, *J. Gen. Virol.* **72:**2197.

Lee, H. J., Shieh, C. K., Gorbalenya, A. E., Koonin, E. V., La-Monica, N., Tuler, J., Bagdzhadzhyan, A., and Lai, M. M., 1991, The complete sequence (22 kilobases) of murine coronavirus gene 1 encoding the putative proteases and RNA polymerase, *Virology* **180:**567.

Leibowitz, J. L., Weiss, S. R., Paavola, E., and Bond, C. W., 1982, Cell-free translation of murine coronavirus RNA, *J. Virol.* **43:**905.

Leibowitz, J. L., Perlman, S., Weinstock, G., DeVries, J. R., Budzilowicz, C., Weissemann, J. M., and Weiss, S. R., 1988, Detection of a murine coronavirus nonstructural protein encoded in a downstream open reading frame, *Virology* **164:**156.

Liu, D. X., and Inglis, S. C., 1991, Association of the infectious bronchitis virus 3c protein with the virion envelope, *Virology* **185:**911.

Liu, D. X., and Inglis, S. C., 1992, Internal entry of ribosomes on a tricistronic mRNA encoded by infectious bronchitis virus, *J. Virol.* **66:**6143.

Liu, D. X., Cavanagh, D., Green, P., and Inglis, S. C., 1991, A polycistronic mRNA specified by the coronavirus infectious bronchitis virus, *Virology* **184:**531.

Liu, D. X., Brierley, I., Tibbles, K., and Brown, T. D. K., 1994, A 100 K polypeptide encoded by ORF 1b of the coronavirus infectious bronchitis virus is processed by ORF 1a products, *J. Virol.* **68:**5772.

Lomniczi, B., and Kennedy, S. I. T., 1977, Genome of infectious bronchitis virus, *J. Virol.* **24:**99.

Luytjes, W., Bredenbeek, P. J., Noten, A. F., Horzinek, M. C., and Spaan, W. J., 1988, Sequence of mouse hepatitis virus A59 mRNA 2: indications for RNA recombination between coronaviruses and influenza C virus, *Virology* **166:**415.

Mahy, B. W. J., Siddell, S., Wege, H., and ter Meulen, V., 1983, RNA-dependent RNA polymerase activity in murine coronavirus-infected cells, *J. Gen. Virol.* **64:**103.

Mounir, S., and Talbot, P. J., 1993, Human coronavirus OC43 RNA 4 lacks two open reading frames located downstream of the S gene of bovine coronavirus, *Virology* **192:**355.

Niesters, H. G. M., Zijderveld, A. J., Seifert, W. F., Lenstra, J. A., Bleumink-Pluym, N. M. C., Horzinek, M. C., and van der Zeijst, B. A. M., 1986, Infectious bronchitis virus RNA D encodes three potential translation products, *Nucleic Acids Res.* **14:**3144.

Oh, C.-S., and Carrington, J. C., 1989, Identification of essential residues in potyvirus protease HC by site-directed mutagenesis, *Virology* **173:**692.

Pachuk, C. J., Bredenbeek, P. J., Zoltick, P. W., Spaan, W. J., and Weiss, S. R., 1989, Molecular cloning of the gene encoding the putative polymerase of mouse hepatitis coronavirus, strain A59, *Virology* **171:**141.

Page, K. W., Mawditt, K. L., and Britton, P., 1991, Sequence comparison of the 5′ end of mRNA 3 from transmissible gastroenteritis virus and porcine respiratory coronavirus, *J. Gen. Virol.* **72:**579.

Pleij, C. N., and Bosch, L., 1989, RNA Pseudoknots—structure, detection, and prediction, *Methods in Enzymology* 180:289.

Raabe, T., and Siddell, S., 1989, Nucleotide sequence of the human coronavirus HCV 229E mRNA 4 and mRNA 5 unique regions, *Nucleic Acids Res.* **17:**6387.

Raabe, T., Schelle-Prinz, B., and Siddell, S. G., 1990, Nucleotide sequence of the gene encoding the spike glycoprotein of human coronavirus HCV 229E, *J. Gen. Virol.* **71:**1065.

Rasschaert, D., Gelfi, J., and Laude, H., 1987, Enteric coronavirus TGEV: Partial sequence of the genomic RNA, its organization and expression, *Biochimie* **69:**591.

Schochetman, G., Stevens, R. H., and Simpson, R. W., 1977, Presence of infectious polyadenylated RNA in the coronavirus infectious bronchitis virus, *Virology* **77:**772.

Schwarz, B., Routledge, E., and Siddell, S. G., 1990, Murine coronavirus nonstructural protein ns2 is not essential for virus replication in transformed cells, *J. Virol.* **64:**4784.

Shieh, C. K., Lee, H. J., Yokomori, K., La-Monica, N., Makino, S., and Lai, M. M., 1989, Identification of a new transcriptional initiation site and the corresponding functional gene 2b in the murine coronavirus RNA genome, *J. Virol.* **63:**3729.

Siddell, S. G., 1983, Coronavirus JHM: Coding assignments of subgenomic mRNAs, *J. Gen. Virol.* **64:**113.

Siddell, S. G., Wege, H., Barthel, A., and ter Meulen, V., 1981, Coronavirus JHM: Intracellular protein synthesis, *J. Gen. Virol.* **53:**145.

Skinner, M. A., and Siddell, S. G., 1985, Coding sequence of coronavirus MHV-JHM mRNA 4, *J. Gen. Virol.* **66:**593.

Skinner, M. A., Ebner, D., and Siddell, S. G., 1985, Coronavirus MHV-JHM mRNA 5 has a sequence arrangement which potentially allows translation of a second, downstream open reading frame, *J. Gen. Virol.* **66:**581.

Smith, A. R., Boursnell, M. E., Binns, M. M., Brown, T. D., and Inglis, S. C., 1990, Identification of a new membrane-associated polypeptide specified by the coronavirus infectious bronchitis virus, *J. Gen. Virol.* **71:**3.

Snijder, E. J., Wassenaar, A. L. M., and Spaan, W. J. M., 1992, The 5′ end of the equine arteritis virus replicase gene encodes a papain-like cysteine protease, *J. Virol.* **66:**7040.

Soe, L. H., Shieh, C. K., Baker, S. C., Chang, M. F., and Lai, M. M., 1987, Sequence and translation of the murine coronavirus 5′-end genomic RNA reveals the N-terminal structure of the putative RNA polymerase, *J. Virol.* **61:**3968.

Somogyi, P., Jenner, A. J., Brierley, I., and Inglis, S. C., 1993, Ribosomal pausing during translation of an RNA pseudoknot, *Mol. Cell. Biol.* **13:**6931.

Strauss, E. G., de Groot, R. J., Levinson, R., and Strauss, J. H., 1992, Identification of the active site residues in the nsP2 proteinase of Sindbis virus, *Virology* **191:**932.

Sutou, S., Sato, S., Okabe, T., Nakai, M., and Sasaki, N., 1988, Cloning and sequencing of genes encoding structural proteins of avian infectious bronchitis virus, *Virology* **165:**589.

ten Damm, E. B., Pleij, C. W. A., and Bosch, L., 1990, RNA pseudoknots, translational frameshifting and readthrough on viral RNAs, *Virus Genes* **4:**121.

Thiel, V., and Siddell, S. G., 1994, Internal ribosome entry in the coding region of murine hepatitis virus mRNA 5, *J. Gen. Virol.* **75:**3041.

Tu, C., Tzeng, T.-H., and Bruenn, J. A., 1992, Ribosomal movement impeded at a pseudoknot required for framshifting, *Proc. Natl. Acad. Sci. USA* **89:**8636.

van der Most, R. G., Bredenbeek, P. J., and Spaan, W. J., 1991, A domain at the 3′ end of the polymerase gene is essential for encapsidation of coronavirus defective interfering RNAs, *J. Virol.* **65:**3219.

Vennema, H., Rossen, J. W., Wesseling, J., Horzinek, M. C., and Rottier, P. J., 1992, Genomic organization and expression of the 3′ end of the canine and feline enteric coronaviruses, *Virology* **191:**134.

Wesley, R. D., Cheung, A. K., Michael, D. D., and Woods, R. D., 1989, Nucleotide sequence of coronavirus TGEV genomic RNA: Evidence for 3 mRNA species between the peplomer and matrix protein genes, *Virus Res.* **13:**87.

Yokomori, K., and Lai, M. M., 1991, Mouse hepatitis virus S RNA sequence reveals that nonstructural proteins ns4 and ns5a are not essential for murine coronavirus replication, *J. Virol.* **65:**5605.

Yu, X., Bi, W., Weiss, S. R., and Leibowitz, J. L., 1994, Mouse hepatitis virus gene 5b protein is a new virion envelope protein, *Virology* **202:**1018.

Zoltick, P. W., Leibowitz, J. L., DeVries, J., Pachuk, C. J., and Weiss, S. R., 1990, Detection of mouse hepatitis virus nonstructural proteins using antisera directed against bacterial fusion proteins, in: *Coronaviruses and Their Diseases, Advances in Experimental Biology and Medicine,* 276 (D. Cavanagh and T. D. K. Brown, eds.), p. 291 Plenum Press, New York.

CHAPTER 11

# The Molecular Biology of Toroviruses

Eric J. Snijder and Marian C. Horzinek

## I. INTRODUCTION

Recently, the International Committee on the Taxonomy of Viruses has included the toroviruses as a new genus in the previously monogeneric family of the *Coronaviridae* (Pringle, 1992; Cavanagh *et al.*, 1994). Toroviruses and coronaviruses share some similarities in their structure and biology (see Chapter 18, this volume); however, it was the molecular characterization of the torovirus prototype, Berne virus (BEV), that produced the evidence for an evolutionary link between the two virus groups.

After the recognition of the link between toroviruses and coronaviruses, the genome organization and replication strategy of a third virus group, the arteriviruses, were found to be strikingly similar to both genera of the *Coronaviridae* (Den Boon *et al.*, 1991b). However, it is evident that the arteriviruses are more distantly related than toroviruses and coronaviruses, and the inclusion of the arteriviruses as a third genus in the *Coronaviridae* was deemed inappropriate. Nevertheless, to indicate the ties between coronaviruses, toroviruses, and arteriviruses the unofficial term "coronaviruslike superfamily" has been put forward (Den Boon *et al.*, 1991b). The similarities and differences between the members of this superfamily will be discussed elsewhere (Chapter 12, this

ERIC J. SNIJDER • Department of Virology, Institute of Medical Microbiology, Leiden University, 2300 AH Leiden, The Netherlands. MARIAN C. HORZINEK • Virology Division, Department of Infectious Diseases and Immunology, University of Utrecht, 3584 CL Utrecht, The Netherlands.

*The Coronaviridae*, edited by Stuart G. Siddell, Plenum Press, New York, 1995.

volume). In this chapter, we will focus on the molecular analysis of the torovirus particle, genome, mRNAs, and structural and nonstructural proteins.

## II. THE TOROVIRION

The equine torovirus BEV was isolated from a diarrheic horse in 1972 at the University of Berne, Switzerland. However, the virus was not studied in detail until particles with the same morphology were observed in the feces of diarrheic cattle in Breda, Iowa, in 1979 (Weiss *et al.*, 1983; Woode *et al.*, 1982). Similar pleiomorphic viruses were also found in human feces collected in Birmingham, UK, and Bordeaux, France (Beards *et al.*, 1984, 1986) and, more recently, in Toronto, Canada (Tellier and Petric, 1993).

To date, BEV is the only torovirus that can be propagated in cultured cells. As a result, it is the best-studied member of the genus. The unique structure of the torovirus particle initially led to the proposal of a new family of enveloped animal viruses, the *Toroviridae* (Horzinek and Weiss, 1984; Horzinek *et al.*, 1987). As illustrated in Figs. 1 and 2, toroviruses are enveloped pleiomorphic particles measuring 120 to 140 nm in their largest axis. Spherical, oval, elongated, and kidney-shaped viruses are observed (Weiss *et al.*, 1983). Drumstick-shaped projections, very similar to the spikes of coronaviruses, decorate the virion surface. However, as illustrated in Fig. 2A, the nucleocapsid structure of the toroviruses is very different from that of coronaviruses: the core of the torovirion is formed by an electron-dense, tubular structure, which may be straight or may display the typical toro-shape, that is prevalent in mature virions. The straight form is observed especially prior to the process of budding, which takes place at intracellular membranes, predominantly those of the Golgi system (Weiss and Horzinek, 1986; Fagerland *et al.*, 1986). The current structural model of BEV, based on electron micrographs and the molecular data described below, is shown in Fig. 2B.

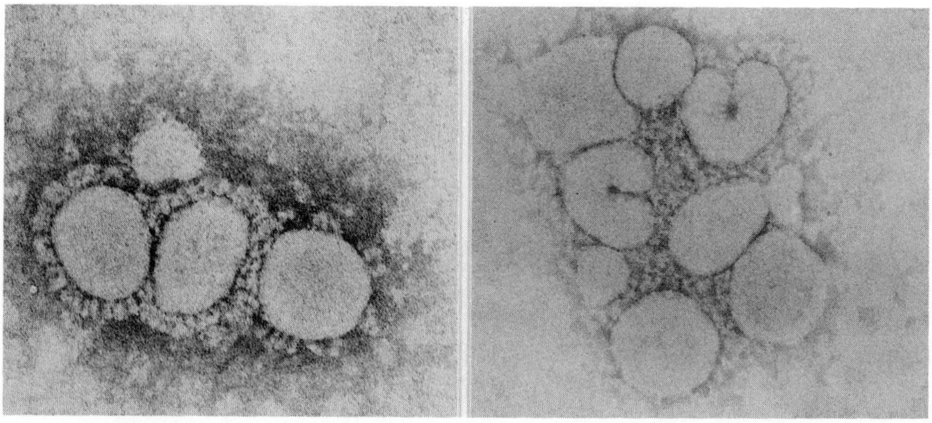

FIGURE 1. Negative staining of Berne virus particles. Magnification: approximately 150,000× (Weiss *et al.*, 1983).

## III. THE TOROVIRUS GENOME

The BEV genome is a single molecule of infectious, polyadenylated RNA (Snijder *et al.*, 1988) with an estimated size of 25–30 kb. This size estimate is based on the sequence analysis of a substantial part of the genome (about 16 kb) and the parallels between the genome organization of BEV and other members of the *Coronaviridae*.

The BEV genome probably contains six open reading frames (ORFs) (Fig. 3). As for all other coronaviruses, the two most 5′ reading frames (ORF 1a and 1b) are translated from the genomic RNA and constitute the viral replicase gene. The four remaining reading frames, of which ORFs 2, 3, and 5 have been identified as structural genes, are expressed by generation of a 3′-coterminal nested set of mRNAs (see Section IV).

The two parts of the BEV genome that have been sequenced are separated by a gap in the ORF 1a region. About 1.5 kb of the 5′ end sequence of the BEV genome was determined from cDNA clones derived from genomic RNA and defective interfering (DI) RNAs (Snijder *et al.*, 1991b). A sequence of about 14.5 kb, starting 1 kb upstream of the ORF 1a/ORF 1b ribosomal frameshift site and ending at the 3′ poly(A) tail, was obtained by analysis of a cDNA library prepared from poly(A)-containing RNA from BEV-infected cells (Snijder *et al.*, 1990a,c).

Assuming that the 5′ ends of the BEV genome and BEV DI RNAs are colinear, the ORF 1a initiation codon is located at nucleotide (nt) position 825–827 in the genome. The region upstream of ORF 1a is probably untranslated since it contains only one other AUG codon (nt 22–24) that is followed by a UGA termination codon three triplets downstream (Snijder *et al.*, 1991b).

The 3′ nontranslated region (NTR) of the BEV genome encompasses 200 nt, excluding the poly(A) tail (Snijder *et al.*, 1989). Interestingly, a computer analysis of this region reveals a large potential stem-loop structure that may be one of the signals for viral RNA replication (Snijder *et al.*, 1989). The 3′ NTR is the only region of the genome for which the sequence of a second torovirus, the bovine Breda virus (BRV), is available. A short 269-nt sequence upstream of the BRV poly(A) tail was obtained from the analysis of a cDNA library prepared from genomic RNA purified from the feces of an infected animal (Koopmans *et al.*, 1991). This sequence, comprising the 3′ end of the BRV nucleocapsid (N) protein gene and the 3′ NTR, is 93% identical to the BEV sequence. Hybridization experiments suggest that there is a high degree of sequence similarity between equine and bovine toroviruses, with the possible exception of the 5′ part of the spike (S) protein gene (Koopmans *et al.*, 1991).

## IV. BERNE VIRUS TRANSCRIPTION AND TRANSLATION

A 3′-coterminal nested set of 5 mRNAs (including the genome) is generated during BEV replication (Snijder *et al.*, 1990c) (Fig. 3). The estimated sizes of these RNAs and their relative abundance late in infection (Snijder *et al.*, 1988) are listed in Table I.

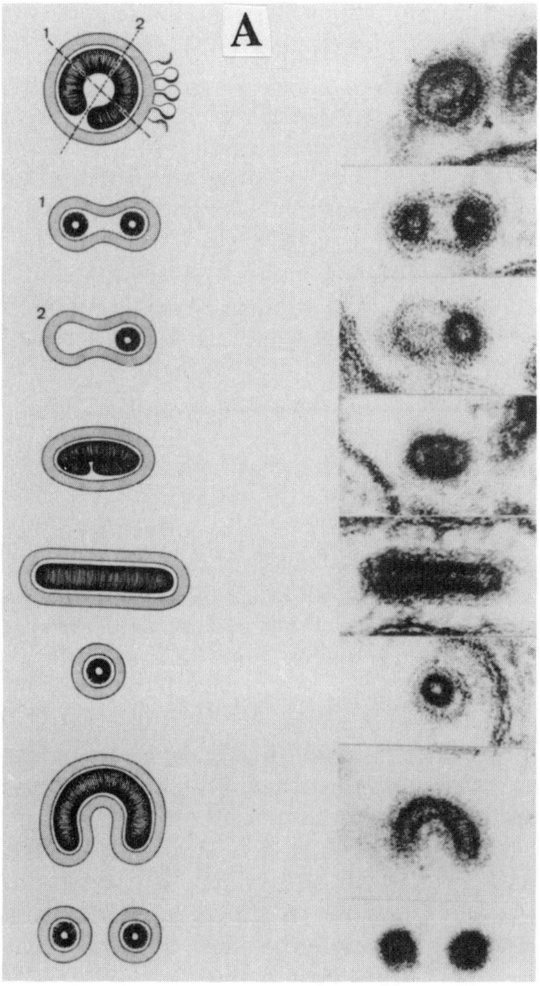
A
1
2
1
2

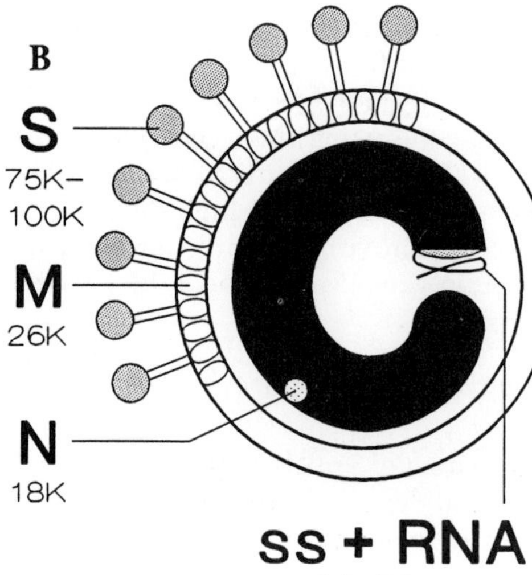
B
S
75K–
100K
M
26K
N
18K
ss + RNA
25–30 kb

The analysis of BEV subgenomic (sg) RNA synthesis by UV transcription mapping (Snijder *et al.*, 1990c) indicates that (late in infection) the BEV mRNAs are transcribed independently, i.e., from separate transcription units. In common with other coronaviruses, the BEV genomic sequence contains conserved intergenic sequences that are probably involved in mRNA transcription. For some coronaviruses, these intergenic motifs [e.g., 5′ AAUCuAuAC 3′ for murine hepatitis virus (MHV)] have been identified as the site where the common 5′ leader sequence is fused to the mRNA bodies (Spaan *et al.*, 1988; Lai, 1990). Although the torovirus genome contains conserved AU-rich intergenic sequences (5′ uaUcUUUAGa 3′) (Fig. 4), there appears to be an important difference, since in the case of torovirus RNAs no evidence for fusion of a common leader to mRNA bodies at this position has been obtained. Thus, none of the clones from a cDNA library prepared from poly(A)-selected intracellular RNA contained a candidate leader sequence. In addition, primer extension analyses of RNA 5 and DI RNAs indicated that BEV mRNAs terminate at, or just upstream of, the conserved intergenic sequence (Snijder *et al.*, 1990c, 1991b). In terms of transcription, however, the consequences of this dissimilarity between toro- and coronaviruses could be limited: direct binding of the polymerase to the various BEV "core promoters" on a negative-stranded template could simply replace a leader-priming mechanism.

In addition to the products of ORFs 2, 3, 4, and 5, which are assumed to be synthesized by monocistronic translation of a nested set of structurally polycistronic mRNAs, the 3′ part of the BEV genome may encode one more protein: ORF 5 completely overlaps with a 264-nt ORF that potentially encodes a hydrophobic 10-kDa protein. Although no such protein has been observed in virions or BEV-infected cells, it is remarkable that a similar situation, a small hydrophobic protein expressed from an ORF that completely overlaps with the N protein gene, has been reported for the bovine coronavirus (BCV) (Senanayake *et al.*, 1992).

BEV ORF 1b is expressed from the genomic RNA by a −1 ribosomal frameshift (Snijder *et al.*, 1990a), which apparently is one of the hallmarks of the coronaviruslike supergroup. As a result, two proteins are produced from the replicase region: the ORF 1a protein and an ORF 1a/1b fusion protein. BEV ORF 1a ends with the nucleotide sequence 5′ U UUA AAC UGU UGA 3′, in which the UGA codon terminates ORF 1a translation. The heptanucleotide sequence 5′ U UUA AAC 3′ is one of the typical "shifty sequences," which have been described to be present at ribosomal frameshift sites of an increasing number of viruses (Jacks *et al.*, 1988; Brierley *et al.*, 1989; Ten Dam *et al.*, 1990). Two versions of the downstream RNA pseudoknot (PK), which usually accompanies a shifty heptanucleotide and which is thought to promote frameshifting by

←

FIGURE 2. (A) Different forms of BEV particles visualized in ultrathin sections of BEV-infected equine dermis cells. On the right, electron micrographs of BEV particles are shown; on the left, schematic interpretations of the viral structures seen in the corresponding photographs are presented. Section planes "1" and "2" cut the nucleocapsid twice and once, respectively (Weiss and Horzinek, 1987). (B) Schematic representation of the architecture of BEV, the torovirus prototype. The localization and sizes of the structural proteins and genome are indicated (Snijder and Horzinek, 1993).

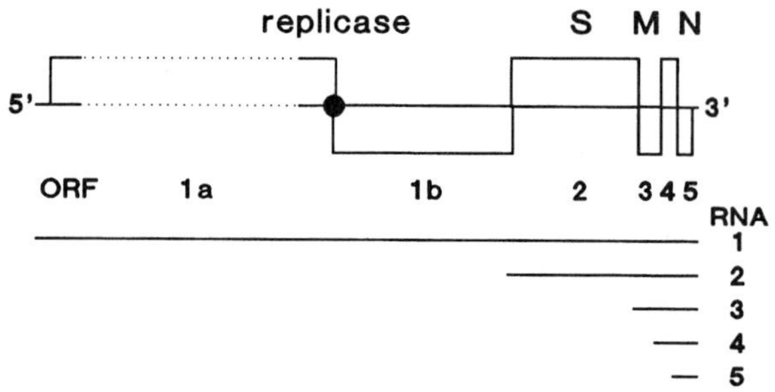


FIGURE 3. Genome organization and expression of BEV. The location of the 6 ORFs is indicated. The dashed lines indicate the ORF 1a region that remains to be sequenced (estimated size: 10 to 12 kb). The lower part of the figure shows the relationship of the genome and the nested set of mRNAs generated in infected cells (Snijder and Horzinek, 1993).

stalling the ribosome at the frameshift site, can be predicted for BEV (Fig. 5). They differ in the size of loop 2 (L2), which is either 11 (PK1) or 69 nt (PK2). The BEV frameshift signal was tested *in vitro* and *in vivo* using a reporter gene construct. Frameshift efficiencies between 20 and 30% were observed. A deletion mutation which obviates the posibility of a PK2 structure in the transcribed RNA did not influence the frameshift efficiency, suggesting that it, in fact, is PK1 that is involved in the regulation of ORF 1b expression (Snijder *et al.*, 1990a).

## V. DI PARTICLES AND RNAs OF BEV

DI RNAs were readily generated during undiluted passage of BEV in embryonic mule skin cells (Snijder *et al.*, 1991b). The sequence of two small BEV DI genomes, which arose and interfered strongly with viral replication within five undiluted passages, has been determined. A DI RNA of about 1 kb was found to contain about 600 nt from the 5′ end of the viral genome and 242 nt [excluding the poly(A) tail] from the 3′ end. A somewhat larger DI RNA (1.4 kb) contained larger parts from the same segments of the genome (about 700 and 441 nt, respectively) and an additional central part of 100 nt, which is thought to be

TABLE I. Characteristics of BEV RNAs and ORFs

| RNA | Size (kb) | Relative molarity (%) | Encoded ORF | Position upstream of 3′ end | Number of aa encoded | Calculated size (kDa) | Protein |
|---|---|---|---|---|---|---|---|
| 1 | 25–30? | 2 | 1a | ?–13,468 | ? | ? | Replicase |
| | | | 1b | 13,477–6,604 | 2291 | 261 | Replicase |
| 2 | 6.9 | 3 | 2 | 6,684–1,941 | 1581 | 178 | Spike |
| 3 | 2.1 | 30 | 3 | 1,909–1,210 | 233 | 26 | Membrane |
| 4 | 1.4 | 13 | 4 | 1,150–724 | 142 | 16 | Pseudogene? |
| 5 | 0.8 | 52 | 5 | 680–200 | 160 | 18 | Nucleocapsid |

```
                                               *******************
RNA 2   G U G C C U A A U U U U U A A A G|U|G|U U U A G|U A C U A G U U U U A
RNA 3   A G A A A A G U U A G U C A C U U|U|C|U U U A G|A A G A A G G U U G C
RNA 4   G U U U G A G U A G C C A C U U A|U|C|U U U A G|A A G A U G U U G A U
RNA 5   G U U A G U G A G A G A C A C U A|U|C|U U U A G|A G A A A G A G C C A
DI                        N N A A C G U A|U|C|U U U A G|A A G U U U A U G U U
                                               *******************
```

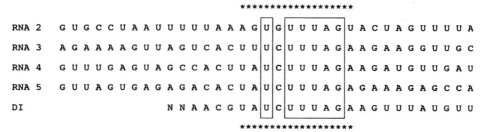

FIGURE 4. Alignment of putative BEV "core promoter" sequences (indicated with asterisks) and their flanking context. Fully conserved nucleotides are boxed. The sequences are derived from sequence analysis of the genome (Snijder *et al.*, 1990c) and of DI RNAs (Snijder *et al.*, 1991b) of BEV (Snijder and Horzinek, 1993).

derived from the ORF1a region, since it does not match any of the known BEV sequences.

The composition of the 1-kb DI genome suggests that the minimal sequences required for BEV RNA replication (and probably also packaging) are located in two small domains present at the termini of the genomic RNA. This suggests a difference with members of the genus coronavirus. The generation and construction (at the cDNA level) of MHV DI genomes that are both replicated and packaged has been described, and DI-based systems are currently being used to study viral transcription and replication (Makino 1990, 1991; van der Most *et al.*, 1991, 1992). However, all viable MHV DI RNAs contain at least three segments from the viral genome. In addition to 5′ and 3′ sequences, sequences from ORF1b, which are assumed to contain a packaging signal, are required for efficient propagation of MHV DI genomes (Makino *et al.*, 1990, 1991; Van der Most *et al.*, 1991).

Although BEV DI genomes and particles were initially generated and used as a tool to study viral replication, they also provide information on the previously decribed heterogeneous nature of BEV preparations (Weiss *et al.*, 1983; Snijder *et al.*, 1988). A second peak of virus-specific material, sometimes referred to as 50S particles, has been observed in both isokinetic and isodensity gradients. Ultracentrifugation experiments, using virus stocks containing the 1-kb DI RNA, confirmed that this second peak represented particles that contained the DI genome and displayed both smaller S values (50–100S instead of 400S) and lower densities (1.07–1.11 g/ml vs. 1.16 g/ml in sucrose) than standard virions. However, the protein composition of BEV DI particles remains to be studied in detail, especially since preliminary data suggested that both the S protein and the N protein were absent, or present only in very small quantities (Snijder *et al.*, 1991b).

## VI. THE BERNE VIRUS REPLICASE GENE

In addition to the discovery that the genome organization and expression strategy of toroviruses are very similar to those of coronaviruses, sequence

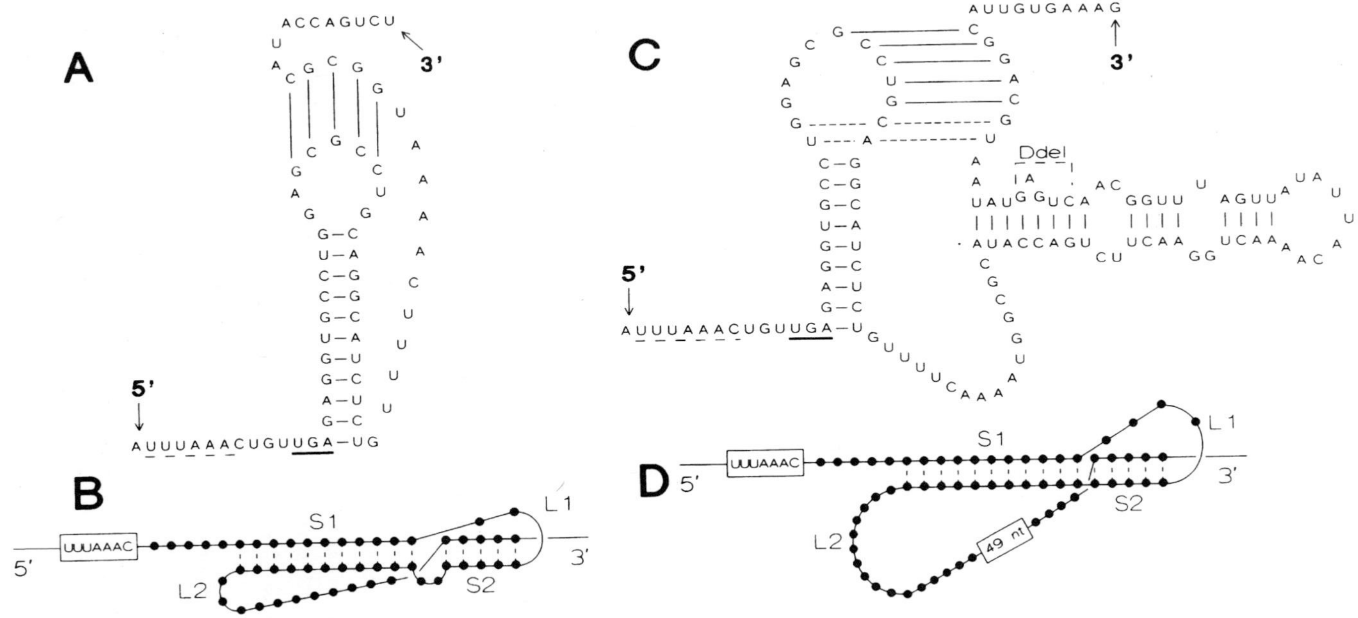


FIGURE 5. Predicted secondary and tertiary RNA structure of the BEV ORF 1a/1b overlap region. (A) Predicted structure of the RNA pseudoknot PK1. The "slippery" sequence U UUA AAC is indicated by a dashed line. The ORF 1a termination codon is underlined. (B) Schematic representation of PK1. The basepaired stem structures S1 and S2 and the connecting loops L1 and L2 are indicated. (C) Alternative model of the RNA pseudoknot PK2. (D) Schematic representation of PK2. The 49-nt box indicates the possible internal stem-loop structure in L2 as shown in panel C (Snijder *et al.*, 1990a).

analysis of the BEV ORF1b region also provides conclusive evidence for an ancestral relationship between the two virus groups (Snijder *et al.*, 1990a).

Two of the three basic enzymatic activities that are usually encountered in the replicase proteins of positive-stranded RNA viruses (polymerase, helicase, and protease) are encoded by the ORF1b sequence of toro- and coronaviruses (Figs. 6 and 7), namely, the polymerase (or SDD) domain (Poch *et al.*, 1989) and the NTP-dependent helicase domain (Gorbalenya and Koonin, 1993). On the basis of the sequence analysis shown in Fig. 7, the percentage of identical amino acids (aa) in toro- and coronavirus polymerase and helicase domains can be calculated at 40 to 45% (70–90% between members of the genus coronavirus). This analysis clearly separates the toro- and coronavirus sequences from the corresponding domains of other virus (super)groups and also illustrates the relatively large evolutionary distance between the two genera of the family *Coronaviridae*.

In addition to the "universal" domains described above, a number of less striking sequence similarities, which nevertheless appear to be *Coronaviridae*-specific, have been detected. The most remarkable example of such a domain is found at the C-terminus of the ORF 1b protein (Snijder *et al.*, 1990a) (Figs. 6 and 7). The last 300 aa of the ORF 1b product are approximately 40% identical when toroviruses and coronaviruses are compared. Additional smaller conserved motifs are a cysteine/histidine-rich domain, located between the polymerase and helicase regions, and a short region of the ORF1a protein, just upstream of the ribosomal frameshift site (Snijder *et al.*, 1990a, 1991a).

Several putative protease domains, belonging to different protease supergroups, have been identified in the ORF1a proteins of the coronaviruses avian infectious bronchitis virus (IBV), MHV, and human coronavirus (HCV) 229E (Gorbalenya *et al.*, 1989; Lee *et al.*, 1991; Herold *et al.*, 1993). Unfortunately, only short sequences from both ends of the BEV ORF1a protein are available and these do not contain any obvious protease domains. Therefore, it remains to be seen whether the proteolytic processing pattern of the torovirus replicase follows the examples provided by the corresponding proteins of coronaviruses and arteriviruses (Snijder and Horzinek, 1993). For these two virus groups, a protease belonging to the chymotrypsin/3C-like supergroup is predicted to be the main protease, responsible for the processing of the C-terminal part of the ORF1a protein and the ORF1b polypeptide. Additional protease domains, belonging to the papainlike family of proteolytic enzymes, direct the processing of the N-terminal part of the ORF1a product (Snijder *et al.*, 1992; Baker *et al.*, 1993).

## VII. THE STRUCTURAL PROTEINS OF TOROVIRUSES

Torovirus proteins were initially analyzed by metabolic labeling of BEV-infected cells (Horzinek *et al.*, 1984). Subsequent sequence analysis of the structural genes and characterization of their products identified three structural proteins: a 19-kDa nucleocapsid (N) protein, a 26-kDa integral membrane protein, and a 180-kDa spike protein (which is posttranslationally cleaved into

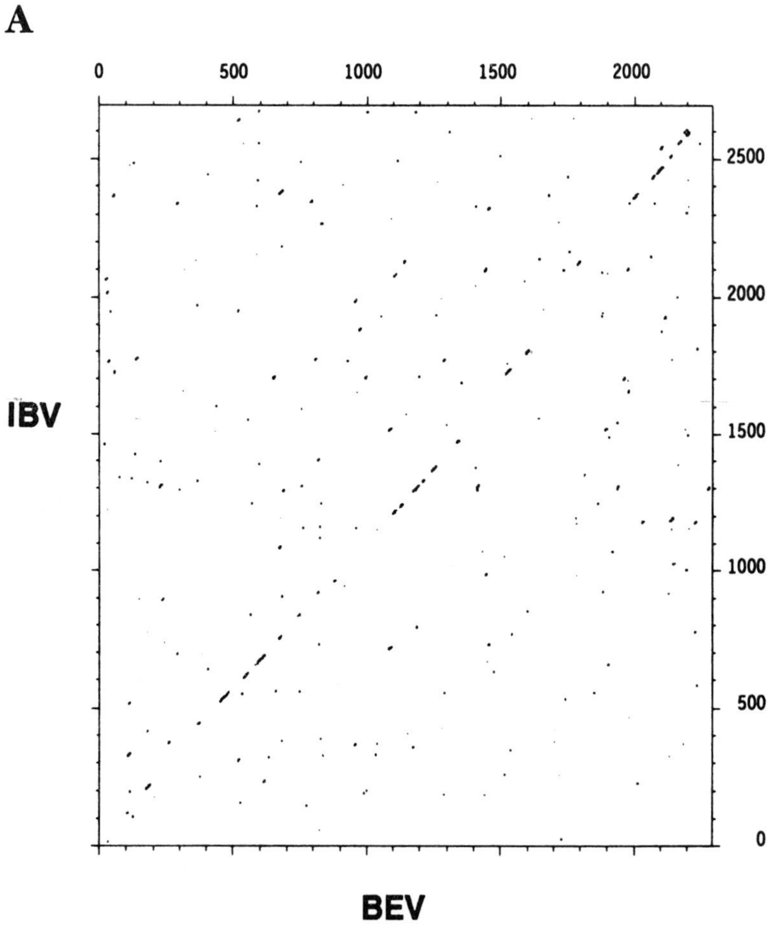


FIGURE 6.

two subunits). These proteins were assigned to ORFs 5, 3, and 2, respectively, in the BEV genomic sequence (Fig. 3) (Den Boon *et al.*, 1991a; Snijder *et al.*, 1989, 1990b). Although the small-membrane protein and the spike protein were initially termed E (for envelope) and P (for peplomer), they have now been renamed to M and S to follow the coronavirus protein nomenclature (Snijder and Horzinek, 1993).

Since BRV cannot be grown in cultured cells, its protein composition was studied by means of surface radioiodination of purified virus (Koopmans *et al.*, 1986). Possible virus-specific polypeptide species of 20, 37, 85, and 105-kDa were identified in this manner.

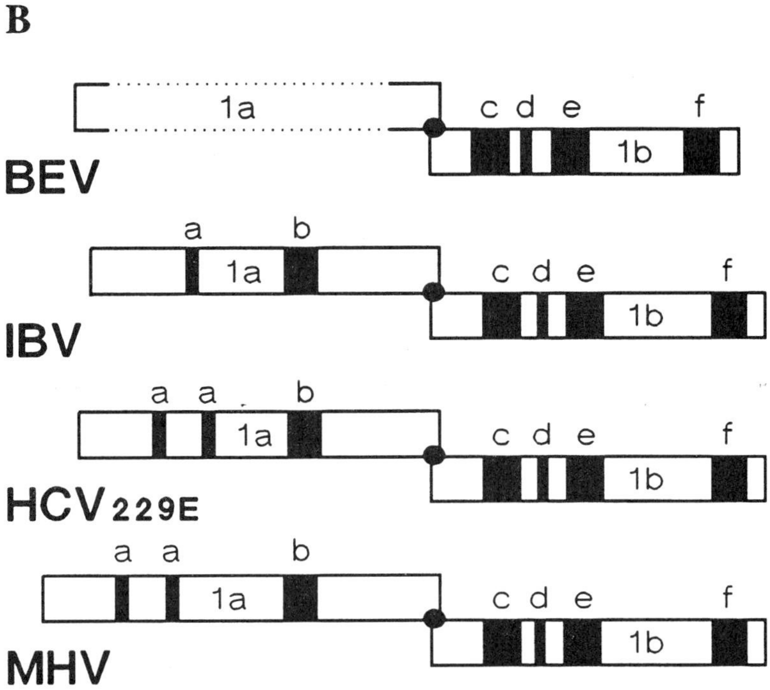


FIGURE 6. (A) Proportional dot matrix comparison of the amino acid sequences of the predicted ORF 1b products of the torovirus BEV and the coronavirus IBV (Snijder *et al.*, 1990a). (B) Position of conserved domains in the replicases of the torovirus BEV and the coronaviruses IBV (Beaudette M42), MHV (strain A59), and HCV 229E: a, papainlike protease; b, 3C-like protease; c, polymerase; d, cysteinc/histidine-rich domain; e, helicase; f, C-terminal 1b domain. Sequence comparisons of domains c, e, and f are presented in Fig. 7.

## A. The N Protein

A 19-kDa protein was present in purified BEV nucleocapsids that were obtained by detergent treatment of purified virus particles (Horzinek *et al.*, 1985). This N protein is the most abundant structural protein of the BEV particle, accounting for about 80% of its protein mass. It is a phosphorylated protein with RNA-binding properties (Horzinek *et al.*, 1985). By *in vitro* translation studies using BEV RNA 5 (Snijder *et al.*, 1988) and sequence analysis of the region upstream of the poly(A) tail (Snijder *et al.*, 1989) the N protein gene was identified as the 3′-terminal gene in the BEV genome.

The N protein consists of 160 aa. It contains 7 (4%) acidic and 22 (14%) basic amino acid residues. The latter are especially clustered in a central domain of the N protein sequence: 15 basic residues are found between aa 34 and 80 (Snijder *et al.*, 1989).

The size of the BEV N protein (18.3kDa) is noticeably smaller than that of the corresponding protein of coronaviruses (45–55kDa) (Spaan *et al.*, 1988). Apparently, the different architecture of the torovirus nucleocapsid, which is

## Polymerase domain

```
BEV  513  lIGvsKyglkfskfLkd-9-vfGsDYtKCDRtfPlsfR-17-Y-L-NE-12-GmllnKPGGTsSGDATTAhsNtfyN-51-yflnfLSDDsfi-fs-37-eEFCSaH-10-L--PsrgRll
           **  *        *       * ** ****  *   *    * * **    *    ***** *******  *   *         ****           **** *    *  *   *
IBV  582  vIGttKfyggwdnmLrn-9-lmGwDYpKCDRamPnllR-21-YrLyNE-12-GgiyvKPGGTsSGDATTAyaNsvfN-57-fslmiLSDDgvvcyn-42-hEFCSqH-12-LpyPdpsRil
MHV  576  vIGttKfyggwddmLrr-9-lmGwDYpKCDRamPnilR-21-YrLaNE-12-GcyyvKPGGTsSGDATTAfaNsvfN-57-fsmmiLSDDgvvcyn-42-hEFCSqH-12-LpyPdpsRil
HCV  575  vIGttKfyggwdnmLkn-9-lmGwDYpKCDRamPsmiR-21-YrLsNE-12-GgfyfKPGGTtSGDATTAyaNsvfN-57-fsmmiLSDDsvvcyn-42-hEFCSqH-12-LpyPdpsRii
```

## Helicase domain

```
BEV 1097  vmGPPGtGKttf-61-cThNtLPfiksavliaDEVSli-15-vVllGDPfQL-sP--v-23-yLtaCYRCPpqIl-67-gLG-dvtTiDSsQGt-18-vNRviVgcsRst-thlv
            **** **         * * **         ****       *  *** **  *        *  *****  *      **    * ** **      **  *   *
IBV 1206  vqGPPGsGKshf-66-sTiNaLPevscdillvDEVSml-16-vVyvGDPaQLpaPrtl-24-fLakCYRCPkelv-78-mLGlnvqTvDSsQGs-18-iNRfnValtRakrgilv
MHV 1199  vqGPPGtGKshl-66-tTiNaLPelvtdiivvDEVSml-16-yVyiGDPaQLpaPrvl-25-fLgtCYRCPkelv-75-vLGlqtqTvDSaQGs-18-vNRfnVaitRakkgilc
HCV 1200  iqGPPGsGKshc-66-sTvNaLPevnadivvvDEVSmc-16-iVyvGDPqQLpaPrvl-25-fLhkCYRCPaelv-75-lLGlqtqTvDSaQGs-18-aNRfnVaitRakkgifc
```

## C-terminal ORF 1b domain

```
BEV 1933  GGvH-31-Krt-TlvD-21-SKVifVniD-17-TfYP-25-NYG-10-NfaKYTQiC-16-GAagvdGcsPGdiVL-33-LivSDiY-16-LALGGtivfKtTEsS-17-FfTagVNtSSSEvF
          ** *    *   *  *    ***  *  *    * **    ***    *  **** *    **    *  **  **    *  ** *    *****    * ** *    * *  ** **** *
IBV 2282  GGlH-39-KqvcTvvD-21-SKVvtVsiD-16-TcYP-29-NYG-11-NvaKYTQlC-17-GAgsdkGvaPGstVL-40-LviSDmY-32-LALGGsfavKvTEtS-17-FcTa-VNaSSSEaF
MHV 2315  GGlH-39-KsvcTviD-18-SKVvnVnvD-16-TfYP-31-NYG-11-NvaKYTQlC-17-GAgsdkGvaPGsaVL-40-LiiSDmY-28-LALGGsvaiKiTEfS-17-FcTn-VNaSSSEgF
HCV 2281  GGlH-40-KtvcTymD-18-SKVheViiD-16-TfYP-30-NYG-11-NvvKYTQlC-19-GAgsdyGvaPGtaVL-40-LlISDmY-28-LALGGslaiKvTEyS-17-FcTs-VNtSSSEaF
```

FIGURE 7. Amino acids sequence comparisons of the three most conserved domains of toro- and coronaviral ORF 1b proteins: polymerase, helicase, and C-terminal domain, respectively. Residues identical in all four sequences are shown in capitals and indicated with asterisks.

tubular instead of the helical core found in the genus coronavirus, dictates the use of a different type of N protein. Therefore, it is not surprising that no significant sequence similarities are detected in the N proteins of toro- and coronaviruses are compared.

## B. The M Protein

The BEV M protein, which is translated from ORF 3 (Snijder *et al.*, 1988), is an unglycosylated polypeptide (Horzinek *et al.*, 1986), accounting for about 13% of the virion protein mass (Horzinek *et al.*, 1985). The M protein is a 26.5-kDa polypeptide of 233 aa. It does not contain an N-terminal signal sequence. In its N-terminal half, the three membrane-spanning α-helixes that are so characteristic for coronavirus M proteins are encountered (Fig. 8) (Den Boon *et al.*, 1991a). The small difference between the calculated and observed sizes of the M protein in sodium dodecyl sulfate–polyacrylamide gel electrophoresis (SDS-PAGE) is explained by aberrant migration, probably due to the extreme hydrophobicity of the protein.

The membrane topology of the M protein has been studied by *in vitro* translation using the M protein itself and a hybrid protein which contained a C-terminal tag (Den Boon *et al.*, 1991a). After *in vitro* translation in the presence of microsomes, about 85% of the M protein was resistant to protease K digestion. Therefore, its disposition in the membrane is thought to be very similar to that of the triple-spanning M proteins of coronaviruses such as MHV and IBV. The N-terminus is located in the lumen of the endoplasmic reticulum (and eventually at the virion surface), a large central part of the protein is embedded in the membrane, and the C-terminus is located at the cytoplasmic face of the endoplasmic reticulum. One of the hydrophobic transmembrane domains is assumed to function as an internal signal sequence. The BEV M hybrid carrying the C-terminal tag accumulated in intracellular membranes during transient expression experiments. Thus the torovirus M protein may play a role in the intracellular budding process, as has been suggested for its coronavirus counterpart (Dubois-Dalcq *et al.*, 1982; Holmes *et al.*, 1981; Rottier *et al.*, 1981).

Despite the striking similarities in size, structure, and functional characteristics, no significant primary sequence similarities can be detected between the M protein sequences of BEV and coronaviruses (Den Boon *et al.*, 1991a). Still, the observed similarities are taken as evidence for a process of divergent evolution during which the properties essential for virus assembly have been conserved.

## C. The S Protein

The BEV envelope is studded with drumstick-shaped projections that measure about 20 nm in length (Weiss *et al.*, 1983) (Fig. 1). In early torovirus studies, the heterogeneous 75- to 100-kDa protein material was assumed to represent the viral spike protein(s), since it was recognized by both neutralizing and

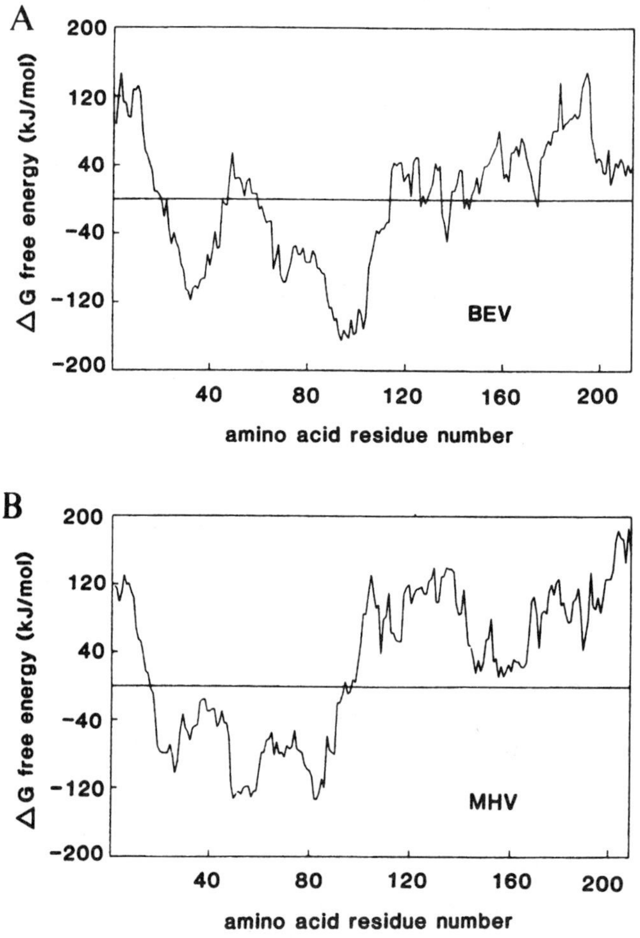


FIGURE 8. Comparison of the hydrophobicity profiles of the M proteins of the torovirus BEV and the coronavirus MHV. Free energy values corresponding to the insertion of successive 21-aa protein segments from an aqueous environment into a lipid bilayer. Calculations were done according to Von Heijne (1981) and plotted as a function of the first amino acid residue in the segment (den Boon *et al.*, 1991a).

hemagglutination-inhibiting monoclonal antibodies (Horzinek *et al.*, 1986; Kaeffer *et al.*, 1989). Pulse-chase labeling experiments revealed that the 75- to 100-kDa material is derived from a 200-kDa precursor that is found in infected cells, but not in virus particles (Horzinek *et al.*, 1986). Extensive N-glycosylation and proteolytic cleavage of the precursor are part of the posttranslational processing of the torovirus S protein. The cross-reactivity between different BRV "serotypes" in radioimmunoprecipitation indicates that the 85- and 105-kDa protein species, which were identified in radioiodination studies, also probably represent the surface structures (Koopmans *et al.*, 1986).

The backbone of the BEV S protein is an apoprotein of 1581 aa ($M_r$ about 178 kDa) encoded by ORF2. Among the features it shares with the S proteins of other coronaviruses are a large number (18) of potential N-glycosylation sites, an N-terminal signal sequence, a putative C-terminal transmembrane anchor, two putative heptad repeat domains, and a possible cleavage site for a "trypsin-like" protease (Fig. 9). The mature S protein consists of two subunits (S1 and S2, according to the coronavirus protein nomenclature). Endoglycosidase F digestion of S1 and S2 strongly suggests that the S protein is, indeed, cleaved at the predicted cleavage site, which consists of five consecutive arginine residues (Snijder *et al.*, 1990b). As postulated previously for other S proteins (De Groot *et al.*, 1987), the heptad repeats in S2 are assumed to be involved in the generation of an intrachain coiled-coil secondary structure. Interchain interactions of the same kind may be involved in protein oligomerization. In a sucrose gradient assay the BEV S protein was shown to form dimers (Snijder *et al.*, 1990b). The intra- and interchain coiled-coil interactions may stabilize the elongated spikes of all *Coronaviridae*.

## VIII. INDICATIONS FOR RECOMBINATION DURING TOROVIRUS EVOLUTION

Although all *Coronaviridae* display the same basic gene order (replicase, S, M, N), the number and location of additional reading frames in the genomic RNA is quite variable (Spaan *et al.*, 1988). RNA recombination, which is considered an important factor in RNA virus evolution (Strauss and Strauss, 1988; Goldbach and Wellink, 1988; Lai, 1992), is the most attractive explanation for the observed genetic heterogeneity. This idea is supported by the fact that MHV

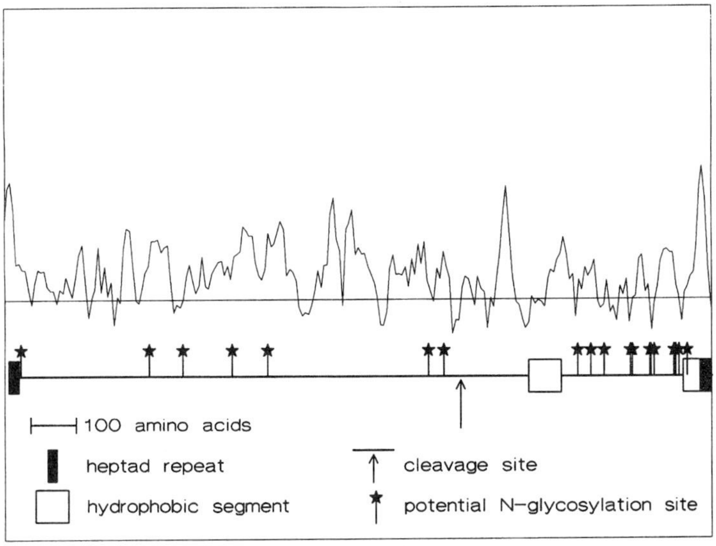


FIGURE 9. Schematic representation of the position of characteristic domains in the BEV peplomer protein. The positions of heptad repeats, hydrophobic segments, potential N-glycosylation sites and the putative cleavage site are indicated. In addition, a hydrophobicity plot of the amino acid sequence is presented. The plot was generated according to the method of Kyte and Doolittle (1982) using a window size of 21 (Snijder *et al.*, 1990b).

displays a high (homologous) recombination frequency (Makino and Lai, 1989; Spaan *et al.*, 1988; Lai, 1990, 1992), which may be a consequence of the replication strategy and replicase properties of coronaviruses.

Although only the BEV genome has been (partially) sequenced, the toroviruses appear to follow the example of the coronaviruses. Thus, indications for two independent recombination events during BEV evolution have been obtained (Snijder *et al.*, 1991a).

The first putative recombination involves BEV ORF 4. Its 142-aa sequence shares 30–35% identical amino acid residues with the C-terminal parts of both the influenza C virus (IVC) and the coronavirus hemagglutinin esterase (HE) protein sequences, which were previously found to be related by common ancestry and heterologous recombination (Luytjes *et al.*, 1988; see also Chapter 8, this volume). Since the sequences corresponding to the 5′ two thirds of the HE gene are lacking in BEV, ORF 4 has been hypothesized to be a pseudogene (Snijder *et al.*, 1991a).

The second putative recombination event that has occurred during BEV evolution is quite similar to the case of ORF 4. The C-terminus of the BEV ORF 1a protein contains 31 to 36% identical amino acid residues compared with the N-terminal 190 aa of the nonstructural coronavirus protein, ns30/32kDa (Bredenbeek *et al.*, 1990; Cox *et al.*, 1989). Like the HE protein gene, this ns protein is found only in the coronaviruses closely related to MHV (Snijder *et al.*, 1991a), where it is located between the polymerase and HE genes (Fig. 10). Apparently a sequence related to the 5′ two thirds of this coronavirus nonstructural gene has been integrated into ORF 1a of BEV and is now expressed as part of its replicase.

Remarkably, both recombinations previously described associate toroviruses with coronaviruses, in particular with the antigenic cluster to which MHV belongs, and it is clear that these sequence similarities must be the result of nonhomologous RNA recombination events. However, they do not imply that toroviruses are more closely related to the MHV cluster than to other

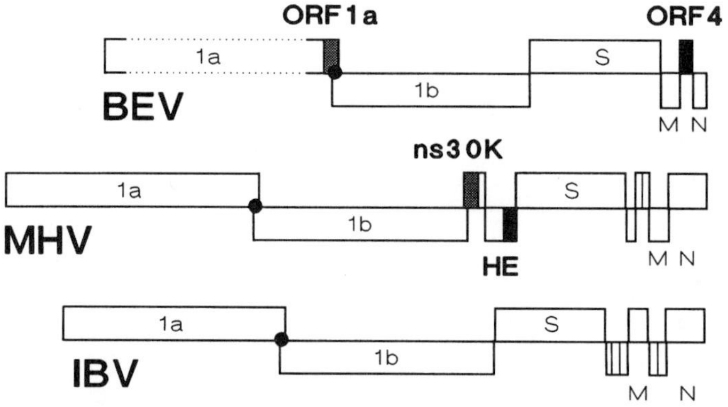


FIGURE 10. Schematic representation of the genome organization of the torovirus BEV and the coronaviruses MHV and IBV. The replicase gene (1a and 1b) and the structural genes S, M, and N are indicated. Filled (ORF 4/HE) and cross-hatched (ORF 1a/ns30kDa) boxes indicate homologous sequences of BEV and MHV that are thought to be derived from independent recombination events (Snijder and Horzinek, 1993).

coronaviruses. First, the homologous sequences are located at different positions in the genome (Fig. 10). Second, the BEV replicase is not more closely related to the MHV replicase than to, e.g., that of IBV. Third, the sequence similarities between the S and M proteins of corona- and toroviruses are so low (see the previous paragraph) that the high similarities discussed above (>30% identical residues) can only be explained if divergence between BEV ORF 4 and the coronavirus HE gene, and between the BEV ORF 1a fragment and the coronavirus ns30/32-kDa gene is a more recent event than the divergence of the other homologous genes of toro- and coronaviruses.

Considering the fact that several extant representatives of both virus groups cause enteric infections, direct recombination between toro- and coronaviruses during coinfection of the same cell is feasible. However, the involvement of a third party of viral or cellular origin cannot be excluded.

## IX. CONCLUDING REMARKS

The history of torovirus research illustrates the impact of molecular virology on the taxonomy of viruses. The taxonomic career of the toroviruses—at first proposed as a new family, then a free-floating genus, and finally a genus in the *Coronaviridae*—was guided by increasingly detailed knowledge of toroviral genes and proteins.

The BEV genome has turned out to be a showcase for the two driving forces in RNA virus evolution: divergence from a common ancestor and RNA recombination. Traces of these processes are revealed in the primary protein sequence homologies in the replicase, conserved structural properties (in spite of diverged primary sequences) in the S and M proteins, the presence of apparently unrelated N proteins coupled to a clearly different nucleocapsid architecture, and finally the recombination events discussed in the previous paragraph. Thus, the comparative analysis of the genomes and proteins of toroviruses and coronaviruses, supplemented with the data from the molecular characterization of the arteriviruses (see Chapter 12, this volume), has increased our understanding of the replication and evolution of all three virus groups.

ACKNOWLEDGMENTS. A number of the figures of this chapter have been published previously (Snijder and Horzinek, 1993). We acknowledge the valuable contributions of Joke Ederveen, Johan den Boon, and Willy Spaan to the torovirus research described in this chapter. We thank Alexander Gorbalenya for his assistance with the replicase sequence alignment, Stuart Siddell for helpful suggestions, and Mareen de Best for her assistance in preparation of the manuscript.

## X. REFERENCES

Baker, S. C., Yokomori, K., Dong, S., Carlisle, R., Gorbalenya, A. E., Koonin, E. V., and Lai, M. M. C., 1993, Identification of the catalytic sites of a papain-like cysteine proteinase of murine coronavirus, *J. Virol.* **67**:6056.

Beards, G. M., Hall, C., Green, J., Flewett, T. H., Lamouliatte, F., and Du Pasquier, P., 1984, An enveloped virus in stools of children and adults with gastroenteritis that resembles the Breda virus of calves, *Lancet* **2**:1050.

Beards, G. M. Brown, D. W. G., Green, J., and Flewett, T. H., 1986, Preliminary characterisation of torovirus-like particles of humans: Comparison with Berne virus of horses and Breda virus of calves, *J. Med. Virol.* **20**:67.

Bredenbeek, P. J., Noten, J. F. H., Horzinek, M. C., and Spaan, W. J. M., 1990, Identification and stability of a 30 Kd nonstructural protein encoded by mRNA 2 of mouse hepatitis virus in infected cells, *Virology* **175**:303.

Brierley, I., Diggard, P., and Inglis, S., 1989, Characterization of an efficient coronavirus ribosomal frameshifting signal: Requirement for an RNA pseudoknot, *Cell* **57**:537.

Cavanagh, D., Brian, D. A., Brinton, M., Enjuanes, L., Holmes, K. V., Horzinek, M. C., Lai, M. M. C., Laude, H., Plagemann, P. G. W., Siddell, S., Spaan, W. J. M., Taguchi, F., Talbot, P. J., 1994, Revision of the taxonomy of the Coronavirus, Torovirus and Arterivirus genera, *Arch. Virol.* **135**:227– 237.

Cox, G. J., Parker, M. D., and Babiuk, L. A., 1989, The sequence of cDNA of bovine coronavirus 32K nonstructural gene, *Nucleic Acids Res.* **17**:5847.

De Groot, R. J., Luytjes, W., Horzinek, M. C., Van der Zeijst, B. A. M., Spaan, W. J. M., and Lenstra, J. A., 1987, Evidence for a coiled-coil structure in the spike proteins of coronaviruses, *J. Mol. Biol.* **196**:963.

Den Boon, J. A., Snijder, E. J., Krijnse Locker, J., Horzinek, M. C., and Rottier, P. J. M., 1991a, Another triple-spanning envelope protein among intracellularly budding RNA viruses: The torovirus E protein, *Virology* **182**:655.

Den Boon, J. A., Snijder, E. J., Chirnside, E. D., De Vries, A. A. F., Horzinek, M. C., and Spaan, W. J. M., 1991b, Equine arteritis virus is not a togavirus but belongs to the coronavirus-like superfamily, *J. Virol.* **65**:2910.

Dubois-Dalcq, M. E., Doller, E. W., Haspel, M. V., and Holmes, K. V., 1982, Cell tropism and expression of mouse hepatitis virus (MHV) in mouse spinal chord cultures, *Virology* **119**:317.

Fagerland, J. A., Pohlenz, J. F. L., and Woode, G. N., 1986, A morphological study of the replication of Breda virus (proposed family Toroviridae) in bovine intestinal cells, *J. Gen. Virol.* **67**:1293.

Goldbach, R., and Wellink, J., 1988, Evolution of plus-strand RNA viruses, *Intervirology* **29**:260.

Gorbalenya, A. E., and Koonin, E. V., 1993, Helicases: Amino acid sequence comparisons and structure–function relationships, *Curr. Opin. Struct. Biol.* **3**:419.

Gorbalenya, A. E., Koonin, E. V., Donchenko, A. P., and Blinov, V. M., 1989, Coronavirus genome: Prediction of putative functional domains in the non-structural polyprotein by comparative amino acid sequence analysis, *Nucleic Acids Res.* **17**:4847.

Herold, J., Raabe, T., Schelle-Prinz, B., and Siddell, S. G., 1993, Nucleotide sequence of the human coronavirus 229E RNA polymerase locus, *Virology* **195**:680.

Holmes, K. V., Doller, E. W., and Sturman, L. S., 1981, Tunicamycin resistant glycosylation of a coronavirus glycoprotein: Demonstration of a novel type of viral glycoprotein *Virology* **115**:334.

Horzinek, M. C., and Weiss, M., 1984, Toroviridae: A taxonomic proposal, *Zentralbl. Vet. Med. [B]* **31**:649.

Horzinek, M. C., Weiss, M., and Ederveen, J., 1984, Berne virus is not "Coronavirus-like," *J. Gen. Virol.* **65**:645.

Horzinek, M. C., Ederveen, J., and Weiss, M., 1985, The nucleocapsid of Berne virus, *J. Gen. Virol.* **66**:1287.

Horzinek, M. C., Ederveen, J., Kaeffer, B., De Boer, D., and Weiss, M., 1986, The peplomers of Berne virus, *J. Gen. Virol.* **67**:2475.

Horzinek, M. C., Flewett, T. H., Saif, L. J., Spaan, W. J. M., Weiss, M., and Woode, G. N., 1987, A new family of vertebrate viruses: *Toroviridae*, *Intervirology* **27**:17.

Jacks, T., Madhani, H. D., Masiarz, F. R., and Varmus, H. E., 1988, Signals for ribosomal frameshifting in the Rous sarcoma virus gag-pol region, *Cell* **55**:447.

Kaeffer, B., Van Kooten, P., Ederveen, J., Van Eden, W., and Horzinek, M. C., 1989, Properties of monoclonal antibodies against Berne virus (Toroviridae), *Am. J. Vet. Res.* **50**:1131.

Koopmans, M., Ederveen, J., Woode, G. N., and Horzinek, M. C., 1986, Surface proteins of Breda virus, *Am. J. Vet. Res.* **47**:1896.

Koopmans, M., Snijder, E. J., and Horzinek, M. C., 1991, cDNA probes for the diagnosis of bovine torovirus (Breda virus) infection, *J. Clin. Microbiol.* **29**:493– 497.

Kyte, J., and Doolittle, R. F., 1982, A simple method for displaying the hydropathic character of a protein, *J. Mol. Biol.* **157**:105.

Lai, M. M. C., 1990, Coronavirus—organization, replication and expression of genome, *Annu. Rev. Microbiol.* **44**:303.

Lai, M. M. C., 1992, RNA recombination in animal and plant viruses, *Microbiol. Rev.* **56**:61.

Lee, H. J., Shieh, C. K., Gorbalenya, A. E., Koonin, E. V., Lamonica, N., Tuler, J., Bagdzhadzhuyan, A., and Lai, M. M. C., 1991, The complete sequence (22 kilobases) of murine coronavirus gene-1 encoding the putative proteases and RNA polymerase, *Virology* **180**:567.

Luytjes, W., Bredenbeek, P. J., Noten, J. F. H., Horzinek, M. C., and Spaan, W. J. M., 1988, Sequence of mouse hepatitis virus A59 mRNA2: Indications for RNA recombination between coronaviruses and influenza C virus, *Virology* **164**:415.

Makino, S., and Lai, M. M. C., 1989, High-frequency leader sequence switching during coronavirus defective interfering RNA replication, *J. Virol.* **63**:5285.

Makino, S., Yokomori, K., and Lai, M. M. C., 1990, Analysis of efficiently packaged defective interfering RNAs of murine coronavirus: Localization of a possible RNA-packaging signal, *J. Virol.* **64**:6045.

Makino, S., Joo, M., and Makino, J. K., 1991, A system for study of coronavirus mRNA synthesis: A regulated, expressed subgenomic defective interfering RNA results from intergenic site insertion, *J. Virol.* **65**:6031.

Poch, O., Sauvaget, I., Delarue, M., and Tordo, N., 1989, Identification of four conserved motifs among the RNA dependent polymerase encoding elements, *EMBO J.* **8**:3867.

Pringle, C. R., 1992, Committee pursues medley of virus taxonomic issues, *ASM News* **58**:475.

Rottier, P. J. M., Horzinek, M. C., and Van der Zeijst, B. A. M., 1981, Viral protein synthesis in mouse hepatitis virus strain A59-infected cells: Effect of tunicamycin, *J. Virol.* **40**:350.

Senanayake, S. D., Hofmann, M. A., Maki, J. L., and Brian, D. A., 1992, The nucleocapsid protein gene of bovine coronavirus is bicistronic, *J. Virol.* **66**:5277.

Snijder, E. J., and Horzinek, M. C., 1993, Toroviruses: Replication, evolution and comparison with other members of the coronavirus-like superfamily, *J. Gen. Virol.* **74**:2305.

Snijder, E. J., Ederveen, J., Spaan, W. J. M., Weiss, M., and Horzinek, M. C., 1988, Characterization of Berne virus genomic and messenger RNAs, *J. Gen. Virol.* **69**:2135.

Snijder, E. J., Den Boon, J. A., Spaan, W. J. M., Verjans, G. M. G. M., and Horzinek, M. C., 1989, Identification and primary structure of the gene encoding the Berne virus nucleocapsid protein, *J. Gen. Virol.* **70**:3363.

Snijder, E. J., Den Boon, J. A. Bredenbeek, P. J., Horzinek, M. C., Rijnbrand, R., and Spaan, W. J. M., 1990a, The carboxyl-terminal part of the putative Berne virus polymerase is expressed by ribosomal frameshifting and contains sequence motifs which indicate that toro- and coronaviruses are evolutionarily related, *Nucleic Acids Res.* **18**:4535.

Snijder, E. J., Den Boon, J. A., Spaan, W. J. M., Weiss, M., and Horzinek, M. C., 1990b, Primary structure and post-translational processing of the Berne virus peplomer protein, *Virology* **178**:355.

Snijder, E. J., Horzinek, M. C., and Spaan, W. J. M., 1990c, A 3′-coterminal nested set of independently transcribed messenger RNAs is generated during Berne virus replication, *J. Virol.* **64**:331.

Snijder, E. J., Den Boon, J. A., Horzinek, M. C., and Spaan, W. J. M., 1991a, Comparison of the genome organization of toro- and coronaviruses: Both divergence from a common ancestor and RNA recombination have played a role in Berne virus evolution, *Virology* **180**:448.

Snijder, E. J., Den Boon, J. A., Horzinek, M. C., and Spaan, W. J. M., 1991b, Characterization of defective interfering Berne virus RNAs, *J. Gen. Virol.* **72**:1635.

Snijder, E. J., Wassenaar, A. L. M., and Spaan, W. J. M., 1992, the 5′ end of the equine arteritis virus genome encodes a papainlike cysteine protease. *J. Virol.* **66**:7040.

Spaan, W. J. M., Cavanagh, D., and Horzinek, M. C., 1988, Coronaviruses: Structure and genome expression, *J. Gen. Virol.* **69**:2939.

Strauss, J. H., and Strauss, E. G., 1988, Evolution of RNA viruses, *Annu. Rev. Microbiol.* **42**:657.

Ten Dam, E. B., Pleij, C. W. A., and Bosch, L., 1990, RNA pseudoknots; translational frameshifting and readthrough on viral RNAs, *Virus Genes* **4**:121.

Tellier, R., and Petric, M., 1993, Human torovirus—purification from faeces, in: *Abstracts of the IXth International Congress of Virology*, p. 47, Glasgow, Scotland.

van der Most, R. G., Bredenbeek, P. J., and Spaan, W. J. M., 1991, A domain at the 3′ end of the polymerase gene is essential for encapsidation of coronavirus defective interfering RNAs, *J. Virol.* **65**:3219.

van der Most, R. G., Heijnen, L., Spaan, W. J. M., and De Groot, R. J., 1992, Homologous RNA recombination allows efficient introduction of site-specific mutations into the genome of coronavirus MHV-A59 via synthetic co-replicating RNAs, *Nucleic Acids Res.* **20**:3375.

Von Heijne, G., 1981, On the hydrophobic nature of signal sequences, *Eur. J. Biochem.* **116**:419.

Weiss, M., and Horzinek, M. C., 1986, Morphogenesis of Berne virus (proposed family Toroviridae), *J. Gen. Virol.* **67**:1305.

Weiss, M., and Horzinek, M. C., 1987, The proposed family *Toroviridae*: Agents of enteric infections, *Arch. Virol.* **92**:1.

Weiss, M., Steck, F., and Horzinek, M. C., 1983, Purification and partial characterization of a new enveloped RNA virus (Berne virus), *J. Gen. Virol.* **64**:1849.

Woode, G. N., Reed, D. E., Runnels, P. L., Herrig, M. A., and Hill, H. T., 1982, Studies with an unclassified virus isolated from diarrhoeic calves, *Vet. Microbiol.* **7**:221.

CHAPTER 12

# The Coronaviruslike Superfamily

ERIC J. SNIJDER AND WILLY J. M. SPAAN

## I. INTRODUCTION

Until recently, the *Coronaviridae* was classified as a monogeneric family of closely related viruses. However, in the past four years, it has become evident that similarities in the genome organization, replication strategies, and nucleotide sequences of coronaviruses, toroviruses, and arteriviruses, require a revision of this taxonomy. The "superfamily" concept (Strauss and Strauss, 1988; Goldbach and Wellink, 1988), which is based on evolution and phylogeny and which has already closed the gaps between other virus groups (e.g., the alphaviruslike and picornaviruslike superfamilies), can now also be applied to a group of "coronaviruslike viruses."

The sequence analysis of the genomes of the coronaviruses infectious bronchitis virus (IBV) (Boursnell *et al.*, 1987) and mouse hepatitis virus (MHV) (Bredenbeek *et al.*, 1990; Lee *et al.*, 1991), the Berne torovirus (BEV) (Snijder *et al.*, 1990a,b), and the (at the time) "unclassified togavirus" equine arteritis virus (EAV) (Den Boon *et al.*, 1991b) revealed unexpected evolutionary links. The common features of these viruses are centered around the coronaviruslike replicase gene and its associated replication and expression strategy. Consequently, the corona- and toroviruses have now been formally joined together in the family *Coronaviridae* (Chapter 11, this volume), and the International Committee on the Taxonomy of Viruses has recently created a new study group to establish a virus family that will comprise the members of the presently free-floating genus arterivirus: EAV (the prototype of the genus), lactate dehydro-

ERIC J. SNIJDER AND WILLY J. M. SPAAN • Department of Virology, Institute of Medical Microbiology, Leiden University, 2300 AH Leiden, The Netherlands.

*The Coronaviridae*, edited by Stuart G. Siddell, Plenum Press, New York, 1995.

genase-elevating virus (LDV), simian hemorrhagic fever virus (SHFV), and the recently discovered porcine reproductive and respiratory syndrome virus (PRRSV; also known as Lelystad virus). The question of how the evolutionary links between the *Coronaviridae* and arteriviruses should be reflected in their taxonomic status is still open to discussion (Snijder and Horzinek, 1993; Cavanagh *et al.*, 1994).

The biological properties of EAV, LDV, and SHFV have recently been reviewed by Plagemann and Moennig (1992). The morphological characteristics and genome size of arteriviruses (12–15 kb) are most comparable to those of the *Togaviridae*. However, as will be described, the arterivirus replicase and replication strategy are strikingly similar to those of the *Coronaviridae*. In this short comparative chapter, we will briefly summarize the molecular characteristics of the arteriviruses and focus on the most important similarities and differences with the *Coronaviridae*.

## II. ARTERIVIRUSES

EAV was first isolated from a fetus aborted during an endemic disease outbreak in pregnant mares (Doll *et al.*, 1957). The sequence of the 12.7-kb positive-stranded, polyadenylated EAV genome was published in 1991 by Den Boon *et al.* (1991b). Two years later, the molecular characterization of PRRSV of swine (Meulenberg *et al.*, 1993a) and LDV of mice (Godeny *et al.*, 1993a) revealed slightly larger genome sizes: 15.1 and 14.2 kb, respectively. At the IXth Congress of Virology in Glasgow, 1993, the first sequence data from the SHFV genome were reported and confirmed its close relationship to the other three arteriviruses (Godeny *et al.*, 1993b).

Due to their morphology, EAV and LDV were initially classified as members of the togavirus family. The spherical arterivirus particle has a diameter of 50 to 70 nm (Hyllseth, 1973) and is comprised of an icosahedral core structure of 35 nm surrounded by an envelope carrying ringlike structures with a diameter of 12 to 15 nm (Horzinek *et al.*, 1971). The identification and characterization of four structural EAV proteins has been reported recently (de Vries *et al.*, 1992): a 12-kDa nucleocapsid (N) protein, an unglycosylated 18-kDa transmembrane protein (M), a 25-kDa glycoprotein $G_S$, and a second glycoprotein, $G_L$, which, due to heterogeneous glycosylation, has sizes between 30 and 42 kDa. The current model of the EAV particle is shown and compared with the *Coronaviridae* in Fig. 1. It is clear that the nucleocapsid architecture (a classic trait for viral taxonomy, with the same ranking as nucleic acid type or the presence of an envelope), is different for arteriviruses (icosahedral) and *Coronaviridae* (helical). Another important difference is the fact that the EAV envelope does not bear the elongated spikes which are so characteristic of the *Coronaviridae*.

## III. ARTERIVIRUS GENOME ORGANIZATION AND EXPRESSION

The arterivirus genome is a polycistronic RNA that contains 8 open reading frames (ORFs) in the case of EAV, PRRSV, and LDV (Fig. 2). Like *Corona-*

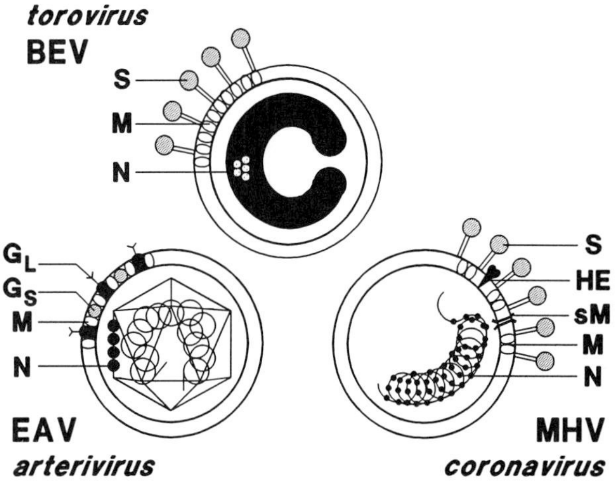


FIGURE 1. Schematic representation of the structure of BEV (a torovirus), EAV (an arterivirus), and MHV (a coronavirus). The major structural proteins of each virus group are illustrated.

*viridae*, arteriviruses produce a 3′-coterminal nested set of viral mRNAs in the infected cell. These mRNAs range in size from genome length to about 0.7 kb and are used to express internally located ORFs (de Vries *et al.*, 1990; Kuo *et al.*, 1991; Meulenberg *et al.*, 1993a). Viral subgenomic (sg) RNAs are composed of leader and body sequences that are not contiguous on the genome, another striking similarity with coronaviruses (but possibly not with toroviruses; see Chapter 11, this volume).

The EAV ORFs 2, 5, 6, and 7 have been shown to encode the structural $G_S$, $G_L$, M, and N proteins, respectively (de Vries *et al.*, 1992). The structural characteristics of the products of ORFs 3 and 4 are typical of membrane proteins, but no information on the function of these proteins has been obtained so far.

The arterivirus replicase gene is between 9.7 (EAV) and 11.6 (PRRSV) kb in

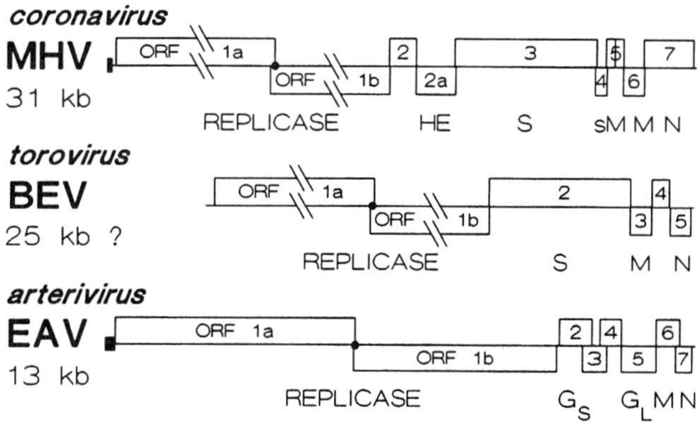


FIGURE 2. Genome organization of MHV (a coronavirus), BEV (a torovirus), and EAV (an arterivirus). The genes encoding the viral replicases and major structural proteins are illustrated.

size and is comprised of two ORFs (1a and 1b) that are expressed from the genomic RNA. As in the case of *Coronaviridae*, the arterivirus ORF 1b is probably expressed by ribosomal frameshifting during the translation of genomic RNA (den Boon *et al.*, 1991b). "Shifty" sequences (5′ UUUAAAC 3′ for LDV and PRSSV; 5′ GUUAAAC 3′ for EAV) and a downstream RNA pseudoknot structure are thought to form the frameshift signals (Jacks *et al.*, 1988; Brierley *et al.*, 1989; Ten Dam *et al.*, 1990). The ORF 1a/ORF 1b frameshift mechanism and the RNA structures involved in this process are remarkably conserved in corona-, toro, and arteriviruses, indicating that translational frameshifting is an ancient and probably essential regulatory step in replicase gene expression.

## IV. mRNA SYNTHESIS

The generation of a 3′-coterminal nested set of mRNAs from a nonsegmented genome distinguishes the members of the coronaviruslike superfamily from other groups of positive-stranded RNA viruses. The common ancestry of corona-, toro-, and arteriviral replicases (see Section V) suggests that this expression strategy is related to the properties of the coronaviruslike replicase and that similar (in fact, homologous) transcription mechanisms may be used by these viruses.

An early analysis of EAV mRNA transcription by means of UV transcription mapping suggested that all sg RNAs are derived from the processing of a genome-length precursor molecule (van Berlo *et al.*, 1982), a situation that would be very different from the independent transcription of coronavirus and torovirus mRNAs, which was deduced using the same experimental approach (Jacobs *et al.*, 1981; Snijder *et al.*, 1990b). To resolve this point, the UV transcription mapping analysis of both EAV and MHV at a late time point in infection has recently been repeated (Den Boon *et al.*, 1995b), and in contrast to the observations of van Berlo *et al.* (1982), a clear correlation between UV target size and physical size for the sg RNAs of EAV was established. However, a quantitative analysis indicated that for both arteriviruses and coronaviruses this correlation does not justify the conclusion that the sg RNA transcription is fully independent from the synthesis of the genomic RNA (Den Boon *et al.*, 1995b). Nevertheless, splicing can be ruled out as the major mechanism for arterivirus sg RNA production late in infection.

The comparative analysis of mRNA transcription of corona- and arteriviruses was extended by preparing replicative form (RF) RNA in order to study the EAV minus strands (Den Boon *et al.*, 1995c). As in coronavirus-infected cells (Sethna *et al.*, 1989; Sawicki and Sawicki, 1990), sg RFs were detected, suggesting that complements of both genomic and sg positive strands are generated. However, it remains to be excluded that the sg negative strands of EAV are derived from cleavage of RNase-sensitive sites in a genome-length RF. For a discussion on the possible role of the sg minus strands of coronaviruses, the reader is referred to Chapter 2 (this volume).

Both corona- and arteriviral mRNAs contain a leader sequence at their 5′ end that is identical to the 5′ end of their respective genomes (de Vries *et al.*,

1990). Thus the generation of corona- and arterivirus subgenomic mRNAs must involve a process of discontinuous transcription. The arterivirus leader RNA has between 170 and 210 nucleotides and is, therefore, two to three times larger than that of coronaviruses. For EAV, LDV, and PRRSV, conserved AU-rich sequences have been identified at the leader-to-body junction (LBJ) sites of the mRNAs (Fig. 3) (de Vries *et al.*, 1990; Den Boon *et al.*, 1995c; Chen *et al.*, 1993; Meulenberg *et al.*, 1993b). Coronavirus genomes contain similar conserved AU-rich boxes (known as intergenic sequences) at the fusion sites of leader and mRNA bodies (reviewed by Spaan *et al.*, 1988; Lai, 1990). However, since most of the arterivirus ORFs 2 to 6 overlap to some extent, the coronavirus term "intergenic sequence" is inappropriate for arteriviruses. Although no evidence for a common leader sequence in torovirus mRNAs has been obtained, a conserved AU-rich sequence, which is thought to function as a promoter sequence for mRNA transcription, is also found upstream of ORFs 2 to 5 of the Berne virus (Fig. 3) (Chapter 11, this volume).

In view of its relatively small size, it is, perhaps, not surprising that the arterivirus LBJ sequences are found at sites not involved in mRNA synthesis. The EAV 5′ UCAAC 3′ LBJ motif, for example, occurs 18 times in the genomic sequence. Thus, it is evident that this sequence element cannot be the sole determinant of mRNA transcription.

The role of the LBJ motif in mRNA synthesis has not yet been fully elucidated. As in coronaviruses, the extreme 3′ end of the arterivirus leader sequence is identical to the LBJ sequence. Thus, it is possible that, in the arterivirus minus strand, the complements of LBJ sites may function as mRNA promoters in a discontinuous leader-primed transcription mechanism involving basepairing between the leader and the promoter region. This process may correspond to the model that has been put forward for coronavirus mRNA transcription (Baric *et al.*, 1983; Lai *et al.*, 1983; Spaan *et al.*, 1983). A basepairing interaction between leader and mRNA promoter sequences on the negative

*arteriviruses*

| | |
|---|---|
| **EAV** | U C A A C u |
| **LDV** | n u a A C C |
| **PRRSV** | u n a A C C |

*coronaviruses*

| | |
|---|---|
| **MHV** | A A U C u A A a C |
| **IBV** | C U U A A C A A |
| **HCV** | u c U c A A C U a A a |

*toroviruses*

| | |
|---|---|
| **BEV** | U c U U U A G a a |

FIGURE 3. Alignment of the conserved AU-rich sequences which are thought to be involved in the transcription of the subgenomic mRNAs of arteriviruses (leader-to-body junction sites), coronaviruses (intergenic regions), and toroviruses ("core promoter"). Fully conserved nucleotides are printed in capitals; for the other positions the predominant nucleotide is shown.

strand is still the most attractive explanation for the discontinuous step during coronavirus mRNA transcription. Since the coronavirus system has been studied more extensively, we would refer the reader to Chapter 2 (this volume) for a more detailed discussion of this subject.

## V. THE CORONAVIRUSLIKE REPLICASE

Genome replication is a fundamental process in the viral life cycle and viral replicase genes are at the hub of viral biology. Replicase proteins are conserved among seemingly disparate groups of plant and animal RNA viruses (Strauss and Strauss, 1988; Goldbach and Wellink, 1988), and this evolutionary link is clearly apparent in the replicases of the coronaviruslike (CVL) superfamily. The comparison of the replication strategy and replicase properties of corona-, toro-, and arteriviruses has clearly distinguished the CVL replicase module from those of alpha, picorna, and flaviviruses. The organization of the CVL replicase in two ORFs, expression of the gene by ribosomal frameshifting, and the arrangement of conserved domains within the gene product are unique.

Although arterivirus replicase genes are considerably smaller (9.5–12 kb) than their toro- and coronavirus counterparts, they contain a number of conserved domains that are present in the same relative positions (Den Boon *et al.*, 1991b). The sequence alignments (Fig. 4) of these domains reveal up to 30% amino acid sequence identity in the most conserved regions, a percentage that cannot be due to convergent evolution. The conservation of two of these domains (polymerase and helicase), which are common to all positive-stranded RNA viruses (Poch *et al.*, 1989; Gorbalenya and Koonin, 1993), is not very surprising; their presence indicates that these viruses have probably all descended from the same RNA virus prototype. It is remarkable, however, that only in CVL replicases the helicase domain is located downstream of the polymerase motif. The polymerase motif also carries another CVL replicase trademark: the substitution of the classic GDD in the core of the motif by an SDD. Also, the conservation of additional replicase domains, for example, the carboxyl-terminal ORF 1b domain, for which no homologue can be found in other viral replicases, clearly indicates that the CVL replicases are more related to each other than to any other group of positive-stranded RNA viruses.

## VI. PROTEOLYTIC PROCESSING OF THE CVL REPLICASE

Proteolytic processing of nonstructural proteins fulfills a key role in the life cycle of most viruses. In the course of virus evolution, highly specific virus-encoded proteases have evolved, and their importance for the regulation of virus replication is becoming more and more evident. For detailed information on this topic, the reader is referred to an excellent review by Dougherty and Semler (1993).

CVL replicase gene expression leads to the production of an ORF 1a/1b fusion protein that is large in the case of arteriviruses (345–420 kDa) (Den Boon *et al.*, 1991b; Meulenberg *et al.*, 1993a; Godeny *et al.*, 1993a) and extremely large in coronaviruses (740–810 kDa) (Boursnell *et al.*, 1987; Lee *et al.*, 1991; Herold *et al.*, 1993) and probably also in toroviruses. The presence of multiple putative protease domains in the ORF 1a proteins of both coronaviruses (Gorbalenya *et al.*, 1989a; Lee *et al.*, 1991; Herold *et al.*, 1993) and arteriviruses (Den Boon *et al.*, 1991b; Meulenberg *et al.*, 1993a; Godeny *et al.*, 1993a) suggests that extensive proteolytic processing is involved in the regulation of CVL replicase function. This assumption is supported by the fact that the characterization of temperature-sensitive mutants of MHV revealed the presence of at least five $RNA^-$ complementation groups (Schaad *et al.*, 1990; Baric *et al.*, 1990).

A number of replicase cleavage products have recently been detected in MHV-infected cells (Denison *et al.*, 1992) and in *in vitro* translation reactions programmed with genomic RNA (Denison *et al.*, 1991). The preliminary processing scheme of the EAV ORF 1a protein comprises at least five proteolytic cleavages carried out by three different viral proteases (Snijder *et al.*, 1994a). Combined with cleavage site predictions based on the putative protease types encountered in corona- and arteriviruses (Gorbalenya *et al.*, 1989a; Lee *et al.*, 1991; Godeny *et al.*, 1993a), these experimental data forecast the generation of at least 10 (and possibly up to 20) cleavage products from the CVL ORF 1a/1b protein. If this estimate is correct, the number of intermediary processing products could be very large. This implies that the CVL replicase processing may be very complex, especially since the processing analysis of other viral replicases has revealed that intermediary products can be functional subunits themselves, e.g., in the replication of poliovirus (Jore *et al.*, 1988; Ypma-Wong *et al.*, 1988) and Sindbis virus (de Groot *et al.*, 1990; Strauss and Strauss, 1990; Shirako and Strauss, 1994).

The functional characterization of the proteases in the CVL ORF 1a sequences, which belong to different protease supergroups, has only just begun. However, a general pattern for coronaviruses and arteriviruses seems to emerge from comparative replicase sequence analysis. In comparable positions, the ORF 1a proteins of both virus groups contain a domain belonging to the protease superfamily that comprises the chymotrypsinlike and picornavirus 3C-like proteolytic enzymes (Fig. 5) (Gorbalenya *et al.*, 1989b; Bazan and Fletterick, 1988). Although the predicted catalytic nucleophile of the coronavirus protease (Cys) differs from that in the arterivirus proteases (Ser), this domain may still be a remnant from a common ancestor of both virus groups. The exchange of Cys for Ser at the active site of the enzyme is considered to be feasible (Gorbalenya *et al.*, 1989b; Bazan and Fletterick, 1988, 1990; Dougherty and Semler, 1993).

The functions encoded from the central region of ORF 1a to the 3′-end of ORF 1b appear to be the core of the CVL replicase polyprotein: the well-conserved domains (protease–polymerase–helicase–C-terminal "unique" ORF 1b domain) are within this area, and only small insertions and deletions in this part of the replicase can be detected within the coronavirus or arterivirus groups (Fig. 5). The chymotrypsin/3C-like protease is most likely to be responsible for the processing of the core replicase.

The N-terminal half of the ORF 1a protein, on the other hand, is quite variable: the variability of the replicase gene in this area is largely responsible

## *Polymerase domain*

```
MHV    576  VIGTTKfYGGWDdMLrr-9-LMGWDYPKCDRAMPnilR-21-YRLaNE-12-GcyYvKPGGTsSGDATTAfANSVFN-57-FSmMILSDDgVVCYN-42-HEFCSQH-12-LPYPDPSRIl
IBV    582  VIGTTKfYGGWDnMLrn-9-LMGWDYPKCDRAMPnllR-21-YRLyNE-12-GgiYvKPGGTsSGDATTAyANSVFN-57-FSlMILSDDgVVCYN-42-HEFCSQH-12-LPYPDPSRIl
HCV    575  VIGTTKfYGGWDnMLkn-9-LMGWDYPKCDRAMPsmiR-21-YRLsNE-12-GgfYfKPGGTtSGDATTAyANSVFN-57-FSmMILSDDsVVCYN-42-HEFCSQH-12-LPYPDPSRIi
             ** *      *      * ** **** *  *     * * **    *   ***** ******* *  *        ****            **** *  *  *  *

BEV    513  lIGvsKyglkfskfLkd-9-vfGsDYtKCDRtfPlsfR-17-Y-L-NE-12-GmllnKPGGTsSGDATTAhsNtfyN-51-yflnfLSDDsfi-fs-37-eEFCSaH-10-L--PsrgRll
                                *  ***  *  *     * * *       * **  ***  *  *          ***              *          *     *

PRRSV  367  w-kSPIaLG--knKFke-9-LEa-DLaSCDRSTPAivR-19-YVL-N-- 9-dgaFtKRGGLSSGDPvTSvSNTvYS-38---mlvYSDDlVL-ya-34--sFlGCR- 6-Lv-pnrDRil
LDV    365  i-gSPIyLG--nnKFtp-9-LEa-DLaSCDRSTPAiiR-19-YVL-N-- 9-sgcFdKRGGLSSGDPvTSvSNTvYS-38---llvYSDDvVL-yd-34--qFpGCR- 6-Lv-pqrDRil
EAV    369  kdgSPIyLG--ksKFdp-8-LEt-DLeSCDRSTPAlvR-19-YVL-N-- 9-svaFtKRGGLSSGDPiTSiSNTiYS-39---vyiYSDDvVL-tt-35--sFlGCR- 9-La-slqDRvt
```

## *Helicase domain*

```
MHV   1199  vQGPPGtGKSHl-66-tTiNALPElvtDIivVDEVSMl-16-yVYiGDPaQLPAPRvL-25-FLgtCYRCPkEIV-75-vLGLqtQTVDSaQGS-18-vNRFNVAiTRAKkGIlc
IBV   1206  vQGPPGsGKSHf-66-sTiNALPEvscDIllVDEVSMl-16-vVYvGDPaQLPAPRtL-24-FLakCYRCPkEIV-78-mLGLnvQTVDSsQGS-18-iNRFNVAlTRAKrGIlv
HCV   1200  iQGPPGsGKSHc-66-sTvNALPEvnaDIvvVDEVSMc-16-iVYvGDPqQLPAPRvL-25-FLhkCYRCPaEIV-75-lLGLqtQTVDSaQGS-18-aNRFNVAiTRAKkGIfc
             **** **       * * **         ****       *  *** **  *        *  *****  *        **    * ** **      **  *  *

BEV   1097  vmGPPGtGKttf-61-cThNtLPfiksavliaDEVSli-15-vVllGDPfQL-sP--v-23-yLtaCYRCPpqIl-67-gLG-dvtTiDSsQGt-18-vNRviVgcsRst-thlv
             * ** **                    **          *** **          *    *              * ** **      *  *  *

PRRSV  788  ivGpPGsGKTtw-51-RLias-GhvpGrvsylDEagYc-13-lvclGDLqQL-------19--LttiyRFGpnic-40-igs--aiTIDSsQGa-16-ksRAlVAiTRArhglfi
LDV    785  itGaPGtGKTty-51-RLira-GfipGrvsylDEaaYc-13-lvcvGDLnQL-------19--LievfRFGpsiv-40-vdg--aiTIDSsQGc-16-saRAlVAiTRArfyvfv
EAV    801  veGpPGsGKTfh-58-RLpqv-GtseGe-tfvDEvaYf-13-vkgyGDLnQL-------22--LrvchRFGaavc-40-glg--hrTIDSiQGc-16-rpRAvVAvTRAsqelyi
```

## *Carboxyl-terminal ORF 1b domain*

```
MHV   2315  GGLH-39-KsVCTviD-18-SKVvnVnvD-16-TfYP-31-NYG-11-NVaKYTQLC-12-NMRV
IBV   2282  GGLH-39-KqVCTvvD-21-SKVvtVsiD-16-TcYP-29-NYG-11-NVaKYTQLC-12-NMRV
HCV   2281  GGLH-40-KtVCTymD-18-SKVheViiD-16-TfYP-30-NYG-11-NVvKYTQLC-12-NMRV
            ** *     *  *  *    ***  *  *    * **      ***     *  **** *      *  *
BEV   1933  GGvH-31-Krt-TlvD-21-SKVifVniD-17-TfYP-25-NYG-10-NfaKYTQiC-11-NalV
            ** *     *     *    *        *     * *                            *  *
PRRSV 1225  GGCH-28-KavCtltD-15-SkcwKlklD-12-TaY--12-dya- 7-davVyidpc- 6-NrkV
LDV   1208  GGCH-28-KelCtvtD-15-SmdyKllvD-12-TaY-- 9-sms- 7-eegVffd--- 6-NakV
EAV   1248  GGsH-28-KaaCsvvD-15-SrvyKimiD-12-TfY--13-avs- 7-nepVsfd-v- 6-NalV
```

FIGURE 4. Amino acid sequence comparisons of the three most conserved domains of CVL ORF 1b proteins: polymerase, helicase, and the C-terminal domain, respectively. Amino acids identical in all three coronaviruses are shown in capitals. Amino acids in the BEV sequence that are identical to the amino acid at that position in all three coronaviruses are also shown in capitals and indicated with asterisks on the line between the coronavirus and BEV sequences. Within the arterivirus group, capitals again indicate absolutely conserved amino acids. The asterisks on the line between the sequences of the arteriviruses and BEV indicate the amino acids conserved in all available CVL ORF 1b sequences.

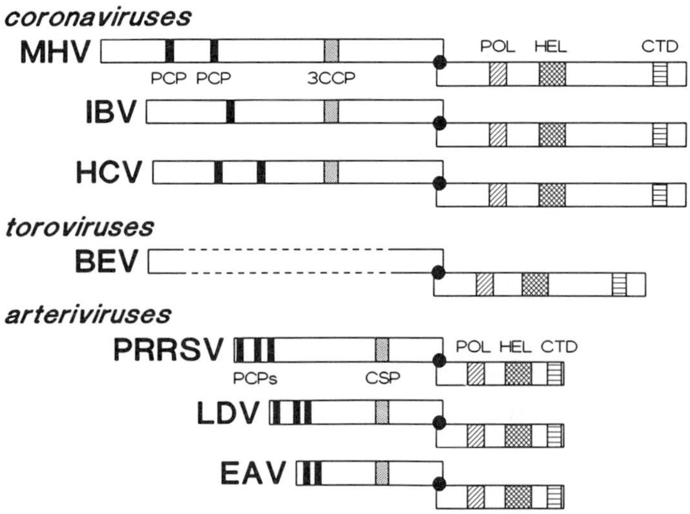


FIGURE 5. Scaled representation of the replicase gene organization of coronaviruses, toroviruses, and arteriviruses. The filled circle indicates the ORF 1a/1b frameshift site. The largest replicase gene (MHV) is about 22 kb, the smallest one (EAV) approximately 9.5 kb. Boxes represent conserved domains: PCP, papainlike cysteine protease; 3CCP, 3C-like cysteine protease; CSP, chymotrypsinlike serine protease; POL, polymerase; HEL, helicase; CTD, carboxyl-terminal ORF 1b domain.

for the size differences in corona- or arterivirus genomes. A comparison between the coronavirus and arterivirus N-terminal ORF 1a protein sequences does not yield any significant similarities, and even within the coronavirus and arterivirus groups there is little conservation in this region.

However, one striking observation can be made. Both corona- and arteriviruses contain putative protease domains in the amino-terminal half of their ORF 1a proteins. These proteases have been shown, or are predicted, to belong to the papainlike cysteine protease superfamily (Fig. 5) (Gorbalenya *et al.*, 1991). A number of these proteases have recently been characterized. In EAV (Snijder *et al.*, 1992) and MHV (Baker *et al.*, 1993), a papainlike protease is responsible for the production of an amino-terminal ORF 1a cleavage product. Furthermore, the production of the next amino-terminal cleavage product from the EAV ORF 1a protein has now been shown to be mediated by an autoproteolytic activity that is most similar to papainlike proteases (Snijder *et al.*, 1995). Two other arteriviruses, PRRSV and LDV, have been found to use even a third, papainlike protease in the processing of the N-terminal region of the ORF 1a protein (Den Boon *et al.*, 1995a). During arterivirus evolution, EAV appears to have lost the protease function that is responsible for the production of an additional amino-terminal cleavage product in PRRSV and LDV (Fig. 5).

Unfortunately, only the 5′- and 3′-sequences of the Berne torovirus ORF 1a region (about 1 kb from each end) have been determined so far and no putative proteases have been identified in these regions. Their future identification would be very useful, especially for a comparative analysis of the CVL chymotrypsin/3C-like proteases.

## VII. STRUCTURAL PROTEINS

Viral structural proteins are known to diverge at a higher rate than nonstructural proteins. This is exemplified by the relationship between the corona- and torovirus replicase, S, and M proteins (Chapter 11, this volume), a set of genes that is thought to be related by common ancestry. Whereas, undisputable sequence homologies are present in the replicase proteins of these viruses, these are absent in the S and M protein sequences, which only display structural similarities. The corona- and torovirus N proteins, on the other hand, are so different in size and sequence that common ancestry is highly unlikely. Also, the nucleocapsid structures that they form together with the genomic RNA are significantly different. It therefore has been proposed that during evolution either the coronavirus or the torovirus group may have acquired a novel N protein gene (Snijder and Horzinek, 1993).

When the arteriviruses are included in the comparison of the structural proteins of CVL replicase-containing viruses, it is hard to detect any similarity at all, either in size, sequence, or structural characteristics (Den Boon *et al.*, 1991b; de Vries *et al.*, 1992). The N protein (12kDa) is even smaller than that of toroviruses and gives rise to an icosahedral nucleocapsid. The presently known surface glycoproteins $G_S$ and $G_L$ are small compared to the S proteins of corona- and toroviruses. Only the presence of a triple-spanning membrane (M) protein, encoded by arterivirus ORF 6, is similar to the *Coronaviridae*, even though this protein (18kDa) is also smaller than its coronaviral counterparts. The triple-spanning membrane proteins of corona- and toroviruses have been implicated to play a crucial role in determining the intracellular budding process of these viruses (Dubois-Dalcq *et al.*, 1982, Holmes *et al.*, 1981; Rottier *et al.*, 1981; Den Boon *et al.*, 1991a). Since arteriviruses also assemble at intracellular membranes, the conservation or acquisition of an M protein with similar structural properties may have been expected during arterivirus evolution (de Vries *et al.*, 1992).

## VIII. THE EVOLUTION OF THE CVL SUPERFAMILY

The common ancestry of the CVL replicase proteins and the associated replication strategy are evident. The features that are shared by the members of the CVL superfamily are the basic genome organization, namely, "replicase-envelope protein–nucleocapsid protein," the production of 3′-coterminal nested sets of mRNAs, and the conserved organization of the replicase gene. However, the N proteins and nucleocapsid structures of the three CVL genera are apparently unrelated. Furthermore, the arterivirus surface proteins show no evidence of a common ancestry.

The coupling of different sets of structural genes to the same replicase gene has been explained by the recombination of complete genes or gene sets (modules) (Zimmern, 1987; Strauss and Strauss, 1988; Goldbach and Wellink, 1988). The RNA recombination frequency during MHV replication has been shown to be remarkably high (reviewed by Spaan *et al.*, 1988; Lai, 1992), a characteristic that may be determined by their replicase properties. The different sets of structural genes (and the varying number of "additional" genes) that are now known to be linked to the CVL replicase gene indicate that this characteristic may be shared by other members of the superfamily ("modular" evolution). Direct evidence for multiple recombination events during torovirus evolution has already been obtained (Snijder *et al.*, 1991). Together with divergent evolution, a high recombination frequency can account for the diverse composition of CVL genomes. It has been hypothesized that the acquisition of a novel nucleocapsid protein gene, giving rise to a helical virus core rather than an icosahedral one, may have created the possibility for the corona/torovirus branch to diverge from the arterivirus branch (Godeny *et al.*, 1993a). The increased genome size of the corona- and toroviruses could in this case be explained by the more relaxed packaging constraints of a helical nucleocapsid structure.

Arteriviruses and, especially, coronaviruses possess exceptionally large genomes in comparison with other positive-stranded animal RNA viruses. The apparent gap between the genome sizes of arteriviruses (EAV:12.7 kb) and coronaviruses (MHV:31.5 kb) is slowly being reduced now that the sequences of larger arterivirus genomes (PRRSV:15.1 kb) and smaller coronavirus genomes [human coronavirus (HCV) 229E:27.3 kb] have been determined. A group of positive-strand RNA viruses that almost bridge the remaining gap are the plant closteroviruses, with genome sizes between 7 and 20 kb. The molecular biology properties of this virus group has recently been reviewed by Dolja *et al.* (1994). A number of remarkable similarities with the CVL group is found in the closterovirus group: e.g., polycistronic genomes, frameshifting during replicase expression, and the generation of a 3′-coterminal nested of mRNAs to express the 3′ proximal genes. Although phylogenetic studies, including closterovirus and CVL replicase sequences, indicated that these similarities are analogies rather than homologies (Koonin and Dolja, 1993), this example of convergent evolution may elucidate some general principles governing the evolution of viruses with large RNA genomes. The size variation and plasticity of both closterovirus and CVL genomes is remarkable, suggesting again the frequent involvement of recombination during the evolution of these virus groups (Dolja *et al.*, 1994). The process of recombination could also partially compensate for the general high error frequency of viral RNA-dependent RNA polymerases, a feature that will greatly influence the stability and evolution of large RNA genomes.

## IX. THE TAXONOMY OF CVL VIRUSES

The "split personality" of the arteriviruses, a CVL replicase gene coupled to a non-CVL set of structural genes, is an interesting test for viral taxonomists. Using traditional taxonomic criteria, e.g., the virion and nucleocapsid structure, there is no reason to propose a taxonomic relationship between arteriviruses and *Coronaviridae*. At the same time, certain characteristics, such as genome organization, replication strategy, and sequence homologies, provide a solid basis for a formal taxonomic status of the CVL superfamily. Any meaningful classification of the members of the CVL group clearly requires four hierarchical levels (Fig. 6) (Snijder and Horzinek, 1993; Cavanagh *et al.*, 1994): the coronavirus and torovirus species have now been classified into two genera that belong to the *Coronaviridae* family. The obvious evolutionary link of this family to the arteriviruses would be reflected most accurately by promoting the present arterivirus genus to the family status and by establishing an *order* (to replace the "superfamily") comprising the *Coronaviridae* and *Arteriviridae* families. Classification of the arteriviruses as a third genus of the coronavirus family is a less attractive alternative because this would not recognize the more distant position of the arteriviruses. However, this problem could be circumvented by establishing two subfamilies (*Coronaviridae* and *Arteriviridae*) (Cavanagh *et al.*, 1994) and changing the family name of the *Coronaviridae*.

The taxonomic fate of the "CVL superfamily" will depend on the future weighing of both old (structural) and new (genetic) criteria for virus classification. The balance between these criteria will decide whether the genetic similarities between coronaviruses and arteriviruses outweigh their structural dissimilarities. If, as in other areas of biology, the taxonomy of viruses should have a genetic basis and should reflect phylogenetic relationships, the creation of an order (Fig. 6) for the *Coronaviridae* and *Arteriviridae* would be the most elegant solution to recognize both their similarities and differences (Snijder and Horzinek, 1993; Cavanagh *et al.*, 1994). A taxonomic system that is unable to ac-

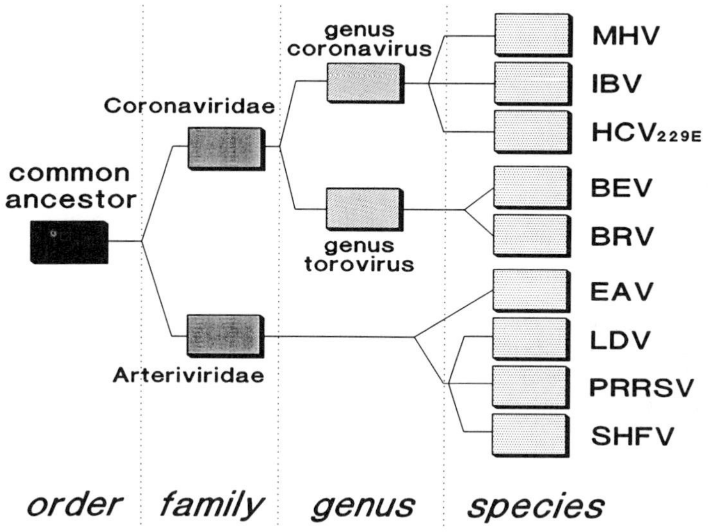


FIGURE 6. Pedigree of the CVL superfamily. The distances between the various CVL viruses and groups are shown arbitrarily. The four hierarchical (taxonomic) levels that can be discerned are indicated. Abbreviations: BEV, Berne virus; BRV, Breda virus; EAV, equine arteritis virus; HCV 229E, human coronavirus 229E; IBV, infectious bronchitis virus; LDV, lactate dehydrogenase-elevating virus; MHV, mouse hepatitis virus; PRRSV, porcine reproductive and respiratory syndrome virus; SHFV, simian hemorrhagic fever virus. For simplicity, the coronavirus clusters that include MHV and HCV 229E are shown as a single species.

commodate the results of virus evolution (mutation, recombination, and selection) will have no future.

ACKNOWLEDGMENTS. We thank Alexander Gorbalenya for his assistance with the replicase sequence alignments, Stuart Siddell and Johan den Boon for critical reading of this chapter, and Mareen de Best for her assistance in preparation of the manuscript.

## X. REFERENCES


Baker, S. C., Yokomori, K., Dong, S., Carlisle, R., Gorbalenya, A. E., Koonin, E. V., and Lai, M. M. C., 1993, Identification of the catalytic sites of a papain-like cysteine proteinase of murine coronavirus, *J. Virol.* **67:**6056.

Baric, R. S., Stohlman, S. A., and Lai, M. M. C., 1983, Characterization of replicative intermediate RNA of mouse hepatitis virus: Presence of leader RNA sequences on nascent chains, *J. Virol.* **48:**633.

Baric, R. S., Fu, K., Schaad, M. C., and Stohlman, S. A., 1990, Establishing a genetic recombination map for murine coronavirus strain A59 complementation groups, *Virology* **177:**646.

Bazan, J. F., and Fletterick, R. J., 1988, Viral cysteine proteases are homologous to the trypsin-like family of serine proteases: Structural and functional implications, *Proc. Natl. Acad. Sci. USA* **85:**7872.

Bazan, J. F., and Fletterick, R. J., 1990, Structural and catalytic models of trypsin-like viral proteases, *Semin. Virol.* **1:**311.

Boursnell, M. E. G., Brown, T. D. K., Foulds, I. J., Green, P. F., Tomley, F. M., and Binns, M. M., 1987, Completion of the sequence of the genome of the coronavirus avian infectious bronchitis virus, *J. Gen. Virol.* **68:**57.

Bredenbeek, P. J., Pachuk, C. J., Noten, J. F. H., Charité, J., Luytjes, W., Weiss, S. R., and Spaan, W. J. M., 1990, The primary structure and expression of the second open reading frame of the polymerase gene of the coronavirus MHV-A59, *Nucleic Acids Res.* **18:**1825.

Brierley, I., Diggard, P., and Inglis, S., 1989, Characterization of an efficient coronavirus ribosomal frameshifting signal: Requirement for an RNA pseudoknot, *Cell* **57:**537.

Cavanagh, D., Brian, D. A., Brinton, M., Enjuanes, L., Holmes, K. V., Horzinek, M. C., Lai, M. M. C., Laude, H., Plagemann, P. G. W., Siddell, S., Spaan, W. J. M., Taguchi, F., and Talbot, P. J., 1994, Revision of the taxonomy of the Coronavirus, Torovirus and Arterivirus genera, *Archives Virology* **135:**227.

Chen, Z., Kuo, L., Rowland, R. R. R., Even, C., Faaberg, K. S., and Plagemann, P. G. W., 1993, Sequence of 3′-end of genome and 5′-end of ORF 1a of lactate dehydrogenase-elevating virus (LDV) and common junction motifs between 5′-leader and bodies of seven subgenomic mRNAs, *J. Gen. Virol.* **74:**643.

de Groot, R. J., Hardy, W. R., Shirako, Y., and Strauss, J. H., 1990, Cleavage-site preferences of Sindbis virus polyproteins containing the non-structural proteinase. Evidence for temporal regulation of polyprotein processing *in vivo*, *EMBO J.* **9:**2631.

Den Boon, J. A., Snijder, E. J., Krijnse Locker, J., Horzinek, M. C., and Rottier, P. J. M., 1991a, Another triple-spanning envelope protein among intracellularly budding RNA viruses: The torovirus E protein, *Virology* **182:**655.

Den Boon, J. A., Snijder, E. J., Chirnside, E. D., De Vries, A. A. F., Horzinek, M. C., and Spaan, W. J. M., 1991b, Equine arteritis virus is not a togavirus but belongs to the coronavirus-like superfamily, *J. Virol.* **65:**2910.

Den Boon, J. A., Faaberg, K. S., Meulenberg, J. J. M., Wassenaar, A. L. M., Plagemann, P. G. W., Gorbalenya, A. E., and Snijder, E. J., 1995a, Processing and evolution of the N-terminal region of the arterivirus ORFIa protein: Identification of two papainlike cysteine proteases, *J. Virol.*, in press.

Den Boon, J. A., Spaan, W. J. M., and Snijder, E. J., 1995b, manuscript in preparation.

Den Boon, J. A. Kleijnen, M. F., Spaan, W. J. M., and Snijder, E. J., 1995c, manuscript in preparation.

Denison, M. R., Zoltick, P. W., Leibowitz, J. L., Pachuk, C. J., and Weiss, S. R., 1991, Identification of polypeptides encoded in open reading frame 1b of the putative polymerase gene of the murine coronavirus mouse hepatitis virus A59, *J. Virol.* **65:**3067.

Denison, M. R., Zoltick, P. W., Hughes, A., Giangreco, B., Olsen, A. L., Perlman, S., Leibowitz, J. L., and Weiss, S. R., 1992, Intracellular processing of the N-terminal ORF1a proteins of the coronavirus MHV-A59 requires multiple proteolytic events, *Virology* **189:**274.

de Vries, A. A. F., Chirnside, E. D., Bredenbeek, P. J., Gravenstein, L. A., Horzinek, M. C., and Spaan, W. J. M., 1990, All subgenomic mRNAs of equine arteritis virus contain a common leader sequence, *Nucleic Acids Res.* **18:**3241.

de Vries, A. A. F., Chirnside, E. D., Horzinek, M. C., and Rottier, P. J. M., 1992, Structural proteins of equine arteritis virus, *J. Virol.* **66:**6294.

Dolja, V. V., Karasev, A. V., and Koonin, E. V., 1994, Molecular biology and evolution of closteroviruses: Sophisticated build-up of large RNA genomes, *Annu. Rev. Phytopathol.* (in press).

Doll, E. R., Bryans, J. T., McCollum, W. H. M., and Wallace, M. E., 1957, Isolation of a filterable agent causing arteritis of horses and abortion by mares. Its differentiation from the equine abortion (influenza) virus, *Cornell Vet.* **47:**3.

Dougherty, W. G., and Semler, B. L., 1993, Expression of virus-encoded proteinases: Functional and structural similarities with cellular enzymes, *Microbiol. Rev.* **57:**781.

Dubois-Dalcq, M. E., Doller, E. W., Haspel, M. V., and Holmes, K. V., 1982, Cell tropism and expression of mouse hepatitis virus (MHV) in mouse spinal chord cultures, *Virology* **119:**317.

Godeny, E. K., Chen, L., Kumar, S. N., Methven, S. L., Koonin, E. V., and Brinton, M. A., 1993a. Complete genomic sequence and phylogenetic analysis of the lactate dehydrogenase-elevating virus, *Virology* **194:**585.

Godeny, E. K., Zeng, L., Smith, S. L., and Brinton, M. A., 1993b, Simian haemorrhagic fever virus: Another member of the coronavirus-like superfamily, in: *Abstracts of the IXth International Congress of Virology*, p. 22, Glasgow, Scotland.

Goldbach, R., and Wellink, J., 1988, Evolution of plus-strand RNA viruses, *Intervirology* **29:**260.

Gorbalenya, A. E., and Koonin, E. V., 1993, Helicases: Amino acid sequence comparisons and structure–function relationships, *Curr. Opin. Struct. Biol.* **3:**419.

Gorbalenya, A. E., Koonin, E. V., Donchenko, A. P., and Blinov, V. M., 1989a, Coronavirus genome: Prediction of putative functional domains in the non-structural polyprotein by comparative amino acid sequence analysis, *Nucleic Acids Res.* **17:**4847.

Gorbalenya, A. E., Donchenko, A. P., Blinov, V. M., and Koonin, E. V., 1989b, Cysteine proteases of positive strand RNA viruses and chymotrypsin-like serine proteases: A distinct protein superfamily with a common structural fold, *FEBS Lett.* **243:**103.

Gorbalenya, A. E., Koonin, E. V., and Lai, M. M. C., 1991, Putative papain-related thiol proteases of positive-stranded RNA viruses, *FEBS Lett.* **288:**201.

Herold, J., Raabe, T., Schelle-Prinz, B., and Siddell, S. G., 1993, Nucleotide sequence of the human coronavirus 229E RNA polymerase locus, *Virology* **195:**680.

Holmes, K. V., Doller, E. W., and Sturman, L. S., 1981, Tunicamycin resistant glycosylation of a coronavirus glycoprotein: Demonstration of a novel type of viral glycoprotein, *Virology* **115:**334.

Horzinek, M. C., Maess J., and Laufs, R., 1971, Studies on the structure of togaviruses. II. Analysis of equine arteritis, rubella, bovine viral diarrhoea, and hog cholera viruses, *Arch. Gesamte Virusforsch.* **33:**306.

Hyllseth, B., Structural proteins of equine arteritis virus, 1973, *Arch. Gesamte Virusforsch.* **40:**177.

Jacks, T., Madhani, H. D., Masiarz, F. R., and Varmus, H. E., 1988, Signals for ribosomal frameshifting in the Rouse sarcoma virus gag-pol region, *Cell* **55:**447.

Jacobs, L., Spaan, W. J. M., Horzinek, M. C., and van der Zeijst, B. A. M., 1981, Synthesis of subgenomic mRNAs of mouse hepatitis virus is initiated independently: evidence from UV transcription mapping, *J. Virol.* **39:**401.

Jore, J, de Geus, B., Jackson, R. J., Pouwels, P. H., and Enger-Valk, B. E., 1988, Poliovirus 3CD is the active protease for processing of the precursor protein P1 *in vitro*, *J. Gen. Virol.* **69:**1627.

Koonin, E. V., and Dolja, V. V., 1993, Evolution and taxonomy of positive-strand RNA viruses: Implications of comparative analysis of amino acid sequences, *Crit. Rev. Biochem. Mol. Biol.* **28:**375.

Kuo, L. L., Harty, J. T., Erickson, L., Palmer, G. A., and Plagemann, P. G. W., 1991, A nested set of eight RNAs is formed in macrophages infected with lactate dehydrogenase-elevating virus, *J. Virol.* **65:**5118.

Lai, M. M. C., 1990, Coronavirus—organization, replication and expression of genome, *Annu. Rev. Microbiol.* **44:**303.

Lai, M. M. C., 1992, RNA recombination in animal and plant viruses, *Microbiol. Rev.* **56:**61.

Lai, M. M. C., Patton, C. D., Baric, R. S., and Stohlman, S. A., 1983, Presence of leader sequences in the mRNA of mouse hepatitis virus, *J. Virol.* **46:**1027.

Lee, H. J., Shieh, C. K., Gorbalenya, A. E., Koonin, E. V., Lamonica, N., Tuler, J., Bagdzhadzhuyan, A., and Lai, M. M. C., 1991, The complete sequence (22 kilobases) of murine coronavirus gene-1 encoding the putative proteases and RNA polymerase, *Virology* **180:**567.

Meulenberg, J. J. M., Hulst, M. M., De Meijer, E. J., Moonen, P. L. J. M., Den Besten, A., De Kluyver, E. P., Wensvoort, G., and Moormann, R. J. M., 1993a, Lelystad virus, the causative agent of porcine epidemic abortion and respiratory syndrome (PEARS), is related to LDV and EAV, *Virology* **192:**62.

Meulenberg, J. J. M., De Meijer, E., and Moormann, R. J. M., 1993b, Subgenomic RNAs of Lelystad virus contain a conserved leader-body junction sequence, *J. Gen. Virol.* **74:**1697.

Plagemann, P. G., and Moennig, V., 1992, Lactate dehydrogenase-elevating virus, equine arteritis virus and simian haemorrhagic fever virus, a new group of positive strand RNA viruses. *Adv. Virus Res.* **41:**99.

Poch, O., Sauvaget, I., Delarue, M., and Tordo, N., 1989, Identification of four conserved motifs among the RNA dependent polymerase encoding elements, *EMBO J.* **8:**3867.

Rottier, P. J. M., Horzinek, M. C., and van der Zeijst, B. A. M., 1981, Viral protein synthesis in mouse hepatitis virus strain A59-infected cells: effect of tunicamycin, *J. Virol.* **40:**350.

Sawicki, S. G., and Sawicki, D. L., 1990, Subgenomic mouse hepatitis virus replicative intermediates function in RNA synthesis, *J. Virol.* **64:**1050.

Schaad, M. C., Stohlman, S. A., Egbert, J., Lum, K., Fu, K., Wei, T., and Baric, R. S., 1990, Genetics of mouse hepatitis virus transcription: Identification of cistrons which may function in positive and negative strand RNA synthesis, *Virology* **177:**634.

Sethna, P. B., Hung, S. L., and Brian, D. A., 1989, Coronavirus subgenomic minus-strand RNAs and the potential for mRNA replicons *Proc. Natl. Acad. Sci. USA* **86:**5626.

Shirako, Y., and Strauss, J. H., 1994, Regulation of Sindbis virus RNA replication: Uncleaved P123 and nsP4 function in minus-strand RNA synthesis, whereas cleaved products from P123 are required for efficient plus-strand synthesis, *J. Virol.* **68:**1874.

Snijder, E. J., and Horzinek, M. C., 1993, Toroviruses: Replication, evolution and comparison with other members of the coronavirus-like superfamily, *J. Gen. Virol.* **74:**2305.

Snijder, E. J., den Boon, J. A., Bredenbeek, P. J., Horzinek, M. C., Rijnbrand, R., and Spaan, W. J. M., 1990a, The carboxyl-terminal part of the putative Berne virus polymerase is expressed by ribosomal frameshifting and contains sequence motifs which indicate that toro- and coronaviruses are evolutionarily related, *Nucleic Acids Res.* **18:**4535.

Snijder, E. J., Horzinek, M. C., and Spaan, W. J. M., 1990b, A 3′-coterminal nested set of independently transcribed messenger RNAs is generated during Berne virus replication, *J. Virol.* **64:**331.

Snijder, E. J., den Boon, J. A., Horzinek, M. C., and Spaan, W. J. M., 1991, Comparison of the genome organization of toro- and coronaviruses: Both divergence from a common ancestor and RNA recombination have played a role in Berne virus evolution, *Virology* **180:**448.

Snijder, E. J., Wassenaar, A. L. M., and Spaan, W. J. M., 1992, The 5′ end of the equine arteritis virus genome encodes a papainlike cysteine protease, *J. Virol.* **66:**7040.

Snijder, E. J., Wassenaar, A. L. M., and Spaan, W. J. M., 1994a, Proteolytic processing of the equine arteritis virus replicase ORF1a protein, *J. Virol.* **68:**5755.

Snijder, E. J., Wassenaar, A. L. M., Spaan, W. J. M., and Gorbalenya, A. E., 1995, The arterivirus nsp2 protease: an unusual cysteine protease with similarities to both papainlike and chymotrypsinlike proteases, *J. Biol. Chem*, in press.

Spaan, W. J. M., Delius, H., Skinner, M., Armstrong, J., Rottier, P. J. M., Smeekens, S., van der Zeijst, B. A. M., and Siddell, S. G., 1983, Coronavirus mRNA synthesis involves fusion of non-contiguous sequences, *EMBO J.* **2:**1839.

Spaan, W. J. M., Cavanagh, D., and Horzinek, M. C., 1988, Coronaviruses: Structure and genome expression, *J. Gen. Virol.* **69:**2939.

Strauss, J. H., and Strauss, E. G., 1988, Evolution of RNA viruses, *Annu. Rev. Microbiol.* **42:**657.
Strauss, J. H., and Strauss, E. G., 1990, Alphavirus proteinases, *Semin. Virol.* **1:**347.
Ten Dam, E. B., Pleij, C. W. A., and Bosch, L., 1990, RNA pseudoknots; translational frameshifting and read-through on viral RNAs, *Virus Genes* **4:**121.
Van Berlo, M. F., Horzinek, M. C., and van der Zeijst, B. A. M., 1982, Equine arteritis virus-infected cells contain six polyadenylated virus-specific RNAs, *Virology* **118:**345.
Ypma-Wong, M. F., Dewalt, P. G., Johnson, V. H., Lamb, J. G., and Semler, B. L., 1988, Protein 3CD is the major poliovirus proteinase responsible for cleavage of the P1 capsid precursor, *Virology* **166:**165.
Zimmern, D., 1987, Evolution of RNA viruses, in *RNA Genetics*, vol. 2 (J. J. Holland, E. Domingo, and P. Ahlquist, eds.), pp. 211–240. Boca Raton, FL, CRC Press.

CHAPTER 13

# Pathogenesis and Diseases of the Central Nervous System Caused by Murine Coronaviruses

SAMUEL DALES AND ROBERT ANDERSON

## I. INTRODUCTION

This chapter is an account of studies of central nervous system (CNS) diseases connected with neurotropic variants of MCV such as J. Howard Muller Virus (JHM) and A59 and deals with animal models that may have relevance to an understanding of human diseases of putative viral etiology such as multiple sclerosis (MS). From the time of (JHMV) isolation from paralyzed mice by Cheever *et al.* (1949) and Bailey *et al.* (1949), this agent and related strains have provided copious data about encephalitic and demyelinating diseases in rodents. To date, however, any possible connection between murine coronavirus (MCV) and MS is tenuous. The reported isolation of coronavirus (CV) particles from MS patients' brain (Burks *et al.*, 1980) or electron microscopic visualization of CV-like particles in brain tissue of one MS patient (Tanaka *et al.*, 1976), require confirmation. An older report of JHMV-induced panencephalitis in monkeys (Kersting and Pette, 1956), however, has been confirmed by Murray *et al.* (1992a) in their description of demyelinative disease in several monkey

SAMUEL DALES • Cytobiology Group, Department of Microbiology and Immunology, The University of Western Ontario, London, Ontario, N6A 5C1 Canada. ROBERT ANDERSON • Department of Microbiology and Immunology, Dalhousie University, Halifax, Nova Scotia, B3H 4H7 Canada.

*The Coronaviridae*, edited by Stuart G. Siddell, Plenum Press, New York, 1995.

species. This finding, coupled with the identification of viral RNA and protein within demyelinative plaques in human brain tissue from MS patients (Murray *et al.*, 1992b; Stewart *et al.*, 1992), are highly provocative data concerning an involvement of CV in the etiology of MS. However, a cautious evaluation of these results may be in order, in view of the recent demonstration of genomic and antigenic relatedness between the N protein of JHMV and the microtubule-associated protein tau (Pasick *et al.*, 1994).

The disease process associated with CV infection is highly variable, both in animals and at the cellular level, depending on the virus, cell types, and host species. In view of the voluminous literature on the subject, this relatively brief review is concerned primarily with disease processes in rodents infected by murine CVs and emphasizes infections of neural cells and pathogenesis in the CNS.

## II. VIRUS–CELL INTERACTIONS RELATED TO PATHOGENESIS AND DISEASE

To comprehend the overall disease process elicited by neurotropic MCV, one should consider three major parameters. First, it is necessary to consider the complex organization of CNS tissue, involving a variety of cell types and lineages, that provides a great diversity of cell–cell interactions. Second, one must consider the highly error-prone synthesis of CV genomes that can undergo a high frequency of recombination, generating rapidly evolving mutants or variants (Lee *et al.*, 1991; Lai, 1990). Such variants manifest altered tropism and virulence, as detailed in Section IV. Third, one must keep in mind the effects of host immunity evoked by CV infections. These responses, primarily of the cellular type, but also to some extent humoral, can modulate pathogenesis and disease, as described in Section III.

### A. Correlations between Infections *in Vivo* and with Explanted Neurons and Glia

Infections of different tissues and organs *in vivo* develop according to the MCV type and the route of inoculation employed. The natural means for initiating infection is through intranasal (IN) instillation of virus, but introduction through intraperitoneal (IP) and intracranial (IC) injections is also commonly employed. Neonates and juveniles usually succumb more rapidly than adults, developing fulminant infections manifested by widespread destruction of tissues. In the CNS an early onset acute encephalitis develops.

In adult mice, responses are variable depending on the MCV type and the genetic constitution of the host, as discussed in more detail in Section IV. When adult mice are challenged intranasally with the highly virulent murine hepatitis virus S (MHV-S), occurrence of a rapidly fatal infection of the CNS is associated with generalized virus dissemination to the bowel, liver, spleen, and other organs (Taguchi *et al.*, 1979; Barthold *et al.*, 1986; Lavi *et al.*, 1984a).

Histopathology in the CNS includes spongiform lesions within the brain stem. Comparable IN inoculation of neonates with neurotropic A59 and JHMV also results in a wide dissemination into the CNS, respiratory, and vascular endothelium and elsewhere (Barthold and Smith, 1984).

By contrast, A59 and JHM, when given IN to adult mice, induce milder forms of disease. The viruses progress slowly through the olfactory tracts before spreading into the brain and spinal cord (SC) (Barthold *et al.*, 1986). Encephalitis or subacute paralytic symptoms frequently ensue (Lavi *et al.*, 1988; Barthold *et al.*, 1986). The viscerotropic MHV3 strain of MCV, when inoculated into resistant A/JX mice, fails to produce disease symptoms, but nevertheless replicates in liver and brain (Tardieu *et al.*, 1986). Evidently MCVs belonging to the neurotropic or viscerotropic categories possess a more general tropism in the mouse than is indicated by their designation.

After IN infection of mice, virus dissemination and progress of disease have been followed within the CNS and other tissues by means of antibodies and nucleic acid probes. The localization of viral antigen and RNA by means of *in situ* hybridization in sections has been highly informative. With C57BL/6 mice, in which acute encephalitis develops upon infection with A59 or JHMV, the virus becomes widespread within the cortical gray matter (GM), brain stem, the white matter (WM) tracts in the regions of the optic chiasma, and the SC (Lavi *et al.*, 1988; Perlman *et al.*, 1988, 1989, 1990). A time-course reconstruction of events leads to the conclusion that virus spreads from the trigeminal olfactory locus along anatomically and functionally interconnected neuronal tracts. Dissemination of virus along such neuronal pathways provides a rationale for the speed at which virus can spread toward and within the SC. Virus expression is detectable at the cellular level in neurons and glia, consistent with the histopathology associated with the acute encephalitic or progressive demyelinative diseases caused by A59 and JHM (Perlman *et al.*, 1988; Fleming *et al.*, 1987; Spaan *et al.*, 1988). In the mouse CNS the neurons and oligodendrocytes (OL) are the prominent cell types containing viral RNA and antigen (Lavi *et al.*, 1987), although astrocytes (AS) may also become targets in primary explants. In such cell cultures from neonatal BALB/c or CD.1 mice, both OL and AS are infectable by MCV A59, JHM, and human coronavirus (HCV) OC43 (Wilson *et al.*, 1986; van Berlo *et al.*, 1989; Pearson and Mims, 1985). Likewise, explanted neurons can be productive host cells for A59 and JHMV (Dubois-Dalq *et al.*, 1982; Knobler *et al.*, 1981).

Although the CNS disease process induced in the rat by MCV has many parallels with that observed in mice, notable differences are evident. They include an age-related refractoriness to infection, apparent by the time of weaning (Sorensen *et al.*, 1987a); a requirement for larger inocula, which are usually administered IC to promote efficiency, although neonates can be challenged successfully IN (Hirano *et al.*, 1980); and an association of the CNS and paralytic forms of disease with JHMV but not MHV3, although the latter can produce inapparent infections (Hirano *et al.*, 1980). Again, the progress of infection can be followed in the CNS by antigen and RNA probes, which demonstrate that dissemination occurs in a temporal sequence, culminating with virus spread within the SC (Sorensen *et al.*, 1980, 1984b). The cell types prominently involved in infection include neurons and OL. *In situ* hybridization by means of cDNA probes has pinpointed the presence of viral RNA in hippocampal and cerebellar neurons. The virus may remain sequestered in these targets for prolonged periods, even during episodes of subacute, paralytic disease evident in the Wistar Furth (WF) rat model (Sorensen and Dales, 1985; Parham *et al.*, 1986). Rat CNS explants, cultured to promote neuronal survival *in vitro*, are likewise infectable with JHMV. In such cultures, replication occurs preferentially or exclusively in neurons (Pasick and Dales, 1991). The susceptibility of primary neuron explants to JHMV is consistent with them being primary targets for the development of acute encephalomyelitis (Knobler *et al.*, 1981; Dubois-Dalq *et al.*, 1982; Buchmeier *et al.*,1984; Sorensen and Dales, 1985; Pasick and Dales, 1991; Matsubara *et al.*, 1991; Zimprich *et al.*, 1991).

Contrary to the dual tropism of JHMV observed with murine cells, in our studies with explanted rat cells, this virus exhibits a reciprocally exclusive tropism for OL compared with MHV3, which infects only type 1 AS (Beushausen and Dales, 1985). This tropism pertains to explants from WF or Wistar Lewis (WL) strains. Our findings were confirmed with Wistar rat glial cultures (van Berlo *et al.*, 1986), although JHMV infection of both OL and AS was detected with glial cultures from WL rat CNS (van Berlo *et al.*, 1989). In this case, the AS lineage (type 1 or 2) was not established.

The inverse tropism to those that we determined has been described in CNS explants from WL rats (Massa *et al.*, 1986). In this study, immunofluorescence labeling indicated selective infection of AS rather than OL, with some modulation in tropism (Massa *et al.*, 1988). These diverse findings on the tropism of JHMV for rat glia have not been reconciled completely, but could be related to the identity of the lineage of infectable AS. The presence of glial acidic fibrillary protein (GFAP) as filament bundles that characterize type 1 AS, also occur, albeit transiently, in the progenitor cell type O2A, from which both OL and type 2 AS differentiate (Raff, 1989; Lillien and Raff, 1990). Since immature, rather than fully differentiated, rat OL appear to be susceptible to JHMV (Beushausen and Dales, 1985), it is likely, as shown by Pasick and Dales (1991), that $GFAP^+$ O2A cells become infected and express virus antigen, accounting for the observations of Massa *et al.* (1986, 1988).

In general, the tropism defined *in vitro* matches the observed distribution of CV infections of the CNS *in vivo*, and is also consistent with the pathogenesis and disease processes that become evident.

## B. Ligand–Receptor Interactions

The surface component of CV that attaches to receptors on the host cell is the surface glycoprotein. This molecule possesses features generally associated with peplomers of enveloped viruses (Schmidt *et al.*, 1987). Activation of the fusogenic and penetration activities of S is enhanced by proteolytic cleavage into the subunits $S_1$ and $S_2$ (Schmidt *et al.*, 1987; Luytjes *et al.*, 1987; Frana *et al.*, 1985; Sturman *et al.*, 1985; Stauber *et al.*, 1993; Taguchi, 1993). However, attachment and neutralization by antibodies are not contingent on the proteolytic processing of S (Luytjes *et al.*, 1989; Collins *et al.*, 1983). The cell-to-cell spread of infection in the human line Medical Research Council (MRC-C) challenged with HCV 229E (Appleyard and Tisdale, 1985) or the murine cell line 17 Cl I challenged with A59 (Sturman *et al.*, 1985; Frana *et al.*, 1985), requires the activation of S by a cellular protease that is inhibited by agents such as leupeptin (Appleyard and Tisdale, 1985). Also, virus dissemination is not absolutely contingent on the formation of syncytia, even after S has been processed into $S_1$/$S_2$. Thus, MHV A59 is highly fusogenic for 17Cl I and L cells, but in contrast to MHV3 and JHMV is nonfusogenic *in vivo* or for explanted rodent glial cells (Lucas *et al.*, 1977; Sturman *et al.*, 1985; Frana *et al.*, 1985; Lavi *et al.*, 1987; Wilson and Dales, 1988). Differences in syncytiogenic activity among CVs remain unexplained.

The receptor specificity for CV has also been studied as a determinant of tropism. In one system using rat glial cultures, one can demonstrate an unambiguous tropism of JHMV for OL and MHV3 for AS, but there is no difference in the binding efficiency of labeled inoculum particles of either virus to either cell type (Beushausen and Dales, 1985). With continuous cell lines such as the rat RN2-2 Schwannoma, which is infectable by JHMV but not by MHV3, both virus types likewise become attached equally well (Lucas *et al.*, 1977, 1978). In the case of the C6 rat astrocytoma cell line, which is resistant to infection by both viruses, receptors for JHMV, MHV3, and other MCV strains are present (van Dinter and Flintoff, 1987; Flintoff and van Dinter, 1989; Kooi *et al.*, 1988). In the case of sublines derived from the prototype L strain murine line, the more restrictive subline LM-K can absorb A59 virus as efficiently as the highly permissive L2 cell (Kooi *et al.*, 1988). A similar situation is found with cells from SJL/J mice, which are highly resistant to JHMV but not to MHV3. Thus, both viruses can be adsorbed with the same efficiency to OL and AS of SJL/J mice and to OL and AS of permissive BALB/c and CD.1 mice (Wilson *et al.*, 1986).

A presumed absence of receptors for A59 on SJL/J hepatocytes and enterocytes, which was assumed to account for differences in resistance of SJL/J mice compared to susceptible BALB/c mice (Boyle *et al.*, 1987), also can now be explained. The MHV receptor is a 110- 120-kDa G-P, a member of the carcinoembryonic antigen family (Williams *et al.*, 1991). Although this G-P is expressed on SJL/J cells, its form is modified (Yokomori and Lai, 1992), presumably affecting the efficiency of JHMV replication. Whatever the deficiency of receptors for JHMV and A59 might be in SJL/J mice, the resistance of neurons and glia from these mice is due to a defect in cell–cell virus spread rather than an absence of receptors per se (Wilson and Dales, 1988; Pasick *et al.*, 1992). Virus dissemination may involve a deficiency in the processing of S because a more fusogenic JHMV variant ATf11, isolated from rat SC (Morris *et al.*, 1989), is not restricted in SJL/J glial cells (Pasick *et al.*, 1992).

The identification of a hemagglutinin esterase (HE) glycoprotein component in BCV (Vlasak *et al.*, 1988) drew attention to the possible function of HE during early interactions with host cells. By sequence comparison, HE shows evolutionary relatedness with the influenza type C hemagglutinin $H_1$ subunit (Kienzle *et al.*, 1990). The esterase of HE has receptor-binding activity and inactivates receptors on erythrocytes by hydrolyzing O-linked acetylsialic acid (Schultze *et al.*, 1991). Since this reaction and virus entry are both blocked with diisopropyl fluorophosphate (Vlasak *et al.*, 1988), it is presumed that HE acts in a manner similar to that of a receptor-destroying enzyme of type C influenza virus (Kienzle *et al.*, 1990). Evidently the esterase is not essential for infectivity because it is absent from MCV A59 and some MCV JHM strains (Parker *et al.*, 1990). Some of the roles suggested for HE are: as a determinant of tropism or virulence in the CNS; as an additional ligand for binding of cell targets; or for facilitating the penetration of inoculum virus (reviewed in Parker *et al.*, 1990; Vlasak *et al.*, 1988).

## C. Penetration

An electron microscopic examination of the interaction between surfaces covering the microvilli in calf intestine inoculated with the calf diarrhea CV was interpreted as showing fusion between the virus envelope and plasma membrane on microvilli (Doughri *et al.*, 1976). Studies on cells in culture, using biochemical and electron microscopic methods, suggested that MCV initiates an infection following sequestration or viropexis of the inoculum (reviewed in Krzystyniak and Dupuy, 1984). The inhibition of early stages of A59 or MHV3 infection, including the eclipse phase (Mizzen *et al.*, 1985), in L2 cells treated with agents such as $NH_4Cl$ and chloroquine (Krzystyniak and Dupuy, 1984), also supports a requirement for engulfment as an obligatory function for initiating infection. More recent unpublished data from our laboratories indicate that $NH_4Cl$ only marginally inhibits the early stages of MHV infection, implying that penetration and release of the genome occurs in "early" endosomes containing a neutral or only slightly acidic milieu due to their intracellular location close to the cell surface. (Schmid *et al.*, 1989).

Studies of penetration by JHM, A59, and MHV3 into continuous neural and nonneural cell lines suggest that interactions with the host may vary. The virions that are absorbed onto and sequestered by C6 astrocytoma cells do not become eclipsed (van Dinter and Flintoff, 1987), due to an unidentified defect that can be circumvented by inducing fusion between viral and cell membranes by means of polyethylene glycol (PEG). PEG by itself does not enhance the infectiousness of permissive cells, among them L, RN2-2 Schwannoma, and HTC hepatoma lines. The failure of a "low pH shock" to promote infectability of MCV in the manner observed with vesicular stomatitis virus (VSV) and A-type influenza viruses also argues against involvement of an acidic endosomal compartment in the penetration sequence (Flintoff and van Dinter, 1989). Host control over virus penetration, examined by means of cell–cell hybrids between the resistant C6 and either partially or fully permissive cell lines showed that C6 × L2 and RN-2 × L2 cell lines could be infected without the use of PEG (Flintoff and van Dinter, 1989; Coulter-Mackie *et al.*, 1984). However, these permissive cell hybrids retain the phenotype of the parent that resists extensive syncytia formation. This demonstrates that penetration and MCV-induced cell–cell fusion are controlled independently.

## D. Uncoating and Initiation of Genome Functions

Although the ssRNA genome penetrates from inoculum virions into the host within a helical coat of N protein (Davies *et al.*, 1981), the infectivity of the naked genomic RNA (Siddell *et al.*, 1982) implies that, in the normal course of penetration, the uncoating step involves dissociation of N from the RNA. The nascent N polypeptide of JHMV has a molecular weight of 50 kDa which increases to 56 kDa upon phosphorylation. The 56-kDa phosphorylated form of N exists in virions (Armstrong *et al.*, 1983; Wilbur *et al.*, 1986; Siddell *et al.*, 1981). From sequence data on N (Armstrong *et al.*, 1983; Skinner and Siddell, 1983) and other evidence (Wilbur *et al.*, 1986), one can deduce that N has two to four, or perhaps more, potential sites for phosphorylation, a modification effected prior to virion assembly by either cellular kinases (PK) or a MCV-specified PK. In a converse sequence of events taking place during penetration, the molecular weight of N is reduced initially from 56 to 50 kDa, then further breakdown into smaller molecular weight polypeptide fragments occurs (Beushausen *et al.*, 1987). Conceivably, dephosphorylation of N promotes the uncoating process, as documented for retroviruses (Beushausen *et al.*, 1987; Mohandas and Dales, 1991).

Our demonstration of the existence, in an endosome fraction of a cellular phosphoprotein phosphatase [(PPPase) belonging to the neutral type 1 class of enzyme, unpublished], with a high specificity for N (Mohandas and Dales, 1990), is consistent with the view that uncoating may either commence or proceed to completion in "early" endosomes and/or close to the cell surface. In this context, it is worth mentioning that endosomal PPPases of a type similar to that acting on N have been recognized in clathrin-coated vesicles, where they may catalyze the rupturing of polymeric clathrin protein cages and facilitate movement of the freed endosomes and fusion of the endosomal membrane with other membrane compartments (Pauloin *et al.*, 1988; Greene and Eisenberg, 1988; Loeb *et al.*, 1989).

## E. Effect on Cellular Protein Synthesis

Viruses have developed diverse strategies to ensure the translation of virus-specific mRNAs in the infected cell. MCV gives rise to mRNAs that share the structural characteristics of most eurkaryotic mRNAs, such as the presence of 5′ terminal caps (Lai *et al.*, 1982) and 3′ poly A tails (Cheley *et al.*, 1981; Cheley and Anderson, 1981). It appears unlikely, therefore, that translational control in MCV infection is mediated by the alteration of host initiation factors such as occurs, for example, in poliovirus infection and results in the selective translation of structurally distinct (e.g., uncapped) mRNAs (Kaufman *et al.*, 1976; Trachsel *et al.*, 1980).

At high multiplicities of infection, MCV inhibits host protein synthesis in certain cell lines. In mouse L2 fibroblasts, infection results in a decline in total protein synthesis to about 7% of controls. This overall decrease is likely due to a partial block of translation at the initiation stage, as indicated by an increase in free 80S ribosomes. There is also a shift in size from heavier to lighter polysomes, reflecting increased competition by a greater number of shorter mRNAs for a limited quantity of available ribosomes. The overall total poly(A) +RNA content of the infected cell is increased approximately threefold by 6 hr postinfection, as a result of the synthesis and accumulation of viral mRNA. Most interestingly, by this time, and later in infection, cellular mRNA transcripts disappear, apparently due to a degradative mechanism (Hilton *et al.*, 1986).

Previously, virus-induced hydrolysis of host mRNA had been noted after infection with complex DNA viruses, such as herpes simplex (Nishioka and Silverstein, 1978) and vaccinia (Rice and Roberts, 1983), and with influenza, an RNA virus. In the case of influenza, breakdown of host mRNA is a consequence of the unique mode of transcapping, i.e., the capture of host-derived 5′-caps (Bouloy *et al.*, 1980; Plotch *et al.*, 1981). Most other RNA viruses studied to date, however, appear to effect translational controls by nondegradative mechanisms (Fernandez-Munoz and Darnell, 1976; Gallwitz *et al.*, 1977; Lodish and Porter, 1981). Evidence has also been presented suggesting that during infections with some viruses, including adenovirus (Lazaridis *et al.*, 1988) and La Crosse virus (Raju and Kolakofsky, 1988), viral functions are expressed that affect the stability of cellular or viral mRNAs.

It is intriguing to speculate that the mechanism of RNA degradation connected with MCV infection (Hilton *et al.*, 1986) is related to the unique replication strategy of coronavirus. It has been suggested that the synthesis of CV subgenomic mRNAs may involve nucleolytic trimming (Lai, 1986). Although this type of nucleolytic activity has yet to be identified, if present, such a MCV-induced nuclease could also presumably recognize as a substrate cleavage sites in cellular mRNAs, thereby accelerating their hydrolysis by host degradative mechanisms.

Another obvious cell-killing function of MCV infection is the inhibition of host translation (e.g., Marvaldi *et al.*, 1977). Differences among host cells in their susceptibility to MCV-mediated translational control could be manifested in the cytopathology that occurs and also determine whether the outcome is an acute or persistent infection.

### F. Effects on Cellular Membranes

The unique sites of CV assembly by budding from smooth endoplasmic reticulum membranes, recognized by electron microscopy (Massalski *et al.*, 1982; Barthold and Smith, 1984; Pasick *et al.*, 1994; Krijnse-Locker *et al.*, 1994; Klumperman *et al.*, 1994), has not yet been explained. Earlier, it was thought that the M protein had an important role in organizing the assembly of virion envelopes on these internal membranes, mainly because this protein has signals for retention in the Golgi (Swift and Machamer, 1991). Recently, however, it has become apparent that the Golgi compartment is beyond the site of budding (Klumperman *et al.*, 1994). One reason suggested for M being anchored and retained on internal membranes is the preferential O-linked glycosylation

(Rottier *et al.*, 1984; Niemann and Klenk, 1981; Niemann *et al.*, 1984). In contrast, the S protein, which possesses N-linked carbohydrates, behaves like the usual peplomer G-P of viruses budding from the cell surface. Thus, although S attaches to the formative CV envelopes within the budding compartment, it can also migrate independently to the plasma membrane where it accumulates and can express fusogenic activity (Rottier *et al.*, 1984; Niemann *et al.*, 1984).

Electron microscopic examination of MCV-infected cells has revealed apparent distensions or amplification of intracellular membranes that may be derived from smooth endoplasmic reticulum (David-Ferreira and Manaker, 1965; Massalski *et al.*, 1982; Dubois-Dalq *et al.*, 1982; van Berlo *et al.*, 1986; Tooze *et al.*, 1984). It has been speculated that such membrane proliferation or reorganization may reflect a cellular response to the removal of membrane lipid as it becomes incorporated into virion envelopes (Tooze *et al.*, 1984). This explanation seems to be unlikely since membrane proliferation has been observed during infection with a temperature-sensitive (*ts*) mutant of MCV, when virion assembly was interrupted (van Berlo *et al.*, 1986). The stimulation of membrane proliferation and lipid biosynthesis resulting from cytopathology have also been observed following infections with other agents, among them nonenveloped polio- and mengovirus (Penman, 1965; Amako and Dales, 1967.

A very obvious manifestation of virus–cell membrane interactions occurring within the replication cycle of MCV is membrane fusion. Fusion occurs both early during virion entry and uncoating and later when syncytiogenesis is evident. The control of cell–cell fusion is closely related to events that contribute to the establishment of a cytolytic versus an attenuated or persistent infection (Kooi *et al.*, 1988; Mizzen *et al.*, 1983, 1985, 1987b).

The S proteint is a key mediator for both cell fusion and pathogenesis. Within the infected CNS, differences among cell types have been noted in relation to virus-induced cell fusion (Sorensen *et al.*, 1980; Dubois-Dalcq *et al.*, 1982; Buchmeier *et al.*, 1984). By inducing cell–cell fusion from within, MCV may facilitate both dissemination of virus and maintenance of persistent infections, as demonstrated in the presence of neutralizing antibodies added to the medium (Buchmeier *et al.*, 1984; Sorensen *et al.*, 1984a; Stohlman and Weiner, 1978). In cultures, such as the Cl-1 line, which is relatively resistant to JHMV-mediated syncytiogenesis, individual cells may survive a cytolytic, acute infection, thereby maintaining a state of virus persistence (Mizzen *et al.*, 1983).

The resistance of murine cells to various strains of MCV appears to be under the control of a single autosomal, recessive gene (Bang and Warwick, 1960; Virelizier and Allison, 1976; Stohlman and Frelinger, 1978). Concomitantly, the complete or partial genetic resistance to disease observed with some strains of mice is inherent in explanted macrophages, hepatocytes, fibroblasts, and other cell types. The infection of cells from such mice results in either an abortive (Virelizier and Allison, 1976) or limited infection, with reduced cytopathology (MacNaughton and Patterson, 1980; Arnheiter *et al.*, 1982; Lamontagne and Dupuy, 1984a). Although cultured macrophages, OL and AS from SJL/J mice, which are resistant to JHMV as primary targets of infection, can support MCV replication, the cell–cell spread of virus due to a fusion-dependent mechanism is restricted (Knobler *et al.*, 1984). Some deficiency of an SJL/J cell protease was suggested as an explanation for the absence of cell–cell spread of virus. In fact, studies on the spontaneous acquisition of resistance by L2 cell mutants, selected for the survival of a usually lethal MCV infection, demonstrate that partial resistance to infection and to S-mediated syncytia formation are comutable characteristics of the host cell (Daya *et al.*, 1989). Homeostatic control of fatty acid and cholesterol metabolism has been shown to be another important regulator of virus-mediated cell fusion both *in vivo* and *in vitro* (Cervin and Anderson, 1991).

It has been assumed that differences in the infectability and susceptibility of cells to MCV-mediated fusion are correlated (Mizzen *et al.*, 1983, 1987b), and depend on proteolytic activation of S, presumably involving a host cell protease (Frana *et al.*, 1985). Quantitative cleavage of S is, however, not necessary for cell–cell fusion, and there is no strict correlation between the proteolysis of S and the extent of cell fusion (Frana *et al.*, 1985), as indicated when S was mutated in the cleavage signal and expressed independently by vaccinia recombinants (Stauber *et al.*, 1993; Taguchi, 1993). Also, one class of fusion (from within) defective MCV mutant, which possesses a single mutation within the cleavage signal of S, has been isolated from persistently infected glial cultures (Gombold *et al.*, 1993). It is possible that uncleaved S protein is fusogenic, but that fusogenicity is enhanced upon proteolytic cleavage.

CV-induced membrane fusion and the severity of infection are also influenced by host cell membrane lipid composition (Daya *et al.*, 1988), particularly the cholesterol content (Daya *et al.*, 1988). Conceivably there may be a specific interaction between peptide sequences of the S protein and cholesterol, in a manner similar to that observed with Sendai virus F protein (Asano and Asano, 1988). Another modulation of MCV infection by cholesterol may pertain to the inhibition of Kupffer cell function during *in vivo* infection (Pereira *et al.*, 1987). The observed association of the MCV N protein with cellular membranes may be another example of lipid-dependent events during replication. N is synthesized on free polysomes but associates posttranslationally with membranes, possibly as a result of its phosphorylation (Stohlman *et al.*, 1983). Since N binds the viral M protein (Sturman *et al.*, 1980) and the M protein is found in association with smooth endoplasmic reticulum (ER) membranes (Rottier and Rose, 1987), the membrane attachment may be mediated by the M. Nevertheless, N protein is able to associate with cellular membranes in the absence of M (Anderson and Wong, 1993), raising the possibility that the components involved are certain membrane phospholipids (Anderson and Wong, 1993). Interestingly, the ability of N to bind both membrane phospholipids and nucleic acid is perhaps analogous to that shown by certain bacterial DNA-binding proteins such as DnaA (Sekimizu and Kornberg, 1988) and RecA (Krishna and van de Sande, 1990). The possibility, therefore, arises that an association of N with viral RNA is influenced by competitive binding to membrane phospholipids.

The involvement of the cytoskeleton in a number of MCV infections has also been investigated. Treatment of L2 fibroblasts with colchicine or cytochalasin B, drugs that disorganize microtubules and actin filaments, alters cell shape without affecting syncytiogenesis or the acute cytopathology (Daya *et al.*, 1988). By contrast, in peritoneal macrophages exposed to cytochalasin B, the infection is changed from an acute to a persistent type (Mallucci and Edwards, 1982). Quite possibly, the cytoskeleton plays a role in regulating MCV assembly and cytopathology, involving membranes that may vary in a cell-dependent manner according to differences in the organization of the cytoskeleton. The close association of N of JHMV with microtubules in neurites of cultured neurons, recently recognized by Pasick *et al.*, (1994), was explained in terms of a specific binding by N, via a region of the protein that has a close sequence homology to the microtubule-binding site of the cytoskeletal protein tau.

As mentioned above, the presence of S at the cell surface is related to cell fusion (Collins *et al.*, 1982). However, it may also affect control of plasma membrane permeability, as evidenced by an influx of sodium ions (Mizzen *et al.*, 1987a). This occurs during both acute and persistent infections (MacIntyre *et al.*, 1989). However, in contrast to certain other virus infections (Carrasco and Smith, 1976), there is evidence that in MCV-infected cells an elevated intracellular sodium ion concentration affects the regulation of translational control (Mizzen *et al.*, 1987a). One should not exclude the possibility that altered membrane permeability can affect other aspects of viral cytopathology.

## G. Factors Related to Latency and Persistence

Infection of the CNS can establish a very prolonged state of CV latency or persistence. Whether the animal remains asymptomatic or undergoes a late-onset progressive paralytic disease with demyelination, virus expression in the form of RNA, antigen, or infectious particles may continue for prolonged periods in mice (Pickel *et al.*, 1981; Lavi *et al.*, 1984a,b, 1987; Perlman *et al.*, 1988; Knobler *et al.*, 1981) or rats (Wege *et al.*, 1983, 1984; Sorensen *et al.*, 1984b; Sorensen and Dales, 1985). Persistent infections are readily established in explanted primary neural cells (Dubois-Dalq *et al.*, 1982; Beushausen and Dales, 1985; Wilson *et al.*, 1986; Pearson and Mims, 1985; van Berlo *et al.*, 1989; Lavi *et al.*, 1987) and continuous cell lines of neural and nonneural derivation (Flintoff and van Dinter, 1989; van Dinter and Flintoff, 1987; Coulter-Mackie *et al.*, 1984, 1985; Lucas *et al.*, 1977; 1978; Collins and Sorensen, 1986; Leibowitz *et al.*, 1984). Evidently, conditions prevail in the CNS of postweanling and adult rodents, as well as in neural cells in culture, which favor maintenance of the state of persistence. Analogous observations have been documented for other neurotropic RNA viruses such as measles virus (Miller and Carrigan, 1982; Yoshikawa and Yamanouchi, 1984), canine distemper virus (Pearce-Kelling *et al.*, 1990; Zurbriggen *et al.*, 1987), rubella virus (van Alstyne and Paty, 1983), picorna viruses (Rodriguez *et al.*, 1990), and alpha viruses (Bruce *et al.*, 1984). A common feature linking MCV with all of the above agents is an initial low rate of replication associated with limited cytopathology. Following a period of adaptation, during which, perhaps, variants are selected and maintained, the infectious process may remain unchanged or switch into a more acute or more attenuated mode (Taguchi and Siddell, 1985) or replication may be entirely suppressed (Collins and Sorensen, 1986). The exceptionally high efficiency of MCV genetic recombination creates numerous variants (Lai *et al.*, 1985; Makino and Lai, 1989), a phenomenon that can account, at least partially, for the varying types of infection encountered, as discussed in Section IV.

The repeated passage of JHMV in murine fibroblastic DBT cell lines generates defective-interfering (DI) particles in which genomes are shortened by deletions (Makino *et al.*, 1984, 1985). In a high multiplicity of infection, DIs interfere with the transcription of all mRNAs species with the exception of mRNA7, coding for N. There is no substantial evidence linking DIs with CV infections of the CNS or neural cell explants (Taguchi and Siddell, 1985).

A regular feature of MCV persistence in continuous cell lines, not exclusively of the neural type, is a wavelike, cyclical production of infectious particles (Lucas *et al.*, 1977, 1978; Coulter-Mackie *et al.*, 1984, 1985). Such periodicity in virus titer also has been reported for measles and rabies viruses (reviewed in Lucas *et al.*, 1978) but is not evident during MCV infections of primary neural cell cultures (Beushausen and Dales, 1985; Wilson *et al.*, 1986). With rabies virus, the rise and fall of titers was explained as due to reciprocal periodic development of an antiviral state related to the autocrine activity of interferon (IFN) (reviewed in Coulter-Mackie *et al.*, 1985). During persistent infection of porcine alveolar macrophages by transmissible gastroenteritis virus (TGEV) of pigs, an induction of α-IFN was observed (Laude *et al.*, 1984). By contrast, our extensive, detailed search for the induction of IFN during a cyclical rise and fall of JHMV titers from RN2-2 Schwannoma cells failed to detect the presence of this lymphokine, although these host cells have the capacity to acquire an antiviral state and make IFN when appropriately induced with reovirus (Coulter-Mackie *et al.*, 1985). Likewise, persistent infections of mouse neuroblastoma C1300 cells with JHMV (Leibowitz *et al.*, 1984) and human glioblastoma U-87 HG cells infected with HCV OC43 (Collins and Sorensen, 1986) do not evoke the production of IFN, demonstrating that in these systems IFN or the antiviral state are not required to maintain persistence. The effectiveness of IFN as a protective factor may also depend on the MCV strain involved. Thus, with DBT cells, which become productively infected, α-IFN fails to arrest replication or the development of cytopathology when the highly virulent MHV-2 is used, whereas IFN treatment can abrogate both virus production and cell killing when MHV-S or JHMV are used as infectious agents (Taguchi and Siddell, 1985).

Another reversible cellular response to MCV infections of primary neural cell explants and continuous cell lines is the suppression of persistent virus production that occurs on a shift up of the ambient temperature from a permissive 32–33 °C to the restrictive 39–40 °C (Lucas *et al.*, 1977, 1978; Coulter-Mackie *et al.*, 1984, 1985; Sorensen *et al.*, 1984a). The MHV3 and JHMV progeny generated during permissive infection of L2 cells do not acquire or possess a thermosensitive or thermolabile phenotype during passage through lymphocytic (Lamontagne and Dupuy, 1984b) or neural cells (Lucas *et al.*, 1978; Leibowitz *et al.*, 1984). Similar results were obtained with measles virus (Lucas *et al.*, 1978). This implies that latency at elevated temperature is due to some controlling factor(s) in the host cells per se that is not IFN (Coulter-Mackie *et al.*, 1985; Leibowitz *et al.*, 1984). By contrast, replication of HCV OC43 in the UH87 HG line generates thermosensitive variants, so that during restriction at high temperature persistent or latent OC43 virus is rapidly eliminated and cells are "cured" (Collins and Sorensen, 1986).

A detailed study of latency was undertaken using the RN2-2 cells infected with JHMV (Coulter-Mackie *et al.*, 1985). The virus formed during persistent infections at 33 °C is not temperature sensitive. Within 24 hr after shift up to 39 °C, replication is arrested, but the latent genomes are maintained in the host cell for prolonged periods and gene expression, measured as translation into N protein, can be detected for several days. If the latently infected cells, which continue to grow and divide at 39 °C, are returned to the permissive temperature following 2 to 3 weeks, virus production slowly resumes. Analyses made by following a time-course in the decline of infectious centers yielded data compatible with a probability model that predicts that due to continual segregation of the limited number of latent JHMV genomes among daughter cells, the virus genome is eventually lost by a dilution effect.

## III. RELATIONSHIP BETWEEN DEVELOPMENT AND DIFFERENTIATION OF THE CENTRAL NERVOUS SYSTEM AND THE DISEASE PROCESS

### A. Characteristics of the Cells Involved

Under normal circumstances, the suceptibility of rats to CV-induced disease extends only to the time of weaning (Sorensen *et al.*, 1980), whereas mice remain susceptible to challenge into adulthood. The idea that the tropism of CV for specific cells in the CNS can explain the basis of encephalitic and demyelinative paralytic cells in the CNS has already been considered in Section II. In the rat model, a temporal coincidence between the completion of myelination and the onset of resistance to JHMV drew our attention to OL maturation as a possible determinant of resistance to JHMV-induced demyelinating disease (Beushausen and Dales, 1985; Wilson *et al.*, 1986; Beushausen *et al.*, 1987). Also, the availability of methods for reproducibly culturing cells explanted from the CNS now makes it possible to follow the process of differentiation and infection among defined types of neural cells.

Soon after explantation from the brain, a mixture of neurons, cells of glial lineage, and other elements are present in culture. The composition of the nutrient media, including a serum supplement, can profoundly affect the cell types that remain during prolonged cultivation and the cell type selection (Fabre *et al.*, 1985; Gard and Pfeiffer, 1989; Muraoka and Takahashi, 1989; Koper *et al.*, 1984). Thus, with respect to glial cultures, type 1 AS, identified by the presence of GFAP, and the O2A bipotential OL and type 2 AS progenitor cells, which have receptors for monoclonal antibody (MAb) A2B5 (and hence are designated as A2B5$^{+}$) (Raff, 1989), are the most numerous elements when neonatal cortex is explanted into nutrient medium with added serum (Beushausen and Dales, 1985; Koper *et al.*, 1984). Moreover, the progression of O2A cells into AS or OL lineages, respectively, can be reconstructed by following the progeny of individual cells in rat optic nerve cultures (Raff, 1989). These elegant analyses demonstrated that, even after their separation from the CNS, glial cells are endowed with a predetermined developmental "time-clock" that probably simulates the program existing *in vivo.* The aquisition of GFAP by daughters of O2A parents identifies them as type 2 AS. Daughters of O2A expressing myelin basic protein (MBP), CNPase activity, and GalC on their surfaces are destined to become OL.

The appearance of the mature OL phenotype is more rapid if cells are sparsely distributed in culture dishes (Aloisi *et al.*, 1988) or are grown on poly L lysine, which promotes tight adherence to the substrate. Factors promoting differentiation favor the onset of resistance to JHMV infection (Beushausen and Dales, 1985). Conversion from the O2A into 2AS or OL types is also influenced by growth factors such as basic fibroblastic growth factor (bFGF) and platelet-derived growth factor (PDGF) (McKinnon *et al.*, 1990; Bögler *et al.*, 1990). For example, PDGF, secreted by type 1 AS, promotes the appearance of OL (Raff, 1989; Levine, 1989). When recently established cultures are switched from a serum containing medium to a defined, serumless medium, the survival of neurons is promoted and an alteration of $A2B5^+$ cells into differentiated OL is enhanced (Raff, 1989; Levine 1989; Koper *et al.*, 1984). This suggests that there are factors in serum that help to maintain O2A cells but may inhibit the survival of neurons. Interactions between neurons and OL are indicated by the myelination of neuronal processes *in vitro*, albeit of a rudimentary type (Muraoka and Takahashi, 1989; Pasick and Dales, 1991). Myelin elaboration in cultures of OL also has been reported in the absence of neurons, implying that axons are not an essential stimulus for triggering myelin formation (Fabre *et al.*, 1985)

## B. The Infectious Process in Differentiating OL

If kept for several days, cultures of mixed rat glial cell types contain tightly adherent type 1 AS on the bottom and more loosely attached O2A and OL above the subjacent AS layer. The mixed glia culture can be infected persistently with JHMV and MHV3. After shaking off the top cell layer, type 1 AS are infectable exclusively by MHV3. After replanting, the shaken cells can support only JHMV replication (Beushausen and Dales, 1985; Wilson *et al.*, 1986). Shaken cells, when kept in culture for 2 to 3 weeks, become resistant to JHMV and switch from $A2B5^+$ into the mature OL phenotype. Shaken cells can be induced to differentiate more rapidly and resist JHMV infection by pretreatment with $N_6,O_2$-dibutyryl 3′:5′ cAMP (dbcAMP) (see Table I). By contrast, AS treated with this metabolite continue to produce abundant quantities of MHV3. Cultured neurons exposed in the same manner to dbcAMP can be infected persistently and produce high titers of JHMV (Pasick and Dales, 1991). With other neurotropic viruses quite different effects of dbcAMP are observed. Infection of AS with rubella virus becomes latent, but it is induced to a productive infection by dbcAMP (van Alstyne and Paty, 1983). In contrast, infection of neuronal and other cell lines with measles virus is reversibly inhibited by dbcAMP, possibly at a late stage in virus expression (Miller and Carrigan, 1982; Yoshikawa and

TABLE I. Effect of dbcAMP on the Replication of JHMV in Primary Oligodendrocytes[a]

| Treatment of culture | Days postinfection 2 | 3 | 4 | 6 | 8 | 12 | 14 |
|---|---|---|---|---|---|---|---|
| Control | 13 | 46.8 | 55 | 38.8 | 720 | 31 | 20 |
| 1 mM postinfection | 33.4 | 100[b] | 15.5 | 100 | 500 | 26 | 30 |
| 1 mM 48 hr before infection | 0 | 0 | 0 | 0 | 0 | 0 | 0 |
| 1 mM 48 hr before infection | 0 | 0[c] | 0 | 0 | 0 | 0 | 0 |

NOTE: All titers expressed × $10^2$ PFU/ml multiplicity of infection 0.5–1.0/cell in each case.
[a]Adapted from Beushausen and Dales (1985).
[b]dbcAMP added.
[c]dbcAMP removed.

Yamanouchi, 1984). These observations indicate that dbcAMP effects may be different for each type of neural cell and virus tested, even among MCV strains.

The resistance to JHMV in cells belonging to the OL lineage can be affected by metabolites other than dbcAMP, as long as they can act to elevate intracellular concentrations of cAMP (Beushausen *et al.*, 1987) (Fig. 1). These data indicate that adenylate cyclase metabolism is generally involved, perhaps through the cAMP-dependent protein kinase(s) (PK). Oligodendrocytes, like brain tissue generally, constitutively possess the PK II activity but lack PK I. A striking effect of induced OL differentiation is a tenfold increase in the amount of PK I regulatory subunit RI despite the absence of constitutive PK I enzyme from OL (Beushausen *et al.*, 1987). Infection by MCV is arrested in mature OL at some stage after internalization of the inoculum but prior to the initiation of genome expression. The critical step affected by dbcAMP treatment, or by the normal differentiation process, could be the uncoating step, defined here as the separation of genome RNA from a phosphorylated form of N constituting the helical capsid (Beushausen *et al.*, 1987). Evidence has been obtained showing that dephosphorylation of N is followed by its proteolytic breakdown. This step, which likely occurs after attachment and uptake, as illustrated in Fig. 2, may be catalyzed by a neutral PPPase concentrated in endosomes (Beushausen *et al.*, 1987; Mohandas and Dales, 1991). A PPPase that is highly specific for N occurs in rat brain tissue, OL ROC-1 cells (a hybrid cell line created from OL × C6), and in the highly permissive L2 cell (Mohandas and Dales, 1990). It is remarkable that the N-specific, endosomal PPPase activity is inhibited by the addition of RI to the reaction, providing evidence for a putative mechanism involved in blocking JHMV infection in OL (Wilson *et al.*, 1990). As might be expected from these findings, the block to the process of MCV uncoating, relating to the dephosphorylation of N in mature OL, can be circumvented by transfection with genomic JHMV RNA (K. Kalicharran and S. Dales, unpublished data).

## C. Immunity as a Factor Regulating Pathogenesis and Disease

Both cellular and humoral components of immunity modulate the disease process due to CNS infections by MCV. Protection against acute, lethal enceph-

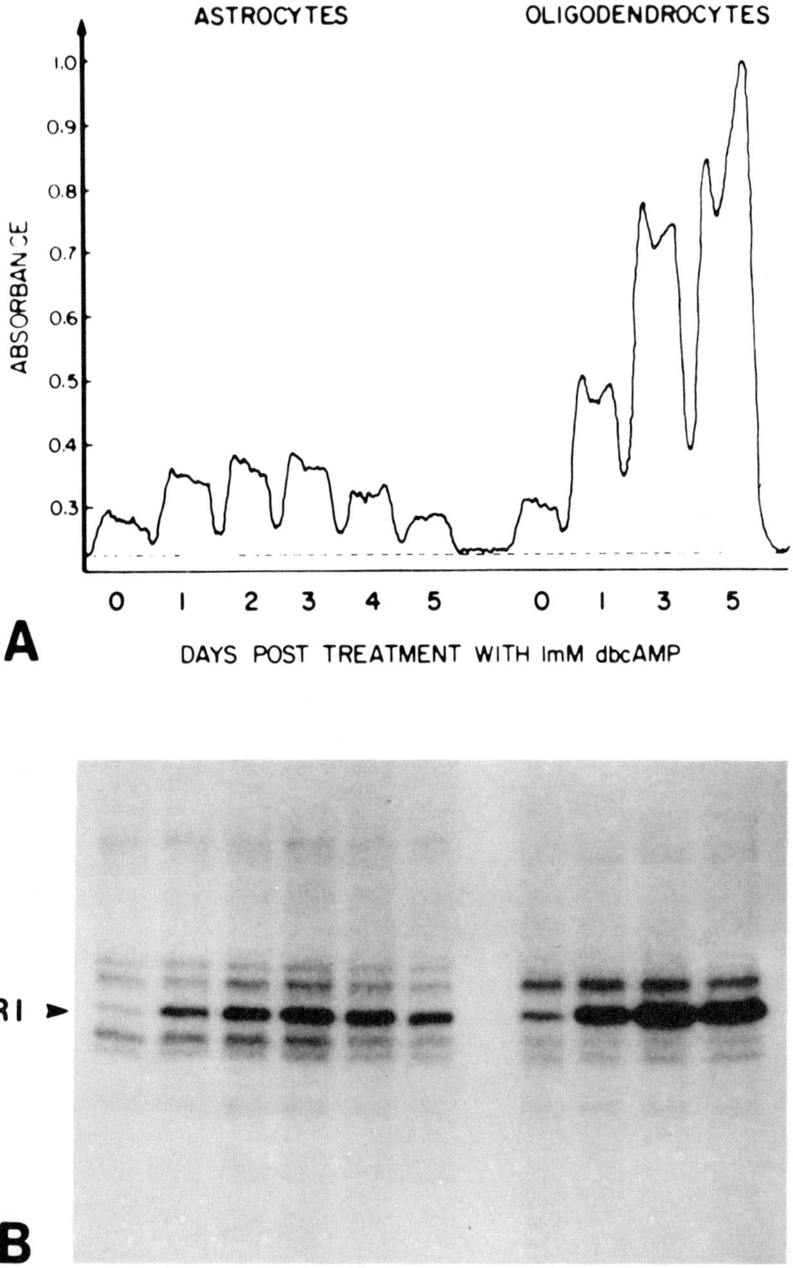


FIGURE 1. Modulation of R I and R II in primary rat astrocytes and oligodendrocytes during treatment with 1 mM dbcAMP. The concentrations of the regulatory subunits in cytosol (100,000 × g) fractions from astrocytes and oligodendrocytes were determined by binding of 8-azido-[$^{32}$P]-cAMP. A densitometer tracing (A) made from an autoradiogram (B), obtained after 10% sodium dodecyl sulfate–polyacrylamide gel electrophoresis, enabled a comparison of the time-related changes in the R I regulatory subunit. Absorbance units have been normalized to the band of greatest density (oligodendrocytes, 5 days posttreatment). (From Beushausen *et al.*, 1987.)

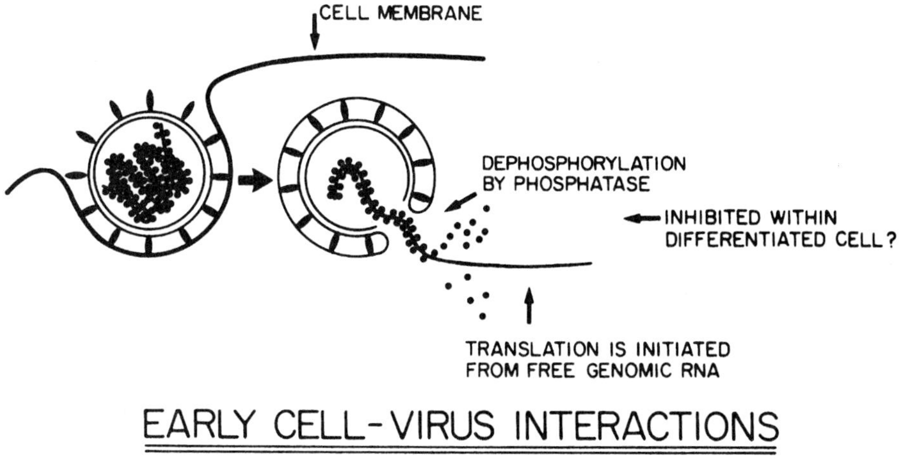


FIGURE 2. Diagrammatic representation of the early events connected with MCV uncoating. Dephosphorylation of NC protein is proposed as a necessary event for initiation of processing of NC to facilitate conformational relaxation of the genome and/or release of NC from the genome to allow the expression of subsequent virus-specified functions.

alitis is achieved by passive immunization with MAb against viral envelope glycoproteins S, M, and HE. Such immunization, however, fails to arrest the delayed, demyelinating form of disease and may in fact promote it (Fleming *et al.*, 1989). Our own studies with rat tetencephalic cultures suggest that the subacute disease with demyelination may occur only after survival of the acute, encephalitic phase of disease (Pasick and Dales, 1991). An antibody-mediated shift in tropism is indicated by the interference with neuronal, but not OL infections, when MAb against S or HE is employed (Buchmeier *et al.*, 1984; Yokomori *et al.*, 1992). This implies that complement-mediated cytolysis or antibody-dependent, cell-mediated cytotoxicity (ADCT) may eliminate infected targets following the mobilization and infiltration of mononuclear cells. In addition, an increase of B and plasma cells as well as IgG$^+$ complement deposits within demyelinating lesions of WL rats (Zimprich *et al.*, 1991) is consistent with the possible role of ADCT.

During acute phases of murine CNS infection by JHMV, the T-cell response may be crucial in mediating clearance of virus and arrest of the disease process. Upon adoptive transfer of CD4$^+$, Lyt2-cloned T cells specific for JHMV, C57BL/6 mice are protected from developing lethal encephalitis by an major histocompatability complex II (MHC-II)-restricted delayed type hypersensitivity (DTH) response (Stohlman *et al.*, 1986, 1988), without a reduction in virus titer in the CNS. Such protected mice subsequently manifest OL demyelinating disease. The CNS concentrations of anti-JHMV antibodies are not altered in the above T-cell recipients or in WF rats with demyelinating lesions (Sorensen *et al.*, 1984b). It can be concluded from these data that the local synthesis of antiviral antibodies within the CNS does not prevent demyelination. The adoptive transfer of T-cell clones specific for JHMV can also protect against encephalitis by reducing virus replication in neurons (Stohlman *et al.*, 1986). Furthermore, some data suggest that the clearance of JHMV from the CNS of mice appears to involve both $CD4^+$ and $CD8^+$ T cells (Sussmann *et al.*, 1989; Williamson and Stohlman, 1990; Yamaguchi *et al.*, 1991). These results are in line with the evocation of a CTL response against the carboxy-terminal end of N (Stohlman *et al.*, 1992). Another study, however, failed to demonstrate cytotoxic lymphocytes (CTLs) against viral components in C57BL/6 mice, although other T-helper cell responses were detected (Mobley *et al.*, 1992). Complementary studies employing athymic, nude mice and rats or animals treated with immunosuppressing agents confirmed the involvement of cellular immunity in the disease process. They showed that, relative to control animals, gray matter and neuronal involvement is enhanced during T-cell deficiency, resulting in acute encephalitic rather than demyelinative, white matter disease (Sorensen *et al.*, 1982, 1987b; Zimmer and Dales, 1989; Pasick *et al.*, 1992).

Postweanling rats challenged with JHMV do not exhibit any disease symptoms, but on histological examination show signs of limited CNS histopathology and some virus expression. The age-limited resistance does not apply to athymic nude (nu/nu) rats or animals treated with the immunosuppressants cyclophophosphamide and cyclosporin A (Sorensen *et al.*, 1987b; Zimmer and Dales, 1989). Rats challenged with virus even beyond 70 days of age develop a late-onset encephalitic form of disease that is rapidly fatal. Histopathology reveals wide dissemination of infection in neuron-rich gray matter but a limited involvement of white matter, where small foci of demyelination develop during the late-onset disease. In adult mice of strains that are normally resistant to A59, MHV3, and JHMV MCV infections, when initiated by the IN and IP routes, immunosuppression promotes a similar, late-onset encephalitis arising from extensive neuronal damage (Sorensen *et al.*, 1982, 1984a,b). It appears that CNS neurons in older rats and mice are fully susceptible to infection but replication and dissemination of MCV progeny is suppressed when cellular immune responses are activated. Presumably, the paucity of susceptible OL targets for MCV infection within the CNS of mature animals results in a limitation of pathogenesis, manifested as very small WM demyelinating lesions (Sorensen *et al.*, 1987b; Zimmer and Dales, 1989).

Consistent with the *in vivo* situation, the susceptibility of neurons can also be demonstrated with cortical explants from rat embryos that are maintained in long-term culture. Unlike explanted OL, which are intrinsically programmed to differentiate within 3 weeks after birth or at an accelerated rate by treatment with dbcAMP (Raff, 1989; Beushausen and Dales, 1985), isolated neurons remain susceptible to persistent JHMV infection for long periods of time and can produce large quantities of virus after pretreatment with dbcAMP (Pasick and Dales, 1991). The late-onset, fulminant infection of neurons in immunosuppressed rats might commence from a reservoir of virus at a foci of latency, perhaps in neurons of the cerebellum and hippocampus (Sorensen and Dales, 1985).

Unlike cellular immunity, the humoral immune responses that occur within the CNS compartment following infection of WF and WL rats fail to suppress virus replication or interrupt the development of disease (Sorensen *et al.*, 1984b). In Brown Norway rats, a vigorous humoral immunity within the

CNS develops after JHMV infection, which could explain why there is a suppression of disease and virus replication in this model (Dörries *et al.*, 1987).

The infection of macrophages and lymphotropism of MHV3 during infection of partially susceptible (C57BL/6 × A/J)$F_1$ mice initially causes a low-grade infection of the liver, then of the brain. Upon spreading to the CNS, the infection produces paralytic symptoms, sometimes lasting for periods as long as 1 yr. Lymphotropism is associated with a progressive decline of immunoglobulin production during the initial 3 months, then a return to normal levels (Leray *et al.*, 1982). After introduction of JHMV by IN inoculation, adult BALB/c mice become protected against challenge for 12 months. This is due to antibodies raised against the initial inoculation (Barthold and Smith, 1989). The virus may incite an autoimmune disease process in WL rats and SJL/J mice, which follows myelin destruction, as described in Section IV. Autoimmunity was postulated to involve expression of the MHC-I associated (Ia) determinant on AS when these cells become differentiated into antigen-presenting cells due to the presence of JHMV (Massa and ter Meulen, 1987; Calder *et al.*, 1988). This phenomenon does not appear to pertain in mice, in which only class I H-2 antigen presentation is observed in AS and OL (Suzumura *et al.*, 1986, 1988). The presentation of class I H-2 required continuous presence or formation of viable A-59 virus and is mediated by factors produced in glial cells, probably AS (Lavi *et al.*, 1989).

## IV. GENETIC VARIABILITY OF VIRUS AND HOST AS DETERMINANTS OF PATHOGENESIS AND DISEASE

Differences in the cytopathology and disease process in animals are greatly affected by the genotypes of both the virus and the host. During interactions with cell targets, the S peplomer is most probably the major determinant of tropism, virulence or attenuation, and cytopathology. The host controls the disease process and susceptibility or resistance to infection at the cellular and organismic level, including responses by the immune system.

### A. Relationship of the CV Replication Strategy to Genetic Variability

An understanding of the CV replication strategy can explain why these agents have the capacity to evolve rapidly into diverse genetic variants. Thus, transcription by means of discontinuous, leader-primed synthesis (Lai, 1986), as well as the occurence of subgenomic replicative intermediates (Sethna *et al.*, 1989; Sawicki and Sawicki, 1990), provide possible mechanisms for highly efficient recombination among the nonsegmented CV genomes (Lai *et al.*, 1985). Based on the currently available CV genomic and mRNA sequences, it is possible to explain the genesis of diversity observed among MCV progeny following passage through cells in culture or during CNS infection of animals. Also, in the case of HCV, circulating among human populations, genetic variability could arise by repeated OC43 and 229E reinfections of the upper respiratory tract (MacNaughton, 1982; Reed, 1984).

The occasional association of OC43 infections of humans with neurological symptoms (Hellevi and Hovi, 1980) and the induction of a demyelinating disease process by MCV in rodents and monkeys have a bearing on the search for an etiologic infectious agent in multiple sclerosis (MS). However, the claim that two CV isolates, designated SK and SD, had originated from the tissue of MS patients (Mendelman *et al.*, 1983) remains questionable. Sensitive antigen detection (Hasony and MacNaughton, 1982) and nucleic acid probes (Weiss, 1983; Weiss and Leibowitz, 1983) that are able to discriminate between HCV and MCV have grouped SK and SD viruses with MCV (Weiss, 1983).

## B. Role of S in Pathogenesis and Disease

The S peplomer, which is involved in attachment, penetration, syncytiogenesis, and other aspects of cytopathology, appears to be one of the most variable component of coronavirions. In addition to molecular diversity in S from the various CV strains infecting mammals, variability in the MCV S protein arises during passage through rodents, generating antigenic and pathogenicity variants, as summarized in Table II. An important location for molecular modulation could be a hypervariable RNA sequence indentified in the S gene (Parker *et al.*, 1989). During replication, this region may either be a preferred crossing over site or a site of strong selection, thereby giving rise to recombinants of S in which modifications, including deletions, are introduced frequently.

The unquestionable role of S in cytopathogenicity and neuroattenuation was demonstrated when S was expressed by means of a vaccinia virus recombinant vector. This experiment of Daya *et al.* (1989) demonstrated unequivocally that S, by itself, can induce the same cytopathology as the MCV from which it was derived. Also, it has been possible to identify separate regions on S associated with pathogenicity and other functions (Table II). These various biological activities of S were mapped using panels of MAbs able to affect individual functions of this G-P. Thus, three topologically distinct antigenic sites, eliciting protective cellular or humoral responses in mice, were identified on the S protein of JHMV by Talbot *et al.* (1984). The survival of mice was associated with decreased virus production and confinement of the infection to neurons, sparing the glia (Buchmeier *et al.*, 1984).

Another comparative study on the MCV strains MHV-1, -3, -5, JHM, and A59 demonstrated the presence of five antigenic sites, A through E, among which B is involved with neurotropism and C with virulence (Talbot and Buchmeier, 1985). Such analyses also uncovered polymorphism in the S protein of JHMV variants isolated after escaping neutralization by MAb (Dalziel *et al.*, 1986; Buchmeier *et al.*, 1988). Some isolates obtained in this manner were attenuated and others more virulent when compared with the parental JHMV. Attenuation was manifested through a requirement for larger inocula to initiate CNS disease and a change in preferential tropism from neurons to OL. Thus, a paralytic, demyelinating disease rather than an acute encephalitis was elicited (Buchmeier *et al.*, 1988). Controlled dual infection with JHMV and A59 has also generated recombinants in the S gene. Epitope mapping carried out with MAb demonstrated that the critical determinant(s) for neuropathogenicity occur at the C terminal one third of the S protein (Makino *et al.*, 1987; Gallagher *et al.*, 1990). In some variants, attenuation is linked with a *ts* phenotype of the virus, as in the case of one A59 isolate that replicated at a normal rate in Sac(−) cells but behaved as *ts* in glial cultures and was mildly neuropathogenic in mice (van Berlo *et al.*, 1986; Koolen *et al.*, 1983). From this and other findings, it becomes evident that a temperature restriction regulating CV infection can be exerted by the neural cell host and is not due to thermosensitivity of the virus, as discussed in Section II.G (Lucas *et al.*, 1978; van Berlo *et al.*, 1986; Koolen *et al.*, 1983).

It is possible that some phenotypic variations of S are due to point mutations alone, but it is known that during passage through neural cells in explants or within the CNS, the S protein of the emerging stable variants undergoes considerable changes, as described in Table II. One JHMV isolate, passaged in C1300 neuroblastoma, possessed a deletion in S and coincidentally became nonsyncytiogenic (Lavi *et al.*, 1990). In another JHMV isolate from the CNS that could replicate more efficiently in AS, the mRNA encoding S was larger (Taguchi *et al.*, 1986). In another study of six JHMV isolates from mouse CNS, four were shown to possess larger and two smaller than normal length S genes. In the smaller S variants, sequence deletions were linked with an attenuated phenotype producing a subacute demyelinating disease in mice (Taguchi and Fleming, 1989).

Similar modulations in S of JHMV were generated by passage through the CNS of rats. One report described a stable change to a larger S polypeptide (Taguchi *et al.*, 1985). As described in Table II, genetically stable JHMV variants have been isolated and different phenotypes for the induction of neurological disease in the white and gray matter of the CNS can be recognized (Baybutt *et al.*, 1984; Morris *et al.*, 1989). One such attenuated isolate from spinal cord (SC), termed AT11f, has a deletion in S, is highly fusogenic for cells in culture, and rather specific for inducing SC demyelination (Morris *et al.*, 1989). AT11f overcomes the resistance of explanted glial cells in adult SJL/J mice, apparently, unlike infection with parental JHMV, because dissemination of the AT11f progeny is not as restricted and high titers are produced (Pasick *et al.*, 1992). Thus, an alteration in S has made this isolate more, not less, virulent for mouse neural cells. A double mutant attenuated in both encephalitic and demyelinating potential contained a point mutation within S2 plus a deletion in the hypervariable region of S1 (Fleming *et al.*, 1986, 1987). This finding led to the conclusion that the role of S in demyelination may involve multiple determinants on this molecule (Wang *et al.*, 1982). Evidently, care must be taken in defining attenuation and virulence as it pertains to diseases produced by these variants.

## C. Influence of the Genetic Constitution of the Host

There is abundant documentation that the resistance or susceptibility to MCV-induced disease process is controlled by the genetic determinants of the host. With mice, the determinants have been recognized at the cellular level.

TABLE II. Relationship between Changes in S of JHMV and CNS Disease[a]

| Parental virus strain | Variant | Basis of selection | Alteration in the molecule | Neuropathology observed | References |
|---|---|---|---|---|---|
| JHMV | AT11f cord | Spinal cord of WF rate with demyelinating encephalomyelitis | 441-nucleotide deletion in hypervariable region of S1 segment; 738-nucleotide deletion in the HE gene | Neuroattenuated; induces acute demyelinating disease | Morris *et al.* (1989); La Monica *et al.* (1991) |
| | 2.2-V-1 | Neutralizing resistance to MAb | Single point mutation at nucleotide 3340 in S2 segment | Neuroattenuated; induces subacute demyelinating disease | Fleming *et al.* (1986); Wang *et al.* (1992) |
| | 7.2-V-1 | Neutralization resistance to MAb | Two point mutations at nucleotides 1766 and 1950 in S1 | Retains encephalitogenic potential | Fleming *et al.* (1986); Wang *et al.* (1992) |
| | 2.2/7.2-V-2 | Double neutralization resistance | Deletion spanning nucleotides 1523 to 1624 in S1 hypervariable region plus point mutation at nucleotide 3340 in S2 | Neuroattenuated for both encephalitic and demyelinating potential | Fleming *et al.* (1986); Wang *et al.* (1992) |
| | c1-2 | Brain of Lewis with acute encephalitis | 15-kDa enlargements of S protein | Enhanced neurovirulence with widespread neuronal and glial involvement in weanling Lewis rats | Matsubara *et al.* (1991); Taguchi *et al.* (1985) |
| | CNSV | Isolated from cultured rat neural cells | Similar though not identical enlargement of S found in c1-2 | Enhanced tropism for astrocytes in vitro, enhanced neurovirulence *in vivo* | Matsubara *et al.* (1991); Taguchi *et al.* (1986) |

| | | | | | |
|---|---|---|---|---|---|
| | N3 | Chemical mutagenesis *ts* mutant | RNA negative | Neuronal infection without encephalitis; induces meningitis and demyelination | Robb *et al.* (1979) |
| | ts43 | Chemical mutagenesis *ts* mutant | — | Persistent infections of rat glial cultures with predominant involvement of astrocytes; demyelinating disease with inflammatory infiltrates | Massa *et al.* (1988) |
| MHV-4 | V5A 13.1 | Neutralization resistance to S MAb | 426-nucleotide deletion in S1 involving nucleotides 1298 to 1723 | Neuroattenuated; increased $LD_{50}S$; demyelination associated with moderate inflammatory cell infiltration; apparent loss of neuronal tropism | Dalziel *et al.* (1986); Gallagher *et al.* (1990); Parker *et al.* (1989) |
| MHV-4 | V4B11.3 | Neutralization resistance to S MAb | 477-nucleotide deletion in S1 involving nucleotides 1285 to 1761 | | |
| | ts8 | Chemical mutagenesis *ts* mutant | — | Neuroattenuated; acute demyelinating disease without associated inflammatory cell infiltration | Haspel *et al.* (1978); Knobler *et al.* (1982) |

[a]Adapted from Pasick and Dales (in press).

Notably, SJL/J mice resist infection by JHMV but not by the related MHV3. This distinction between virus strains becomes clear when different primary cell types are challenged, including neurons, macrophages (Knobler *et al.*, 1981), AS, and OL (Wilson and Dales, 1988). In contrast to SJL/J mice, A/J adults and their peritoneal exudate macrophages are resistant to MHV3. By challenging IN the hybrid mice crosses SJL/J × CD.1 and A/J × DBA and explanted cells from these crosses, it can be demonstrated (as shown in Table III) that resistance is manifested as a recessive trait (reviewed in Wilson and Dales, 1988). This result is consistent with the idea that a specific, cellular protease is required to activate S by cleavage to $S_1$ and $S_2$, and thus promote the spread of infection. However, the disease pattern produced with the virus when inoculated into $F_1$ mice is unlike that expected from fully susceptible mouse strains. For example, IP challenge of the $F_1$ A/J × DBA cross with MHV3, rather than inducing a fulminant visceral and liver disease, instead elicits late-onset neurological symptoms with paralysis (Sorensen *et al.*, 1982; Le Prévost *et al.*, 1975). Evidently some additional factor(s) is controlling virus expression and influencing the *in vivo* disease process.

It has been suggested that any variable that influences survival of the animal host, following virally induced encephalitis, results in the subsequent manifestation of a demyelinating form of disease (Fleming *et al.*, 1989). Our own studies indicate that subacute demyelination may occur after the host has undergone a transitory and limiting form of encephalitis (Pasick and Dales, 1991).

Genetic variability of the cellular immune response could be one of such factor determining the outcome of the disease process. Peritoneal macrophages

TABLE III. Comparison of Replication in Glial Cultures from Purebred and Hybrid Mice[a,b]

| Mouse strain from which culture originated | Time after inoculation (hr) | Titer ($10^2$ PFU/ml)[c] | | |
|---|---|---|---|---|
| | | JHMV[c] | MHV3[d] | A59[d] |
| SJL | 12 | 0.17 ± 0.29 (3) | 590 ± 114 (3) | 6 ± 2 (6) |
| | 24 | 23 ± 12 (14) | 435 ± 292 (14) | 13 ± 0.5 (6) |
| | 48 | 48 ± 20 (14) | 886 ± 62 (10 | 2 ± 0.6 (3) |
| | 72 | 22 ± 9 (12) | 475 ± 119 (6) | ND[e] |
| CD.1 | 12 | 350 ± 216 (5) | 77 ± 12 (4) | ND |
| | 24 | 3,260 ± 2,110 (13) | 778 ± 30 (11) | 21,300 ± 19,600 (3) |
| | 48 | 9,400 ± 500 (9) | 3,340 ± 2,300 (7) | 95,800 ± 2,700 (3) |
| (CD.1 × SJL)$F_1$ | 24 | 961 ± 34 (5) | 1,120 ± 730 (5) | ND |
| | 48 | 6,260 ± 4,320 (5) | 2,400 ± 2,000 (5) | ND |
| | 72 | 943 ± 480 (5)[f] | 213 ± 52 (5)[f] | ND |

[a]From Wilson and Dales (1988).
[b]Primarily oligodendrocytes and astrocytes (35).
[c]The values are means with standard deviations. The number of cultures tested is shown in parentheses.
[d]MOI, 1 PFU per cell.
[e]ND, not determined.
[f]Decrease in titers was attributed to rapid cell killing.

from a substrain of $C_3H$ mice, normally resistant to MCV, became susceptible if the host had been pretreated with hydrocortisone, implying that a lymphokine-induced resistance was involved (Taylor *et al.*, 1981). Infection of susceptible C57BL/6 mice with MHV3 may abrogate or depress cellular immunity directly because the virus is lymphotropic, hence immunosuppressive, and causes the spleen and thymus to atrophy (Lamontagne *et al.*, 1989). The role of macrophages in the MHV3 disease process can be demonstrated in the adoptive transfer of resistance by means of bone marrow cells from resistant into susceptible X-irradiated mouse recipients (Lamontagne *et al.*, 1989; Dupuy *et al.*, 1984).

Among rats, strain WF has a more prolonged age-related susceptibility to inoculation with JHMV than WL and other strains (Sorensen *et al.*, 1982). The tropism of this virus for immature OL *in vitro* (Beushausen and Dales, 1985; Wilson *et al.*, 1986; Beushausen *et al.*, 1987) and the documented, extended duration of susceptibility to JHMV in postnatal, preweanling WF rats may be connected. Since the maturation of the CNS in this strain is slower, presumably due to genetic deficiency of circulating growth hormone (reviewed in Sorensen *et al.*, 1980), a greater number of infectable OL progenitors might be present for a longer duration in WF versus WL rats. As is the case with the mouse, rats also manifest genetic variation of immune responses, which are superimposable on virus–target cell interactions. An autoimmune type of disease, similar to experimental allergic encephalomyelitis (EAE), was recognized following JHMV infection of WL rats (Watanabe *et al.*, 1983; Wege *et al.*, 1984). This strain is readily inducible for EAE by inoculation with myelin components. However, despite similarities in appearances of demyelinating lesions evident in WL and WF rats, EAE cannot be induced with myelin antigens in WF rats (Zimmer and Dales, 1989). These combined data indicate that JHMV infection of WL rats may elicit a primary demyelination due to OL infection, as well as an EAE-like disease, provoked by circulating myelin antigens released by cytolysis of infected OL. This duality of disease processes may occur in WL rats, perhaps because they are endowed with a high frequency of T-cell responders to myelin basic protein.

In rats, naturally or deliberately made deficient in cellular immunity, susceptibility to JHMV-related CNS disease becomes extended indefinitely beyond weaning. Athymic nu/nu rats undergo a late-onset acute encephalitis with only a limited white matter involvement, evident by the presence of small lesions (Sorensen *et al.*, 1987b). The disease process is primarily connected with fulminant gray matter neuronal infections, which cannot be kept in check due to a genetic deficiency of cellular immunity. Humoral immune responses, although they do occur locally within the CNS of JHMV-infected WL and WF rats, appear not to influence the demyelinative disease process (Sorensen *et al.*, 1984b; Dörries *et al.*, 1986). On the other hand, a vigorous humoral immunity detected in the CNS of BN rats could explain why the disease is suppressed in this strain (Dörries *et al.*, 1987).

From the above considerations, it is evident that genetic variability of the virus and the host can result in differing diseases in animals and varying cytopathic responses when infections are carried out *in vitro*. Within the animal there are additional important factors, most notably the immune system, that also determine the nature of pathogenesis and type of disease that develops.

## V. SUMMARY

Based on the information reviewed in this chapter, it becomes clear that both the virus and host are involved in determining the outcome of CNS infections by MCV. The current information is consistent with the idea that virus tropism for specific cell types within the CNS is profoundly influenced by: (1) the state of maturation of the CNS, i.e, by cell lineages and parameters controlling differentiation; (2) immune responses; and (3) possibly, evolution of genetic variants of the virus. Thus, pathogenesis and neurological disease in this complex rodent model of virus infection become a manifestation of an interplay between the above three, equally important, parameters.

## VI. REFERENCES

Aloisi, F., Agresti, C., D'Urso, D., and Levi, G., 1988, Differentiation of bipotential glial precursors into oligodendrocytes is promoted by interaction with type 1 astrocytes in cerebellar cultures, *Proc. Natl. Acad. Sci. USA* **85:**6167.

Amako, K., and Dales, S., 1967, Cytopathology of mengovirus infection. II. Proliferation of membranous cisternae, *Virology* **32:**201.

Anderson, R., and Wong, F., 1993, Membrane and phospholipid binding by murine coronaviral nucleocapsid N protein, *Virology* **194:**224.

Appleyard, G., and Tisdale, M., 1985, Inhibition of the growth of human coronavirus 229E by leupeptin, *J. Gen. Virol.* **66:**363.

Armstrong, J., Smeekens, S., and Rottier, P., 1983, Sequence of the nucleocapsid gene from murine coronavirus MHV-A59. *Nucleic Acids Res.* **11:**883.

Arnheiter, H., Baechi, T., and Haller, O., 1982, Adult mouse hepatocytes in primary monolayer culture express genetic resistance to mouse hepatitis virus type 3, *J. Immunol.* **129:**1275.

Asano, K., and Asano, A., 1988, Binding of cholesterol and inhibitory peptide derivatives with the fusogenic hydrophopic sequence of F-glycoprotein of HVJ (Sendai virus): Possible implication in the fusion reaction, *Biochemistry* **27:**1321.

Bailey, O. T., Pappenheimer, A. M., Cheever, F. S., and Daniels, J. B., 1949, A murine virus (JHM) causing disseminated encephalomyelitis with extensive destruction of myelin. II. Pathology, *J. Exp. Med.* **90:**195.

Bang, F. B., and Warwick, A., 1960, Mouse macrophages as host cells for the mouse hepatitis virus and the genetic basis for their susceptibility, *Proc. Natl. Acad. Sci. USA* **46:**1065.

Barthold, S. W., and Smith, A. L., 1984, Mouse hepatitis virus strain-related patterns of tissue tropism in suckling mice, *Arch. Virol.* **81:**103.

Barthold, S. W., and Smith, A. L., 1989, Duration of challenge immunity to coronavirus JHM in mice, *Arch. Virol.* **107:**171.

Barthold, S. W., Beck, D. S., and Smith, A. L., 1986, Mouse hepatitis virus nosoencephalopathy is dependent upon virus strain and host genotype, *Arch. Virol.* **91:**247.

Baybutt, H. N., Wege, H., Carter, M. J., and Ter Meulen, V., 1984, Adaptation of coronavirus JHM to persistent infection of murine sac(−) cells, *J. Gen. Virol.* **65:**915.

Beushausen, S., and Dales, S., 1985, *In vivo* and *in vitro* models of demyelinating disease. IX. Tropism and differentiation regulate the infectious process of coronaviruses in primary explants of the rat CNS, *Virology* **141:**89.

Beushausen, S., Narindrasorasak, K., Sanwal, B. D., and Dales, S., 1987, *In vivo* and *in vitro* models of demyelinating disease: Activation of the adenylate cyclase system influences JHM virus expression in explanted rat oligodendrocytes, *J. Virol.* **48:**3795.

Bögler, O., Wren, D., Barnett, S. C., Land, H., and Noble, M., 1990, Cooperation between two growth factors promotes extented self-renewal and inhibits differentiation of ologodendrocyte-type 2 astrocyte (O2-A) progenitor cells, *Proc. Natl. Acad. Sci. USA* **87:**6368.

Bouloy, M., Plotch, S. J., and Krug, R. M., 1980, Both in the 7-methyl and the 2′-O-methyl groups in the cap of mRNA strongly influence its ability to act as primer for influenza virus RNA transcription, *Proc. Natl. Acad. Sci. USA* **77:**3952.

Boyle, J. F., Weismiller, D. G., and Holmes, K. C., 1987, Genetic resistance to mouse hepatitis virus correlates with absence of virus-binding activity on target tissue, *J. Virol.* **61:**185.

Bruce, C. B., Chapman, J., Suckling, A. J., and Rumsby, M. G., 1984, Infection of rat brain primary cell cultures with an avirulent A7 strain of Semliki forest virus, *J. Neuro. Sci.* **66:**77.

Buchmeier, M. J., Lewicki, H. A., Talbot, P. J., and Knobler, R. L., 1984, Murine hepatitis virus-4 (strain JHM)-induced neurologic disease is modulated *in vivo* by monoclonal antibody, *Virology* **132:**261.

Buchmeier, M. J., Dalziel, R. G., and Koolen, M. J. M., 1988, Coronavirus-induced CNS disease: A model for virus-induced demyelination, *J. Neuroimmun.* **20:**111.

Burks, J. S., Devald, B. L., Jankovsky, L. D., and Gerdes, J. C., 1980, Two coronaviruses isolated from central nervous system tissue of two multiple sclerosis patients, *Science* **209:**933.

Calder, V. L., Wolswijk, G., and Noble, M., 1988, The differentiation of O-2A progenitor cells into oligodendrocytes is associated with loss of inducibility of la antigens, *Eur. J. Immunol.* **18:**1195.

Carrasco, L., and Smith, A. E., 1976, Sodium ions and the shut off of host cell protein synthesis by picornaviruses, *Nature* **264:**807.

Cavanagh, D., Brian, D. A., Enjuanes, L., Holmes, K., Lai, M., Laude, S., Siddell, S., Spaan, W., Taguchi, F., and Talbot, P., 1990, Recommendations of the coronavirus study group for the nomenclature of the structural proteins, mRNAs, and genes of coronaviruses, *Virology* **176:**306.

Cervin, M., and Anderson, R., 1991, Modulation of coronavirus-mediated cell fusion by homeostatic control of cholesterol and fatty acid metabolism, *J. Med. Virol.* **35:**142.

Cheever, F. S., Daniels, J. B., Pappenheimer, A. M., and Bailey, O. T., 1949, A murine virus causing disseminated encephalomyelitis with extensive destruction of myelin I. Isolation and biological properties of the virus, *J. Exp. Med.* **90:**181.

Cheley, S., and Anderson, R., 1981, Cellular synthesis and modification of murine hepatitis virus polypeptides, *J. Gen. Virol.* **54:**301.

Cheley, S., Morris, V. L., Cuppels, M. J., and Anderson, R., 1981, RNA and polypeptide homology among murine coronaviruses, *Virology* **115:**310.

Collins, A. R., and Sorensen, O., 1986, Regulation of viral persistence in human glioblastoma and rhabdomyosarcoma cells infected with coronavirus OC43, *Microb. Pathol.* **1:**573.

Collins, A. R., Knobler, R. L., Powell, H., and Buchmeier, M. J., 1982, Monoclonal antibodies to murine hepatitis virus-4 (strain JHM) define the viral glycoprotein responsible for attachment and cell–cell fusion, *Virology* **119:**358.

Collins, A. R., Runison, A. L., and Knobler, R. L., 1983, Mouse hepatitis virus type 4 infection of primary glial cultures from genetically susceptible and resistant mice, *Infect. Immun.* **40:**1192.

Coulter-Mackie, M. B., Flintoff, W. F., and Dales, S., 1984, *In vivo* and *in vitro* models of demyelinating disease. X. A Schwannoma-L-2 somatic cell hybrid persistently yielding high titres of mouse hepatitis virus strain JHM, *Virus Res.* **1:**477.

Coulter-Mackie, M., Adler, R., Wilson, G., and Dales, S., 1985, *In vivo* and *in vitro* models of demyelinating diseases. XII. Persistence and expression of corona JHM virus functions in RN2-2 schwannoma cells during latency, *Virus Res.* **3:**245.

Dalziel, R. G., Lampert, P. W., Talbot, P. J., and Buchmeier, M. J., 1986, Site-specific alteration of murine hepatitis virus type 4 peplomer glycoprotein E2 results in reduced neurovirulence, *J. Virol.* **59:**463.

David-Ferreira, J. F., and Manaker, R. A., 1965, An electron microscope study of the development of a mouse hepatitis virus in tissue culture cells, *J. Cell. Biol.* **24:**57.

Davies, H. A., Dourmashkin, R. R., and MacNaughton, M. R., 1981, Ribonucleoprotein of avian infectious bronchitis virus, *J. Gen. Virol.* **53:**67.

Daya, M., Cervin, M., and Anderson, R., 1988, Cholesterol enhances mouse hepatitis virus-mediated cell fusion, *Virology* **163**:276.

Daya, M., Wong, F., Cervin, M., Evans, G., Vennema, H., Spaan, W., and Anderson, R., 1989, Mutation of host cell determinants which discriminate between lytic and persistent mouse hepatitis virus infection results in a fusion-resistant phenotype, *J. Gen. Virol.* **70**:3335.

Dörries, R., Watanabe, R., Wege, H., and Ter Meulen, V., 1986, Murine coronavirus-induced encephalomyelitides in rats: Analysis of immunoglobulins and virus-specific antibodies in serum and cerebrospinal fluid, *J. Neuroimmun.* **12**:131.

Dörries, R., Watanabe, R., Wege, H., and Ter Meulen, V., 1987, Analysis of the intrathecal humoral response in Brown Norway (BN) rats, infected with murine coronavirus JHM, *J. Neuroimmun.* **14**:305.

Doughri, A. M., Storz, J., Hajer, I., and Fernando, H. S., 1976, Morphology and morphogenesis of a coronavirus infecting intestinal epithelial cells of newborn calves, *Exp. Mol. Pathol.* **25**:355.

Dubois-Dalq, M. E., Doller, E. W., Haspel, M. V., and Holmes, K. V., 1982, Cell tropism and expression of mouse hepatitis virus (MHV) in mouse spinal cord cultures, *Virology* **119**:317.

Dupuy, J. M., Dupuy, C., and Decarie, D., 1984, Genetically determined resistance to mouse hepatitis virus 3 is expressed in hematopoeietic donor cells in radiation chimeras, *J. Immunol.* **133**:1609.

Fabre, M., Langley, O. K., Bologia, L., Delaundy, J-P., Lowenthal, A., Ferret-Jena, V., Vincendon, G., and Sarliève, L., 1985, Cellular development and myelin production in primary cultures of embryonic mouse brain, *Dev. Neurosci.* **7**:323.

Fernandez-Munoz, R., and Darnell, J. E., 1976, Structural differences between the 5′-termini of viral and cellular mRNA in poliovirus-infected cells, *J. Virol.* **18**:719.

Fleming, J. O., Trousdale, M. D., El-Zaatari, F. A. K., Stohlman, S. A., and Weiner, L. P., 1986, Pathogenicity of antigenic variants of murine coronavirus JHM selected with monoclonal antibodies, *J. Virol.* **58**:869.

Fleming, J. O., Trousdale, M. D., Bradbury, J., Stohlman, S. A., and Weiner, L. P., 1987, Experimental demyelination induced by coronavirus JHM (MHV-4): Molecular identification of a viral determinant of paralytic disease, *Microb. Pathol.* **3**:9.

Fleming, J. O., Shubin, R. A., Sussman, M. A., Casteel, N., and Stohlman, S. A., 1989, Monoclonal antibodies to the matrix (E1) glycoprotein of mouse hepatitis virus protect mice from encephalitis, *Virology* **168**:162.

Flintoff, W. F., and van Dinter, S., 1989, Several rat cell lines share a common defect in their inability to internalize murine coronaviruses efficiently, *J. Gen. Virol.* **70**:1713.

Frana, M. F., Behnke, J. N., Struman, L. S., and Holmes, K. V., 1985, Proteolytic cleavage of the E2 glycoprotein of murine coronavirus: Host-dependent differences in the proteolytic cleavage and cell fusion, *J. Virol.* **56**:912.

Gallagher, T. M., Parker, S. E., and Buchmeier, M. J., 1990, Neutralization-resistant variants of a neurotropic coronavirus are generated by deletions with the amino-terminal half of the spike glycoprotein, *J. Virol.* **64**:731.

Gallwitz, D., Traub, U., and Traub, P., 1977, Fate of histone messenger RNA in mengovirus-infected Ehrlich ascites tumor cells, *Eur. J. Biochem.* **81**:387.

Gard, A. L., and Pfeiffer, S. E., 1989, Oligodendrocyte progenitors isolated directly from developing telencephalon at a specific phenotypic stage: Myelinogenic potential in a defined environment, *Development* **106**:119.

Gombold, J. L., Hingley, S. T., and Weiss, S. R., 1993, Fusion-defective mutants of mouse hepatitis virus A59 contain a mutation in the spike protein cleavage signal, *J. Virol.* **67**:4504.

Greene, L. E., and Eisenberg, E., 1988, Effect of phosphatase on the ability of the uncoating ATPase to dissociate clathrin from coated vesicles, *J. Cell. Biol.* **107A**:773.

Hasony, H. J., and MacNaughton, M. R., 1982, Serological relationships of the subcomponents of human coronavirus strain HCV 229E and mouse hepatitis virus 3, *J. Gen. Virol.* **58**:449.

Haspel, M. V., Lampert, P. W., and Oldstone, M. B. A., 1978, Temperature-sensitive mutants of mouse hepatitis virus produce a high incidence of demyelination, *Proc. Natl. Acad. Sci. USA,* **75**:4033.

Hellevi, R., and Hovi, T., 1980, Coronavirus infections of man associated with diseases other than the common cold, *J. Med. Virol.* **6**:259.

Hilton, A., Mizzen, L., MacIntyre, G., Cheley, S., and Anderson, R., 1986, Translational control in murine hepatitis virus infection, *J. Gen. Virol.* **67:**923.

Hirano, N., Goto, N., Ogawa, T., Ono, K., Murakani, T., and Fujiwara, K., 1980, Hydrocephalus in suckling rats infected intracerebrally with mouse hepatitis virus MHV-A59, *Microb. Immunol.* **24:**825.

Kaufman, Y., Goldstein, E., and Penman, S., 1976, Poliovirus-induced inhibition of polypeptide initiation *in vitro* on native polysomes, *Proc. Natl. Acad. Sci. USA* **73:**1834.

Kersting, G., and Pette, E., 1956, Pathohistologie und Pathogenese der experimentellen JHM-Virusencephalomyelitis des Affen, *Dtsch. Nervenheilkund.* **174:**283.

Kienzle, T. E., Abraham, S., Hogue, B. G., and Brian, D., 1990, Structure and orientation of expressed bovine coronavirus hemagglutinin-esterase protein, *J. Virol.* **64:**1834.

Klumperman, J., Krijnse-Locke, J., Meijer, A., Horzinek, M. C., Geuze, H., and Rottier, P. J. M., 1994, Coronavirus M proteins accumulate in the Golgi complex beyond the site of virion budding, *J. Virol.* **68:**6523.

Knobler, R. L., Haspel, M. V., and Oldstone, M. B. A., 1981, Mouse hepatitis virus type 4 (JHM strain)-induced fatal central nervous system disease, *J. Exp. Med.* **153:**832.

Knobler, R. L., Lampert, P. W., and Oldstone, M. B. A., 1982, Virus persistence and recurring demyelination produced by a temperature-sensitive mutant of MHV, *Nature* **298:**279.

Knobler, R. L., Tunison, L. A., and Oldstone, M. B. A., 1984, Host genetic control of mouse hepatitis virus type 4 (JHM strain) replication. I. Restriction of virus amplification and spread in macrophages from resistant mice, *J. Gen. Virol.* **65:**1543.

Kooi, C., Mizzen, L., Alderson, C., Daya, M., and Anderson, R., 1988, Early events of importance in determining host cell permissiveness to mouse hepatitis virus infection, *J. Gen. Virol.* **69:**1125.

Koolen, M. J. M., Osterhaus, A. D. M. E., Van Steenis, G., Horzinek, M. C., and van der Zeijst, B. A. M., 1983, Temperature-sensitive mutants of mouse hepatitis virus strain A59: Isolation, characterization and neuropathogenic properties, *Virology* **125:**393.

Koper, J. W., Lopes-Cardozo, M., Romijn, H. J., and van Golde, E. M. G., 1984, Culture of rat cerebellar oligodendrocytes in a serum-free, chemically defined medium, *J. Neurosci. Meth.* **10:**157.

Krijnse-Locker, J., Ericsson, M., Rottier, P. J. M., and Griffiths, G., 1994, Characterization of the budding compartment of mouse hepatitis virus: Evidence that transport from the RER to the Golgi complex requires only one vesicular transport step, *J. Cell. Biol.* **124:**55.

Krishna, P., and van de Sande, J. H., 1990, Interaction of RecA protein with acidic phospholipids inhibits DNA-binding activity of RecA, *J. Bacteriol.* **172:**279.

Krzystyniak, K., and Dupuy, J. M., 1984, Entry of mouse hepatitis virus 3 into cells, *J. Gen. Virol.* **65:**227.

La Monica, N., Banner, L. R., Morris, V. L., and Lai, M. M. C., 1991, Localization of extensive deletions in the structural genes of two neurotropic variants of murine coronavirus JHM, *Virology* **182:**883.

Lai, M. M. C., 1986, Coronavirus leader-RNA-primed transcription: An alternative mechanism to RNA splicing, *Bioessays* **5:**257.

Lai, M. M. C., 1990, Coronavirus: Organization, replication and expression of genome, *Annu. Rev. Microbiol.* **44:**303.

Lai, M. M. C., Patton, C. D., and Stohlman, S. A., 1982, Further characterization of mRNAs of mouse hepatitis virus: Presence of common 5′ end nucleotides, *J. Virol.* **41:**557.

Lai, M. M. C., Baric, R. S., Makino, S., Keck, J. G., Egbert, J., Leibowitz, J. L., and Stohlman, S. A., 1985, Recombination between nonsegmented RNA genomes of murine coronavirus, *J. Virol.* **56:**449.

Lamontagne, L., and Dupuy, J. M., 1984a, Persistent infection with mouse hepatitis virus 3 in mouse lymphoid cell lines, *Infect. Immun.* **44:**716.

Lamontagne, L., and Dupuy, J. M., 1984b, Natural resistance of mice to mouse hepatitis virus type 3 infection is expressed in embryonic cells, *J. Gen. Virol.* **65:**1165.

Lamontagne, L., Descoteux, J. P., and Jolicoeur, P., 1989, Mouse hepatitis virus replication in T and B lymphocytes correlate with viral pathogenicity, *J. Immunol.* **142:**4458.

Laude, H., Charley, B., and Gelfi, J., 1984, Replication of transmissible gastroenteritis coronavirus (TGEV) in swine alveolar macrophages, *J. Gen. Virol.* **65:**327.

Lavi, E., Gilden, D. H., Highkin, M. K., and Weiss, S. R., 1984a, Persistence of mouse hepatitis virus A59 RNA in a slow virus demyelinating infection of mice as detected by in situ hybridization, *J. Virol.* **51**:563.

Lavi, E., Gilden, D. H., Wroblewska, Z., Rorke, L. B., and Weiss, S. B., 1984b, Experimental demyelination produced by the A59 strain of mouse hepatitis virus, *Neurology* **34**:597.

Lavi, E., Suzumura, A., Hiragama, M., Highkin, M. K., Dambach, D. M., Silberberg, D. H., and Weiss, S. R., 1987, Coronavirus mouse hepatitis virus (MHV)-A59 causes persistent, productive infection in primary glial cell cultures, *Microb. Pathol.* **3**:79.

Lavi, E., Fishman, P. S., Highkin, M. K., and Weiss, S. R., 1988, Limbic encephalitis after inhalation of a murine coronavirus, *Lab. Invest.* **58**:31.

Lavi, E., Suzumura, A., Murray, E. M., Silberberg, B. H., and Weiss, S. R., 1989, Induction of MHC class I antigens on glial cells is dependent on persistent mouse hepatitis virus infection, *J. Neuroimmunol.* **22**:107.

Lavi, E., Murray, E. M., Makino, S., Stohlman, S. A., and Lai, M. M. C., 1990, Determinants of coronavirus MHV pathogenesis are localized to 3′ portions of the genome as determined by ribonucleic acid-ribonucleic acid recombination, *Lab. Invest.* **62**:570.

Lazaridis, I., Babich, A., and Nevins, J. R., 1988, Role of adenovirus 72-kDa DNA binding protein in the rapid decay of early viral mRNA, *Virology* **165**:438.

Lee, H.-J., Shieh, C.-K., Gorbalenya, A. E., Koonin, E. V., La Monica, N., Tuler, J., Bagdzhadzhyan, A., and Lai, M. M. C., 1991, The complete sequence (22 kilobases) of murine coronavirus gene 1 encoding the putative protease and RNA polymerase, *Virology* **180**:567.

Leibowitz, J. L., Bond, C. W., Anderson, K., and Goos, S., 1984, Biological and macromolecular properties of murine cells persistently infected with MHV-JHM, *Arch. Virol.* **80**:315.

Le Prévost, C., Verlizier, J. L., and Dupuy, J. M., 1975, Immunpathology of mouse hepatitis virus type 3 infection, *J. Immunol.* **115**:640.

Leray, D., Dupuy, C., and Dupuy, J. M., 1982, Immunpathology of mouse hepatitis virus type 3 infection. IV. MHV3-induced immunodepression. *Clin. Immun. Immunopathol.* **23**:539.

Levine, J. M., 1989, Neuronal influences on glial progenitor cell development, *Neuron* **5**:103.

Lillien, L. E., and Raff, M. C., 1990, Differentiation signals in the CNS: Type-2 astrocyte development *in vitro* as a model system, *Neuron* **5**:111.

Lodish, H. F., and Porter, M., 1981, Vesicular stomatitis virus mRNA and inhibition of translation of cellular mRNA-is there a P function in vesicular stomatitis virus, *J. Virol.* **38**:504.

Loeb, J. E., Cantournet, B., Vartanian, M.-P., Goris, J., and Mervelde, W., 1989, Phosphorylation/dephosphorylation of β light chain of clathrin from rat liver coated vesicles, *Eur. J. Biochem.* **182**:195.

Lucas, A., Flintoff, W., Anderson, R., Percy, D., Coulter, M., and Dales, S., 1977, *In vivo* and *in vitro* models of demyelinating diseases: Tropism of JHM strain of murine hepatitis virus for cells of glial origin, *Cell* **12**:553.

Lucas, A., Coulter, M., Anderson, R., Dales, S., and Flintoff, W., 1978, *In vivo* and *in vitro* models of demyelinating diseases. II. Persistence and host regulated thermosensitivity in cells of neural derivation infected with mouse hepatitis and measles viruses, *Virology* **88**:325.

Luytjes, W., Sturman, L., Bredenbeek, P. J., Charite, J., van der Zeijst, B. A. M., Horzinek, M. C., and Spaan, W. J. M., 1987, Primary structure of glycoprotein E2 of coronavirus MHV-A59 and identification of the trypsin cleavage site, *Virology* **161**:479.

Luytjes, W., Geerts, D., Posthumus, W., Meleon, R., and Spaan, W., 1989, Amino acid sequence of conserved neutralizing epitope of murine coronaviruses, *J. Virol.* **63**:1408.

MacIntyre, G., Wong, F., and Anderson, R., 1989, A model for persistent murine coronavirus infection involving maintenance via cytopathically infected cell centres, *J. Gen. Virol.* **70**:763.

MacNaughton, M. R., 1982, Occurrence and frequency of coronavirus infections in humans as determined by enzyme-linked immunosorbent assay, *Infect. Immun.* **38**:419.

MacNaughton, M. R., and Patterson, S., 1980, Mouse hepatitis virus strain 3 infection of C57, A/Sn and A/J strain mice and their macrophages, *Arch. Virol.* **66**:71.

Makino, S., and Lai, M. M. C., 1989, Evolution of the 5′-end of genomic RNA of murine coronaviruses during passage *in vitro, Virology* **169**:227.

Makino, S., Taguchi, F., and Fujiwara, L., 1984, Defective interfering particles of mouse hepatitis virus, *Virology* **133**:9.

Makino, S., Fujioka, N., and Fujiwara, K., 1985, Structure of the intracellular defective viral RNAs of defective interfering particles of mouse hepatitis virus, *J. Virol.* **54:**329.
Makino, S., Fleming, J. O., Keck, J. G., Stohlman, S. A., and Lai, M. M. C., 1987, RNA recombination of coronaviruses: Localization of neutralizing epitopes and neuropathogenic determinants on the caroxyl terminus of peplomers, *Proc. Natl. Acad. Sci. USA* **84:**6567.
Mallucci, L., and Edwards, B., 1982, Influence of cytoskeleton on the expression of a mouse hepatitis virus (MHV-3) in peritoneal macrophages: Acute and persistent infection, *J. Gen. Virol.* **63:**217.
Marvaldi, J. L., Lucas-Lenard, J., Sekellick, M. J., and Marcus, P., 1977, Cell killing by viruses. IV. Cell killing and protein synthesis inhibition by vesicular stomatitis virus require the same gene functions, *Virology* **79:**267.
Massa, P. T., and ter Meulen, V., 1987, Analysis of Ia induction on Lewis rat astrocytes *in vitro* by virus particles and bacterial adjuvants, *J. Neuroimmun.* **13:**259.
Massa, P. T., Wege, H., and ter Meulen, V., 1986, Analysis of murine hepatitis virus (JHM strain) tropism towards Lewis rat glial cells *in vitro* Type I astrocytes and brain macrophages (microglia) as primary glial target cells, *Lab. Invest.* **55:**318.
Massa, P. T., Wege, H., and ter Meulen, V., 1988, Growth pattern of various JHM coronavirus isolates in primary rat glial cell cultures correlates with differing neurotropism *in vivo, Virus Res.* **9:**133.
Massalski, A., Coulter-Mackie, M., Knobler, R. L., Buchmeier, M. J., and Dales, S., 1982, *In vivo* and *in vitro* models of demyelinating diseases. V. Comparison of the assembly of mouse hepatitis virus, strain JHM, in two murine cell lines, *Intervirology* **18:**135.
Matsubara, Y., Watanabe, R., and Taguchi, F., 1991, Neurovirulence of six murine coronavirus JHMV variants of rats, *Virus Res.* **20:**45.
McKinnon, R. D., Matsui, T., Dubois-Dalcq, M., and Aaronson, S. A., 1990, FGF modulates the PDGF-driven pathway of oligodendrocyte development, *Neuron* **5:**603.
Mendelman, P. M., Jankovsky, L. D., Murray, R. S., Licari, P., Devald, B., Gerdes, J. C., and Burks, J. S., 1983, Pathogenesis of coronavirus SD in mice. I. Prominent demyelination in the absence of infectious virus production, *Arch. Neurol.* **40:**493.
Miller, C. A., and Carrigan, D. B., 1982, Reversible repression and activation of measles virus infection in neural cells, *Proc. Natl. Acad. Sci. USA* **79:**1629.
Mizzen, L., Cheley, S., Rao, M., Wolf, R., and Anderson, R., 1983, Fusion resistance and decreased infectability as major host cell determinants of coronavirus persistence, *Virology* **128:**407.
Mizzen, L., Hilton, A., Cheley, S., and Anderson, R., 1985, Attenuation of murine coronavirus infection by ammonium chloride, *Virology* **142:**378.
Mizzen, L., MacIntyre, G., Wong, F., and Anderson, R., 1987a, Translational regulation in mouse hepatitis virus infection is not mediated by altered intracellular ion concentrations, *J. Gen. Virol.* **68:**2143.
Mizzen, L., Daya, M., and Anderson, R., 1987b, The role of protease-dependent cell membrane fusion in persistent and lytic infections of murine hepatitis virus, *Adv. Exp. Med. Biol.* **218:**175.
Mobley, J., Evans, G., Dailey, M. O., and Perlman, S., 1992, Immune response to a murine coronavirus: Identification of a homing receptor-negative $CD4^+$ T cell subset that responds to viral glycoproteins, *Virology* **187:**443.
Mohandas, D. V., and Dales, S., 1990, *In vivo* and *in vitro* models demyelinating disease: A phosphoprotein phosphatase in host cell endosomes dephosphorylating the nucleocapsid protein of coronavirus JHM, in: *Coronavirus and Their Diseases* (D. Cavanagh and T. D. K. Brown, eds.), pp. 255–260, Plenum Press, New York.
Mohandas, D. V., and Dales, S., 1991, Endosomal association of a protein phosphatase with high dephosphorylating activity against a coronavirus nucleocapsid protein, *FEBS Lett.* **282:**419.
Morris, V. L., Tieszer, C., MacKinnon, J., and Percy, D., 1989, Characterization of coronavirus JHM variants isolated from Wistar Furth rats with a viral-induced demyelinating disease, *Virology* **169:**127.
Muraoka, S., and Takahashi, T., 1989, Primary dissociated cell culture of fetal rat central nervous tissue. II. Immunocytochemical and ultrastructural studies of myelinogenesis, *Dev. Brain Res.* **49:**63.
Murray, R. S., Cai, G.-Y., Hoel, K., Zhang, J.-Y., Soike, K. F., and Cabirac, G. F., 1992a, Coronavirus infects and causes demyelination in primate central nervous system, *Virology* **188:**274.

Murray, R. S., Brown, B., Brian, D., and Cabirac, G. F., 1992b, Detection of coronavirus RNA and antigen in multiple sclerosis brain, *Ann. Neurol.* **31:**525.

Niemann, H., and Klenk, H.-D., 1981, Coronavirus glycoprotein El, a new type of viral glycoprotein, *J. Mol. Biol.* **153:**993.

Niemann, H., Geyer, R., Klenk, H.-D., Lindner, D., Strim, S., and Wirth, M., 1984, The carbohydrates of mouse hepatitis virus (MHV) A59: Structures of the O-glycosidically linked oligosaccharides of glycoprotein El, *EMBO J.* **3:**665.

Nishioka, Y., and Silverstein, S., 1978, Requirement of protein synthesis for the degrading of host mRNA in Friend erytholeukemia cells infected with herpes simplex virus, *J. Virol.* **27:**619.

Parham, D., Tereba, A., Talbot, P. J., Jackson, D. P., and Morris, V. L., 1986, Analysis of JHM central nervous system infections in rats, *Arch. Neurol.* **43:**702.

Parker, M. D., Yoo, D., and Babiuk, L. A., 1990, Expression and secretion of the bovine coronavirus hemagglutinin-esterase glycoprotein by insect cells infected with recombinant baculoviruses, *J. Virol.* **64:**1625.

Parker, S. E., Gallagher, T. M., and Buchmeier, M. J., 1989, Sequence analysis reveals extensive polymorphism and evidence of deletions within the E2 glycoprotein gene of several strains of murine hepatitis virus, *Virology* **173:**664.

Pasick, J. M. M., and Dales, S., 1991, Infection by coronavirus JHM of rat neurons and oligodendrocyte-type-2 astrocyte lineage cells during distinct development stages, *J. Virol.* **65:**5013.

Pasick, J. M. M., Wilson, G. A. R., Morris, V. L., and Dales, S., 1992, SJL/J resistance to mouse hepatitis virus JHM-induced neurologic disease immunosuppression, *Microbiol. Pathol.* **13:**1.

Pasick, J. M. M., Kalicharran, K., and Dales, S., 1994, Distribution and traffiking of JHM coronavirus structural proteins and virions in primary the OBL-21 neuronal cell line, *J. Virol.* **68:**2915.

Pauloin, A., Thurieau, C., and Jolles, P., 1988, Cyclic phosphorylation/dephosphorylation cascade in bovine brain coated vesicles, *Biochem. Biophys. Acta* **968:**91.

Pearce-Kelling, S., Mitchell, W. J., Summers, B. A., and Appal, M. J. G., 1990, Growth of canine distemper virus in cultured astrocytes: Relationship to *in vivo* persistence and disease, *Microb. Pathol.* **8:**71.

Pearson, J., and Mims, C. A., 1985, Differential susceptibility of cultured neural cells to the human coronavirus OC43, *J. Virol.* **53:**1016.

Penman, S., 1965, Stimulation of the incorporation of choline in poliovirus-infected cells, *Virology* **25:**148.

Pereira, C. A., Steffan, A. M., Koehren, F., Douglas, C. R., and Kirn, A., 1987, Increased susceptibility of mice to MHV 3 infection induced by hypercholesterolemic diet: Impairment of Kupffer cell function, *Immunobiology* **174:**253.

Perlman, S., Jacobsen, G., and Moore, S., 1988, Regional localization of virus in the central nervous system of mice persistently infected with murine coronavirus JHM, *Virology* **166:**328.

Perlman, S., Jacobsen, G., and Afifi, A., 1989, Spread of a neurotropic coronavirus into the CNS via the trigeminal and olfactory nerves, *Virology* **170:**556.

Perlman, S., Jacobsen, G., Olson, A. L., and Afifi, A., 1990, Identification of the spinal cord as a major site of persistence during chronic infection with a murine coronavirus, *Virology* **175:**418.

Pickel, K., Muller, M. A., and ter Meulen, V., 1981, Analysis of age-dependent resistance to murine coronavirus JHM infection in mice, *Infect. Immun.* **34:**648.

Plotch, S. J., Bouloy, M., Ulmanen, I., and Krug, R. M., 1981, A unique cap (m7Gpppm)-dependent influenza virion endonuclease cleaves capped RNAs to generate the primers that initiate viral RNA transcription, *Cell* **23:**847.

Raff, M. C., 1989, Glial cell diversification in the rat optic nerve, *Science* **243:**1450.

Raju, R., and Kolakofsky, D., 1988, La Crosse virus infection of mammalian cells induces mRNA instability, *J. Virol.* **62:**27.

Reed, S., 1984, The behaviour of recent isolates of human respiratory coronavirus *in vitro* and in volunteers: Evidence of heterogeneity among 229E-related strains, *J. Med. Virol.* **13:**179.

Rice, A. P., and Roberts, B. E., 1983, Vaccinia virus induces cellular mRNA degradation, *J. Virol.* **47:**529.

Robb, J. A., Bond, C. W., and Leibowitz, J. L., 1979, Pathogenic murine coronaviruses. III. Biological and biochemical characterization of temperature-sensitive mutants of JHMV, *Virology* **94:**385.

Rodriguez, M., Kenny, J. J., Thiemann, R. L., and Woloschak, G. E., 1990, Theiler's virus-induced demyelination in mice immunosuppressed with anti-IgM and in mice expressing the xid gene, *Microb. Pathol.* **8**:23.

Rottier, P. J. M., and Rose, J. K., 1987, Coronavirus El glycoprotein expressed from cloned cDNA localizes in the Golgi region, *J. Virol.* **61**:2042.

Rottier, P. J. M., Brandenburg, D., Armstrong, J., van der Zeijst, B., and Warren, G., 1984, Assembly *in vitro* of a spinning membrane protein of the endoplasmic reticulum: The El glycoprotein of coronavirus mouse hepatitis virus A59, *Proc. Acad. Natl. Sci. USA* **81**:1421.

Sawicki, S. G., and Sawicki, D. L., 1990, Coronavirus transcription: Subgenomic mouse hepatitis virus replicative intermediates function in mRNA synthesis, *J. Virol.* **64**:1050.

Schmid, S., Fuchs, R., Kielian, M., Helenius, A., and Mellman, I., 1989, Acidification of endosome subpopulations in wild-type Chinese hamster ovary cells and temperature-sensitive acidification-defective mutants, *J. Cell. Biol.* **108**:1291.

Schmidt, I., Skinner, M., and Siddell, S., 1987, Nucleotide sequence of the gene encoding the surface projection glycoprotein of coronavirus MHV-JHM, *J. Gen. Virol.* **68**:47.

Schultze, B., Wahn, K., Klenk, H.-D., and Herrler, G., 1991, Isolated HE protein from hemagglutinating encephalomyelitis virus and bovine coronavirus has receptor-destroying and receptor-binding activity, *Virology* **180**:221.

Sekimizu, K., and Kornberg, A., 1988, Cardiolipin activation of dnaA protein, the initiation protein of replication in *Escherichia coli, J. Biol. Chem.* **263**:7131.

Sethna, P. B., Hung, S.-L., and Brian, D. A., 1989, Coronavirus subgenomic minus-strand RNAs and the potential for mRNA replicons, *Proc. Natl. Acad. Sci. USA* **86**:5626.

Siddell, S. G., Barthel, A., and ter Meulen, V., 1981, Coronavirus JHM: A virion associated protein kinase, *J. Gen. Virol.* **52**:235.

Siddell, S. G., Wege, H., and ter Meulen, V., 1982, The structure and replication of coronaviruses, in: *Current Topics in Microbiology and Immunology*, Vol. 99 (M. Cooper *et al.*, eds.), pp. 131–163, Springer, New York.

Skinner, M. A., and Siddell, S. G., 1983, Coronavirus JHM: Nucleotide sequence of the mRNA that encodes nucleocapsid protein, *Nucleic Acid Res.* **11**:5045.

Sorensen, O., and Dales, S., 1985, *In vivo* and *in vitro* models of demyelinating disease: JHM virus in the rat central nervous system localized by *in situ* cDNA hybridization and immunofluorescent microscopy, *J. Virol.* **56**:434.

Sorensen, O., Percy, D., and Dales, S., 1980, *In vivo* and *in vitro* models of demyelinating diseases. III. JHM virus infection of rats, *Arch. Neurol.* **37**:478.

Sorensen, O., Dugre, R., Percy, D., and Dales, S., 1982, *In vivo* and *in vitro* models of demyelinating disease: Endogenous factors influencing demyelinating disease caused by mouse hepatitis virus in rats and mice, *Infect. Immun.* **37**:1248.

Sorensen, O., Beushausen, S., Puchalski, S., Cheley, S., Anderson, R., Coulter-Mackie, M., and Dales, S., 1984a, *In vivo* and *in vitro* models of demyelinating diseases. VIII. Genetic, immunologic and cellular influences on JHM virus infection on rats, in: *Molecular Biology and Pathogenesis of Coronaviruses* (P. J. M. Rottier *et al.*, eds.), pp. 279–297, Plenum Press, New York.

Sorensen, O., Coulter-Mackie, M. B., Puchalski, S., and Dales, S., 1984b, *In vivo* and *in vitro* models of demyelinating disease. IX. Progression of JHM virus infection in the central nervous system of the rat during overt and asymptomatic phases, *Virology* **137**:347.

Sorensen, O., Beushausen, S., Coulter-Mackie, M., Adler, R., and Dales, S., 1987a, *In vivo* and *in vitro* models of demyelinating disease, in: *Viruses, immunity and mental disorders* (K. Kurstak, Z. J. Lipowski, and P. V. Morozow, eds.), pp. 199–210, Plenum Press, New York.

Sorensen, O., Saravani, A., and Dales, S., 1987b, *In vivo* and *in vitro* models of demyelinating disease. XVII. The infectious process in athymic rats inoculated with JHM virus, *Microb. Pathol.* **2**:79.

Spaan, W., Cavanagh, D., and Horzinek, M. C., 1988, Coronaviruses: Structure and genome expression, *J. Gen. Virol.* **69**:2939.

Stauber, R., Pleiderera, M., and Siddell, S., 1993, Proteolytic cleavage of murine coronavirus surface glycoprotein is not required for fusion activity, *J. Gen. Virol.* **74**:183.

Stewart, J. N., Mounir, S., and Talbot, P. J., 1992, Human coronavirus gene expression in brains of multiple sclerosis patients, *Virology* **191**:502.

Stohlman, S. A., and Frelinger, J. A., 1978, Resistance to fatal central nervous system disease by mouse hepatitis virus, strain JHM. I. Genetic analysis, *Immunogenetics* **6**:271.

Stohlman, S. A., and Weiner, L. P., 1978, Stability of neurotropic mouse hepatitis virus (JHM strain) during chronic infection of neuroblastoma cells, *Arch. Virol.* **57**:53.

Stohlman, S. A., Baric, R. S., Nelson, G. N., Soe, L. H., Welter, L. M., and Deans, R. J., 1983, Synthesis and subcellular localization of the murine coronavirus nucleocapsid protein, *Virology* **130**:527.

Stohlman, S. A., Matsushima, G. K., Casteel, N., and Weiner, L. P., 1986, *In vivo* effects of coronavirus-specific T cell clones: DTH inducer cells prevent a lethal infection but do not inhibit virus replication, *J. Immunol.* **136**:3052.

Stohlman, S. A., Sussman, M. A., Matsushima, G. K., Shubin, R. A., and Erlich, S. S., 1988, Delayed-type hypersensitivity response in the central nervous system during JHM virus infection requires viral specificity for protection, *J. Neuroimmunol.* **19**:255.

Stohlman, S. A., Kyuwa, S., Cohen, M., Bergmann, C., Polo, J. M., Yeh, J., Anthony, R., and Keck, J. G., 1992, Mouse hepatitis virus nucleocapsid protein-specific cytotoxic T lymphocytes are Ld restricted and specific for the carboxy terminus, *Virology* **189**:217.

Sturman, L. S., Holmes, K. V., and Behnke, J., 1980, Isolation of coronavirus envelope glycoproteins and interaction with the viral nucleocapsid, *J. Virol.* **33**:449.

Sturman, L. S., Ricard, C. S., and Holmes, K. V., 1985, Proteolytic cleavage of the EL glycoprotein of murine coronavirus: Activation of cell-fusing activity of virion by trypsin and separation of two different 90 K cleavage fragments, *J. Virol.* **56**:904.

Sussman, M. A., Shubin, R. A., Kyuwa, S., and Stohlman, S. A., 1989, T-cell-mediated clearance of mouse hepatitis virus strain JHM from the central nervous system, *J. Virol.* **63**:3051.

Suzumura, A., Lavi, E., Weiss, S. B., and Silberberg, D. H., 1986, Coronavirus infection induces H-2 antigen expression on oligodendrocytes and astrocytes, *Science* **232**:991.

Suzumura, A., Lavi, E., Bhat, S., Murasko, D., and Weiss, S. A., 1988, Induction of the glial cell MHC antigen expression in neurotropic coronavirus infections. Characterization of the H-2 inducing soluble factor elaborated by infected brain cells, *J. Immun.* **140**:2068.

Swift, A. M., and Machamer, C. E., 1991, A Golgi retention signal in a membrane-spanning domain of coronavirus El protein, *J. Cell. Biol.* **115**:19.

Taguchi, F., 1993, Fusion formation by the uncleaved spike protein of murine coronavirus JHMV variant cl-2, *J. Virol.* **67**:1195.

Taguchi, F., and Fleming, J. O., 1989, Comparison of six different murine coronavirus JHM variants by monoclonal antibodies against E2 glycoprotein, *Virology* **169**:233.

Taguchi, F., and Siddell, S. G., 1985, Difference in sensitivity to interferon among mouse hepatitis viruses with high and low virulence for mice, *Virology* **147**:41.

Taguchi, F., Goto, Y., Aiuchi, M., Hayshi, T., and Fujiwara, K., 1979, Pathogenesis of mouse hepatitis virus infection. The role of nasal epithelial cells as a primary target of low-virulence virus, MHV-S, *Microb. Immunol.* **23**:249.

Taguchi, F., Siddell, S., Wege, H., and ter Meulen, V., 1985, Characterization of a variant virus selected in rat brains after infection by coronavirus mouse hepatitis virus JHM, *J. Virol.* **54**:429.

Taguchi, F., Massa, P. T., and ter Meulen, V., 1986, Characterization of a variant isolated from neural cell culture after infection of mouse coronavirus JHMV, *Virology* **155**:267.

Talbot, P. J., and Buchmeier, M. J., 1985, Antigenic variation among murine coronaviruses: Evidence for polymorphism on the peplomer glycoprotein E2, *Virus Res.* **2**:317.

Talbot, P. J., Salmi, A. A., Knobler, R. L., and Buchmeier, M. J., 1984, Topographical mapping of epitopes on the glycoproteins of murine hepatitis virus-4 (strain-JHM): Correlation with biological activities, *Virology* **132**:250.

Tanaka, R., Iwasaki, Y., and Koprowski, H., 1976, Ultrastructural studies of perivascular cuffing cells in multiple sclerosis brain, *J. Neurol. Sci.* **28**:121.

Tardieu, M., Boespflug, O., and Barbe, T., 1986, Selective tropism of a neurotropic coronavirus for ependymal cells, neurons, and meningeal cells, *J. Virol.* **60**:574.

Taylor, C. E., Weiser, W. Y., and Bang, F. B., 1981, *In vitro* macrophage manifestation of cortisone-induced decrease in resistance to mouse hepatitis virus, *J. Exp. Med.* **153**:732.

Tooze, J., Tooze, S., and Warren, G., 1984, Replication of coronavirus MHV-A59 in sac-cells: Determination of the first site of budding of progeny virions, *Eur. J. Cell. Biol.* **33**:281.

Trachsel, H., Sonnenberg, N., Shatkin, A. J., Rose, J. K., Leong, K., Bergmann, J. E., Gordon, J., and Baltimore, D., 1980, Purification of a factor that restores translation of vesicular stomatitis virus mRNA in extracts from poliovirus-infected HeLa cells, *Proc. Natl. Acad. Sci. USA* **77**:770.

van Alstyne, D., and Paty, D. W., 1983, The effect of dibutyryl cyclic AMP on restricted replication of rubella virus in rat glial cells in culture, *Virology* **124**:173.

van Berlo, M. F., Wolswijk, G., Calafat, J., Koolen, M. J. M., Horzinek, M. C., and van de Zeijst, B. A. M., 1986, Restricted replication of mouse hepatitis virus A59 in primary mouse brain astrocytes correlates with reduced pathogenicity, *J. Virol.* **58**:426.

van Berlo, M. F., Warringa, R., Wolswijk, G., and Lopez-Cardoso, M., 1989, Vulnerability of rat and mouse brain cells to murine hepatitis virus (JHM-strain): Studies *in vivo* and *in vitro*, *Glia* **2**:85.

van Dinter, S., and Flintoff, W. F., 1987, Rat glial C6 cells are defective in murine coronavirus internalization, *J. Gen. Virol.* **68**:1677.

Virelizier, J. L., and Allison, A. C., 1976, Correlation of persistent mouse hepatitis (MHV-3) infection with its effects on mouse macrophage cultures, *Arch. Virol.* **50**:279.

Vlasak, R., Luytjes, W., Leider, J., Spaan, W., and Palese, P., 1988, The E3 protein of bovine coronavirus is a receptor-destroying enzyme with acetyl esterase activity, *J. Virol.* **62**:4686.

Wang, F.-I., Fleming, J. O., and Lai, M. M. C., 1992, Sequence analysis of the spike protein gene of murine coronavirus variants: Study of genetic sites affecting neuropathogenicity, *Virology* **186**:742.

Watanabe, R., Wege, H., and ter Meulen, V., 1983, Adoptive transfer of EAE-like lesions from rats with coronavirus-induced demyelinating encephalomyelitis, *Nature* **305**:150.

Wege, H., Koga, M., Watanabe, R., Nagashina, K., and ter Meulen, V., 1983, Neurovirulence of murine coronavirus JHM temperature-sensitive mutants in rats, *Infect. Immun.* **39**:1316.

Wege, H., Watanabe, H., and ter Meulen, V., 1984, Relapsing subacute demyelinating encephalomyelitis in rats during the course of coronavirus JHM infection, *J. Neuroimmun.* **6**:325.

Weiss, S. B., 1983, Coronaviruses SD and SK share extensive nucleotide homology with murine coronavirus MHV-A59, more than that shared between human and murine coronaviruses, *Virology* **126**:669.

Weiss, S. B., and Leibowitz, J. L., 1983, Characterization of murine coronavirus RNA by hybridization with virus-specific cDNA probes, *J. Gen. Virol.* **64**:127.

Wilbur, S. M., Nelson, G. W., Lai, M. C., McMillan, M., and Stohlman, S. A., 1986, Phosphorylation of the mouse hepatitis virus nucleocapsid protein, *Biochem. Biophys. Res. Commun.* **141**:7.

Williams, R. K., Gui-Sen, J., and Holmes, K. V., 1991, Receptor for mouse hepatitis virus is a member of the carcinoembryonic antigen family of glycoproteins, *Proc. Natl. Acad. Sci. USA* **88**:5533.

Williamson, J. S. P., and Stohlman, S. A., 1990, Effective clearance of mouse hepatitis virus from the central nervous system requires both $CD4^+$ and $CD8^+$ T cells, *J. Virol.* **64**:4589.

Wilson, G. A. R., and Dales, S., 1988, *In vivo* and *in vitro* models of demyelinating disease: Efficiency of virus spread and formation of infectious centers among glial cells is genetically determined by the murine host, *J. Virol.* **62**:3371.

Wilson, G. A. R., Mohandas, D. V., and Dales, S., 1990, *In vivo* and *in vitro* models of demyelinating disease. Possible relationship between induction of regulatory subunit form cAMP dependent protein kinases and inhibition of JHMV replication in cultured oligodendrocytes, in: *Coronaviruses and Their Diseases* (D. Cavanagh and T. D. K. Brown, eds.), pp. 261–266, Plenum Press, New York.

Yamaguchi, K., Goto, N., Kyuwa, S., Hayami, M., and Toyoda, Y., 1991, Protection of mice from a lethal coronavirus infection in the central nervous system by adoptive transfer of virus-specific T cell clones, *J. Neuroimmunol.* **32**:1.

Yokomori, K., and Lai, M. M. C., 1992, Mouse hepatitis virus utilizes two antigens as alternative receptors, *J. Virol.* **66**:6194.

Yokomori, K., Bakers, S. C., Stohlman, S. A., and Lai, M. M. C., 1992, Hemagglutinin-esterase-specific monoclonal antibodies alter the neuropathogenicity of mouse hepatitis virus, *J. Virol.* **66**:2865.

Yoshikawa, Y., and Yamanouchi, K., 1984, Effect of papaverine treatment on replication of measles virus in human neural and nonneural cells, *J. Virol.* **50**:489.

Zimmer, M. J., and Dales, S., 1989, *In vivo* and *in vitro* models of demyelinating diseases. XXIV. The infectious process in cyclosporin A treated Wistar Lewis rats inoculated with JHM virus, *Microb. Pathol.* **6**:7.

Zimprich, F., Winter, J., Wege, H., and Lassmann, H., 1991, Coronavirus induced primary demyelination: Indications for the involvement of a humoral immune response, *Neuropathol. Appl. Neurobiol.* **17**:469.

Zurbriggen, A., Vandevelde, M., and Vollo, E., 1987, Demyelinating, non-demyelinating and attenuated canine distemper virus strains induce oliogodendroglial cytolysis *in vitro*, *J. Neuro. Sci.* **79**:33.

CHAPTER 14

# Feline Infectious Peritonitis

RAOUL J. DE GROOT AND MARIAN C. HORZINEK

## I. INTRODUCTION

Feline infectious peritonitis (FIP) is one of the most intriguing diseases caused by a coronavirus. It involves immune-mediated phenomena such as antibody-dependent enhancement of virus infection- and immune complex-induced pathology. Furthermore, there is increasing evidence for the existence of a carrier state. FIP and feline coronaviruses have been extensively reviewed (Pedersen, 1976b, 1983a,b, 1987a; Barlough and Stoddart, 1986, 1990; Olsen, 1993). Most of these reviews emphasized the pathology, epidemiology, and classical virology. During the last 10 years, our knowledge of the molecular biology of feline coronaviruses has increased considerably. We will summarize clinicopathological findings and the history of FIP research only briefly. The focus will be on the molecular aspects of feline coronaviruses and of FIP pathogenesis.

## II. CLINICAL SIGNS AND PATHOLOGY

FIP is a progressive, debilitating lethal disease of domestic and wild Felidae. The disease is characterized by disseminated perivascular pyogranulomatous inflammation and exudative fibrinous serositis in the abdominal and thoracic cavities. The initial signs of naturally occurring FIP are not very characteristic. The affected cats show anorexia, chronic fever, and malaise. Occasionally, ocular and neurological disorders occur. In classical "wet" or effusive FIP these signs are accompanied by a gradual abdominal distension due to the accumulation of a viscous yellow ascitic fluid. The quantity of fluid can vary from a few

RAOUL J. DE GROOT AND MARIAN C. HORZINEK • Virology Division, Department of Infectious Diseases and Immunology, University of Utrecht, 3584 CL Utrecht, The Netherlands.

*The Coronaviridae*, edited by Stuart G. Siddell, Plenum Press, New York, 1995.

milliliters to well over a liter. There is also the "dry" or noneffusive form of FIP where little or no exudate is present. The wet and dry forms of FIP are different manifestations of the same infection (Montali and Strandberg, 1972). Gross FIP lesions appear as multiple grayish-white nodules (<1 to 10 mm) in the serosal membranes, liver, lungs, spleen, omentum, intestines, and kidneys (Wolfe and Griesemer, 1971; Montali and Strandberg, 1972). Microscopic lesions consist of disseminated foci of necrosis and pyogranulomatous inflammation, frequently located around smaller vessels. These lesions are characterized by accumulations of fibrin and necrotic debris and by perivascular infiltrations of macrophages, neutrophils, and lymphocytes (Wolfe and Griesemer, 1971; Montali and Strandberg, 1972; Hayashi *et al.*, 1977; Weiss *et al.*, 1980).

## III. DISCOVERY AND EARLY STUDIES

The early 1960s are usually quoted as the period when FIP was first recognized (Holzworth, 1963; Feldmann and Jortner, 1964). However, the disease was probably seen earlier. In 1912/13, a case of conspicuous abdominal distension due to ascites formation in a domestic cat was reported (Jakob, 1914); the retrospective diagnosis is supported by the description of fever, high specific gravity of the abdominal fluid containing many granulocytes, and ophthalmological signs in this animal (Fig. 1).

The infectious nature of FIP was established by Wolfe and Griesemer (1966). These authors showed that FIP could be produced in specific pathogen-free cats by intraperitoneal inoculation of ascitic fluid collected from diseased cats. Zook *et al.* (1968) and Ward *et al.* (1968) were the first to present evidence supporting a viral etiology. Filtration studies indicated that the causative agent could pass 200-nm pores. In areas of inflammation, viral particles were observed by electron microscopy within or budding into the endoplasmic reticulum of macrophagelike cells (Zook *et al.*, 1968; Ward *et al.*, 1968; Ward, 1970; Pedersen, 1976a). The viral etiology was formally proven by transmission experiments using virus grown in autochthonous peritoneal macrophage cultures (Pedersen, 1976a) and virus suspensions purified by density gradient techniques (Horzinek *et al.*, 1977).

The morphology and morphogenesis of FIP virus (FIPV) was typical of a coronavirus (Zook *et al.*, 1968; Ward, 1970). Serological studies revealed an antigenic relationship between FIPV and established members of the Coronaviridae family, such as porcine transmissible gastroenteritis virus (TGEV), canine coronavirus (CCV), and the human coronavirus (HCV) 229E (Reynolds *et al.*, 1977; Witte *et al.*, 1977; Pedersen *et al.*, 1978; Horzinek *et al.*, 1982). In fact, CCV, TGEV, and FIPV are antigenically so similar that they may be regarded as host range mutants rather than as separate species (Horzinek *et al.*, 1982). Consistent with this view, inoculation of FIPV causes TGE-like lesions in the small intestine of piglets (Woods *et al.*, 1981). Furthermore, FIP-like pyogranulomatous lesions were found in cats experimentally infected with the CCV strain Insvac-1 (McArdle *et al.*, 1992). Another coronavirus of pigs, the porcine epidemic diarrhea virus (PEDV), was found to be antigenically related to FIPV;

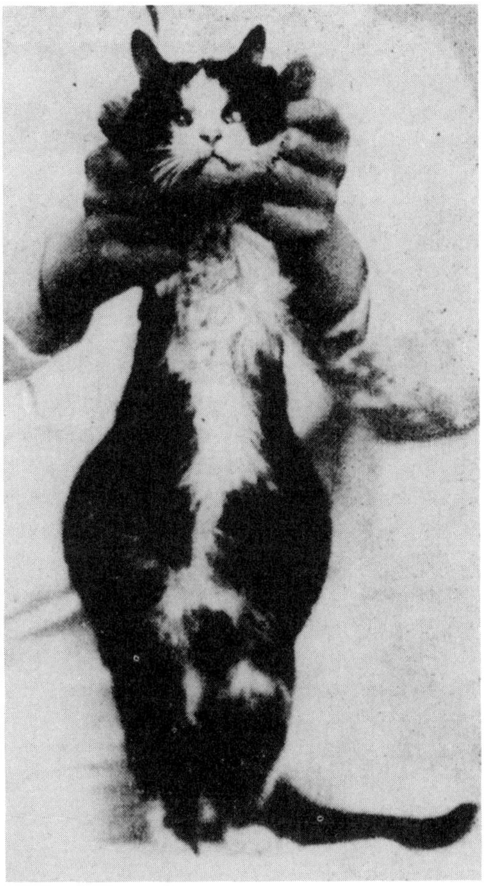

FIGURE 1. Probably the earliest case of FIP documented in the literature (at the State Veterinary School in Utrecht, 1912/13); the retrospective diagnosis is plausible from the description of a chronic exudative peritonitis, dyspnea (pleuritis?), fever, and eye symptoms (Jakob, 1914).

serological cross-reactions were confined to the nucleocapsid protein (Zhou *et al.*, 1988).

The isolation of feline coronaviruses (FCoVs) from naturally infected cats is notoriously difficult, probably due to their poor *in vitro* growth; FCoV strains isolated so far are listed in Table I. Since initial attempts to grow the virus in embryonated eggs, primary cells, and continuous cell lines were not successful, FCoV strains were adapted to and propagated in the brains of suckling mice, rats, and hamsters (Osterhaus *et al.*, 1978a,b). *In vitro* propagation was first reported in macrophage cultures (Pedersen, 1976a). A breakthrough was the finding that FCoVs could be grown in continuous lines and primary cultures of feline fetal cells (O'Reilly *et al.*, 1979; Black, 1980; Evermann *et al.*, 1981; Pedersen *et al.*, 1981a). Currently, FCoVs are mostly grown in Crandell feline kidney (CrFK) cells, feline embryonic lung (FEL) cells (O'Reilly *et al.*, 1979), and the line of *Felis catus* whole fetus (fcwf) cells developed by Niels Pedersen; the latter exhibit macrophagelike features at early passage levels (Jacobse Geels and Horzinek, 1983).

TABLE I. Feline Coronavirus Isolates

| Strain | Presumptive serotype | Reference |
|---|---|---|
| FIPV UCD1 (NW1) | I | Pedersen *et al.* (1981a) |
| FIPV UCD2 | I | Pedersen and Floyd (1985) |
| FIPV UCD3 | I | Pedersen and Floyd (1985) |
| FIPV UCD4 | I | Pedersen and Floyd (1985) |
| FIPV UCD5 | ? | Pedersen (personal communication) |
| FIPV UCD6 | ? | Pedersen (personal communication) |
| FECV UCD | I | Pedersen *et al.* (1981b) |
| FIPV TN-406 | I | Black (1980) |
| FIPV Yayoi | I | Hayashi *et al.* (1981) |
| FIPV Dahlberg | ? | Osterhaus *et al.* (1978a) |
| FIPV KU-2 | I | Hohdatsu *et al.* (1991b) |
| FIPV 79-1146 | II | McKeirnan *et al.* (1981) |
| FIPV NOR15 (DF2) | II | Everman *et al.* (1981) |
| FECV 79-1683 | II | McKeirnan *et al.* (1981) |
| FIPV Cornell-1 | II | Scott (1987) |
| FIPV KU-1 | II | Hohdatsu *et al.* (1991b) |
| FIPV Wellcome | ? | O'Reilly *et al.* (1979) |

## IV. EPIZOOTIOLOGY AND EPIDEMIOLOGY

Although FIP is a disease mainly of domestic cats, it has been reported in several wild felids, such as the lion (Colby and Low, 1970), leopard (Tuch *et al.*, 1974), European wildcats (Watt *et al.*, 1993), caracal, and lynx (Poelma *et al.*, 1971). The cheetah (*Acinonyx jubatus*) is extremely sensitive FIP, a peculiarity attributed to the genetic uniformity in this species (O'Brien *et al.*, 1985).

FIP occurs in domestic cats of both sexes at about the same frequency (Pedersen, 1976c). The disease occurs most commonly in young cats between 6 months and 2 years of age. In cats between 5 and 13 years of age, FIP is less prevalent; but there appears to be an increased incidence in cats older than 14 years (Pedersen, 1976c, 1983a,b; Addie and Jarrett, 1992a). Close contact between cats is required for effective transmission. Virus is shed from the oropharynx and in the feces (Stoddart *et al.*, 1988a,b). Therefore, the infection most likely results from ingestion and/or inhalation of the virus.

Epidemiological studies showed that FCoVs have a worldwide distribution (Horzinek and Osterhaus, 1979) and are widespread in the cat population. Antibodies directed against FIPV or a closely related coronavirus are found in 80 to 90% of the cats in catteries and in 10 to 50% of the cats in single cat households (Pedersen, 1976c; Loeffler *et al.*, 1978; Sparkes *et al.*, 1991, 1992; Addie and Jarrett, 1992a,b). The disease itself, however, occurs sporadically: only 5 to 10% of the seropositive cats actually develop FIP (Pedersen, 1976b; Addie and Jarrett, 1992a,b). These findings seem at odds with the fact that certain FIPV strains such as 79-1146 cause FIP in almost 100% of the experimentally infected cats. To explain this discrepancy, it was initially proposed that most cats in the field become infected with an avirulent virus closely related to FIPV. The isolation of two strains, UCD (Pedersen *et al.*, 1981b) and 79-1683

(McKeirnan *et al.*, 1981), that only cause a mild enteric infection but no FIP seemed to support this view. These avirulent viruses were designated feline enteric coronaviruses (FECVs) (Pedersen *et al.*, 1981b, 1984; Pedersen, 1983a). There is now, however, ample evidence that FECV and FIPV are merely virulence variants of the same virus. The FECV isolates are antigenically (Pedersen *et al.*, 1983; Boyle *et al.*, 1984; Fiscus and Teramoto, 1987a; Hohdatsu *et al.*, 1991a,b; W. V. Corapi, personal communication) and genetically (Herrewegh, Horzinek, Rottier, and de Groot, in preparation; see Section VI) indistinguishable from FIPV strains. The antigenic differences between FIPV 79-1146 and FECV 79-1683 (Fiscus *et al.*, 1987; Hohdatsu *et al.*, 1991b) simply reflect FCoV strain variations and do not correlate with their respective pathogenic properties. FECV 79-1683 is antigenically (Pedersen *et al.*, 1984; Hohdatsu *et al.*, 1991a,b; W. V. Corapi, personal communication) and genetically (Herrewegh, Vennema, Horzinek, Rottier, and de Groot, in preparation; see Section VI) more similar to FIPV 79-1146 than to FECV strain UCD. Moreover, attenuated FIPV strains with properties similar to FECV are readily obtained (Pedersen and Black, 1983; Barlough and Stoddart, 1990), and it is now recognized that even FIP-inducing strains vary dramatically in their pathogenicity (Pedersen and Floyd, 1985; Pedersen, 1987b). In the field, highly virulent strains may be as uncommon as completely avirulent strains. It is therefore more appropriate to consider all strains under the general category of FCoV (Barlough and Stoddart, 1990; Addie and Jarrett, 1992b).

The outcome of an FCoV-infection not only depends on the virus strain but also on the infective dose and the route of inoculation (Pedersen *et al.*, 1981a; Pedersen and Floyd, 1985), the age of the host, concurrent viral infections [e.g., with feline leukemia virus (FeLV)] (Hardy and Hurvitz, 1971; Cotter *et al.*, 1975; Pedersen and Floyd, 1985; Pedersen, 1987b), "stress" (Pedersen, 1976b), and, possibly, the genetic predisposition of the host (O'Brien *et al.*, 1985; Addie and Jarrett, 1992a,b). The combined data suggest that FIP is an infrequent manifestation of a common, inapparent infection (Loeffler *et al.*, 1978; Horzinek and Osterhaus, 1979; Pedersen, 1976b; Addie and Jarrett, 1992a,b). From an evolutionary point of view, viruses that cause low morbidity and low mortality have a survival advantage, especially if they can establish persistent infections. There is increasing evidence for the occurrence of asymptomatic carriers of FCoV. Kittens exposed to healthy, FCoV-seropositive cats seroconvert within 2 to 10 weeks (Pedersen *et al.*, 1981b; Addie and Jarrett, 1992a); a number of these kittens subsequently develop FIP. Strong support for the existence of a FCoV carrier state stems from a series of experiments in which FIPV-immune cats were immunosuppressed by superinfection with FeLV. Thus, FIP could be induced in cats that were kept in strict isolation up to 4 months after exposure to FIPV (Pedersen and Floyd, 1985; Pedersen, 1987b). The reverse transcription–polymerase chain reaction (RT-PCR) detection of FCoV in the feces and plasma of healthy, seropositive cats (Herrewegh *et al.*, 1995) provides further support for a FCoV carrier state.

The epidemiological findings fit a model in which most FCoVs, though in principle pathogenic, behave as innocent commensals. The host's immune system (presumably the cellular arm) (Pedersen, 1987b) keeps the infection in check but is unable to clear the virus completely. Thus, upon natural infection disease would only ensue under exceptional circumstances, i.e., when (1) the host is exposed to a high virus dose, (2) the host is unable to develop a protective immune response, or (3) an immunocompetent host is exposed to factors that compromise its immune system.

## V. MOLECULAR BIOLOGY OF FCoV

As for most other coronaviruses, FCoV virions are composed of three main structural protein species: a 45-kDa nucleocapsid protein N, a 25- to 30-kDa matrix protein M, and a 180- to 210-kDa peplomer protein S (Horzinek *et al.*, 1982; Boyle *et al.*, 1984; de Groot *et al.*, 1987a). Virions of TGEV also contain a small membrane protein (sM) of 10 kDa (Godet *et al.*, 1992). Recent data obtained in our laboratory indicate that sM is present in FCoV virions as well (Vennema, Godeke, Horzinek, and Rottier, in preparation). Further details on the characteristics of the structural proteins and coronavirus assembly will be presented elsewhere in this volume (Chapters 5–9).

The FCoV genome is a positive-stranded RNA molecule with an estimated length of about 30 kb. Thus far, the FCoV strain best characterized at the molecular level is FIPV 79-1146. Initially, six virus-specific, poly(A)-containing RNAs were found in infected cells, with lengths ranging from 1.4 to more than 20 kb. An identical set of RNAs was found in cells infected with FIPV strain NOR15 (de Groot *et al.*, 1987a). Recently, we have obtained evidence for a seventh virus-specific RNA species (see below).

By *in vitro* translation, subgenomic RNAs of 3.8 and 2.8 kb were shown to encode the M and N protein, respectively (de Groot *et al.*, 1987a). Sequence analysis of cDNA clones (de Groot *et al.*, 1987b,c) identified the 9.6-kb RNA species to encode the S protein. Of the FIPV 79-1146 genome, the nucleotide sequence downstream of the polymerase gene has now been completed (de Groot *et al.*, 1987b, 1988; Vennema, de Groot, and Spaan, unpublished results). The gene organization is identical to that of CCV strain Insavc-1 (Horsburgh *et al.*, 1992). In addition to the genes encoding N, M, S, and sM, there are five open reading frames (ORFs) (Fig. 2). The sequence 5′ AACUAAAC 3′ is interspersed between the ORFs; the complementary sequence, 3′ GUUUAGUU 5′, is thought to function as a promoter for subgenomic RNA synthesis (see Chapter 2).

Interestingly, the sM gene is preceded by the partial promoter sequence 3′ GUUUAG 5′ (Vennema, de Groot, and Spaan, unpublished results), suggesting that it may be translated from a separate, hitherto undetected RNA species (de Groot *et al.*, 1987a, 1988). Upon separation of FIPV RNAs in 1.5% agarose gels, a minor 4.0-kb RNA species indeed was found. The identity of this RNA was confirmed by RT-PCR using oligonucleotide primers located in ORF4 and the FIPV leader RNA, followed by sequence analysis (Mijnes and de Groot, unpublished results). The mRNA for sM is made in very low amounts, and in 1% agarose gels it is masked by the far more predominant 3.8-kb RNA species. In view of these data it seems appropriate to change the nomenclature of the FIPV

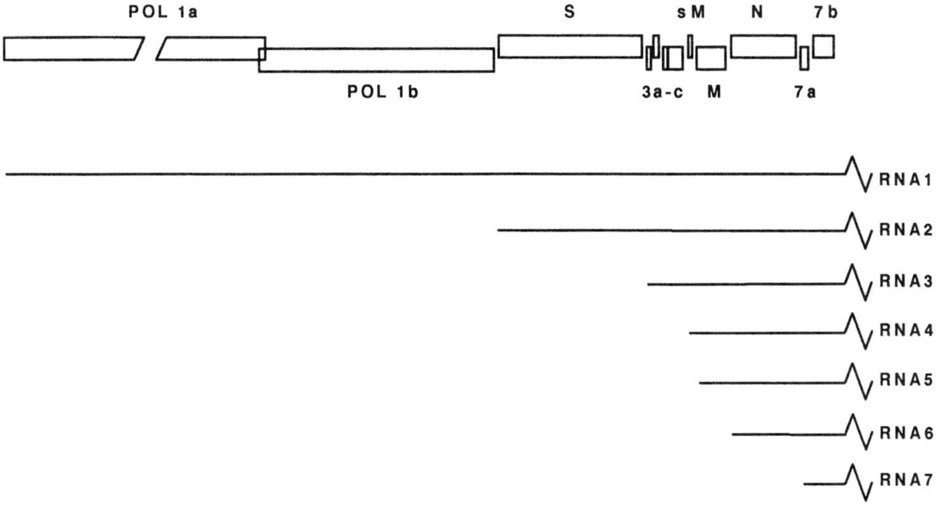


FIGURE 2. Organization and expression of the genome of FIPV strain 79-1146. Open reading frames are indicated by boxes. The structure of the seven FIPV RNAs is also shown.

RNAs as indicated in Fig. 2. The ORFs at the 3′ end of the FIPV genome should thus be referred to as 7a and 7b (formerly 6a and 6b), according to nomenclature recommended by the Coronavirus Study Group (Cavanagh *et al.*, 1990).

As illustrated in Fig. 2, RNAs 3 and 7 are potentially polycistronic. ORFs 3a–c and ORFs 7a and 7b are thought to encode nonstructural proteins. It is still unclear, however, whether all of these ORFs, in particular ORFs 3b and c, are actually expressed (Horsburgh *et al.*, 1992). The ORF7a homologue of TGEV contains a 69 nucleotide (nt) deletion compared to ORF7a of FCoV and CCV (de Groot *et al.*, 1988; Vennema *et al.*, 1992b; Horsburgh *et al.*, 1992). Its product, a 9.1-kDa hydrophobic protein, has been detected in TGEV-infected cells (Garwes *et al.*, 1989; Tung *et al.*, 1992). It therefore is likely that ORF7a is expressed in FCoV-infected cells as well.

ORF7b is present only in the genomes of FCoV and CCV (Kapke and Brian, 1986; de Groot *et al.*, 1988; Vennema *et al.*, 1992b; see Chapter 3). It codes for a nonvirion secretory glycoprotein, gp7b (Vennema *et al.*, 1992a). The glycoprotein gp7b is not required for virus replication *in vivo* and *in vitro*: FECV strain 79-1683 contains a large deletion in ORF7b (Vennema *et al.*, 1992b), but replicates to high titers in tissue culture cells and causes enteritis upon experimental infection of cats. Also, ORF7b is readily lost upon *in vitro* passage of other FCoV strains: large deletions in this gene were found in high passage stocks of strains TN406 and UCD4, in UCD2 and in the DF2-derived *Primucell* vaccine strain (Herrewegh, Vennema, Horzinek, Rottier, and de Groot, in preparation). Still, with the exception of strain 79-1683, an intact ORF7b is present, at least at low passage numbers, in all nine FCoV laboratory strains studied thus far, including the nonvirulent FECV strain UCD. Moreover, no deletions were found in ORF7b of 16 field strains as determined by RT-PCR analysis on fecal samples and tissues (Herrewegh *et al.*, 1995). It therefore appears that although the 7b gene is dispensible for replication in tissue culture cells and enterocytes, its presence provides a selective advantage during natural infection.

The function of gp7b remains enigmatic. In the infected cell, the protein is present only in the lumen of the endoplasmic reticulum (ER) and the Golgi (Vennema *et al.*, 1992a). An involvement in cytoplasmic processes such as RNA replication can therefore be excluded. Presumably, gp7b exerts its function extracellularly, perhaps as a virus-encoded modifier of the immune response and/or inflammatory reaction (Gooding, 1992). The loss of gp7b may well contribute to the reduced virulence of FECV 79-1683, high passage TN406 (Pedersen and Black, 1983), FIPV UCD2, and the *Primucell* vaccine strain.

## VI. SEROLOGICAL AND GENETIC RELATIONSHIPS OF FCoV ISOLATES

As first noted by Pedersen *et al.* (1983), FCoVs can be allocated to at least two serotypes on the basis of *in vitro* neutralization. Sera from cats experimentally infected with so-called type I FCoVs (FIPV TN-406 or FECV UCD) neutralized other type I FCoVs but not type II FCoVs (79-1146 and 79-1683) and vice versa. The type II strains appeared to be more closely related to CCV and TGEV, i.e., immunodominant neutralization epitopes shared by the spike proteins of TGEV, CCV, and the type II FCoV strains seemed to be absent in the type I strains (Pedersen *et al.*, 1983). These findings were confirmed and extended in studies using monoclonal antibodies (MAbs) (Hohdatsu *et al.*, 1991a,b, 1992; W. V. Corapi, personal communication). Strains DF-2 (NOR15), CU1 and KU-1 were tentatively identified as type II viruses, whereas UCD-1, -2, -3, and -4, and Yayoi and KU-2 were assigned to serotype I (Table I) (Hohdatsu *et al.*, 1991a,b; W. V. Corapi, personal communication). Strain UCD-2 is interesting in that a number of epitopes on the spike protein, shared by other group I strains, are absent (Fiscus and Teramoto, 1987a,b; W. V. Corapi, personal communication). This led Fiscus and Teramoto to suggest that UCD-2 represents a separate antigenic type (Fiscus and Teramoto, 1987a).

There is some confusion in regard to the classification of FCoV strain UCD1. According to some reports, UCD1 belongs to type I (Pedersen *et al.*, 1983; Hohdatsu *et al.*, 1991a,b). Other studies suggest a closer antigenic relationship to type II strains (Corapi *et al.*, 1992). As it now appears, virus stocks may have been interchanged. The alleged UCD1 strain used by Scotts' group (Corapi *et al.*, 1992; Olsen *et al.*, 1992), is antigenically (W. V. Corapi, personal communication) and genetically more similar to 79-1146 (type II) than to the original type I UCD1 (NW1) strain from Niels Pedersen.

Although the data are limited and rather sketchy, type I FCoV strains seem to predominate in the field. Of 20 sera from natural FIP cases in the United States, only one contained neutralizing antibodies directed against FIPV strain 79-1146 (type II). The remaining 19 antisera neutralized TN-406 (type I) with titers ranging from 40 to 3200 (Pedersen *et al.*, 1983). Hohdatsu *et al.* (1992)

screened 237 FCoV-positive sera, as determined by immunofluorescent assay (IFA), by using a neutralization assay with FIPV strain 79-1146 as the test virus and by a competition enzyme-linked immunosorbent assay (ELISA) using a type II-specific MAb. Their results suggest that type II viruses account for only 20–30% of the natural FCoV infections in Japan.

The type I and II viruses also appear to differ in their *in vitro* growth characteristics. The type I strains grow poorly in tissue culture, yielding titers of cell-free virus of only 0.5 to $1 \times 10^5$ TCID50/ml. More than 90% of the infectivity remains cell-associated (Pedersen *et al.*, 1983; Pedersen and Floyd, 1985; Hohdatsu *et al.*, 1991b). In contrast, type II FCoVs grow to high titers of cell-free virus (up to $5 \times 10^7$ TCID50/ml) and also in these properties closely resemble CCV (Pedersen *et al.*, 1983).

One interpretation of the data is that the FCoV are a heterogeneous group, with the type II viruses being genetically more closely related to CCV than to type I strains (Pedersen *et al.*, 1983). The fact that some CCV strains induce FIP upon experimental inoculation of cats (McArdle *et al.*, 1992) would lend further credence to this idea. To study this issue, we compared the 3′-most transcription unit comprising ORFs 7a, 7b, and the 3′ nontranslated sequences (NTR) of nine different FCoV isolates. Nucleotide sequence similarities ranged from 87 to 99% (Herrewegh, Vennema, Horzinek, Rottier, and de Groot, in preparation). As shown in Fig. 3, types I and II cannot be distinguished in this part of the genome. Clearly, they are much more related to each other than to the CCV strains sequenced thus far.

From serological studies, it appears that the differences between type I and II FCoVs are mostly located in the S-protein (Fiscus and Teramoto, 1987a,b; Hohdatsu *et al.*, 1991a,b; W. V. Corapi, personal communication). Very recently, these findings were confirmed by sequence analysis. The spike genes of type II FCoVs bear much greater resemblance to those of TGEV and CCV than to the spike genes of type I isolates KU-2 (Motokawa *et al.*, 1995) and UCD3 (Vennema *et al.*, 1995). How can these observations be reconciled with those made for the ORF7a/7b transcription unit? During coronavirus replication, homologous RNA recombination occurs at a high frequency. It is quite conceivable that the type II FCoV strains have arisen by RNA recombination between a type I FCoV and CCV (Vennema *et al.*, 1995; Herrewegh *et al.*, in preparation).

Figure 3 also shows that the avirulent "FECV" strains UCD and 79-1683 are more similar to pathogenic "FIPV" strains than to each other. These findings again indicate that the virulent and avirulent FCoVs are not separate virus species.

## VII. ANTIBODY-DEPENDENT ENHANCEMENT

The key pathogenic event in FIP is the infection of monocytes and macrophages (Ward, 1970; Pedersen, 1976a). Avirulent FCoV strains remain confined to the digestive tract and usually do not spread beyond the intestinal epithelium and regional lymph nodes (Pedersen *et al.*, 1981b, 1984). The virulent strains,

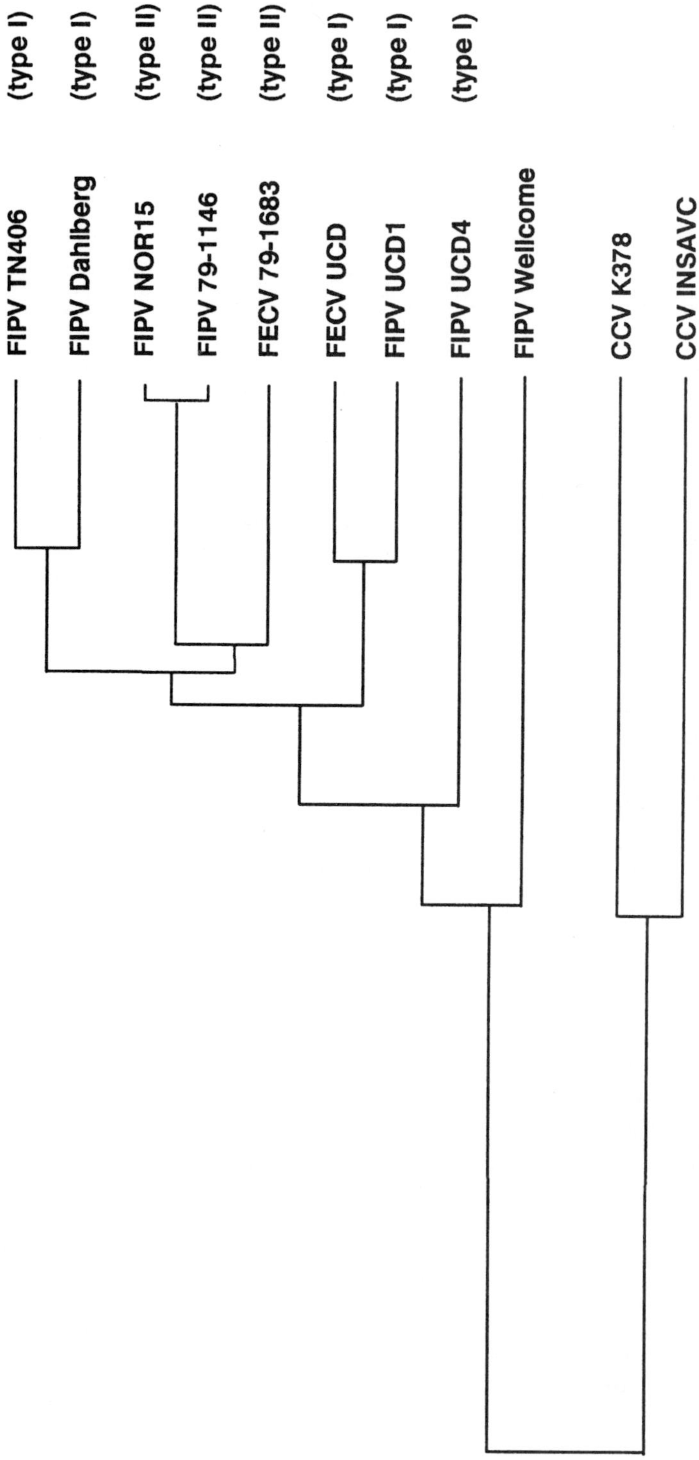


FIGURE 3. A dendrogram showing the sequence similarities between FCoV and CCV strains. The nucleotide sequence of the 3′-most transcription unit, containing ORFs 7a and 7b, was determined for nine different FCoV strains (Herrewegh *et al.*, in preparation). The FCoV and CCV sequences were aligned and the tree was constructed using unweighted pair-group method using arithmetic averages (Sneath and Sokal, 1973). The distance along the horizontal axis is proportional to the differences between sequences.

however, disseminate to other organs most likely via blood-borne monocytes (Weiss and Scott, 1981a). The virulence of FCoV strains appears to correlate with their ability to infect feline peritoneal macrophages *in vitro*. When virulent and avirulent strains were compared, the latter infected fewer macrophages and produced lower virus titers than the virulent strains. Moreover, the avirulent strains were less able to sustain viral replication and to spread to other macrophages (Stoddart and Scott, 1989).

There is ample evidence for an involvement of the immune system in the pathogenesis of FIP. Several authors have emphasized the similarities between FIP and immune-mediated diseases such as Aleutian disease of mink and Dengue hemorrhagic fever (Horzinek and Osterhaus, 1979; Pedersen and Boyle, 1980; Weiss and Scott, 1981b; Halstead, 1988). Humoral immunity is obviously not protective. FCoV-seropositive cats that are experimentally infected with FIPV often develop an accelerated, fulminating course of the disease. Clinical signs and lesions develop earlier, and the mean survival time is dramatically reduced compared to seronegative cats (early death syndrome) (Pedersen and Boyle, 1980; Weiss *et al.*, 1980; Weiss and Scott, 1981a–c). Direct evidence for the involvement of antibodies in early death syndrome was obtained by passive immunization of cats with the purified immunoglobulin IgG fraction of feline αFCoV-antisera. These cats also developed enhanced FIP upon an experimental challenge infection (Pedersen and Boyle, 1980).

Heterologous expression of S, M, and N provided a means to assess the role of each individual protein in FIP pathogenesis (de Groot *et al.*, 1989; Vennema *et al.*, 1990b, 1991). S was unequivocally identified as the main, if not the sole, factor in the induction of early death (Vennema *et al.*, 1990b, 1991). Kittens that had been infected with a recombinant vaccinia virus (recVV) expressing S died within 9 days after an FIPV challenge infection. In contrast, kittens that had been infected with wild-type vaccinia virus or with recVVs expressing M or N either survived FIPV infection or succumbed after a protracted course of the disease (21–41 days) (Vennema *et al.*, 1990b, 1991).

The early death syndrome can be explained by an antibody-dependent enhancement of virus infection (ADE), a phenomenon described for a wide variety of viruses, including the human immunodeficiency virus (Takeda *et al.*, 1988; for reviews, see Porterfield, 1986; Halstead, 1988). Binding of antibody to virus produces infectious immune complexes that attach to the surface of the target cell (via Fc or complement receptors) with higher efficiency than virus alone (Chanas *et al.*, 1982; Burstin *et al.*, 1983; Gollins and Porterfield, 1984, 1985; Halstead and O'Rourke, 1977; Peiris and Porterfield, 1979; Peiris *et al.*, 1981).

Cheryl Stoddart (1989) provided the first *in vitro* evidence for an involvement of ADE in FIP by using feline peritoneal macrophage cultures (Stoddart and Scott, 1988). The number of macrophages infected with FIPV UCD-1* *in vitro* increased up to 12-fold when the virus had been preincubated with serum from an FIPV-sensitized cat. This enhancement also occurred when using the high-pressure liquid chromatography (HPLC)-purified IgG-fraction of the serum. ADE could be prevented by blocking the macrophage Fc receptor with aggregated IgG or by blocking the Fc portion of the enhancing IgG with protein

A (Stoddart, 1989). The levels of enhancement varied greatly between experiments with different batches of macrophages; similar difficulties have been experienced in ADE studies with peripheral blood leukocyte preparations (Porterfield, 1986).

Recently, these observations were confirmed and extended by Hohdatsu *et al.* (1991c) and by Fred Scott's group (Olsen *et al.*, 1992; Corapi *et al.*, 1992). In these studies, ADE in primary feline macrophages was demonstrated both with feline antisera (Olsen *et al.*, 1992; Olsen and Scott, 1993) and with murine monoclonal antibodies (Hohdatsu *et al.*, 1991c; Olsen *et al.*, 1992; Corapi *et al.*, 1992). The variation between experiments was minimized by using batches of macrophages obtained from a single cat (Olsen *et al.*, 1992). Upon screening of 67 MAbs specific for S, M, and N, 17 MAbs were identified that enhanced FIPV infection (Olsen *et al.*, 1992, 1993; Corapi *et al.*, 1992). Consistent with the data of Vennema *et al.* (1990b, 1991), the ADE-inducing MAbs were all directed against the S protein (Olsen *et al.*, 1992; Corapi *et al.*, 1992). There was a distinct correlation between the ability of the MAbs to neutralize infection of CrFK cells and the induction of ADE: 15 of 19 neutralizing antibodies enhanced FIPV infection. All but one of the enhancing MAbs were of the immunoglobulin G2a subclass; the four neutralizing MAbs that did not induce ADE were of the G1 subclass. From these findings Corapi *et al.* (1992) suggested a restriction in the immunoglobulin subclasses mediating ADE. The findings of Hohdatsu *et al.* (1991c) seem to conflict with those of all other authors in that nonneutralizing MAbs directed against both M and S induced ADE. However, the enhancement levels obtained with these MAbs were very low (two- to threefold), while in the studies of Olsen *et al.* (1992) and Corapi *et al.* (1992) some MAbs produced up to a 100-fold increase in the number of infected cells. The most convincing enhancement reported by Hohdatsu *et al.* (1991c) (a sixfold increase in the number of infected cells) notably occurred with a neutralizing S-specific MAb of the immunoglobulin G2a subclass.

There are at least five distinct neutralization sites on the S protein of FIPV, four of which have been implicated in ADE (Corapi *et al.*, 1992). By kinetics-based competitive ELISA, two of the enhancement sites were found to correlate with the previously defined sites A and E/F of the TGEV S protein (Olsen *et al.*, 1992). It was also shown that αTGEV MAbs directed against these sites enhanced FIPV infection *in vitro* (Olsen *et al.*, 1993). By sequence analysis of MAb-resistant mutants and PEPSCAN epitope mapping, amino acid residues involved in the formation of TGEV site A have been localized between amino acid residues 538 and 591 (Correa *et al.*, 1990; Gebauer *et al.*, 1991). The epitopes of the enhancing FIPV site A MAbs are probably located in the homologous region (residues 543–597) (Jacobs *et al.*, 1987) on the FIPV spike (Olsen *et al.*, 1993).

Olsen *et al.* (1992) noted strain variations of ADE in that certain MAbs would enhance FIPV UCD-1* more than 79-1146 and vice versa. This is most likely related to the epitopes recognized by the enhancing MAbs and their relative avidity. These observations may in part explain why FIPV does not always cause antibody-mediated early death in seropositive cats; thus, early death did not occur in kittens that had been inoculated with either FECV UCD, FECV 79-1683, or FIPV UCD-2 prior to a lethal challenge with FIPV 79-1146

(Pedersen *et al.*, 1984; Pedersen and Floyd, 1985; Fiscus *et al.*, 1987). Of the two kittens vaccinated with FIPV UCD-4 and challenged with a lethal dose of FIPV 79-1146, only one developed enhanced FIP (Pedersen and Floyd, 1985; Pedersen, 1987b). Similarly, of four kittens immunized with FIPV UCD-4, only two died of accelerated FIP after a challenge infection with FIPV UCD-1; one kitten died after the normal, protracted course of the disease, and the other one survived. The combined data indicate that a particular virus–antibody combination is required for the development of ADE and early death syndrome.

The mechanism of ADE in FIPV remains to be elucidated. An important question that needs to be addressed is whether the dedicated cell surface receptor of FIPV is involved. Scott and co-workers have argued that the enhanced infection occurs via a very efficient form of FcγR-mediated endocytosis that is independent from the FIPV receptor (Corapi *et al.*, 1992; Olsen *et al.*, 1992). This hypothesis is based upon (1) the "sheer magnitude" of the increase in the number of infected cells due to ADE, (2) the observation that the macrophagelike murine IC-21 cells could be infected with FIPV only in the presence of an enhancing MAb (Corapi *et al.*, 1992), and (3) the as yet unconfirmed claim that ADE can be blocked by lysosomotropic amines (Stoddart, 1989). In this model, the nucleocapsid would be released into the cytoplasm by fusion of the viral envelope and the endosomal membrane (illustrated in Fig. 4B).

The above model is difficult to reconcile with other observations made for coronavirus entry. During normal entry, membrane fusion is mediated by the viral spikes; it occurs readily at neutral pH, but is inhibited at lower pH values. Therefore, coronaviruses most likely enter the host cell by fusion at the plasma membrane rather than via the endosomal route (see also Chapter 5). Furthermore, heterologously expressed spike proteins induce membrane fusion in a host cell-dependent manner: the spike protein of FIPV only induced cell fusion in cells of feline origin. Similarly, fusion by the S protein of mouse hepatitis virus was restricted to murine cells (de Groot *et al.*, 1989; Vennema *et al.*, 1990a). These data indicate that membrane fusion is a highly specific process, most likely requiring recognition of the cell surface receptor by the viral spikes (Fig. 4A).

Corapi's experiments with the murine IC-21 cells (Corapi *et al.*, 1992) would suggest that during ADE FcγRs suffice for viral entry. However, it cannot be excluded that IC-21 cells carry a low affinity receptor for FIPV that is only recognized efficiently when virus–antibody complexes are fixed by the FcγRs at the plasma membrane. Support for the existence of FIPV receptors on murine cells stems from studies in which FIPV strains were adapted to murine Sac$^-$ cells (de Groot, unpublished results) or propagated in brains of suckling mice (Osterhaus *et al.*, 1978a).

We currently favor a model for ADE in which viral entry is mediated by interaction between the spike and the virus receptor, where the probability of this event is increased due to a more efficient binding of the opsonized virus to the cell surface (explained in Fig. 4C). This mechanism would be similar to that found for human immunodeficiency virus (HIV): ADE of HIV infection requires both the FcγR and the HIV cell receptor, CD4 (Takeda *et al.*, 1990; Perno *et al.*, 1990; Zeira *et al.*, 1990; Connor *et al.*, 1991).

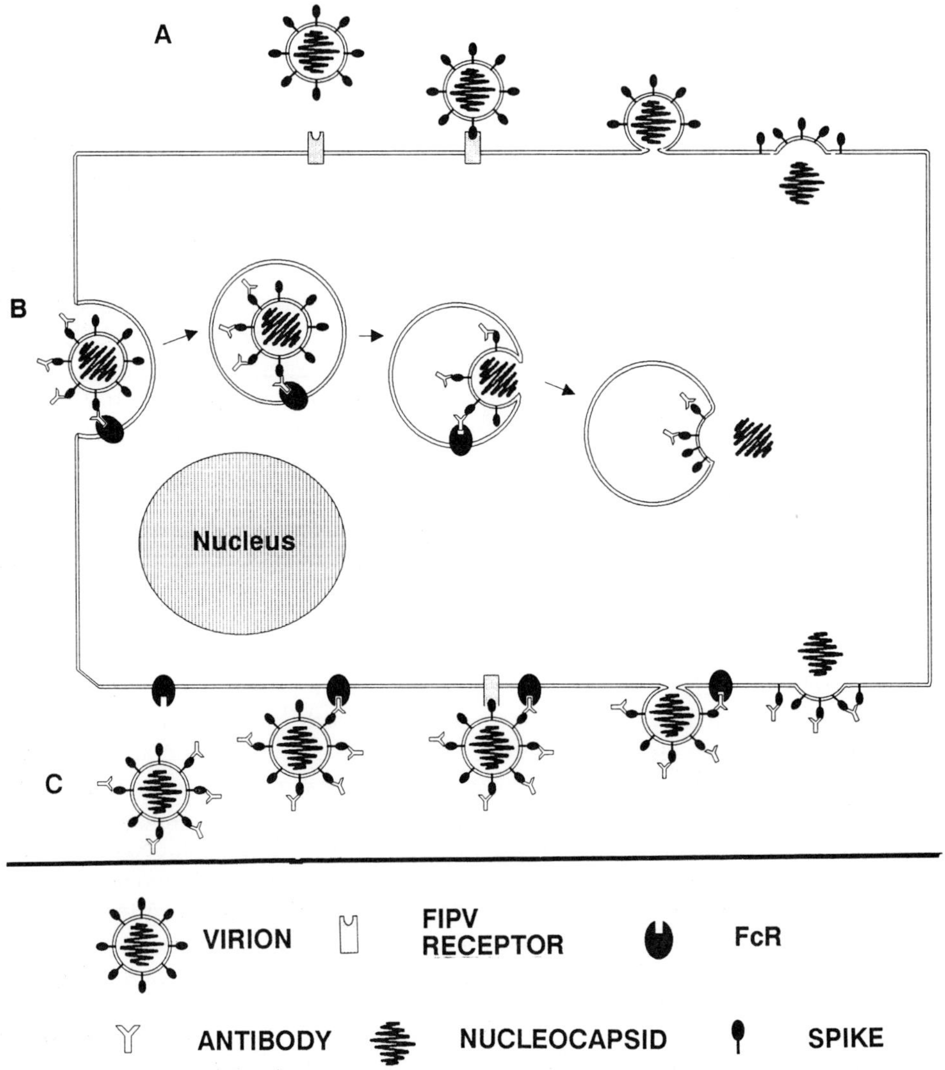


FIGURE 4. Models for viral entry during (A) normal infection and (B,C) ADE. (B) Model for ADE, in which entry occurs via the endosomal route. Adsorption and penetration requires only the Fc receptor. (C) Model for ADE, in which viral entry requires both the Fc receptor and the virus receptor.

Clearly, the receptor for FIPV must be identified before this issue can be resolved. Two closely related viruses, TGEV and HCV 229E, adsorb to and enter their host cell via the aminopeptidase N (APN) molecule, a membrane-bound metalloprotease (Delmas *et al.*, 1992; Yeager *et al.*, 1992). The tissue distribution of APN coincides with the cell tropism of FIPV; it is present on enterocytes and macrophages (Look *et al.*, 1989). It is thus quite likely that FIPV utilizes APN as a receptor as well.

## VIII. IMMUNE-MEDIATED LESIONS IN FIP

Most authors consider the vascular and perivascular lesions in FIP to be immune-mediated, but there is uncertainty about the actual pathogenic mechanism. At least some vascular injury may be attributed to immune-mediated lysis of infected cells (Weiss and Scott, 1981a): FIPV-infected white blood cells were detected in the lumen, intima and wall of veins and in perivascular locations (Weiss and Scott, 1981a; Weiss *et al.*, 1988). Furthermore, inflammatory mediators such as cytokines (Goitsuka *et al.*, 1988, 1990; Hasegawa and Hasegawa, 1991), leukotrienes, and prostaglandins (Weiss *et al.*, 1988) that are released by infected macrophages could play a pivotal role in the development of the perivascular pyogranulomata. As hypothesized by Weiss *et al.* (1988), these products could induce vascular permeability changes and provide additional chemotactic stimuli for neutrophils and monocytes. In response to the inflammation, the attracted cells may release additional mediators and cytotoxic substances; the monocytes would also serve as new targets for FIPV. The end result of this would be enhanced local virus production and increased tissue damage.

Other observations point toward an immune complex pathogenesis. Deposition of immune complexes and subsequent complement activation is thought to cause an intense inflammatory response that may extend across blood vessel walls. The resulting vascular damage would permit leakage of fluid into the intercellular space and eventually lead to the accumulation of thoracic and abdominal exudate (Hayashi *et al.*, 1977; Horzinek and Osterhaus, 1979; Jacobse-Geels *et al.*, 1980, 1982; Pedersen and Boyle, 1980; Weiss *et al.*, 1980; Weiss and Scott, 1981a–c). The morphological features of the vascular lesions (necrosis, polymorphonuclear cell infiltration associated with small veins and venules) strongly indicate an Arthus-type reaction (Hayashi *et al.*, 1977; Weiss and Scott, 1981c). The lesions contain focal deposits of virus, IgG, and C3 (Pedersen and Boyle, 1980; Weiss and Scott, 1981b). Moreover, complement depletion and circulating immune complexes were demonstrated in cats with terminal FIP (Pedersen and Boyle, 1980). In a horizontal study of six experimentally infected cats, first clinical signs were accompanied by increased C3 concentrations in the plasma; subsequently, antibody titers and circulating immune complexes (ICX) increased with concomitant decrease of complement concentrations. At the time of death, maximum ICX and minimum C3 concentrations were measured (Jacobse-Geels *et al.*, 1982). Glomerular ICX deposits were found in FIPV-infected cats but not in uninfected controls (Jacobse-Geels *et al.*, 1980).

## IX. VACCINE DEVELOPMENT

The development of vaccines against FIPV is cumbersome and has been frustrated by the occurrence of ADE. Vaccination with heterologous live virus vaccines (TGEV, CCV, HCV 229E) has failed to provide protection (Witte *et al.*, 1977; Toma *et al.*, 1979; Woods and Pedersen, 1979; Barlough *et al.*, 1984, 1985).

Occasionally, protective immunity can be induced by administering FECV, low virulent FIPV (e.g., FIPV UCD-3), or sublethal amounts of virulent FIPV (Pedersen and Black, 1983; Pedersen *et al.*, 1984; Pedersen and Floyd, 1985; Pedersen, 1987b). However, as quoted from Pedersen and Black (1983, p. 20), the results of immunization with homologous live virus vaccines were "too inconsistent and hazardous to have clinical relevance": while some cats became solidly immune, others developed disease or became sensitized. Also, cats immunized with recombinant vaccinia viruses expressing the spike gene of FIPV 79-1146 (Vennema *et al.*, 1990b) or of FECV 79-1683 (Bhogal, Martinez, Reed, KLepfer, Jones, Pfeiffer, and Miller, in preparation) developed early death syndrome upon an FIPV challenge infection.

A temperature-sensitive strain of FIPV has been shown to stimulate protective immunity in cats. Upon intranasal inoculation into cats, virus replication appeared to be restricted to the upper respiratory tract. The mutant did not differ from the wild-type virus with respect to RNA and protein synthesis, but produced little progeny at the nonpermissive temperature (Christianson *et al.*, 1989; Gerber *et al.*, 1990). There is still controversy over the safety and efficaciousness of this vaccine in that protection may depend on the strain and dose of the challenge virus (Olsen, 1993). Furthermore, even as avirulent and immunizing modified live FIPV strains become available, their widespread use as vaccines is not encouraged in view of the notorious genetic flexibility of coronaviruses. Thus it has been demonstrated that new epidemic strains of avian infectious bronchitis virus have arisen by point mutations in the genomes of attenuated vaccine strains (Kusters *et al.*, 1987). While this may be acceptable in production animals where new vaccine strains can be employed at intervals, the high individual value of a companion animal would plead against the use of live coronavirus vaccines. The possible risks are best illustrated by outbreaks of an unprecedented FIP-like disease in dogs after the widespread use of a modified live CCV vaccine (Martin, 1985).

Designing safe and effective FIP vaccines that avoid immunopathology will probably require a better understanding of the complex interplay between the virus and the host immune system and, perhaps just as important, identification of the viral proteins involved. The molecular cloning of the FIPV genome and the isolation of viral genes provides powerful tools to tackle these problems. Obviously, the S protein is not the most attractive candidate for vaccine development (Vennema *et al.*, 1990b; Bhogal, Martinez, Reed, Klepfer, Jones, Pfeiffer, and Miller, in preparation). The role of M in immunity against FIPV is still unclear. Upon immunization of eight cats with a vaccinia virus recombinant expressing M, three kittens survived, while two other kittens showed protracted survival times compared to the controls (Vennema *et al.*, 1991). Clearly, these findings deserve further scrutiny.

## X. CONCLUDING REMARKS

Substantial progress has been made in FIP research. Our knowledge of the molecular genetics of FCoVs has increased considerably as has our understanding of the mechanism of ADE and early death syndrome. Other areas would require closer examination, for instance, the role of cellular immunity in protection against FIP. As feline cytokine genes are cloned and expressed, new tools become available to modulate the feline immune response and possibly also FIP pathogenesis. Another interesting field is the presumptive FCoV carrier state. It remains to be determined whether FCoVs can cause a true persistent infection, and if so, in which cells and tissues of the host the virus resides and how it manages to escape immunosurveillance. Finally, FCoVs produce a number of proteins that are dispensable for replication in tissue culture cells, most notably gp7b. What is the function of these proteins during natural infection? Though the last ten years of FIP research have been exciting, the coming years promise to be better still.

ACKNOWLEDGMENTS. The authors thank Arnold Herrewegh, Harry Vennema, Tim Miller, and Wayne Corapi for sharing their unpublished results and Dr. Alfred M. Legendre for critically reading the manuscript. Ad Rijnberk from the Small Animal Clinic of the Utrecht State University drew our attention to the historic publication of Fig. 1, which to our knowledge is the earliest case of FIP documented in veterinary literature.

## XI. REFERENCES

Addie, D. D., and Jarrett, J. O., 1992a, A study of naturally occurring feline coronavirus infections in kittens, *Vet. Rec.* **130:**133.

Addie, D. D., and Jarrett, J. O., 1992b, Feline coronavirus antibodies in cats, *Vet. Rec.* **131:**202.

Barlough, J. E., and Stoddart, C. A., 1986, Feline infectious peritonitis, in: *Contemporary Issues in Small Animal Practice 3* (F. W. Scott, ed.), pp. 93–108, Churchill Livingstone, New York.

Barlough, J. E., and Stoddart, C. A., 1990, Feline coronaviral infections, in: *Infectious Diseases of the Dog and Cat* (C. E. Greene, ed.), pp. 300–311, WB Saunders Co., Philadelphia.

Barlough, J. E., Stoddart, C. A., Sorresso, G. P., Jacobson, R. H., and Scott, F. W., 1984, Experimental inoculation of cats with canine coronavirus and subsequent challenge with feline infectious peritonitis virus, *Lab. Anim. Sci.* **34:**592.

Barlough, J. E., Johnson-Lussenburg, C. M., Stoddart, C. A., Jacobson, R. H., and Scott, F. W., 1985, Experimental inoculation of cats with human coronavirus 229E and subsequent challenge with feline infectious peritonitis virus, *Can. J. Comp. Med.* **49:**303.

Black, J. W., 1980, Recovery and *in vitro* cultivation of a coronavirus from laboratory-induced cases of feline infectious peritonitis (FIP), *Vet. Med. Small. Anim. Clin.* **75:**811.

Boyle, J. F., Pedersen, N. C., Evermann, J. F., McKeirnan, A. J., Ott, R. L., and Black, J. W., 1984, Plaque assay, polypeptide composition and immunochemistry of feline infectious peritonitis virus and feline enteric coronavirus isolates, in *Molecular Biology and Pathogenesis of Coronaviruses* (P. J. M. Rottier, B. A. M. van der Zeijst, W. J. M. Spaan, and M. C. Horzinek, eds.), pp. 133–147, Plenum Press, New York.

Burstin, S. J., Brandriss, M. W., and Schlessinger, J. J., 1983, Effects of immune ascitic fluids and monoclonal antibodies on neutralization and on enhancement of viral growth, *J. Immunol.* **130:**2951.

Cavanagh, D., Brian, D. A., Enjuanes, L., Holmes, K. V., Lai, M. M. C., Laude, H., Siddell, S. G., Spaan, W., Taguchi, F., and Talbot, P. J., 1990, Recommendations of the coronavirus study group for the structural proteins, mRNAs and genes of coronaviruses, *Virology* **176:**306.

Chanas, A. C., Gould, E. A., Clegg, J. C., and Varma, M. G. R., 1982, Monoclonal antibodies to Sindbis virus glycoprotein E1 can neutralize, enhance infectivity, and independently inhibit haemagglutination or haemolysis, *J. Gen. Virol.* **58:**37.

Christianson, K. K., Ingersoll, J. D., Landon, R. M., Pfeiffer, N. E., and Gerber, J. D., 1989, Characterization of a temperature sensitive feline infectious peritonitis coronavirus, *Arch. Virol.* **109:**185.

Colby, E. D., and Low, R. J., 1970, Feline infectious peritonitis, *Vet. Med. Small Anim. Clin.* **65:**783.

Connor, R. I., Dinces, N. B., Howell, A. L., Romet-Lemonne, J.-L., and Pasquali, J.-L., 1991, Fc receptors for IgG on human monocytes and macrophages are not infectivity receptors for human immunodeficiency virus type 1 (HIV-1): Studies using bispecific antibodies to target HIV-1 to various myeloid cell surface molecules, including the FcγR, *Proc. Natl. Acad. Sci. USA* **88:**9593.

Corapi, W. V., Olsen, C. W., and Scott, F. W., 1992, Monoclonal antibody analysis of neutralization and antibody-dependent enhancement of feline infectious peritonitis, *J. Virol.* **11:**6695.

Correa, I., Gebauer, F., Bullido, M. J., Suñé, C., Baay, M. F. D., Zwaagstra, K. A., Posthumus, W. P. A., Lenstra, J. A., and Enjuanes, L., 1990, Localization of antigenic sites of the E2 glycoprotein of transmissible gastroenteritis coronavirus, *J. Gen. Virol.* **71:**271.

Cotter, S. M., Hardy, W. D., and Essex, M., 1975, The association of feline leukemia virus with lymphosarcoma and other disorders, *J. Am. Vet. Med. Assoc.* **168:**448.

de Groot, R. J., Ter Haar, R. J., Horzinek, M. C., and van der Zeijst, B. A. M., 1987a, Intracellular RNAs of the feline infectious peritonitis coronavirus strain 79-1146, *J. Gen. Virol.* **68:**995.

de Groot, R. J., Maduro, J., Lenstra, J. A., Horzinek, M. C., van der Zeijst, B. A. M., and Spaan, W. J. M. 1987b, cDNA cloning and sequence analysis of the gene encoding the peplomer protein of feline infectious peritonitis virus, *J. Gen. Virol.* **68:**2639.

de Groot, R. J., Luytjes, W., Horzinek, M. C., van der Zeijst, B. A. M., Spaan, W. J. M., and Lenstra, J. A., 1987c, Evidence for a coiled-coil structure in the spike proteins of coronaviruses, *J. Mol. Biol.* **196:**963.

de Groot, R. J., Andeweg, A. C., Horzinek, M. C., and Spaan, W. J. M., 1988, Sequence analysis of the 3′ end of the feline coronavirus FIPV 79-1146 genome: Comparison with the genome of porcine coronavirus TGEV reveals large insertions, *Virology* **167:**370.

de Groot, R. J., van Leen, R. W., Dalderup, M. J. M., Vennema, H., Horzinek, M. C., and Spaan, W. J. M., 1989, Stably expressed FIPV peplomer protein induces cell fusion and elicits neutralizing antibodies, *Virology* **171:**493.

Delmas, B., Gelfi, J., L'Harridon, R., Vogel, L. K., Sjöström, H., Norén, O., and Laude, H., 1992, Aminopeptidase N is a major receptor for the enteropathogenic coronavirus TGEV, *Nature* **357:**417.

Evermann, J. F., Baumgartner, L., Ott, R. L., Davis, E. V., and McKeirnan, A. J., 1981, Characterization of a feline infectious peritonitis virus isolate, *Vet. Pathol.* **18:**256.

Feldmann, B. F., and Jortner, B. S., 1964, Clinico-pathology conference, *J. Am. Vet. Med. Assoc.* **144:**1409.

Fiscus, S. A., and Teramoto, Y. A., 1987a, Antigenic comparison of feline coronavirus isolates: Evidence for markedly different peplomer glycoproteins, *J. Virol.* **61:**2607.

Fiscus, S. A., and Teramoto, Y. A., 1987b, Functional differences in the peplomer glycoproteins of feline coronavirus isolates, *J. Virol.* **61:**2655.

Fiscus, S. A., Rivoire, B. L., and Teramoto, Y. A., 1987, Humoral immune response of cats to virulent and avirulent feline infectious peritonitis virus isolates, in: *Coronaviruses* (M. M. C. Lai and S. Stohlman, eds.), pp. 559–568, Plenum Press, New York.

Garwes, D. J., Stewart, F., and Britton, P., 1989, The polypeptide of $M_r$ 14000 of porcine transmissible gastroenenteritis virus: Gene assignment and intracellular location, *J. Gen. Virol.* **70:**2495.

Gebauer, F., Posthumus, W. P. A., Correa, I., Suñé, C., Smerdou, C., Sánchez, C. M., Lenstra, J. A., Meloen, R. H., and Enjuanes, L., 1991, Residues involved in the antigenic sites of transmissible gastroenteritis coronavirus S glycoprotein, *Virology* **183:**225.

Gerber, J. D., Ingersoll, J. D., Gast, A. M., Christianson, K. K., Selzer, N. L., Landon, R. M., Pfeiffer, N. E., Sharpee, R. L., and Beckenhauer, W. H., 1990, Protection against feline infectious peritonitis by intranasal inoculation of a temperature-sensitive FIPV vaccine, *Vaccine* **8:**536.

Godet, M., L'Haridon, R., Vautherot, J.-F., and Laude, H., 1992, TGEV coronavirus ORF4 encodes a membrane protein that is incorporated into virions, *Virology* **188:**666.

Goitsuka, R., Onda, C., Hirota, Y., Hasegawa, A., and Tomoda, I., 1988, Feline interleukin 1 production induced by feline infectious peritonitis, *Jpn. J. Vet. Sci.* **50:**209.

Goitsuka, R., Ohashi, T., Ono, K., Yasukawa, K., Koishibara, Y., Fukui, H., Oshugi, Y., and Hasegawa, A., 1990, IL-6 activity in feline infectious peritonitis, *J. Immunol.* **144:**2599.
Gollins, S. W., and Porterfield, J. S., 1984, Flavivirus infection enhancement in macrophages: Radioactive and biological studies on the effect of antibody on viral fate, *J. Gen. Virol.* **65:**1261.
Gollins, S. W., and Porterfield, J. S., 1985, Flavivirus infection enhancement in macrophages: An electron microscopic study of viral cellular entry, *J. Gen. Virol.* **66:**1969.
Gooding, L. R., 1992, Virus proteins that counteract host immune defenses, *Cell* **71:**5.
Halstead, S. B., 1988, Pathogenesis of Dengue: Challenges to molecular biology, *Science* **239:**476.
Halstead, S. B., and O'Rourke, E. J., 1977, Dengue viruses and mononuclear phagocytes. I. Infection enhancement by non-neutralizing antibody, *J. Exp. Med.* **146:**201.
Hardy, W. D., and Hurvitz, A. I., 1971, Feline infectious peritonitis: Experimental studies, *J. Am. Vet. Med. Assoc.* **158:**994.
Hasegawa, T., and Hasegawa, A., 1991, Interleukin 1 alpha mRNA-expressing cells on the local inflammatory response in feline infectious peritonitis, *J. Vet. Med. Sci.* **53:**995.
Hayashi, T., Goto, N., Takahashi, R., and Fujiwara, K., 1977, Systemic vascular lesions in feline infectious peritonitis, *Jpn. J. Vet. Sci.* **39:**365.
Hayashi, T., Yanai, T., Tsurudome, M., Nakayama, H., Watabe, Y., and Fujiwara, K., 1981, Serodiagnosis for feline infectious peritonitis by immunofluorescence using infected suckling mouse brain sections, *Jap. J. Vet. Sci.* **43:**669.
Herrewegh, A. A. P. M., De Groot, R. J., Cepica, A., Egberink, H. F., Horzinek, M. C., and Rottier, P. J. M., 1995, Detection of feline cornavirus RNA in feces, tissues, and body fluids of naturally infected cats by reverse transcriptase PCR. *J. Clin. Microbiol.* **33**(3), in press.
Hohdatsu, T., Okada, S., and Koyama, H., 1991a, Characterization of monoclonal antibodies against feline infectious peritonitis virus type II and antigenic relationship between feline, porcine, and canine coronaviruses, *Arch. Virol.* **117:**85.
Hohdatsu, T., Sasamoto, T., Okada, S., and Koyama, H., 1991b, Antigenic analysis of feline coronaviruses with monoclonal antibodies (MAbs): Preparation of MAbs which discriminate between FIPV strain 79-1146 and FECV strain 79-1683, *Vet. Microbiol.* **28:**13.
Hohdatsu, T., Nakamura, Y., Yamada, H., and Koyama, H., 1991c, A study on the mechanism of antibody-dependent enhancement of feline infectious peritonitis virus infection in feline macrophages by monoclonal antibodies, *Arch. Virol.* **120:**207.
Hohdatsu, T., Okada, S., Ishizuka, Y., Yamada, H., and Koyama, H., 1992, The prevalence of types I and II feline coronavirus infections in cats, *J. Vet. Med. Sci.* **54:**557.
Holzworth, J., 1963, Some important disorders of cats, *Cornell Vet.* **53:**157.
Horsburgh, B. C., Brierley, I., and Brown, T. D. K., 1992, Analysis of a 9.6 kb sequence from the 3′ end of canine coronavirus genomic RNA, *J. Gen. Virol.* **73:**2849.
Horzinek, M. C., and Osterhaus, A. D. M. E., 1979, Feline infectious peritonitis: A world-wide serosurvey, *Am. J. Vet. Res.* **40:**1487.
Horzinek, M. C., Osterhaus, A. D. M. E., and Ellens, D. J., 1977, Feline infectious peritonitis, *Zbl. Vet. Med.* [B] **24:**398.
Horzinek, M. C., Lutz, H., and Pedersen, N. C., 1982, Antigenic relationships among homologous structural polypeptides of porcine, feline and canine coronaviruses, *Infect. Immun.* **37:**1148.
Jacobs, L., de Groot, R. J., van der Zeijst, B. A. M., Horzinek, M. C., and Spaan, W. J. M., 1987, The nucleotide sequence of the peplomer gene of porcine transmissible gastroenteritis virus (TGEV): Comparisosn with the sequence of the peplomer protein of feline infectious peritonitis virus (FIPV), *Virus Res.* **8:**363.
Jacobse-Geels, H. E. L., and Horzinek, M. C., 1983, Expression of feline infectious peritonitis coronavirus antigens on the surface of feline macrophage-like cells, *J. Gen. Virol.* **64:**1859.
Jacobse-Geels, H. E. L., Daha, M. R., and Horzinek, M. C., 1980, Isolation and characterization of feline C3 and evidence for the immune complex pathogenesis of feline infectious peritonitis virus, *J. Immunol.* **125:**1606.
Jacobse-Geels, H. E. L., Daha, M. R., and Horzinek, M. C., 1982, Antibody, immune complexes and complement activity fluctuations in experimental feline infectious peritonitis, *Am. J. Vet. Res.* **43:**666.
Jakob, H., 1914, Therapeutische, kasuistische und statistische Mitteilungen aus der Klinik für kleine Haustiere an der Reichstierarzneischule in Utrecht (Holland). Jahrgang 1912/13, *Zschr. Tiermed.* **18:**193.

Kapke, P. A., and Brian, D. A., 1986, Sequence analysis of the porcine transmissible gastroenteritis coronavirus nucleocapsid protein gene, *Virology* **151**:41.

Kusters, J. G., Niesters, H. G. M., Bleumink-Pluym, N. M. C., Davelaar, F. G., Horzinek, M. C., and van der Zeijst, B. A. M., 1987, Molecular epidemiology of infectious bronchitis virus in the Netherlands, *J. Gen. Virol.* **68**:343.

Loeffler, D. G., Ott, R. L., Evermann, J. F., and Alexander, J. E., 1978, The incidence of naturally occurring antibodies against feline infectious peritonitis in selected cat populations, *Feline Pract.* **8**:43.

Look, A. T., Ashmun, R. A., Shapiro, L. H., and Peiper, S. C., 1989, Human myeloid plasma membrane glycoprotein CD13 (gp150) is identical to aminopeptidase N, *J. Clin. Invest.* **83**:1299.

Martin, M. L., 1985, Canine coronavirus enteritis and a recent outbreak following modified live virus vaccination, *Compend. Cont. Educ. Pract. Vet.* **7**:1013.

McArdle, F., Bennet, M., Gaskell, R. M., Tennant, B., Kelly, D. F., and Gaskell, C. J., 1992, Induction and enhancement of feline infectious peritonitis by canine coronavirus, *Am. J. Vet. Res.* **53**:1500.

McKeirnan, A. J., Evermann, J. F., Hargis, A., Miller, L. M., and Ott, R. L., 1981, Isolation of feline coronaviruses from two cats with diverse disease manifestations, *Feline Pract.* **11**:16.

Montali, R. J., and Strandberg, J. D., 1972, Extraperitoneal lesions in feline infectious peritonitis, *Vet. Pathol.* **9**:109.

Motokawa, K., Hohdatsu, T., Aizawa, C., Koyama, H., and Hashimoto, H., 1995, Molecular cloning and sequence determination of the peplomer protein gene of feline infectious peritonitis virus type I, *Arch. Virol.* **140**:469.

O'Brien, S. J., Roelke, M. E., Marker, L., Newman, A., Winkler, C. A., Meltzer, D., Colly, L., Evermann, J. F., Bush, M., and Wildt, D. E., 1985, Genetic basis for species vulnerability in the Cheetah, *Science* **227**:1428.

Olsen, C. W., 1993, A review of feline infectious peritonitis virus: Molecular biology, immunopathogenesis, clinical aspects, and vaccination, *Vet. Microbiol.* **36**:1.

Olsen, C. W., and Scott, F., 1993, Evaluation of antibody-dependent enhancement of feline infectious peritonitis virus infectivity using *in situ* hybridization, *Microbial Pathogen.* **14**:275.

Olsen, C. W., Corapi, W. V., Ngichabe, C. K., Baines, J. D., and Scott, F. W., 1992, Monoclonal antibodies to the spike protein of feline infectious peritonitis virus mediate antibody-dependent enhancement of infection of feline macrophages, *J. Virol.* **66**:956.

Olsen, C. W., Corapi, W. V., Jacobson, R. H., Simkins, R. A., Saif, L. J., and Scott, F. W., 1993, Identification of antigenic sites mediating antibody-dependent enhancement of feline infectious peritonitis virus infectivity, *J. Gen. Virol.* **74**:745.

O'Reilly, K. J., Fishman, L. M., and Hitchcock, L. M., 1979, Feline infectious peritonitis: Isolation of a coronavirus, *Vet. Rec.* **104**:348.

Osterhaus, A. D. M. E., Horzinek, M. C., and Wirahadiredja, R. M. S., 1978a, Feline infectious peritonitis (FIP) virus. II. Propagation in suckling mouse brain, *Zbl. Vet. Med.* [B] **25**:301.

Osterhaus, A. D. M. E., Horzinek, M. C., and Wirahadiredja, R. M. S., 1978b, Feline infectious peritonitis (FIP) virus. IV. Propagation in suckling rat and hamster brain, *Zbl. Vet. Med.* [B] **25**:816.

Pedersen, N. C., 1976a, Morphologic and physical characteristics of feline infectious peritonitis virus and its growth in autochtonous peritoneal cell cultures, *Am. J. Vet. Res.* **37**:567.

Pedersen, N. C., 1976b, Feline infectious peritonitis: Something old, something new, *Feline Pract.* **6**:42.

Pedersen, N. C., 1976c, Serologic studies of naturally occurring feline infectious peritonitis, *Am. J. Vet. Res.* **37**:1449.

Pedersen, N. C., 1983a, Feline infectious peritonitis and feline enteric coronavirus infections, part 1: Feline enteric coronaviruses, *Feline Pract.* **13**(4):13.

Pedersen, N. C., 1983b, Feline infectious peritonitis and feline enteric coronavirus infections, part 2: Feline infectious peritonitis, *Feline Pract.* **13**(5):5.

Pedersen, N. C., 1987a, Feline infectious peritonitis virus, in: *Virus Infections of Carnivores* (M. C. Horzinek and M. Appel, eds.), pp. 267–286, Elsevier Science Publishers, Amsterdam.

Pedersen, N. C., 1987b, Virologic and immunologic aspects of feline infectious peritonitis virus infection, in: *Coronaviruses* (M. M. C. Lai and S. Stohlman, eds.), pp. 529–550, Plenum Press, New York.

Pedersen, N. C., and Black, J. W., 1983, Attempted immunization of cats against feline infectious peritonitis, using avirulent live virus or sublethal amounts of virulent virus, *Am. J. Vet. Res.* **44:**229.

Pedersen, N. C., and Boyle, J. F., 1980, Immunologic phenomena in the effusive form of feline infectious peritonitis, *Am. J. Vet. Res.* **41:**868.

Pedersen, N. C., and Floyd, K., 1985, Experimental studies with three new strains of feline infectious peritonitis virus: FIPV-UVD2, FIPV-UCD3, and FIPV-UCD4, *Compend. Contin. Educ. Pract. Vet.* **7:**1001.

Pedersen, N. C., Ward, J. M., and Mengeling, W. L., 1978, Antigenic relationship of the feline infectious peritonitis virus to coronaviruses of other species, *Arch. Virol.* **58:**45.

Pedersen, N. C., Boyle, J. F., and Floyd, K., 1981a, Infection studies in kittens utilizing feline infectious peritonitis virus propagated in cell culture, *Am. J. Vet. Res.* **42:**363.

Pedersen, N. C., Boyle, J. F., Floyd, K., Fudge, A., and Barker, J., 1981b, An enteric coronavirus infection of cats and its relationship to feline infectious peritonitis, *Am. J. Vet. Res.* **42:**368.

Pedersen, N. C., Black, J. W., Boyle, J. F., Evermann, J. F., McKeirnan, A. J., and Ott, R. L., 1983, Pathogenic differences between various feline coronavirus isolates, in: *Molecular Biology and Pathogenisis of Coronaviruses* (P. J. M. Rottier, B. A. M. van der Zeijst, W. J. M. Spaan, and M. C. Horzinek, eds.), pp. 365–380, Plenum Press, New York.

Pedersen, N. C., Evermann, J. F., Alison, J., McKeirnan, A. J., and Ott, R. L., 1984, Pathogenicity studies of feline coronavirus isolates 79-1146 and 79-1683, *Am. J. Vet. Res.* **45:**2580.

Peiris, J. S. M., and Porterfield, J. S., 1979, Antibody-mediated enhancement of flavivirus replication in macrophage-like cell lines, *Nature* **282:**509.

Peiris, J. S. M., Gordon, S., Unkeless, J. C., and Porterfield, J. S., 1981, Monoclonal anti-Fc receptor IgG blocks antibody enhancement of viral replication in macrophages, *Nature* **289:**189.

Perno, C.-F., Baseler, M. W., Broder, S., and Yarchoan, R., 1990, Infection of monocytes by human immunodeficiency virus type 1 blocked by inhibitors of CD4-gp120 binding, even in the presence of enhancing antibodies, *J. Exp. Med.* **171:**1043.

Poelma, F. G., Peters, J. C., Mieog, W. H. M., and Zwart, P., 1971, Infektiöse Peritonitis bei Karakal (*Felis caracal*) und Nordluchs (*Felis lynx lynx*), in: *Erkrankungen der Zootiere*, pp. 249–283, 13th International Symposium, Helsinki.

Porterfield, J. S., 1986, Antibody-dependent enhancement of viral infectivity, *Adv. Virus Res.* **31:**335.

Reynolds, D. J., Garwes, D. J., and Gaskell, C. J., 1977, Detection of transmissible gastroenteritis virus neutralizing antibody in cats, *Arch. Virol.* **55:**77.

Scott, F. W., 1987, Immunization against feline cornaviruses, in: *Cornaviruses* (M. M. C. Lai and S. Stohlman, eds.), pp. 569–576, Plenum Press, New York.

Sneath, P. H. A., and Sokal, R. R., 1973, in: *Numeral Taxonomy*, pp. 230–234, W. H. Freeman and Company, San Francisco.

Sparkes, A. H., Gruffydd-Jones, T. J., and Harbour, D. A., 1991, Feline infectious peritonitis: A review of clinical pathological changes in 65 cases and a critical assessment of their diagnostic value, *Vet. Rec.* **129:**209.

Sparkes, A. H., Gruffydd-Jones, T. J., Howard, P. E., and Harbour, D. A., 1992, Coronavirus serology in healthy pedigree cats, *Vet. Rec.* **131:**35.

Stoddart, C. A., 1989, PhD thesis, Cornell University, Ithaca, NY.

Stoddart, C. A., and Scott, F. W., 1988, Isolation and identification of feline peritonteal macrophages for *in vitro* studies of coronavirus–macrophage interactions, *J. Leukocyte Biol.* **44:**319.

Stoddart, C. A., and Scott, F. W., 1989, Intrinsic resistance of feline peritoneal macrophages to coronavirus infection correlates with *in vivo* virulence, *J. Virol.* **63:**436.

Stoddart, M. E., Gaskell, R. M., Harbour, D. A., and Gaskell, C. J., 1988a, Virus shedding and immune responses in cats inoculated with cell culture-adapted feline infectious peritonitis virus, *Vet. Microbiol.* **16:**145.

Stoddart, M. E., Gaskell, R. M., Harbour, D. A., and Pearson, G. R., 1988b, The sites of early viral replication in feline infectious peritonitis, *Vet. Microbiol.* **18:**259.

Takeda, A., Tuazon, C. U., and Ennis, F. A., 1988, Antibody-enhanced infection by HIV-1 via Fc receptor-mediated entry, *Science* **242:**580.

Takeda, A., Sweet, R. W., and Ennis, F. A., 1990, Two receptors are required for antibody-dependent enhancement of human immunodeficiency virus type 1 infection: CD4 and FcγR, *J. Virol.* **64:**5605.

Toma, B., Duret, C., Chappuis, G., and Pellerin, B., 1979, Échec de l'immunisation contre la péritonite infectieuse féline par injection de virus de la gastro-entérite transmissible du porc, *Rec. Med. Vet.* **155:**799.

Tuch, K., Witte, K. H., and Wüller, H., 1974, Feststellung der felinen infektiösen Peritonitis (FIP) bei Hauskatzen und Leoparden in Deutschland, *Zbl. Vet. Med.* [B] **21:**426.

Tung, F. Y., Abraham, S., Sethna, M., Hung, S.-L., Sethna, P., Hogue, B. G., and Brian, D. A., 1992, The 9-kDa hydrophobic protein encoded at the 3′ end of the porcine transmissible gastroenteritis coronavirus genome is membrane-associated, *Virology* **186:**676.

Vennema, H., Heijnen, L., Zijderveld, A., Horzinek, M. C., and Spaan, W. J. M., 1990a, Intracellular transport of recombinant coronavirus spike proteins: Implications for virus assembly, *J. Virol.* **64:**339.

Vennema, H., de Groot, R. J., Harbour, D., Daalderup, M., Gruffydd-Jones, T., Horzinek, M. C., and Spaan, W. J. M., 1990b, Early death after feline infectious peritonitis virus challenge due to recombinant vaccinia virus immunization, *J. Virol.* **64:**1407.

Vennema, H., de Groot, R. J., Harbour, D. A., Horzinek, M. C., and Spaan, W. J. M., 1991, Primary structure of the membrane and nucleocapsid protein genes of feline infectious peritonitis virus and immunogenicity of recombinant vaccinia viruses in kittens, *Virology* **181:**327.

Vennema, H., Heijen, L., Rottier, P. J. M., Horzinek, M. C., and Spaan, W. J. M., 1992a, A novel glycoprotein of feline infectious peritonitis coronavirus contains a KDEL-like endoplasmic reticulum retention signal, *J. Virol.* **66:**4951.

Vennema, H., Rossen, J. W., Wesseling, J. Horzinek, M. C., and Rottier, P. J. M., 1992b, Genomic organization and expression of the 3′ end of the canine and feline enteric coronaviruses, *Virology* **191:**134.

Vennema, H., Poland, A., Floyd Hawkins, K., and Pedersen, N. C., 1995, A comparison of the genomes of FECVs and FIPVs and what they tell us about the relationships between feline coronaviruses and their evolution, *Feline Practice,* in press.

Ward, J. M., 1970, Morphogenesis of a virus in cats with experimental feline infectious peritonitis, *Virology* **41:**191.

Ward, J. M., Munn, R. J., Gribble, D. H., and Dungworth, D. L., 1968, An observation of FIP, *Vet. Res.* **83:**416.

Watt, N. J., MacIntyre, N. J., and McOrist, S., 1993, An extended outbreak of infectious peritonitis in a closed colony of European wildcats (*Felis silvestris*), *J. Comp. Pathol.* **108:**73.

Weiss, R. C., and Scott, F. W., 1981a, Pathogenesis of feline infectious peritonitis: Nature and development of viremia, *Am. J. Vet. Res.* **42:**382.

Weiss, R. C., and Scott, F. W., 1981b, Antibody-mediated enhancement of disease in feline infectious peritonitis: Comparisons with Dengue hemorrhagic fever, *Comp. Immun. Microbiol. Infect. Dis.* **4:**175.

Weiss, R. C., and Scott, F. W., 1981c, Pathogenesis of feline infectious peritonitis: Pathologic changes and immunofluorescence, *Am. J. Vet. Res.* **42:**2036.

Weiss, R. C., Dodds, W. J., and Scott, F. W., 1980, Dissiminated intravascular coagulation in experimentally induced feline infectious peritonitis, *Am. J. Vet. Res.* **41:**663.

Weiss, R. C., Vaughn, D. M., and Cox, N. R., 1988, Increased plasma levels of leukotriene B4 and prostaglandin E2 in cats experimentally inoculated with feline infectious peritonitis virus, *Vet. Res. Commun.* **12:**313.

Witte, K. H., Tuch, K., Dubenkropp, H., and Walther, C., 1977, Untersuchungen über die Antigenverwandtschaft der Viren der Felinen infektiösen Peritonitis und der transmissibelen Gastroenteritis des Schweines, *Berl. Münch. Tierärztl. Wschr.* **90:**396.

Wolfe, L. G., and Griesemer, R. A., 1966, Feline infectious peritonitis, *Pathol. Vet.* **3:**255.

Wolfe, L. G., and Griesemer, R. A., 1971, Feline infectious peritonitis: Review of gross and histopathologic lesions, *J. Am. Vet. Med. Assoc.* **158:**987.

Woods, R. D., and Pedersen, N. C., 1979, Cross-protection studies between feline infectious peritonitis and porcine transmissible gastroenteritis viruses, *Vet. Microbiol.* **4**:11.

Woods, R. D., Cheville, N. F., and Gallagher, J. E., 1981, Lesions in the small intestine of newborn pigs inoculated with porcine, feline and canine coronaviruses, *Am. J. Vet. Res.* **42**:1163.

Yeager, C. L., Ashmun, R. A., Williams, R. K., Cardellichio, C. B., Shapiro, L. H., Look, A. T., and Holmes, K. V., 1992, Human aminopeptidase N is a receptor for human coronavirus 229E, *Nature* **357**:420.

Zeira, M., Byrn, R. A., and Groopman, J. E., 1990, Inhibition of serum-enhanced HIV-1 infection of U937 monocytoid cells by recombinant soluble CD4 and anti-CD4 monoclonal antibody, *AIDS. Res. Hum. Retroviruses* **6**:629.

Zhou, Y., Ederveen, J., Egberink, H., Pensaert, M., and Horzinek, M. C., 1988, Porcine epidemic diarrhea virus (CV 777) and feline infectious peritonitis virus (FIPV) are antigenically related, *Arch. Virol.* **102**:63.

Zook, B. C., King, N. W., Robinson, R. L., and McCombs, H. L., 1968, Ultrastructural evidence for the viral etiology of feline infectious peritonitis, *Pathol. Vet.* **5**:91.

CHAPTER 15

# Epidemiology of Infectious Bronchitis Virus

JANE K. A. COOK AND A. P. A. MOCKETT

## I. INTRODUCTION

Avian infectious bronchitis virus (IBV) is the type species of the genus *Coronavirus* of the family Coronaviridae. It is of considerable economic importance to the poultry industry worldwide, causing a highly contagious disease affecting the respiratory, reproductive, and renal systems.

## II. HOST RANGE

The chicken is believed to be the only natural host for IBV; all ages can be infected and both sexes appear equally susceptible. Vindevogel *et al.* (1976) found that pigeons were resistant to IBV infection. However, Biondi and Schiavo (1966) successfully infected quail but not seven other avian species experimentally with IBV, and Cumming (1969) was able to recover the virus from the kidneys of one of six magpies inoculated via the conjunctiva. Barr *et al.* (1988) isolated IBV from a flock of racing pigeons with respiratory signs, but failed to transmit infection to pigeons experimentally. Based solely on serological data, Weisman *et al.* (1987) claim to have demonstrated the presence of IBV in healthy turkeys, although Cumming (1969) was unable to infect turkeys experimentally with IBV. Spackman and Cameron (1983) isolated IBV from pheasants with a history of respiratory signs and aberrant egg production, and

JANE K. A. COOK • Intervet UK Ltd., Huntingdon, Cambridgeshire PE17 2BQ, England. A. P. A. MOCKETT • Intervet, Inc., Millsboro, Delaware 19966.

*The Coronaviridae*, edited by Stuart G. Siddell, Plenum Press, New York, 1995.

the disease problem in the pheasants was successfully controlled by the use of an oil-based inactivated IB vaccine. This and a further report (Lister *et al.*, 1985) suggest that the pheasant could be a second natural host for IBV and is the only evidence of the ability of IBV to infect avian species other than the chicken.

## III. INCIDENCE

Avian infectious bronchitis is reported from all countries where an intensive poultry industry has developed; in most countries it appears as a serious disease problem shortly after the development of the industry in that area. The relative importance of IBV in a country appears to fluctuate from time to time, being dependent on such factors as the percentage of the poultry population that is vaccinated and on the emergence of virus types against which current vaccines are not wholly effective (see Section VIIA.). Another factor governing the reported incidence of IBV is the efficacy of diagnostic and research facilities in different countries. A detailed review of the incidence of IBV worldwide can be found in the proceedings of the first and second international symposia on infectious bronchitis (Kaleta and Heffels-Redwan, 1988, 1991).

## IV. TRANSMISSION

The most important route of spread of IBV is from the respiratory tract, probably by aerosol. During the clinical phase of the disease and for approximately 28 days after infection, virus can be recovered intermittently from the respiratory tract (Cook, 1968). However, IBV replicates in many tissues throughout the chicken (Hofstad and Yoder, 1966; Doherty, 1967), and once chickens have recovered clinically, the feces are possibly the best site from which to recover virus (Cook, 1984), the intestinal tract (El Houadfi *et al.*, 1986; Ambali and Jones, 1990) and specifically the cecal tonsil (Cook, 1968) being suggested as a likely reservoir for IBV. These findings emphasize that the importance of fecal shedding in the spread of the virus should not be overlooked; although inhalation of airborne virus may be the most common method of transmission, ingestion of virus via contaminated food, drinking water, or litter plays an important role in the epidemiology of IBV.

While lateral transmission is undoubtedly the most important method of IBV spread, it has been shown that virus can be recovered from eggs laid by experimentally inoculated hens for up to 6 weeks after inoculation with a high virus dose (Cook, 1971) and also from a small number of day-old chicks hatched from those hens. However, the progeny showed no signs of IB infection. Earlier work (Cook and Garside, 1967) had shown that it was possible to hatch and rear IB-free chicks from dams showing a high incidence of infection; the conclusion was drawn that egg transmission is of negligible importance in the epizootiology of IBV infections. One interesting finding in that work was that IBV could be recovered from semen of infected cockerels for up to 14 days after inoculation, a finding that could have implications for the spread of IBV.

## A. Carriers/Persistent Infections

### 1. In the Chicken

There is no conclusive evidence that a true carrier state exists with IB, but virus clearly persists for a considerable time in experimentally inoculated chicks. Cook (1968) recovered virus for 49 days from experimentally inoculated chicks housed in strict isolation, and could recover it for over 4 months when isolation was less effective. It was suggested in the latter case that continual reinfection from shedders had occurred. Alexander *et al.* (1978) isolated IBV from the feces of chickens up to 227 days after inoculation of two different strains of IBV. These birds were not housed in isolation and again reinfection was a possibility. Therefore, although IBV can be isolated in the presence of high concentrations of humoral antibody, it was concluded that a true carrier state does not occur. However, the demonstration by Jones and Ambali (1987) of reexcretion of an enterotropic IBV strain at point of lay, with virus being recovered from both tracheal and cloacal swabs, suggests that a state of latency may exist.

### 2. In the Environment

The question of persistence of IBV in the environment is a complex one because of conflicting evidence on the survival of IBV under different environmental conditions. Otsuki *et al.* (1979) demonstrated variation in the resistance of ten IBV strains to chemical and physical treatments. During epidemics the virus can certainly be transmitted between farms. For example, the same IBV serotype has been isolated from four different farms within a 10-mile radius over a period of 13 months (Cook, 1988). Airborne spread is likely to have been involved, but movement of personnel and equipment probably aided transmission. The climatic and other conditions that govern spread are not fully understood, and many aspects of IBV transmission still require elucidation. Moreover, the importance of strain variation in the transmission and persistence of IBV has not received the attention which the subject merits.

# V. PATHOGENESIS

## A. Clinical Signs

IBV can induce respiratory signs, affect the reproductive tract, and cause nephritis. Young chickens up to 4 weeks of age appear to be most susceptible to respiratory disease. The clinical signs, which appear two to three days after infection, include nasal discharge, sneezing, coughing, and tracheal rales. Caseous or catarrhal plugs may be present in the trachea or bronchi. Swollen sinuses may occur in some chickens. IBV respiratory infection predisposes chickens to infection with secondary pathogens (see Section V.C.), although chickens will usually recover quickly from an uncomplicated viral infection.

IBV infection of susceptible layers can result in a drop in egg production or failure to lay at full potential. The extent of the drop varies, depending on such factors as time of infection in relation to stage of egg production and the general health of the flock. Drops in production of 5 to 20% are not uncommon, together with the production of increased numbers of downgraded eggs with thin, misshapen, rough, and soft shells. Watery whites, in which there is no clear demarcation between the thick and thin albumins, are often produced. Although layers usually recover from the disease, their egg production rarely returns to preinfection levels. IBV infection of chickens less than 2 weeks of age can cause damage to the developing reproductive tract, resulting in "false layers" that do not lay normally at sexual maturity (Broadfoot *et al.*, 1956; Jones and Jordan, 1970).

Nephritis has been reported as a result of IBV infection. Early reports were from Australia and showed that certain strains, such as the T strain, had a predilection for the kidney (Cumming, 1963). However, the incidence of nephritis associated with IBV has increased and has been reported from other countries, including Italy (Zanella *et al.*, 1988), Belgium (Froyman *et al.*, 1985), France (Picault *et al.*, 1987), Japan (Otsuki, 1988), and the United States (Kinde *et al.*, 1991). The reason for this increased incidence of renal involvement remains unclear, and particular IBV serotypes do not necessarily appear to be responsible. Swollen, pale kidneys with the tubules and ureters enlarged, due to the presence of urates, are often observed. Polydipsia may also be present. In some outbreaks in Belgium, mortality rates of up to 20% have been reported (Nauwynck and Pensaert, 1988).

## B. Histopathology

This has been covered in previous reviews, for example, King and Cavanagh (1991).

## C. Secondary Infections

The severity of the infection caused by IBV can be increased both by environmental factors and by the presence of other microorganisms that may or may not be pathogens in their own right. In such cases, severe air sacculitis, often accompanied by peritonitis and pericarditis, can result and mortality may occur. In adult chickens combined infection with IBV and *Mycoplasma gallisepticum* has been shown to cause a more severe effect on egg production and quality than inoculation with either agent alone (Blake, 1962).

Other viruses, bacteria, and mycoplasmas have all been shown to exacerbate respiratory disease caused by IBV (Jordan, 1972), although most studies have incriminated *Escherichia coli* and *Mycoplasma synoviae*. Fabricant and Levine (1962) and Springer *et al.* (1974) found that while synergism, as based on severity of air sac lesions, could be demonstrated following inoculation with IBV and either *E. coli* or *M. synoviae*, the most severe infection resulted from inoculation of the three agents together. However, Smith *et al.* (1985) were able to reproduce a highly lethal infection, closely resembling the natural disease, by intranasal inoculation of IBV and invasive serotypes of *E. coli*. Infection of similar severity was caused if the *E. coli* were given at any time between 7 days before and 7 days after the virus. Some insight into the mechanism involved was provided by the observation that, when inoculated intranasally alone, the *E. coli* replicated poorly in the upper respiratory tract and did not invade. However, when IBV was included in the inoculum, *E. coli* multiplied to very high titer in the respiratory tract with subsequent penetration of the invasive *E. coli* strains but not of commensal coliforms. These authors suggested that the *E. coli* were able to invade as a result of the damage that IBV had caused in the respiratory tract, a finding confirmed by Nakamura *et al.* (1992), who showed that histopathological lesions in the respiratory tract were more severe when IB and *E. coli* were inoculated together than when either agent was inoculated alone.

## D. Immunity

It is well established that the chicken responds to IBV infection by producing specific antibodies of the immunoglobulin IgM, IgG, and IgA classes, although the main serum antibody is IgG. Chickens with high levels of specific humoral antibodies can still be susceptible to infection, however, suggesting that immune factors other than humoral antibody are involved in protection.

IgM is the first class of antibody to be detected in the serum following IBV infection; peak concentration is reached about 8 days after infection and levels then decline (Mockett and Cook, 1986). As the presence of this antibody is only transitory, it may be useful in the diagnosis of recent infection (Martins *et al.*, 1991). Since plasma cells soon switch from producing specific IgM to IgG, the latter immunoglobulin, although present at relatively low concentrations 7 days after infection, reaches highest concentrations by about 10 to 14 days, and these levels are maintained for a considerable time. The antiviral IgG response, shown by enzyme-linked immunosorbent assay (ELISA), is similar to that shown by the hemagglutination inhibition test, although the former is much more sensitive. Virus-neutralizing (VN) antibodies are detectable from 9 days after infection, but take longer to reach a peak (Mockett and Darbyshire, 1981). Similar antibody profiles to both the spike and membrane protein have been demonstrated using ELISA (Mockett, 1985). It is generally believed that the spike protein is responsible for inducing VN antibodies and protective immunity (Mockett, 1985; Tomley *et al.*, 1987; Cavanagh *et al.*, 1984, 1988), although recent evidence suggests that such immunity may also be induced by the nucleocapsid protein (Boots *et al.*, 1992).

High concentrations of specific VN antibodies are thought to be important in preventing dissemination of the virus from its primary replication site in the respiratory tract (Box, 1988; Otsuki, 1988). Hence, such antibodies may prevent viral replication in and damage to the reproductive tract and the kidneys. It is also known that high concentrations of maternally derived antibodies, which are transmitted from the hen to the chick via the yolk, can help prevent infection, although for only until 1 to 2 weeks of age (Mockett *et al.*, 1987).

IBV-specific local antibodies have been detected in lachrymal fluid and in nasal and tracheal washings. The antibodies are thought to be produced by plasma cells in lymphoid tissue such as the Harderian gland. The production of both IgG and IgA has been reported (Holmes, 1977; Davelaar *et al.*, 1982; Hawkes *et al.*, 1983; Lütticken *et al.*, 1988). Recent work has suggested that the amount of IgA produced in saliva and lachrymal fluid after infection may be related to the reduced pathogenicity of IBV for certain inbred lines of chickens (Cook *et al.*, 1992). It is interesting to note that aerosol vaccination of day-old chicks results in protective immunity, presumably because local immune mechanisms have been stimulated (Davelaar and Kouwenhoven, 1980).

Although the subject has received little attention, cell-mediated immune responses have been reported following IBV infection. Systemic cell-mediated immunity (CMI) has been detected using a lymphoblast transformation technique (Timms *et al.*, 1980). There was no direct correlation between CMI and humoral antibody production (Timms and Bracewell, 1981). CMI has been demonstrated to the S1, N, and M proteins of IBV (Ignjatovic and Galli, 1993). Although in that work only the S1 glycoprotein protected against challenge, Boots *et al.* (1992) demonstrated that a recombinant product expressing the N protein induced a CMI response and some protective immunity.

## E. Host Factors

### 1. Age Resistance

Although chickens of all ages are susceptible to infection, very young chicks tend to show more severe respiratory signs than older ones and morbidity is generally only observed in very young chicks. Smith *et al.* (1985) found that chicks rapidly became more resistant to a mixed IBV and *E. coli* infection with increasing age. Whereas approximately 90% mortality was recorded following day-old inoculation, this fell to 20% when inoculation was delayed until 6 weeks.

### 2. Genetic Resistance

The availability of genetically defined, inbred lines of chicken has enabled genetic differences in the response to IBV infection to be studied. Purchase *et al.* (1966) demonstrated some variation in mortality following IBV inoculation of embryos from different inbred lines. Using a mixed IBV and *E. coli* infection, Smith *et al.* (1985) and Bumstead *et al.* (1989) have demonstrated large differences in mortality, ranging from 3 to 87%, in different inbred and partially inbred lines of chicken. However, even in the more resistant lines clinical signs of respiratory infection were recorded and pericarditis and peritonitis observed in some of the chicks, indicating that the severity of the IBV infection was sufficient to allow *E. coli* to invade, although not to cause a lethal infection. As a result of challenging with IBV or *E. coli* alone, Bumstead *et al.* (1989) concluded that the variation in mortality in different lines reflected differences in resistance to IBV rather than to *E. coli*. Examination of F1 crosses suggested that mortality differences could not be attributed to maternal effects and that inheritance of resistance was fully dominant. The mortality pattern in F2 and backcross generations was compatible with the inheritance of a dominant autosomal resistance gene and showed no evidence of association with the major histocompatibility complex.

During this study, it was found that up to 87% mortality occurred in some inbred lines of chicken following intranasal inoculation of IBV alone. This enabled a comparison to be made of different parameters of an IBV infection in highly resistant (White Leghorn line C) and highly susceptible (White Leghorn line 15I) chickens. It was found (Otsuki *et al.*, 1990; Nakamura *et al.*, 1991) that the respiratory infection and degree of damage to the trachea were more severe and longer lasting in the susceptible line. While similar amounts of virus were recovered initially from tissues of both lines, overall more was recovered and for longer from the susceptible line. However, tissues from each line were equally susceptible to *in vitro* inoculation of IBV, suggesting that resistance to IBV infection is not related to a lack of specific viral receptors by target cells of the resistant line, as is the case with murine hepatitis virus (Boyle *et al.*, 1987). This finding and the observation of some increased susceptibility to IB infection in bursectomized chicks (Cook *et al.*, 1991a) suggests that some other mechanism must play a role in resistance to IBV infections. The nature of this mechanism has yet to be identified.

## F. Viral Factors

The ability of IBV to vary antigenically has been known for many years, based mainly on the identification of different serological types using virus neutralizing tests (Hopkins, 1974; Darbyshire *et al.*, 1979). As techniques for performing such tests have improved, it has became increasingly easy to define serotypes clearly. Many new serotypes have now been identified, particularly in Europe where most of such work has been done (Davelaar *et al.*, 1983; Cook, 1984; Cook and Huggins, 1986), but also in other countries, for example, Chile (Cubillos *et al.*, 1991) and the United States (Gelb *et al.*, 1991) (Table I). One reason for the increase in numbers of antigenic types of IBV may be the widespread use of live attenuated IB vaccines, leading to the preferential multiplication of virus populations able to escape neutralization by vaccine-induced antibodies (Kusters *et al.*, 1987). However, nucleotide sequencing has indicated that some IBV isolates are genetic recombinants. The possession by two serologically distinct IBV strains of very similar M but very distinct S1 proteins was suggested as evidence of recombination (Cavanagh and Davis, 1988). Kusters *et al.* (1989) compared S1 sequence data from several IBV isolates with T1 fingerprints of the entire IBV genome and reached the same conclusion. Using nucleotide sequencing, Cavanagh *et al.* (1992a) showed that although the S

TABLE I. Cross-Neutralization between IBV Isolates from Different Countries[a]

| Antiserum | UK | Holland | France | Germany | Italy | Portugal | Israel | Morocco | S. Africa | Chile | USA | Australia |
|---|---|---|---|---|---|---|---|---|---|---|---|---|
| UK | *380*[b] | —[c] | — | — | — | — | — | — | — | — | — | — |
| Holland | — | *870* | — | — | — | — | — | — | — | — | — | — |
| France | 20 | 20 | *3200* | — | 26 | — | 50 | — | 26 | — | — | — |
| W. Germany | — | — | — | *300* | NE | — | — | — | — | — | — | — |
| Italy | — | — | — | — | *420* | — | — | — | — | — | — | 20 |
| Portugal | — | — | — | — | — | *180* | — | — | — | — | — | — |
| Israel | — | — | — | — | — | — | *870* | — | — | — | — | — |
| Morocco | — | — | — | — | — | — | — | *400* | — | — | — | — |
| S. Africa | — | — | — | — | — | — | — | — | *490* | — | — | — |
| Chile | — | — | — | NE | — | — | — | — | NE | *1100* | NE | NE |
| USA | — | — | — | — | — | — | — | — | — | — | *244* | — |
| Australia | — | — | — | — | 20 | — | — | — | — | — | — | *1300* |

[a]Virus ($\log_{10}$ 2.0 $CD_{50}$).
[b]Reciprocal neutralization titer; homologous reaction in italics.
[c]Titer <1:10.
NE, Not examined.

sequence of a Massachusetts serotype strain (322/82) was very similar to Massachusetts, the 3C and M gene sequences showed greater homology to those of European non-Massachusetts serotypes, thereby providing evidence of recombination between IBV strains. Similarly, Wang *et al.* (1993) have shown a US field isolate of IBV to be a recombination between the Massachusetts and Arkansas serotypes, the recombination apparently having occurred within S1.

The finding of sharply defined serological differences between IBV strains, however, does not imply lack of *in vivo* cross-protection between them. Using as a challenge system a mixed IBV and *E. coli* infection, it has been shown (Cook *et al.*, 1986) that IBV strains may protect against challenge with antigenically distinct isolates and that protection is much more broadly based than might be inferred from serological data. It may eventually be shown that the grouping of strains into antigenic groups by the use of, for example, monoclonal antibody (MAb)-based ELISAs (Ignjatovic *et al.*, 1991) will correlate better with cross-protection data than do the results of virus neutralizing tests.

It is known that serotype-specific VN antibodies are induced by the highly variable amino-terminal S1 subunit of the spike glycoprotein (Mockett *et al.*, 1984; Cavanagh *et al.*, 1986; Karaca *et al.*, 1992). A surprising finding from sequence analysis of S1 of several British isolates belonging to different serotypes was the similarity of their S1 sequence, the strains differing by only 2% of their S1 residues (Cavanagh *et al.*, 1992b). This provides further evidence that only a few amino acids on S1 form epitopes responsible for inducing VN antibody. Thus, a few key amino acid substitutions can define a new IBV serotype. Analysis of MAb-resistant mutants has shown that an area near the N-terminus of S1 is associated with neutralization epitopes (Cavanagh *et al.*, 1988), and other data (Cavanagh *et al.*, 1992b; Kant *et al.*, 1992) suggest that the region of the first 300 N-terminal residues is an important hypervariable region. As has been shown with transmissible gastroenteritis virus (Jimenez *et al.*, 1986), it is probable that the tertiary structure of the IBV spike protein is important in defining antigenic sites (Koch and Kant, 1990), some of which may be formed by juxtapositioning of linearly well separated amino acids. Recent work (Boots *et al.*, 1991), using a mouse model, has suggested that antigenic variation may also occur in the nucleocapsid of IBV.

The availability of MAbs (Mockett *et al.*, 1984; Koch *et al.*, 1990; Ignjatovic and McWaters, 1991) and the application of molecular techniques to epidemiological studies are permitting the evolution of IBV strains to be studied. It is already clear that the situation is complex; Cavanagh *et al.* (1988) showed that strains of the Massachusetts serotype isolated over many years have retained an S1 protein with over 94% homology, while strains with only 50% homology coexist (Kusters *et al.*, 1989). It has also been shown (Cavanagh *et al.*, 1992b), using neutralizing MAbs, that isolates belonging to different serotypes (defined using pooled, polyclonal chicken antibodies) can share common neutralizing antibody-inducing epitopes. Thus we have examples of S gene homology both within and between serotypes. Much work remains to be done to clarify IBV evolution, a subject further complicated by genetic recombination, but it appears the IBV strains are evolving in different geographical areas by more than one line simultaneously (Cavanagh and Davis, 1992) but at varying rates, depending on many factors such as poultry density, vaccination procedures, and so on.

Progress in studying IBV evolution is now coming both from the application of MAb analysis and the use of the polymerase chain reaction (PCR), which has recently become a useful additional tool to study IBV. Lin *et al.*(1991a) identified a pair of oligonucleotides spaced 400 bases apart in a conserved region of S2. These should identify all IB viruses and indeed they have been used successfully to rapidly confirm a viral isolate as IBV (Parsons *et al.*, 1992). Oligonucleotides have now been identified that enable the DNA of only S1 to be synthesized (Kwon *et al.*, 1993a). PCR products have also been taken, the DNA cut using appropriate restriction enzymes, and IBV strains typed more rapidly than by using conventional virus neutralizing tests (Lin *et al.*, 1991b; Kwon *et al.*, 1993a).

## VI. DIAGNOSIS

### A. Clinical

Avian infectious bronchitis infection may be suspected in a flock because of the clinical signs, which include respiratory signs, drops in egg production, or nephritis. However, other diseases such as Newcastle disease, infectious laryngotracheitis, and turkey rhinotracheitis can produce similar clinical signs. Hence, in order to diagnose an IBV infection it is important to isolate the virus and to demonstrate a specific antibody response.

### B. Virus Isolation

There are a number of ways to propagate IBV, principally tracheal organ cultures, embryonated eggs, and tissue culture. Passage of IBV in embryonated eggs inoculated via the allantoic route may produce no effects on the embryo initially, but after two or three passages dwarfing or death of the embryo is usually observed and high titers of infectious virus can be present in the allantoic fluid. Probably all IBV strains grow in tracheal organ cultures or embryonated eggs but adaption is required for growth in tissue culture. Hence, this substrate is not used for primary virus isolation. As mentioned above, swabs of nasal discharge collected in the early stages of the disease are a rich source of virus as, in the later stages, is fecal material.

Infected allantoic fluid and tracheal organ culture supernatant fluid frequently contain enough virus particles to be visualised by electron microscopy. The typical size (about 80–100 nm) and shape (usually round with projecting 20-nm spikes) of the coronavirus (Berry *et al.*, 1964) enables a presumptive diagnosis to be made. However, the electron microscopy facility is usually found only in research establishments.

The immunofluorescence test is useful in confirming the identity of isolates as IBV, and a technique for classification based on neutralization of immunofluorescent foci (Csermelyi *et al.*, 1988) has been used successfully. Polyclonal antisera, which react with antigenic determinants common to all IBV strains, were used and smears of infected chorioallantoic membranes provided the substrate (Clarke *et al.*, 1972).

Antigen assays, based on MAbs, are now becoming more widely used. A MAb-based immunoperoxidase procedure has been found to be highly sensitive for detecting IBV in infected tissues or chorioallantoic membranes of infected embryos (Naqi, 1990). Yagyu and Ohta (1990) used an indirect immunofluorescence assay with a MAb specific for the IBV nucleocapsid protein and were able to detect IBV in tracheas of infected chickens for longer than was possible by virus isolation. Surprisingly, a streptavidin–biotin immunohistochemical assay was less sensitive than virus isolation (Owen *et al.*, 1991).

By means of the PCR and a biotin-labeled DNA probe it has been possible to amplify and detect IBV in allantoic fluid (Jackwood *et al.*, 1992), in tracheal swabs taken from chickens inoculated with different IBV serotypes, and even in allantoic fluid of embryos inoculated with field samples from IB outbreaks (Kwon *et al.*, 1993b). These authors claim that this technique detects IBV more rapidly and efficiently than more conventional methods.

With the availability of MAbs that are either group-specific or serotype-specific (Koch *et al.*, 1990; Karaca *et al.*, 1992), various ELISAs, such as antigen capture assays, have now been developed (Cavanagh *et al.*, 1992b; Naqi *et al.*, 1993) that allow detection of either all IBVs or particular serotypes. However, the sensitivity of such assays needs improvement (Naqi *et al.*, 1993). The large panel of MAbs now available in The Netherlands has enabled extensive characterization of existing IB strains to be carried out (Cavanagh *et al.*, 1992b).

## C. Serological Methods

A range of serological tests is available to detect IBV antibodies, including agar gel precipitation test, immunofluorescence, hemagglutination inhibition, ELISA, and virus neutralization. The agar gel precipitation test requires the presence of high concentrations of specific antibodies (precipitins) in order to produce a precipitin line; thus the test is relatively insensitive. Precipitins are only present for a short period after infection; MacDonald *et al.* (1982) suggested that their presence indicates recent exposure to the virus. The immunofluorescence test is a useful, relatively sensitive assay, but it has not been used extensively as a serological test for IBV.

The hemagglutination inhibition test gained favor when it was found that IBV could be made to hemagglutinate chicken red blood cells following treatment with crude preparations of the enzyme phospholipase C (Bingham *et al.*, 1975), although more recently neuraminidase has been shown to be the active constituent (Schultze *et al.*, 1992). Interestingly, Davelaar *et al.* (1983) have shown that certain IBV strains may spontaneously hemagglutinate chicken red blood cells. The hemagglutination inhibition test detects antibodies as early as 7 days after infection, and these antibodies persist for some time. This test is used routinely for flock monitoring and in most research or diagnostic laboratories. However, considerable care is required in the performance and interpretation of the test (Cook *et al.*, 1987), which should always be carried out under carefully controlled conditions such as those described by Alexander *et al.* (1983).

In some laboratories, the ELISA has gained acceptance. The test is very sensitive (Garcia and Bankowski, 1981) and can detect antibodies as early as 3 days after infection (Mockett and Darbyshire, 1981). ELISAs that use purified virus (from isopycnic sucrose gradients) as antigen produce highly specific results. As with the hemagglutination inhibition test, antibodies reach highest concentrations 14 to 17 days after infection, after which there is a gradual decline. It is probable that the ELISA will become the serological test of choice for monitoring responses to vaccination since it is especially useful for the large-scale screening of poultry flocks. However, neither the ELISA nor the hemagglutination inhibition test identify a particular serotype of IBV. For this, the virus neutralizing test performed in embryonated eggs (King and Cavanagh, 1991), cell culture after adaptation of the virus to that system (Hopkins, 1974), or tracheal organ cultures (Darbyshire *et al.*, 1979) was until recently the only available assay. However, the existence of strain-specific MAbs has led to the recent development of serotype-specific ELISAs (Karaca and Naqi, 1993). It seems likely that the use of such techniques will increase in the future.

Since a humoral antibody response is produced following vaccination, it is frequently necessary to be able to differentiate between the response to vaccination and to field challenge. An infection might be indicated by a rise in antibody titer to well above the anticipated vaccinal response level. In order to demonstrate evidence of a field challenge, it is necessary to take sequential sets of serum samples and demonstrate a rise in antibody titer between them.

## VII. CONTROL

### A. Vaccines

In most situations the only practical means of preventing IB is to vaccinate against the infection, and both live attenuated and inactivated vaccines are available. Live attenuated vaccines may be given by eyedrop, aerosol, or in the drinking water and are effective in the face of maternally derived immunity (Cook *et al.*, 1991b). With live vaccines it may be possible to prevent disease but not prevent infection. The more highly passaged virus strains, which have reduced pathogenicity, also have reduced immunogenicity. Furthermore, live vaccines may only induce short-lived immunity, and hence repeated vaccinations may be required. It is advisable to monitor regularly the specific antibody status of the flock.

Initially, most of the virulent IBV strains isolated were of the Massachusetts type. Indeed, it was thought for many years that there was only one serotype of IBV. However, other serologically distinct serotypes were subsequently isolated: Connecticut (Jungherr *et al.*, 1956) followed by others such as Arkansas (Johnson *et al.*, 1973) or 06 (Gelb *et al.*, 1991) in the United States;

D212 and D274 in Holland (Davelaar *et al.*, 1981); A to J in England (Cook and Ellis, 1988); PL84084 in France (Picault *et al.*, 1986); Az-23/74 in Italy (Zanella, 1988) and G in Morocco (El Houadfi *et al.*, 1986) (Table I). Different IBV serotypes appear to be present in distinct geographical areas. Thus the Arkansas serotype has not been isolated in Europe, while the D274 serotype has not been isolated in the United States. Even in particular countries distinct serotypes may be confined to local areas although more than one serotype can coexist in the same area (Gelb *et al.*, 1991).

Live vaccines have been developed to combat some of these serotypes, although it is clear that a new IB vaccine is not required for every new serotype since protection is more broadly based than serological tests might suggest (Cook *et al.*, 1986). From time to time, however, new serotypes emerge against which existing vaccines are not fully effective (Picault *et al.*, 1986; Cubillos *et al.*, 1991; Parsons *et al.*, 1992; Lambrechts *et al.*, 1993; Pensaert and Lambrecht, 1994); only in those areas where a known serotype regularly causes disease should a live vaccine containing that serotype be recommended.

It has been assumed that high concentrations of specific IgG antibody can protect the kidney and the reproductive tract from the effects of IBV. The aim therefore is to induce high concentrations of such antibody. This can be achieved by the use of an inactivated IB vaccine given as an oil-in-water emulsion to chickens that have already been "primed" by vaccination with a live-attenuated IB vaccine. The slow release of antigen stimulates and maintains the specific immune response for a considerable time. The timing of administration of inactivated vaccine is important. Chickens that have high concentrations of circulating antibody do not respond well to inactivated vaccine, possibly because the antibody reacts with the antigen and prevents stimulation of the B-cell response. If the correct time interval is allowed, however, a chicken already primed with a live IB vaccine produces a vigorous B-cell response to the inactivated vaccine.

## B. Management of Environmental Factors

Avian infectious bronchitis is a very labile but highly infectious virus. Thus, commonly used disinfectants can kill the virus, and material containing IBV is rapidly made noninfectious usually within 24 hr at room temperature. However, because IBV is readily spread by fomites and is airborne, careful procedures must be undertaken to keep premises clean, for example, by providing foot baths and restricted access to the premises.

The large numbers of chickens commonly housed together provide ideal conditions for the spread of IBV, and since the virus is highly infectious it spreads rapidly from chicken to chicken in the aerosols created by the coughing and sneezing of infected birds. Management should avoid any factors that might irritate the chicken's respiratory tract, for example, by preventing the build up of ammonia and ensuring adequate ventilation in poultry houses. It cannot be stressed too strongly that good vaccination programs go hand in hand with good management.

# VIII. REFERENCES


Alexander, D. J., Gough, R. E., and Pattison, M., 1978, A long-term study of the pathogenesis of infection of fowls with three strains of avian infectious bronchitis virus, *Res. Vet. Sci.* **24:**228.

Alexander, D. J., Allan, W. H., Biggs, P. M., Bracewell, C. D., Darbyshire, J. H., Dawson, P. S., Harris, A. H., Jordan, F. T. W., Macpherson, I., McFerran, J. B., Randall, C. J., Stuart, J. C., Swarbrick, O., and Wilding, G. P., 1983, A standard technique for haemagglutination inhibition tests for antibodies to avian infectious bronchitis virus, *Vet. Rec.* **113:**64.

Ambali, A. G., and Jones, R. C., 1990, Early pathogenesis in chicks of infection with an enterotropic strain of infectious bronchitis virus, *Avian Dis.* **34:**809.

Barr, D. A., Reece, R. L., O'Rourke, D., Button, C., and Faragher, J. T., 1988, Isolation of infectious bronchitis virus from a flock of racing pigeons, *Aust. Vet. J.* **65:**228.

Berry, D. M., Cruickshank, J. G., Chu, H. P., and Wells, R. J. H., 1964, The structure of infectious bronchitis virus, *Virology* **23:**403.

Bingham, R. W., Madge, M. H., and Tyrrell, D. A. J., 1975, Haemagglutination by avian infectious bronchitis virus—a coronavirus, *J. Gen. Virol.* **28:**381.

Biondi, E., and Schiavo, A., 1966, Susceptibility of various bird species to avian infectious bronchitis, *Acta Med. Vet. Napoli* **12:**537.

Blake, J. T., 1962, Effects of experimental chronic respiratory disease and infectious bronchitis on pullets, *Am. J. Vet. Res.* **23:**847.

Boots, A. M. H., Van Lierop, M. J., Kusters, J. G., Van Kooten, P. J. S., Van der Zeijst, B. A. M., and Hensen, E. J., 1991, MHC class II-restricted T-cell hybridomas recognizing the nucleocapsid protein of avian coronavirus IBV, *Immunology* **72:**10.

Boots, A. M. H., Benaissa-Trouw, B. J., Hesselink, W., Rijke, E., Schrier, C., and Hensen, E. J., 1992, Induction of anti-viral immune responses by immunization with recombinant-DNA encoded avian coronavirus nucleocapsid protein, *Vaccine* **10:**119.

Box, P., 1988, Infectious bronchitis—protection of layers and breeders, in: *Proceedings of the First International Symposium on infectious bronchitis* (E. F. Kaleta and U. Heffels-Redmann, eds.), pp. 289–297, Rauischholzhausen, West Germany.

Boyle, J. F., Weismiller, D. G., and Holmes, K. V., 1987, Genetic resistance to mouse hepatitis virus correlates with absence of virus binding activity on target tissues, *J. Virol.* **61:**185.

Broadfoot, D. I., Pomeroy, B. S., and Smith, W. M. Jr., 1956, Effects of infectious bronchitis in baby chicks, *Poultry Sci.* **35:**757.

Bumstead, N., Huggins, M. B., and Cook, J. K. A., 1989, Genetic differences in susceptibility to a mixture of avian infectious bronchitis virus and *Escherichia coli*, *Br. Poultry Sci.* **30:**39.

Cavanagh, D., and Davis, P. J., 1988, Evolution of avian coronavirus IBV: Sequence of the matrix glycoprotein gene and intergenic region of several serotypes, *J. Gen. Virol.* **69:**621.

Cavanagh, D., and Davis, P. J., 1992, Sequence analysis of strains of avian infectious bronchitis coronavirus isolated during the 1960s in the UK. (Brief report), *Arch. Virol.* **130:**471.

Cavanagh, D., Darbyshire, J. H., Davis, P., and Peters, R. W., 1984, Induction of humoral neutralising and haemagglutination-inhibiting antibody by the spike protein of avian infectious bronchitis virus, *Avian Pathol.* **13:**573.

Cavanagh, D., Davis, P. J., Darbyshire, J. H., and Peters, R. W., 1986, Coronavirus IBV: Virus retaining spike glycopolypeptide S2 but not S1 is unable to induce virus-neutralizing or haemagglutination-inhibiting antibody or induce chicken tracheal protection, *J. Gen. Virol.* **67:**1435.

Cavanagh, D., Davis, P. J., and Mockett, A. P. A., 1988, Amino acids within hypervariable region 1 of avian coronavirus IBV (Massachusetts serotype) spike glycoprotein are associated with neutralization epitopes, *Virus Res.* **11:**141.

Cavanagh, D., Davis, P. J., and Cook, J. K. A., 1992a, Infectious bronchitis virus: Evidence for recombination within the Massachusetts serotype, *Avian Pathol.* **21:**401.

Cavanagh, D., Davis, P. J., Cook, J. K. A., Li, D., Kant, A., and Koch, G., 1992b, Location of the amino acid differences in the S1 spike glycoprotein subunit of closely related serotypes of infectious bronchitis virus, *Avian Pathol.* **21:**33.

Clarke, J. K., McFerran, J. B., and Gay, F. W., 1972, Use of allantoic cells for the detection of avian infectious bronchitis virus, *Arch. Gesamte Virusforschung* **36:**62.

Cook, J. K. A., 1968, Duration of experimental infectious bronchitis in chickens, *Res. Vet. Sci.* **9**:506.
Cook, J. K. A., 1971, Recovery of infectious bronchitis virus from eggs and chicks produced by experimentally inoculated hens, *J. Comp. Pathol.* **81**:203.
Cook, J. K. A., 1984, The classification of new serotypes of infectious bronchitis virus isolated from poultry flocks in Britain between 1981–1983, *Avian Pathol.* **13**:733.
Cook, J. K. A., 1988, Epidemiology of IBV in Great Britain, in: *Proceedings of the First International Symposium on Infectious Bronchitis* (E. F. Kaleta and U. Heffels-Redmann, eds.), pp. 27–30, Rauischholzhausen, West Germany.
Cook, J. K. A., and Ellis, M. M., 1988, Long-term studies on serotyping European IBV strains, in: *Proceedings of the First International Symposium on Infectious Bronchitis* (E. F. Kaleta and U. Heffels-Redmann, eds.), pp. 224–228, Rauischholzhausen, West Germany.
Cook, J. K. A., and Garside, J. S., 1967, A study of the infectious bronchitis status of a group of chicks hatched from infectious bronchitis infected hens, *Res. Vet. Sci.* **8**:74.
Cook, J. K. A., and Huggins, M. B., 1986, Newly isolated serotypes of infectious bronchitis virus: Their role in disease, *Avian Pathol.* **15**:129.
Cook, J. K. A., Smith, H. W., and Huggins, M. B., 1986, Infectious bronchitis immunity: Its study in chickens experimentally infected with mixtures of infectious bronchitis virus and *Escherichia coli*, *J. Gen. Virol.* **67**:1427.
Cook, J. K. A., Brown, A. J., and Bracewell, C. D., 1987, Comparison of the haemagglutination inhibition test and the serum neutralisation test in tracheal organ cultures for typing infectious bronchitis virus strains, *Avian Pathol.* **16**:505.
Cook, J. K. A., Davison, T. F., Huggins, M. B., and McLaughlan, P., 1991a, Effect of *in ovo* bursectomy on the course of an infectious bronchitis virus infection in line C white leghorn chickens, *Arch. Virol.* **118**:225.
Cook, J. K. A., Huggins, M. B., and Ellis, M. M., 1991b, Use of an infectious bronchitis virus and *Escherichia coli* model infection to assess the ability to vaccinate successfully against infectious bronchitis in the presence of maternally-derived immunity, *Avian Pathol.* **20**:619.
Cook, J. K. A., Otsuki, K., Martins, N. R. da Silva, Ellis, M. M., and Huggins, M. B., 1992, The secretory antibody response of inbred lines of chicken to avian infectious bronchitis virus infection, *Avian Pathol.* **21**:681.
Csermelyi, M., Thijssen, R., Orthel, F., Burger, A. G., Kouwenhoven, B., and Lütticken, D., 1988, Serological classification of recent infectious bronchitis virus isolates by the neutralisation of immunofluorescent foci, *Avian Pathol.* **17**:139.
Cubillos, A., Ulloa, J., Cubillos, V., and Cook, J. K. A., 1991, Characterisation of strains of infectious bronchitis virus isolated in Chile, *Avian Pathol.* **20**:85.
Cumming, R. B., 1963, Infectious avian nephrosis (uraemia) in Australia, *Aust. Vet. J.* **39**:145.
Cumming, R. B., 1969, Studies on avian infectious bronchitis virus. 3. Attempts to infect other avian species with the virus, *Aust. Vet. J.* **45**:312.
Darbyshire, J. H., Rowell, J. G., Cook, J. K. A., and Peters, R. W., 1979, Taxonomic studies on strains of avian infectious bronchitis virus using neutralisation tests in tracheal organ cultures, *Arch. Virol.* **61**:227.
Davelaar, F. G., and Kouwenhoven, B., 1980, Vaccination of 1-day-old broilers against infectious bronchitis by eye drop application or coarse droplet spray and the effect of revaccination by spray, *Avian Pathol.* **9**:499.
Davelaar, F. G., Kouwenhoven, B., and Burger, A. G., 1981, Investigations into the significance of infectious bronchitis virus (IBV) variant strains in broiler and egg production, in: *Proceedings of the World Veterinary Poultry Association*, p. 44, International congress of WVPA, Oslo.
Davelaar, F. G., Noordzij, A., and van der Donk, J. A., 1982, A study on the synthesis and secretion of immunoglobulins by the Harderian gland of the fowl after eyedrop vaccination against infectious bronchitis at 1-day-old, *Avian Pathol.* **11**:63.
Davelaar, F. G., Kouwenhoven, B., and Burger, A. G., 1983, Experience with vaccination against infectious bronchitis in broilers and significance of and vaccination against infectious bronchitis variant viruses in breeders and layers in the Netherlands, *Clin. Vet.* **106**:7.
Doherty, P. C., 1967, Titration of avian infectious bronchitis virus in the tissues of experimentally infected chickens, *Aust. Vet. J.* **43**:575.

El-Houadfi, Md., Jones, R. C., Cook, J. K. A., and Ambali, A. G., 1986, The isolation and characterisation of six avian infectious bronchitis viruses isolated in Morocco, *Avian Pathol.* **15:**93.

Fabricant, J., and Levine, P. P., 1962, Experimental production of complicated chronic respiratory disease infection ("air sac" disease), *Avian Dis.* **6:**13.

Froyman, R., Derijcke, J., Meulemans, G., and Vaudermeersch, R., 1985, Infectious bronchitis-associated nephritis in broilers, *Vlaams Diergeneeskundig Tijdschrift* **54:**78.

Garcia, Z., and Bankowski, R. A., 1981, Comparison of a tissue-culture virus-neutralization test and the enzyme-linked immunosorbent assay for measurement of antibodies to infectious bronchitis, *Avian Dis.* **25:**121.

Gelb Jr., J., Wolff, J. B., and Moran, C. A., 1991, Variant serotypes of infectious bronchitis virus isolated from commercial layer and broiler chickens, *Avian Dis.* **35:**82.

Hawkes, R. A., Darbyshire, J. H., Peters, R. W., Mockett, A. P. A., and Cavanagh, D., 1983, Presence of viral antigens and antibody in the trachea of chickens infected with avian infectious bronchitis virus, *Avian Pathol.* **12:**331.

Hofstad, M. S., and Yoder, H. W., 1966, Avian infectious bronchitis—virus distribution in tissues of chicks, *Avian Dis.* **10:**230.

Holmes, H. C., 1977, A study of the local immune response of the chicken to viruses causing respiratory diseases, Ph.D. thesis, University of Surrey.

Hopkins, S. R., 1974, Serological comparisons of strains of infectious bronchitis virus using plaque-purified isolants, *Avian Dis.* **18:**231.

Ignjatovic, J., and Galli, L., 1993, Immune responses to structural proteins of avian infectious bronchitis virus, in: *Avian Immunology in Progress* (Les Colloques, No. 62). (INRA, Paris, F. Coudert, ed.), pp. 237–242, Institut National de la Recherche Agronomique, Paris.

Ignjatovic, J., and McWaters, P. G., 1991, Monoclonal antibodies to three structural proteins of avian infectious bronchitis virus: Characterization of epitopes and antigenic differentiation of Australian strains, *J. Gen. Virol.* **72:**2915.

Ignjatovic, J., McWaters, P. G., and Galli, L., 1991, Antigenic relationship of Australian infectious bronchitis viruses: Analysis using polyclonal and monoclonal antibodies, in: *Proceedings of the Second International Symposium on Infectious Bronchitis* (E. F. Kaleta and U. Heffels-Redmann, eds.), pp. 161–167, Rauischholzhausen, West Germany.

Jackwood, M. W., Kwon, H. M., and Hilt, D. A., 1992, Infectious bronchitis virus detection in allantoic fluid using the polymerase chain reaction and a DNA probe, *Avian Dis.* **36:**403.

Jimenez, G., Correa, I., Melgosa, M. P., Bullido, M. J., and Enjuanes, L., 1986, Critical epitopes in transmissible gastroenteritis virus neutralization, *J. Virol.* **60:**131.

Johnson, R. B., Marquardt, W. W., and Newman, J. A., 1973, A new serotype of infectious bronchitis virus responsible for respiratory disease in Arkansas broiler flocks, *Avian Dis.* **17:**518.

Jones, R. C., and Ambali, A. G., 1987, Re-excretion of an enterotropic infectious bronchitis virus by hens at point of lay after experimental infection at day-old, *Vet. Rec.* **120:**617.

Jones, R. C., and Jordan, F. T. W., 1970, The exposure of day-old chicks to infectious bronchitis and the subsequent development of the oviduct, *Vet. Rec.* **87:**504.

Jordan, F. T. W., 1972, The epidemiology of disease of multiple aetiology: The avian respiratory disease complex, *Vet. Rec.* **90:**556.

Jungherr, E. L., Chomiak, T. W., and Luginbuhl, R. E., 1956, Immunologic differences in strains of infectious bronchitis virus, in: *Proceedings of 60th Annual Meeting of U.S. Livestock Sanitary Association*, pp. 203–209, Chicago.

Kaleta, E. F., and Heffels-Redmann, U., eds., 1988, *Proceedings of the First International Symposium on Infectious Bronchitis*. Rauischholzhausen, Germany.

Kaleta, E. F., and Heffels-Redmann, U., eds., 1991, *Proceedings of the Second International Symposium on Infectious Bronchitis*. Rauischholzhausen, Germany.

Kant, A., Koch, G., van Roozelaar, D. J., Kusters, J. G., Poelwijk, F. A. J., and van der Zeijst, B. A. M., 1992, Location of antigenic sites defined by neutralising monoclonal antibodies on the S1 avian infectious bronchitis virus glycopolypeptide, *J. Gen. Virol.* **73:**591.

Karaca, K., and Naqi, S., 1993, A monoclonal antibody-blocking ELISA to detect serotype-specific infectious bronchitis virus antibodies, *Vet. Microbiol.* **34:**249.

Karaca, K., Naqi, S., and Gelb, J., 1992, Production and characterization of monoclonal antibodies to three infectious bronchitis virus serotypes, *Avian Dis.* **36:**903.

Kinde, H., Daft, B. M., Castro, A. E., Bickford, A. A., Gelb, J., and Reynolds, B., 1991, Viral pathogenesis of a nephropathogenic infectious bronchitis virus isolated from commercial pullets, *Avian Dis.* **35**:415.

King, D. J., and Cavanagh, D., 1991, Infectious bronchitis, in: *Diseases of Poultry* (B. W. Calnek, H. J. Barnes, C. W. Beard, W. M. Reid, and H. W. Yoder, eds.), 9th ed. pp. 471-484, Iowa State University Press, Ames.

Koch, G., and Kant, A., 1990, Binding of antibodies that strongly neutralize infectious bronchitis virus is dependent on the glycosylation of the viral peplomer protein, in: *Coronaviruses and Their Diseases* (D. Cavanagh and T. D. K. Brown, eds.), pp. 143–150, Plenum Press, New York.

Koch, G., Hartog, L., Kant, A., and van Roozelaar, D. J., 1990, Antigenic domains on the peplomer protein of avian infectious bronchitis virus: Correlation with biological functions, *J. Gen. Virol.* **71**:1925.

Kusters, J. G., Niesters, H. G. M., Bleumink-Pluym, N. M. C., Davelaar, F. G., Horzinek, M. C., and van der Zeijst, B. A. M., 1987, Molecular epidemiology of infectious bronchitis virus in the Netherlands, *J. Gen. Virol.* **68**:343.

Kusters, J. G., Niesters, H. G. M., Lenstra, J. A., Horzinek, M. C., and van der Zeijst, B. A. M., 1989, Phylogeny of antigenic variants of avian coronavirus IBV, *Virology* **169**:217.

Kwon, H. M., Jackwood, M. W., and Gelb, J., 1993a, Differentiation of infectious bronchitis virus serotypes using polymerase chain reaction and restriction fragment length polymorphism analysis, *Avian Dis.* **37**:194.

Kwon, H. M., Jackwood, M. W., Brown, T. P., and Hilt, D. A., 1993b, Polymerase chain reaction and a biotin-labelled DNA probe for detection of infectious bronchitis virus in chickens, *Avian Dis.* **37**:149.

Lambrechts, C., Pensaert, M., and Ducatelle, R., 1993, Challenge experiments to evaluate cross-protection induced at the trachea and kidney level by vaccine strains and Belgian nephropathogenic isolates of avian infectious bronchitis virus, *Avian Pathol.* **22**:577.

Lin, Z., Kato, A., Kudou, Y., and Ueda, S., 1991a, A new typing method for the avian infectious bronchitis virus using polymerase chain reaction and restriction enzyme fragment length polymorphism, *Arch. Virol.* **116**:19.

Lin, Z., Kato, A., Kudou, Y., Umeda, K., and Ueda, S., 1991b, Typing of recent infectious bronchitis virus isolates causing nephritis in chicken, *Arch. Virol.* **120**:145.

Lister, S. A., Beer, J. V., Gough, R. E., Holmes, R. G., Jones, J. M. W., and Orton, R. G., 1985, Outbreaks of nephritis in pheasants (*Phasianus colchicus*) with a possible coronavirus aetiology, *Vet. Rec.* **117**:612.

Lütticken, D., Rijke, E. O., Loeffen, T., and Hesselink, W. G., 1988, Aspects of local immune response to IBV, in: *Proceedings of the First International Symposium on Infectious Bronchitis* (E. F. Kaleta and U. Heffels-Redmann, eds.), pp. 173–181, Rauischholzhausen, West Germany.

MacDonald, J. W., Dagless, M. D., McMartin, D. A., Randall, C. J., Pattison, M., Early, J. L., and Aubrey, S., 1982, Field observations on serological responses to vaccine strains of infectious bronchitis virus administered by coarse spray and via the drinking water, *Avian Pathol.* **11**:537.

Martins, N. R. da Silva, Mockett, A. P. A., Barrett, A. D. T., and Cook, J. K. A., 1991, IgM responses in chicken serum to live and inactivated infectious bronchitis virus vaccines, *Avian Dis.* **35**:470.

Mockett, A. P. A., 1985, Envelope proteins of avian infectious bronchitis virus: Purification and biological properties, *J. Virol. Methods* **12**:271.

Mockett, A. P. A., and Cook, J. K. A., 1986, The detection of specific IgM to infectious bronchitis virus in chicken serum using an ELISA, *Avian Pathol.* **15**:437.

Mockett, A. P. A., and Darbyshire, J. H., 1981, Comparative studies with an enzyme-linked immunosorbent assay (ELISA) for antibodies to avian infectious bronchitis virus, *Avian Pathol.* **10**:1.

Mockett, A. P. A., Cavanagh, D., and Brown, T. D. K., 1984, Monoclonal antibodies to the S1 spike and membrane proteins of avian infectious bronchitis coronavirus strain Massachusetts M41, *J. Gen. Virol.* **65**:2281.

Mockett, A. P. A., Cook, J. K. A., and Huggins, M. B., 1987, Maternally derived antibody to infectious bronchitis virus: Its detection in chick trachea and serum and its role in protection, *Avian Pathol.* **16**:407.

Nakamura, K., Cook, J. K. A., Otsuki, K., Huggins, M. B., and Frazier, J. A., 1991, Comparative study of respiratory lesions in two chicken lines of different susceptibility infected with infectious bronchitis virus: Histology, ultrastructure and immunohistochemistry, *Avian Pathol.* **20:**241.

Nakamura, K., Cook, J. K. A., Frazier, J. A., and Narita, M., 1992, *Escherichia coli* multiplication and lesions in the respiratory tract of chickens inoculated with infectious bronchitis virus and/or *E. coli, Avian Dis.* **36:**881.

Naqi, S. A., 1990, A monoclonal antibody-based immunoperoxidase procedure for rapid detection of infectious bronchitis virus in infected tissues, *Avian Dis.* **34:**893.

Naqi, S. A., Karaca, K., and Bauman, B., 1993, A monoclonal antibody-based antigen capture enzyme-linked immunosorbent assay for identification of infectious bronchitis virus serotypes, *Avian Pathol.* **22:**555.

Nauwynck, H., and Pensaert, M., 1988, Studies on the pathogenesis of infections with a nephropathogenic variant of infectious bronchitis virus in chickens, in: *Proceedings of the First International Symposium on Infectious Bronchitis* (E. F. Kaleta and U. Heffels-Redmann, eds.), pp. 113–119, Rauischholzhausen, West Germany.

Otsuki, K., 1988, Epidemiology of avian infectious bronchitis in Japan, in: *Proceedings of the First International Symposium on Infectious Bronchitis* (E. F. Kaleta and U. Heffels-Redmann, eds.), pp. 76–83, Rauischholzhausen, West Germany.

Otsuki, K., Yamamoto, H., and Tsubokura, M., 1979, Studies on avian infectious bronchitis virus (IBV) 1. Resistance of IBV to chemical and physical treatments, *Arch. Virol.* **60:**25.

Otsuki, K., Huggins, M. B., and Cook, J. K. A., 1990, Comparison of the susceptibility to avian infectious bronchitis virus infection of two inbred lines of White Leghorn chickens, *Avian Pathol.* **19:**467.

Owen, R. L., Cowen, B. S., Hattel, A. L., Naqi, S. A., and Wilson, R. A., 1991, Detection of viral antigen following exposure of one-day-old chickens to the Holland 52 strain of infectious bronchitis virus, *Avian Pathol.* **20:**663.

Parsons, D., Ellis, M. M., Cavanagh, D., and Cook, J. K. A., 1992, Characterisation of an infectious bronchitis virus isolated from vaccinated broiler breeder flocks, *Vet. Rec.* **131:**408.

Pensaert, M., and Lambrechts, C., 1994, Vaccination of chickens against a Belgian nephropathogenic strain of infectious bronchitis virus B1648 using attenuated homologous and heterologous strains, *Avian Pathology* **23:**631.

Picault, J. P., Drouin, P., Guittet, M., Bennejean, G., Protais, J., L'Hospitalier, R., Gillet, J. P., Lamande, J., and Le Bachelier, A., 1986, Isolation, characterisation and preliminary cross-protection studies with a new pathogenic avian infectious bronchitis virus (strain PL-84084), *Avian Pathol.* **15:**367.

Picault, J. P., Duée, J. P., Gillet, J. P., Cook, J. K. A., Guittet, M., Bennejean, G., and Lamande, J., 1987, Etude d'un noveau coronavirus néphropathogène (CR-84221) isolé chez des poulets et des poules dans le nord de la France, *Recueil Méd. Vét.* **163:**269.

Purchase, H. G., Cunningham, C. H., and Burmester, B. R., 1966, Genetic differences among chicken embryos in response to inoculation with an isolate of infectious bronchitis virus, *Avian Dis.* **10:**162.

Schultze, B., Cavanagh, D., and Herrler, G., 1992, Neuraminidase treatment of avian infectious bronchitis coronavirus reveals a haemagglutinating activity that is dependent on sialic acid-containing receptors on erythrocytes, *Virology* **189:**792.

Smith, H. W., Cook, J. K. A., and Parsell, Z. E., 1985, The experimental infection of chickens with mixtures of infectious bronchitis virus and *Escherichia coli, J. Gen. Virol.* **66:**777.

Spackman, D., and Cameron, I. R. D., 1983, Isolation of infectious bronchitis virus from pheasants, *Vet. Rec.* **113:**354.

Springer, W. T., Luskus, C., and Pourciau, S. S., 1974, Infectious bronchitis and mixed infections of *Mycoplasma synoviae* and *Escherichia coli* in gnotobiotic chickens. 1. Synergistic role in the airsacculitis syndrome, *Infect. Immun.* **10:**578.

Timms, L. M., and Bracewell, C. D., 1981, Cell mediated and humoral immune response of chickens to live infectious bronchitis vaccines, *Res. Vet. Sci.* **31:**182.

Timms, L. M., Bracewell, C. D., and Alexander, D. J., 1980, Cell mediated and humoral immune response in chickens infected with avian infectious bronchitis, *Br. Vet. J.* **136:**349.

Tomley, F. M., Mockett, A. P. A., Boursnell, M. E. G., Binns, M. M., Cook, J. K. A., Brown, T. D. K., and Smith, G. L., 1987, Expression of the infectious bronchitis virus spike protein by recombinant vaccinia virus and induction of neutralising antibodies in vaccinated mice, *J. Gen. Virol.* **68:**2291.

Vindevogel, H., Gouffaux, M., and Duchatel, J. P. 1976, Resistance in the pigeon to avian infectious bronchitis virus, *Ann. Med. Vet.* **120:**45.

Wang, L., Junker, D., and Collisson, E. W., 1993, Evidence of natural recombination within the S1 gene of infectious bronchitis virus, *Virology* **192:**710.

Weisman, Y., Aronovici, A., and Malkinson, M., 1987, Prevalence of IBV antibodies in turkey breeding flocks in Israel, *Vet. Rec.* **120:**94.

Yagyu, K., and Ohta, S., 1990, Detection of infectious bronchitis virus antigen from experimentally infected chickens by indirect immunofluorescent assay with monoclonal antibody, *Avian Dis.* **34:**246.

Zanella, A., 1988, Avian infectious bronchitis: Properties and application of attenuated vaccine prepared with nephropathogenic strain AZ-23/74, in: *Proceedings of the First International Symposium on Infectious Bronchitis* (E. F. Kaleta and U. Heffels-Redmann, eds.), pp. 335–342, Rauischholzhausen, West Germany.

Zanella, A., Marchi, R., Mellano, D., and Ponti, W., 1988, Avian infectious bronchitis: Nephropathogenic and respiratory virus isolates and their spreading in Italy, in: *Proceedings of the First International Symposium on Infectious Bronchitis* (E. F. Kaleta and U. Heffels-Redmann, eds.), pp. 245–255, Rauischholzhausen, West Germany.

CHAPTER 16

# Molecular Basis of Transmissible Gastroenteritis Virus Epidemiology

LUIS ENJUANES AND BERNARD A. M. VAN DER ZEIJST

## I. HISTORY OF TRANSMISSIBLE GASTROENTERITIS AND CLOSELY RELATED CORONAVIRUSES

A disease with the characteristics of transmissible gastroenteritis (TGE) was first reported in 1935 (Smith, 1956). The viral etiology of TGE was demonstrated 11 years later by Doyle and Hutchings (1946) in the United States. During the next 20 years TGE was reported in all other continents (Table I). Apparently, the disease occurred first in those countries that had imported North American stock and then was also introduced by European stock (Woode, 1969). There is evidence that TGE was not new to the pig, but became important to the pig industry concurrently with its intensification (Woode, 1969).

Until two decades ago, TGE outbreaks in the United States and Europe occurred mainly as acute epizooties on breeding farms, particularly during the winter months. The disease incidence appeared to have a cyclic course. After an outbreak, the disease and the virus disappeared and the herd immunity gradually waned in the next 2 to 3 years. A new outbreak was the consequence of the reintroduction of transmissible gastroenteritis virus (TGEV). A change in this epizootic pattern has occurred in recent decades together with the further intensification of swine breeding. TGEV has become more enzootic, partic-

LUIS ENJUANES • Centro Nacional de Biotecnología, CSIC, Campus Universidad Autónoma, Cantoblanco, 28049 Madrid, Spain. BERNARD A. M. VAN DER ZEIJST • Institute of Infectious Diseases and Immunology, School of Veterinary Medicine, University of Utrecht, 3508 TD Utrecht, The Netherlands.

*The Coronaviridae*, edited by Stuart G. Siddell, Plenum Press, New York, 1995.

TABLE I. Initial Reports of Transmissible Gastroenteritis in Various Countries

| Year | Country or Continent | References |
|---|---|---|
| 1946 | United States | Doyle and Hutchings (1946); R. Wesley (personal communication) |
| 1956 | England | Pritchard (1987) |
| 1956 | Japan | Sasahara *et al.* (1958) |
| 1956 | France | Jestin *et al.* (1987a,b) |
| 1957 | Denmark | Woode (1969) |
| 1958 | China | Woode (1969) |
| 1959 | Germany | Woode (1969) |
| 1960 | Spain | Concellón Martínez (1960) |
| 1961 | Russia | Woode (1969) |
| 1961 | Poland | Woode (1969) |
| 1962 | Romania | Woode (1969) |
| 1962 | The Netherlands | Woode (1969) |
| 1966 | Australia | Woode (1969); not confirmed |
| 1966 | Africa | Woode (1969); not confirmed |
| 1966 | Czechoslovakia | Woode (1969) |
| 1967 | Canada | Woode (1969) |
| 1967 | Belgium | Woode (1969) |

ularly in large breeding farms in which, after a primary outbreak, the virus persists in infected weaned nursery pigs. This enzootic form of TGE cannot be recognized clinically and it has, therefore, become difficult to determine how widespread TGEV is in a country, unless serological surveys were performed (Pensaert *et al.*, 1993; Hill, 1989).

At present, TGE has dominantly been reported in developed areas, which could be caused by a higher incidence in these countries, but could also reflect that the disease is not an obligatory noticeable disease in developing countries (FAO, 1984). In Europe, the prevalence of TGE is decreasing. However, there are still pockets with a high incidence of TGE. For instance, in England only 3% of the national herd was seropositive for TGEV up to 1984 (Pritchard, 1987), but in East Anglia TGEV has been diagnosed virtually every year since 1956 (Pritchard, 1987); the last serious epizootic occurred from 1980 to 1982 (Pritchard and Cartwright, 1982; D. Paton, Central Veterinary Laboratory, UK, personal communication). Evidence of enzootic TGEV infection was found in 50% of pigs studied between 1981 and 1983 in East Anglia. The region, therefore, is a reservoir containing the virus. In general, TGEV outbreaks have diminished, but the virus is still present in Central Europe. In Belgium the number of seropositive fattening pigs has decreased from 15% (Callebaut *et al.*, 1989) to 7.6%, according to a seroepizootiological study carried out between 1989 and 1990. In this study 5% of the farms appeared to harbor seropositive animals (Pensaert *et al.*, 1992). In South European countries such as Spain, the incidence of TGE is low. Two geographical zones, Central and East Spain, can be differentiated. In Central Spain (Castilla and León), sera collected in 1985 and 1988 were seronegative for TGEV (Rubio *et al.*, 1987; Lanza *et al.*, 1993a) and the virus was only occasionally diagnosed, generally associated with swine imports (Laviada *et al.*, 1988; Lanza *et al.*, 1990). Along the Mediterranean Coast, epizootic outbreaks of TGE-like disease have been observed every 3 years since 1980 (Plana *et al.*, 1982; Anon, 1989). TGEV-specific antibodies have been found in sows of two different intensive pig-breeding areas: Murcia and Catalonia. A survey made in 1987 in the East Coast showed that 1.3% of the breeding pigs were seropositive and there was a prevalence of infection in 5% of the farms (Cubero *et al.*, 1990, 1993b).

In the United States, TGE is more of a problem. In 1987 and 1988, TGE was reported in about half of the swine herds (Hill, 1989). It was also a major cause of viral enteritis, since 26% of the cases with neonatal diarrhea submitted to diagnostic laboratories were caused by TGEV, a figure equal to the enteritis caused by *Escherichia coli* (Hoefling, 1989). In the United States, TGEV has created greater economic vulnerability resulting from growth retardation and increased susceptibility to other infectious diseases (Hoefling, 1989). But recently the incidence of TGEV has also decreased in the United States. The percentage of farms with at least one sample testing positive for TGEV was 35.8% [National Animal Health Monitoring System (NAHMS), 1992], lower than the 50% found in a survey of Midwestern swine in the early 1980s (Egan *et al.*, 1982). In the United States, 24.3% of the farms still vaccinate the sows or gilts against TGE (NAHMS, 1992).

Recently, a variant of TGEV, the porcine respiratory coronavirus (PRCV), has been described (Pensaert *et al.*, 1986; Callebaut *et al.*, 1988). The discovery of this virus was based on a survey carried out in 1984 in Belgium, which showed an increase of animals with antibodies to TGEV (up to 68%), with no increase in the incidence of TGE in the preceeding winter, and in the absence of vaccination (Pensaert *et al.*, 1987). In 1986 and 1987, the virus spread to 100% of the swine farms in Belgium, 67% of which are still infected. PRCV-free farms became infected during the autumn (Pensaert *et al.*, 1992). This pattern, showing higher incidence of PRCV in the cold seasons, has repeated itself in later years both in Belgium and in France (Laude *et al.*, 1993). The virus has been observed in many European countries (The Netherlands, Denmark, England, Spain, and France), where it has now spread to almost the whole swine population (Brown and Cartwright, 1986; Pensaert *et al.*, 1987; Jestin *et al.*, 1987a,b; Laviada *et al.*, 1988; Have, 1991; Sánchez *et al.*, 1990). PRCV was detected in Spain for the first time in September 1986 (Martin-Alonso *et al.*, 1992). Thirty-one percent of sera collected in Castilla-León (Central Spain) and 64% of the farms in this area were seropositive for PRCV (Lanza *et al.*, 1993a,b). In England, a survey in 1987 of over 300 elite herds with a high health status revealed that 80.1% of the herds contained seropositive pigs (D. Paton, personal communication). In the United States, serological tests detected PRCV for the first time in Indiana in 1989 (Wesley *et al.*, 1990). Since then, other PRCV isolates have been reported in the United States and Canada (Halbur *et al.*, 1993; Jabrane and Elazhary, 1993). PRCV strongly cross-reacts with its ancestor TGEV (Callebaut *et al.*, 1988; Rasschaert *et al.*, 1990; Sánchez *et al.*, 1990, 1992; Wesley *et al.*, 1991a). PRCV may evoke a protective response to TGEV (Hooyberghs *et al.*, 1988; Paton and Brown, 1990; De Diego *et al.*, 1992; Laude *et al.*, 1993; Wesley and Woods, 1993). PRCV has also been diagnosed in Asia and Eastern European countries (M. Pensaert and J. Musilova, personal communications).

A second enteric coronavirus, the porcine epidemic diarrhea virus (PEDV), serologically unrelated to TGEV, was isolated for the first time in Belgium in 1977 (Pensaert and Debouck, 1978) and then in England (Chasey and Cartwright, 1978) and Spain (Jiménez *et al.*, 1986a). PEDV is now present in all European countries. In 1983, two surveys performed in England, in the spring and in the autumn, showed 19.2 and 8.3% of the animals positive, respectively, which reflects the seasonal incidence of PED. In 1987, 9.2% of the sows tested were seropositive for PED in England. Currently, about 1–2% of the fecal samples submitted to the Central Veterinary Laboratory from all over England are PED positive. PEDV has not been found in the United States (R. Wesley, National Animal Disease Center, Iowa, personal communication). It is intriguing that during 35 years of research into TGE, PEDV had not been recognized before, suggesting that the infection is a recent event or that in the past its diagnosis was less efficient because TGEV-specific reagents were used to detect coronaviruses-causing enteritis.

## II. SUSCEPTIBILITY TO TGEV INFECTION

### A. Tropism and Host Cells

TGEV is basically an enteropathogen, although natural TGEV isolates can also replicate in the respiratory tract (Underdahl *et al.*, 1974; Kemeny *et al.*, 1975). The virus has occasionally been isolated from tonsils, trachea, and lungs. Signs of respiratory disorders were not observed despite the detection of pulmonary lesions (Kemeny *et al.*, 1975; Underdahl *et al.*, 1974, 1975). During infections with virulent isolates, the highest virus concentrations are found in the enteric tract. The jejunum and, to a lesser extent, the ileum and duodenal epithelium are the areas with most pathology. TGEV isolates that have been passaged in cell cultures gradually lose their tropism for the enteric tract, while they gain tropism for respiratory tissues (Harada *et al.*, 1969; Furuuchi *et al.*, 1978, 1979) and for alveolar macrophages, where TGEV replication is partially restricted (Laude *et al.*, 1984). Highly attenuated strains of TGEV, which replicate in the upper respiratory tract, including tonsils and lungs, do not replicate in the intestine of newborn pigs (Furuuchi *et al.*, 1979). TGEV replicates in the apical tubovascular system of villous absorptive cells in newborn pigs; this system is absent in pigs older than 3 weeks (Wagner *et al.*, 1973). The virus also grows in the mammary tissue of lactating sows (Saif and Bohl, 1983); infected sows shed virus in their milk (Kemeny and Woods, 1977). Intrafetal inoculation results in the production of villous atrophy (Redman *et al.*, 1978). The susceptibility of cells from different organs to TGEV should be carefully defined for each situation, since it is a function of the virus dose, the status of the swine used (gnotobiotic, colostrum deprived, or conventional), temperature, and age of the animal, among other factors (Furuuchi *et al.*, 1976; L. Saif, personal communication; M. L. Ballesteros and L. Enjuanes, unpublished results).

In infected animals, the virus can be recovered from macrophages and other cells of the reticuloendothelial system (Underdahl *et al.*, 1974). Macrophages from the intestinal mucosa (Chu *et al.*, 1982) and probably Küppfer cells are also infected, while porcine blood monocytes are not infected by TGEV (Laude *et al.*, 1984). The virus has also been isolated from mesenteric lymph nodes and Peyer's patches up to 9 days after infection. However, the infectious virus persisted only for 3–4 days in T-cell lines derived from these organs (M. J. Bullido and L. Enjuanes, unpublished results).

Even though TGEV usually causes an acute enteric disease, it also leads to persistent infections, in which the virus is not detected in the enteric tract (M. Pensaert, personal communication). Instead, the virus has been found in the respiratory tract of recovered pigs for more than 100 days after infection (Underdahl *et al.*, 1974, 1975). TGEV mutants can persistently infect adult swine with continuous virus shedding in the gut (R. Wesley and R. Roods, personal communication).

After oronasal inoculation, strain BEL85 of PRCV replicates in the nasal mucosa, tonsils, trachea, bronchi, bronchioles, and alveoli and alveolar macrophages, while in the intestine a few cells stained positively with immunofluorescence. Higher virus titers ($>10^8$ $TCID_{50/g}$ of tissue) have been detected in the lungs (O'Toole *et al.*, 1989; Cox *et al.*, 1990a; Laude *et al.*, 1993). Other tissues, including plasma (viremia was observed), mesenteric lymph nodes, and colon were consistently positive, while the virus was sporadically isolated from other lymph nodes, spleen, liver, and thymus. Even when the virus was inoculated directly to the lumen of the small intestine, a limited degree of virus replication, in a few enterocytes located in the transition from the crypts to the villi in the jejunum, was observed (Pensaert *et al.*, 1987). The cells were identified as villous enterocytes by electron microscopy (Popischil *et al.*, 1990) and immunocytochemistry (O'Toole *et al.*, 1989). PRCV is not found in rectal swab samples (Vancott *et al.*, 1993) unless it is administered to the pig artificially via a stomach tube (Wesley and Woods, 1993). A minimum of $10^3$ $TCID_{50}$ of PRCV were needed to start the infection in the intestinal tract (Cox *et al.*, 1990b). This result indicates that the gut is not the target organ for PRCV and that the ability of these coronaviruses to infect a tissue is not an all-or-nothing phenomenon.

It has been debated whether PRCV causes respiratory disease or not. Signs of respiratory disorders caused by PRCV have not been observed by Pensaert's group, either after field infections or after experimental aerosol inoculation, using procedures that produce clinical signs using respiratory viruses such as influenza viruses and Aujeszky disease virus (Pensaert *et al.*, 1987). Also, experimental inoculation of pigs with the British PRCV has not produced disease, other than very occasional and mild rhinitis, although histopathological lesions are present (O'Toole *et al.*, 1989). In England, isolation of PRCV has been associated with respiratory problems, frequently with concurrent swine influenza (Lanza *et al.*, 1992). In contrast, two groups (Duret *et al.*, 1988; Van Nieuwstadt and Pol, 1989) have reported that PRCV can cause pneumonia. The pathology caused by various PRCV strains isolated in the United States also range from inapparent to severe bronchointerstitial pneumonia, which has been observed more recently in gnotobiotic pigs. Lesions in conventional pigs were less severe, however. The pigs developed mild disease and fever (Halbur *et al.*, 1992). The Canadian isolate PRCV IQ90 produced morbidity and mortality rates reaching 100% and 60%, respectively (Jabrane and Elazhary, 1993). Conflicting results have been obtained with different PRCV isolates which differ in their genetic structure (see Section III.A.). This may explain the different results.

## B. Virus Receptors

Two cell surface proteins appeared to be relevant to the entry of TGEV into susceptible cells: aminopeptidase N (APN) and a recently described 200-kDa protein. APN, an N-terminal exopeptidase with preference for neutral amino acids, has been clearly shown to act as a major receptor for TGEV and PRCV in cultured cells (Delmas *et al.*, 1992, 1993). APN is identical to CD13, a surface protein abundantly expressed in the brush border membranes of intestinal epithelial cells and fibroblast, on the apical surfaces of lung and renal cells, as well as in granulocytes, monocytes, and their bone marrow progenitors (Norén *et al.*, 1986; Look *et al.*, 1989). APN plays a role in the digestion of peptides in the gut (Kenny *et al.*, 1987). The distribution of APN strongly suggests that it is also major receptor for TGEV and PRCV *in vivo*. Two key experiments provide evidence for the involvement of APN as a receptor for TGEV and PRCV. First, cells refractory to these viruses became susceptible after being transfected with the cDNA encoding porcine APN. Second, monoclonal antibodies (MAbs) specific for this protein efficiently block the multiplication of these viruses in cell culture. APN also is the major receptor for human coronavirus 229E (Yeager *et al.*, 1992) and for cytomegalovirus (Söderberg *et al.*, 1993).

Additional factors are probably involved in the susceptibility of intestinal epithelial cells both for TGEV and PRCV. Since intestinal cells are not infected (or only infected with extremely low efficiency) by PRCV, the existence of an additional factor involved in virus entry or in later steps of replication has been proposed (Sánchez *et al.*, 1992; Delmas *et al.*, 1993; Laude *et al.*, 1993). A similar situation exists in murine coronaviruses (Yokomori *et al.*, 1993). On the viral side, there is evidence that two regions of the genome might be involved: the 5′-terminus of the S gene is deleted or altered in viruses that have lost enteric tropism (Sánchez *et al.*, 1992; Delmas *et al.*, 1993; Wesley *et al.*, 1991a; Britton *et al.*, 1991); alternatively, the open reading frame (ORF) 3a that codes for a 71–72 amino acid nonstructural polypeptide (Godet *et al.*, 1992), which has been converted to a pseudo-gene in many respiratory virus (Rasschaert *et al.*, 1990; Wesley *et al.*, 1989; Laude *et al.*, 1993), might also be a determinant of tropism and virulence for coronaviruses related to TGEV. It was suggested that the high susceptibility of newborn piglets to TGEV infection and the tropism of the virus for villous enterocytes may be related to a 200-kDa protein. Experimental evidence for the existence of a second receptor binding site for TGEV has recently been provided (Weingartl and Derbyshire, 1993a,b). A saturable, specific binding of TGEV to the plasma membrane of the villous enterocytes in neonate swine has been shown. This binding is inhibited by MAbs that recognize a 200-kDa protein but not APN (Weingartl and Derbyshire, 1993a,b; 1994). This protein was present in tiny or undetectable amounts in cryptal enterocytes of newborn swine or villous enterocytes from weaned pigs, less susceptible to TGEV. At early times post-birth (less than 3-day old piglets) the villi are covered with APN containing epithelial cells, while the 200-kDa protein is only present on the tip of the villi and in a few epithelial cells of the villi.

## C. Effect of Age on Infection

A clear relationship between TGEV pathogenicity and the age of the infected animals has been established. It has been shown that the dose of virus needed to infect an adult swine is $10^4$-fold greater than that required to infect a neonate (Witte and Walther, 1976). Epizootic outbreaks affect all age groups and spread through the herd in 2 or 3 days. Only pigs up to 3 weeks of age often vomit and develop watery diarrhea that leads to dehydration, rapid weight loss, and death in 2 to 5 days. Mortality in these young pigs often approaches 100%. In weanlings (3 to 8 weeks old), mortality is usually less than 10 to 20%, but impaired feed adsorption causes growth retardation. Adult pigs become inappetent and develop diarrhea that usually lasts only 2 to 4 days. Although morbidity may approach 100%, mortality in this age group is usually less than 5% (Hill, 1989).

A sharp difference in the susceptibility of newborn swine to TGEV, depending on their access to colostrum and milk, even when the sows were seronegative for TGEV, has been observed (Furuuchi *et al.*, 1976; M. L. Ballesteros, C. M. Sánchez, J. Plana, and L. Enjuanes, unpublished results). These data suggest a specific effect of the colostrum on intestinal epithelial cell differentiation toward a TGEV-resistant state. This colostrum factor may influence the expression of the 200-kDa or APN TGEV receptors. Alternatively, factors present in the colostrum may interfere with binding of the virus to the receptors. Conventional colostrum-deprived newborn swine were fully susceptible (100% mortality) to high cell passage TGEV (PUR46-MAD strain), while 86% of their littermates that had access to colostrum and milk for 7 hr showed no disease symptoms (M. L. Ballesteros, C. M. Sánchez, J. Plana, and L. Enjuanes, unpublished results). Similar results have been obtained with gnotobiotic pigs and newborn animals that had access to colostrum, using attenuated strains of TGEV (M. Welter, personal communication). Thus, susceptibility to TGEV is a function of several host factors: age of animals, access to colostrum, and environmental conditions, in addition to the dose and virulence of the virus strain.

## D. Transmission of the Virus

The main reservoir for TGEV and related viruses is probably the pig. Both fattening swine and lactating sows are responsible for transmission of TGEV to young animals. Contamination originates from feces, the milk, and aerosols generated in the respiratory tract (Kemeny *et al.*, 1975; Kemeny and Woods,

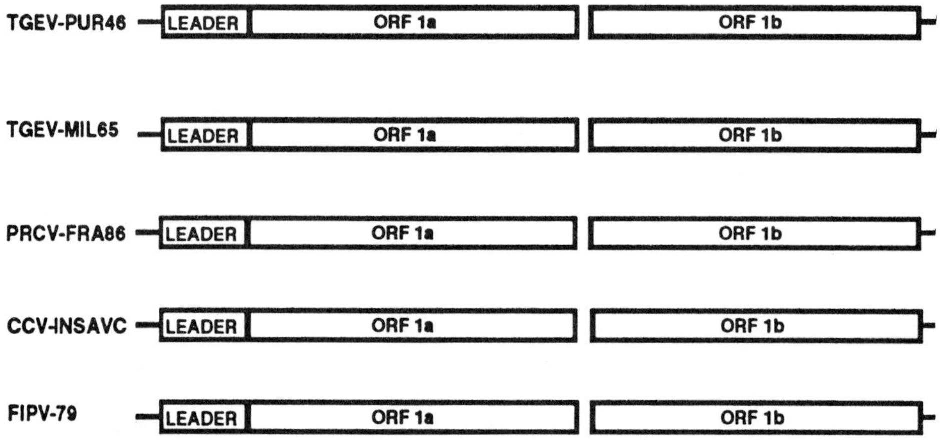


FIGURE 1.

1977; Kemeny, 1978). TGEV can also replicate in dogs, cats, foxes, starlings, domestic flies, and possibly in skunks, opossums, and muskrats (Saif and Wesley, 1992). No signs of infection have been detected in mice, rats, gerbils, or guinea pigs (Garwes, 1982).

TGEV replicates in the mammary gland of lactating sows, and infectious virus can be recovered from their milk and possibly be transmitted to piglets (Saif and Bohl, 1983). Intrauterine transfer of PRCV and TGEV has not been observed (Paton and Brown, 1990; Saif and Bohl, 1983). Transmission by the aerogenic route in TGEV epizootiology is under debate. Some authors suggest that it is relevant since infectious virus has been isolated 100 days after infection in the breath of infected pigs (Underdahl *et al.*, 1975; Torres-Medina, 1975). Other authors suggest that the aerogenic route is not important for TGEV, since the efficiency of spreading by this route was much lower than that for PRCV (Laude *et al.*, 1993). By contrast, PRCV spreads via air, based on the following findings: (1) aerosolized virus initiates the infection; (2) the virus disseminates very rapidly; (3) the virus transferred between farms with no apparent links; and (4) the virus spreads in countries with high hygienic standards, free of TGEV (Pensaert *et al.*, 1986; Jestin *et al.*, 1987a,b; Henningsen *et al.*, 1988; Have, 1991). PRCV is secreted orally nasally. There is no indication that the fecal–oral transmission plays a role in the epizootiology of the natural infection. The transmission of PEDV in infected farms in the center of Spain could not be easily explained, other than by the aerogenic route (Jiménez *et al.*, 1986a).

## III. ANTIGENIC AND GENETIC VARIATION

### A. Genome Organization and Virus Structure

TGEV contains a single-stranded positive-sense RNA genome of around 30 kb, which is infectious (Brian *et al.*, 1980), and generates eight mRNAs, includ-

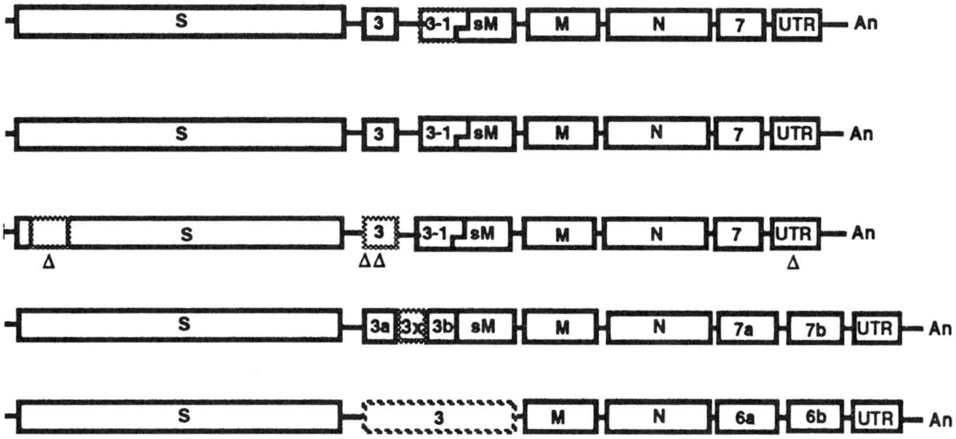


FIGURE 1. Comparison of the genomes of TGEV and related coronaviruses. ORFs are indicated. CCV ORF 3x is most likely not expressed. ORF 3 of FIPV is not well defined, although a 5.2-kb mRNA corresponding to this ORF has been described (de Groot *et al.*, 1987; Vennema *et al.*, 1992). Solid rectangles indicate identified or predicted ORFs; dotted rectangles indicate pseudogenes for which no mRNA has been detected; Δ, indicates presence of deletions; UTR, untranslated regions; An, poly A. The sequences used have been previously reported: TGEV-PUR46 (Rasschaert *et al.*, 1987), TGEV-MIL65 (Wesley *et al.*, 1989; Wesley, 1990); PRCV-FRA86 (Rasschaert *et al.*, 1990); CCV (Horsburgh *et al.*, 1992); FIPV (de Groot *et al.*, 1987; Jacobs *et al.*, 1987; Vennema *et al.*, 1992). The drawing is not to scale.

ing the genomic size mRNA (mRNA 1) and a nested set of seven subgenomic mRNAs. The mRNAs have a common leader of about 90 nucleotides (nt)-long (Page *et al.*, 1990; Sethna *et al.*, 1989) and poly A on the 3′ end (Jacobs *et al.*, 1986). These RNAs are named mRNAs 2 or S, 3, 3-1, 4 or sM, 5 or M, 6 or N, and 7, after the ORFs that are encoded by them (Fig. 1). The nomenclature used is according to the Coronavirus Study Group (Cavanagh *et al.*, 1990, 1994). These genes code for four structural proteins (S, sM, M, and N) (Jacobs *et al.*, 1986; Rasschaert *et al.*, 1987; Godet *et al.*, 1992). mRNA 7 codes for a potential fifth membrane-associated structural protein of 9 kDa (Tung *et al.*, 1992). A comparison of TGEV genome organization with that of closely related coronaviruses is shown (Fig. 1). TGEV, PRCV, canine coronavirus (CCV), and feline infectious peritonitis virus (FIPV) have a similar genome organization with some differences. FIPV and CCV have two ORFs at the 3′ end of the genome, instead of the single ORF7 in TGEV.

The most abundant structural proteins in TGEV are S, M, and N. S protein is the main inducer of neutralizing antibodies (Laude *et al.*, 1986; Jiménez *et al.*, 1986b). Four antigenic sites (A, B, C, and D) have been recognized on the spike protein; site A can be subdivided in three subsites (Aa, Ab, and Ac) (Correa *et al.*, 1988, 1990). These sites have been mapped on the S protein (Fig. 2) in the order C, B, D, and A, starting at the N-terminal end. Site C is located between amino acid 49 to 52; site B is between residues 97 to 144; site D between residues 382 to 389; and site A, around residues 538 (Aa), 543 (Ac), 586 (Aa-Ab), and 591 (Ab) (Correa *et al.*, 1990; Enjuanes *et al.*, 1990; Gebauer *et al.* 1991). Three of these

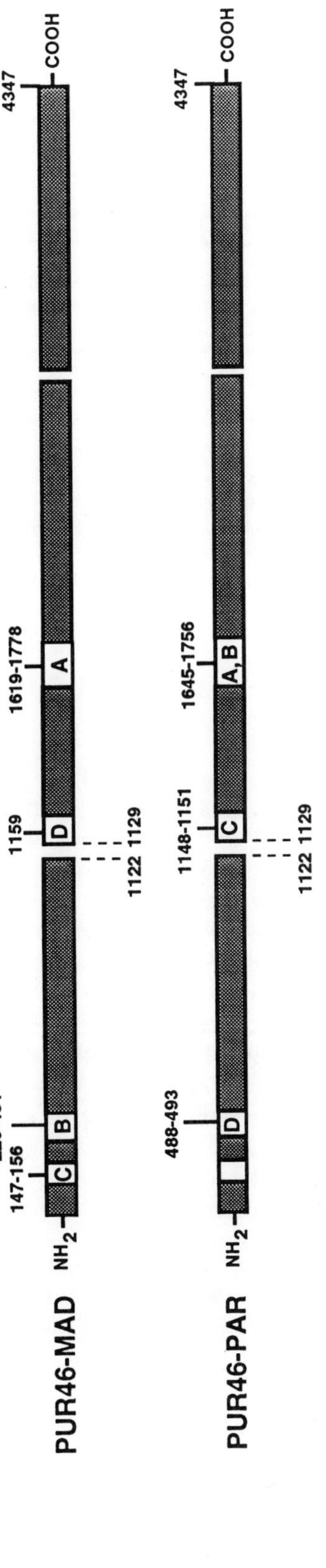


FIGURE 2. Correlation between the antigenic and the physical structure of TGEV S-glycoprotein. The nomenclature used for the antigenic sites by the groups at Paris (Delmas *et al.*, 1990a,b) and Madrid (Correa *et al.* 1988; Gebauer *et al.*, 1991) is indicated. The four antigen sites (A, B, C, and D) were defined by competitive radioimmunoassays using MAbs and the antigenic subsites by characterization of TGEV MAb-resistant mutants (Correa *et al.*, 1988). Numbers above the bars indicate changed residues responsible for altered amino acids in escape mutants selected with MAbs. Nucleotide numbers have been given according to the numbers for the PUR46 virus, adding six residues to compensate the deletion present between residues 1122 and 1129.

sites (B, D, and A) overlap with sites defined by Delmas *et al.* (1990a,b), who named them D, C, and A-B, respectively (Fig. 2). By competitive binding studies using MAbs from different laboratories, more than ten antigenic sites could be differentiated, five of which were involved in the neutralization of the virus, although site A was the major inducer of neutralizing antibodies in all laboratories (D. Garwes, H. Laude, R. Wesley, R. Woods, and L. Enjuanes, unpublished data). The second important sites involved in *in vitro* neutralization are sites D and B. Only some of the MAbs specific for these sites are neutralizing (Laude *et al.*, 1986; Jiménez *et al.*, 1986b; Correa *et al.*, 1988; Posthumus *et al.*, 1990a,b; Van Nieuwstadt *et al.*, 1988). Using neutralizing MAbs, Hohdatsu *et al.* (1987) have defined at least six different epitopes involved in the neutralization of TGEV isolates, which these authors classified into four groups. Interestingly, some critical epitopes for neutralization by MAbs appeared at high (176th) passage levels but were not present at a low (17th) passage of the Toyama strain of TGEV.

## B. Evolution of TGEV Structural Proteins S, M, N, and sM

The binding of MAbs to TGEV, PRCV, and related coronaviruses demonstrated antigenic diversity in the three major viral proteins (Fig. 3). Among TGEV isolates, there is more diversity in the S protein than in the M or N protein. This might be related to the important biological activities of the coronavirus S protein. It is involved in the fusion of infected cells (Frana *et al.*, 1985; Fazakerley *et al.*, 1992); it binds to receptors on the cell surface (Holmes *et al.*, 1981, 1989; Delmas *et al.*, 1992, 1993); it is involved in the pathogenesis of murine hepatitis virus (MHV) (Fleming *et al.*, 1986); and it is the major inducer of complement-independent neutralizing antibodies (Jiménez *et al.*, 1986b; Delmas *et al.*, 1986). Of the four antigenic sites defined (C, B, D, and A) (Correa *et al.*, 1988, 1990; Delmas *et al.*, 1986; Gebauer *et al.*, 1991), sites C and B were particularly variable. Extending the comparison to PRCV, FIPV, feline enteric coronavirus (FECV) and CCV, these isolates also showed a wide variation in the sites C and B. Site D has an intermediate level of conservation, while site A was most conserved, particularly subsite Ac, which is present in porcine, feline, and canine coronaviruses (Fig. 3). From an antigenic point of view, the main difference observed between the attenuated PRCVs and the virulent TGEVs is the absence of the B site on the virulent isolates: MIL65, SHI56, and MAD88 (Fig. 3) (Sánchez *et al.*, 1990), indicating that this site is different in the virulent strains.

Comparison of S protein sequences of TGEVs and PRCVs (Fig. 4) has shown that the S protein from three TGEV strains, MIL65, BRI70-FS772, and TOY56, has 1449 amino acids, two more residues than three clones of the high passage PUR46 strain and two vaccine strains (Sánchez *et al.*, 1990; Register and Wesley, 1994; C. M. Sánchez and L. Enjuanes, unpublished results). The two amino acid deletion is not present in the European PRCVs. An identity of 97% between the S-proteins of MIL65 and the PUR46 strains was observed at both amino acid and nucleotide level. Seventy-two percent of the amino acid changes were located within the N-terminal half of the S protein, which comprises the exposed globular-shaped portion of the peplomer. Only nine residue differences occur

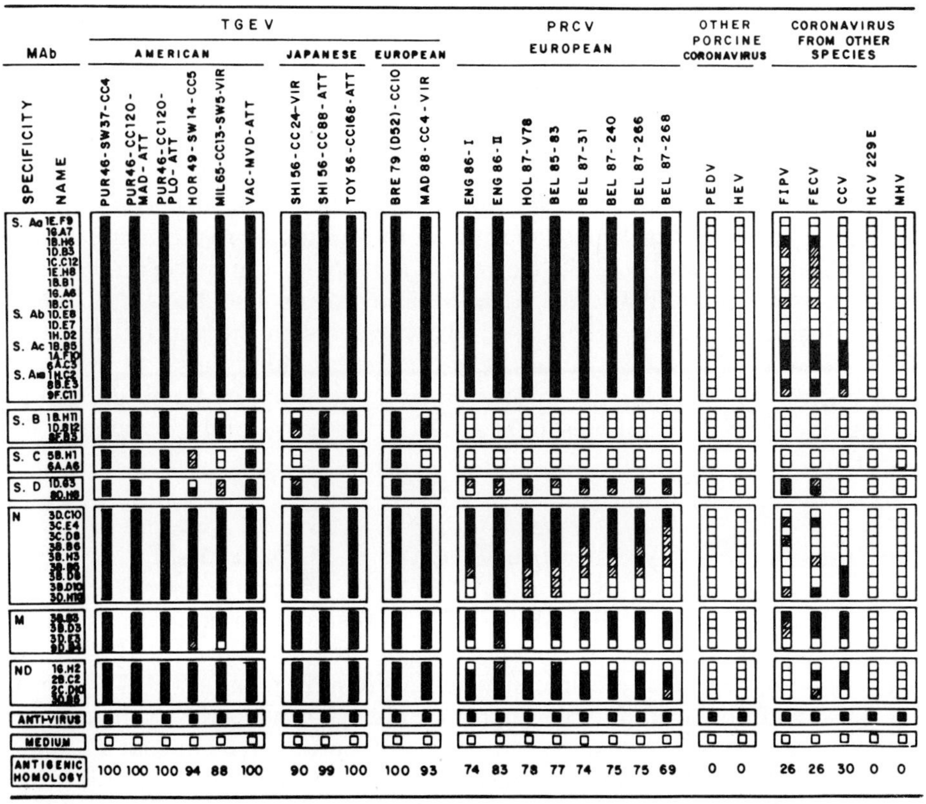


FIGURE 3. Binding of MAbs to coronaviruses. The value of the MAb binding to the PUR46-CC120-MAD strain, determined by RIA, was taken as the reference value (100). The characteristics of the viruses used and the specificity of the MAbs have been reported (Sánchez *et al.*, 1990). The antigenic homology of each virus isolate relative to the reference virus PUR46-CC120-MAD was expressed as a percentage (Sánchez *et al.*, 1990). Symbols: □, 0 to 30; ▨, 31 to 50; ■, 51 to 100. The antivirus sera were TGEV specific in the case of TGEV, PRCV, FIPV, FECV, and CCV and specific for the homologous virus in the case of PEDV, HEV, HCV 229E, and MHV. ND, not determined.

within the C-terminal half of the peplomer, which encompasses the stalk structure and the membrane anchoring domain. The N-terminal amino acid variation in the S-protein has also been detected in MHV and avian infectious bronchitis virus (IBV) (Luytjes *et al.*, 1987; Kusters *et al.*, 1989). In this region (S1) of IBV virus, there are two areas of high amino acid variability (Niesters *et al.* 1986; Cavanagh *et al.*, 1988). No clustering of amino acid changes indicative of highly variable domains were apparent in the S protein of TGEV. The amino acid changes in the S protein of MIL65 and PUR46 strains of TGEV are apparently substitutions that do not affect epitopes involved in *in vitro* TGEV neutralization, since neutralizing MAbs representing five different noncompeting sites were unable to distinguish between these TGEV strains (Wesley, 1990). All these observations on the S protein of TGEV indicate that the peplomer protein is highly conserved among TGEVs.

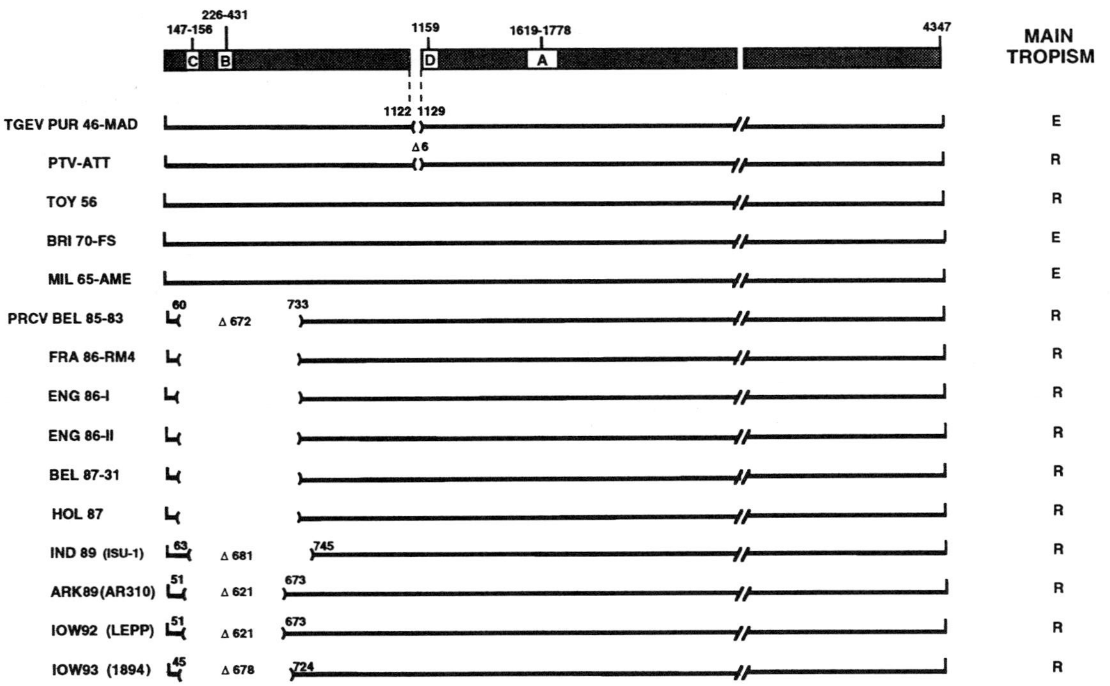


FIGURE 4. Simplified comparison of TGEV and PRCV S genes from European and American isolates. The bar indicates the S gene of PUR46-MAD. Antigenic sites have been identified by letters. Numbers above the bar indicate the location of nucleotides related to the antigenic sites on S protein. The lines represent S genes from other strains. The position of the deletions are indicated by brackets. Numbers above brackets indicate residues flanking the deletion. E and R indicate the predominant tropism of the isolate: enteric or respiratory, respectively. Sequence data were taken from different authors: PUR46-MAD, PTV-ATT (a Purdue type virus, attenuated, formerly named NEB72), TOY56, BEL85-83; ENG86-I, ENG86-II, BEL87-31, and HOL87 (Sánchez *et al.*, 1992); BRI70-FS (Britton and Page, 1990), MIL65-AME (Wesley, 1990); FRA86-RM (Rasschaert *et al.*, 1990); IND89 (Wesley *et al.*, 1991a,b), ARK89, IOW92, IOW93 (Halbur *et al.*, 1993; P. Paul, personal communication).

The genetic relationship among enteric and respiratory isolates of the TGEV cluster has been determined, based on the RNA sequences of the S protein (Sánchez *et al.*, 1992). An evolutionary tree relating these isolates indicated that a main virus lineage evolved from a recent progenitor virus that was circulating around 1941 (Fig. 5). From this, secondary lineages originated, PUR46, TOY56, MIL65, BRI70, and the PRCVs, in this order. Least-squares estimation of the origin of TGEV-related coronaviruses showed a significant constancy in the fixation mutation rate with time, that is, the existence of a well-defined molecular clock. A mutation fixation rate of $7 \pm 2 \times 10^{-4}$ nucleotide substitutions per site and per year was calculated for TGEV-related viruses. This rate falls into the range reported for other RNA viruses. Sequencing data of the S gene of PRCV virus indicate that there is a high similarity, as compared with TGEV, in the areas encoding antigenic sites A and D. A deletion of 672 nucleotides in the 5′ region, which in TGEV codes for sites C and B, leaves PRCV without these sites (Callebaut *et al.*, 1988; Rasschaert *et al.*, 1990; Sánchez *et al.*, 1990, 1992; Wesley *et al.*, 1990; Britton *et al.*, 1991). PRCV was detected about 40 years later than TGEV in Europe, where it spread very fast. All the European PRCVs have an identical deletion of 224 amino acids in the same position within the amino terminal half of the spike protein, suggesting that they were all derived from the same precursor (Sánchez *et al.*, 1992). In contrast, American PRCVs, which were detected for the first time in 1989 (Wesley *et al.*, 1990), have deletions of different sizes (207 to 227 amino acids) located in slightly different positions (Fig. 4) (Vaughn *et al.*, 1994; Halbur *et al.*, 1992, 1993; P. Paul, personal communication), suggesting that they originated independently.

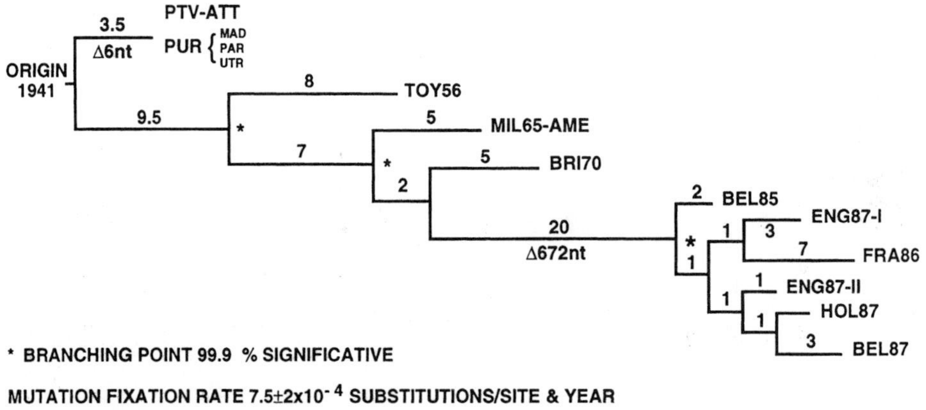


FIGURE 5. Evolutionary tree of TGEV related coronaviruses. Neighbor-joining and least-squares methods of tree reconstruction procedures were applied to the first 1956 nt of 13 virus isolates (Sánchez *et al.*, 1992). Numbers in the diagram indicate residue substitutions between branching points. Δ, Indicates the introduction of a deletion between branching points; *, indicates that all the descendents of this fork have, with a probability of 99.9%, a recent common ancestor.

A comparison of the primary structure of the peplomer proteins of TGEV and FIPV and of the 3′ end of their genome indicates that they are closely related, both in their structural and nonstructural proteins (de Groot *et al.*, 1987; Jacobs *et al.*, 1987). Sequence analysis of the S protein genes revealed one domain, from amino acids 1 to 274, in which the nucleotide sequences were 39% similar, whereas in the second domain, from residues 275 to 1447, the identity was 93%. Comparison of the 3′ ends of the FIPV and TGEV genomes revealed that the highest sequence similarity (98.5%) is in the 3′-noncoding sequences. Large insertions and deletions in the FIPV and TGEV genomes were detected that could be the result of RNA recombination events (de Groot *et al.*, 1988; Vennema *et al.*, 1992). The amino acid sequence of the CCV S protein has 91.1%, 81.6%, and 24% identity with FIPV, TGEV, and MHV, indicating a closer relationship with FIPV. This conclusion is reversed, however, if the M and N proteins are compared (Horsburgh *et al.*, 1992). The CCV S protein has 1452 amino acids, slightly larger that the 1447 or 1449 of TGEVs.

The identity between the M proteins of two TGEV strains (ENG70 and PUR46) is 98% at the nucleotide and amino acid level (Britton *et al.*, 1988a). There are 11 or 12 amino acid substitutions between the British and the PUR46 strain, as determined by Kapke *et al.* (1987) or by Laude *et al.* (1987), respectively.

In the TGEV N protein, three antigenic domains (A, B, and C) have been defined using MAbs (Martín-Alonso *et al.*, 1992). By studying the binding of these MAbs to virus isolates, it was shown that the three domains were highly conserved in TGEV isolates, but that B domain differs between TGEVs and PRCVs (Sánchez *et al.*, 1990). The amino acid sequences of the N protein genes from the British (ENG70) and the PUR46 strains of TGEV are 97.9% identical, and most of the changes are relatively conservative (Kapke and Brian, 1986; Britton *et al.*, 1988b). The data available on the three structural proteins of TGEV isolates indicate that there is also high conservation among isolates collected from remote areas and analyzed after different passage numbers in cell culture.

The sM protein of TGEV has been recently described (Godet *et al.*, 1992). A comparison of the nucleotide sequences of sM genes from virulent (BRI70, MIL65) or attenuated (PUR46, PTV-ATT) TGEVs and PRCVs (ENGII, FRA86) showed a high degree of conservation (Fig. 6) (C. M. Sánchez, M. L. Ballesteros, and L. Enjuanes, unpublished results).

## C. The Nonstructural Proteins

The genes coding for three nonstructural proteins of TGEV (nsp 3, 3-1, and 7) of seven strains of TGEV, including two European PRCVs (ENGII and FRA86) and one American isolate (IND89), have been aligned (Fig. 6). Gene 3 is one of the areas of major genetic variability among TGEVs. There are major differences between the noncoding areas of these genes (Fig. 6) (Wesley *et al.*, 1989, 1990, 1991a,b; Britton *et al.*, 1989, 1991; Rasschaert *et al.*, 1990; Laude *et al.*,

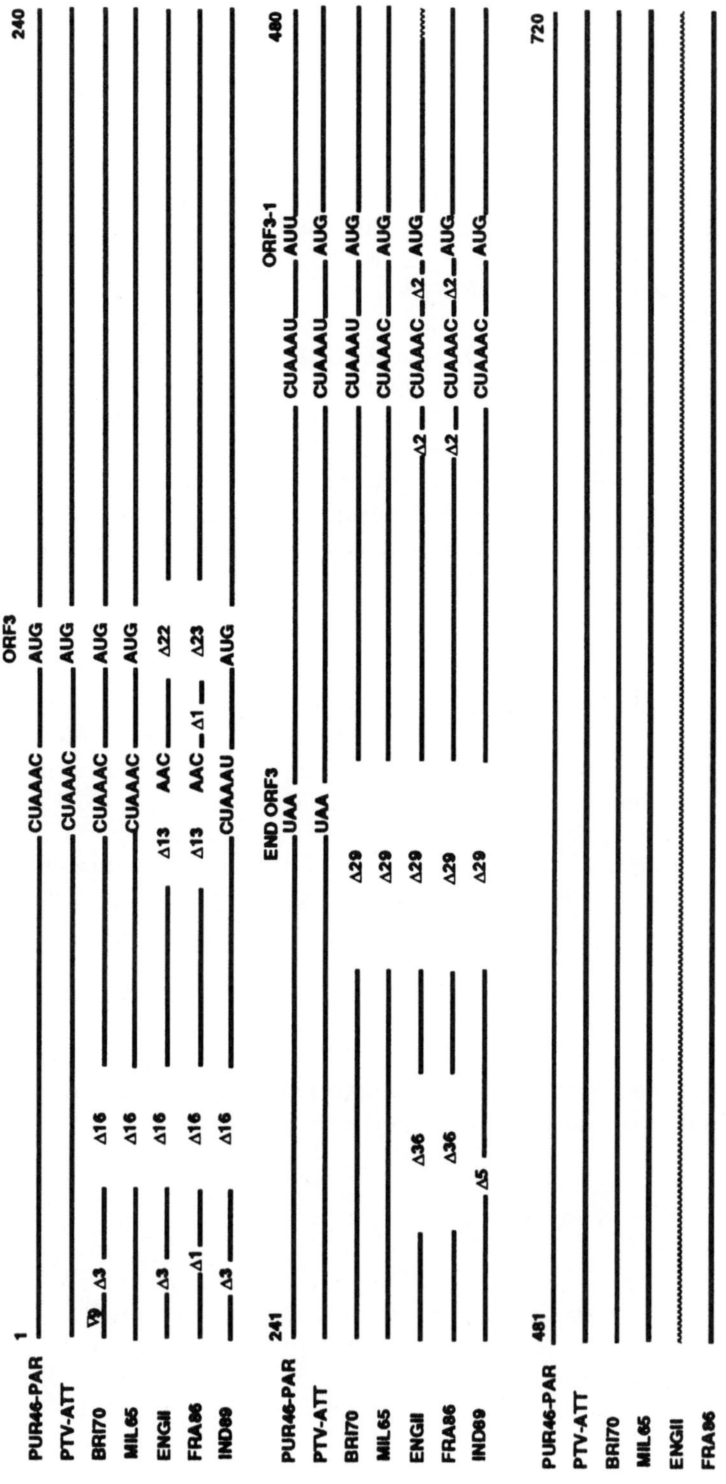

PUR46-PAR
PTV-ATT
BRI70
MIL65
ENGII
FRA86
IND89
1
240
241
480
481
720
ORF3
CUAAAC
AUG
CUAAAU
AAC
Δ3
Δ1
Δ16
Δ13
Δ22
Δ23
END ORF3
UAA
Δ5
Δ36
Δ29
Δ2
ORF3-1
AUU

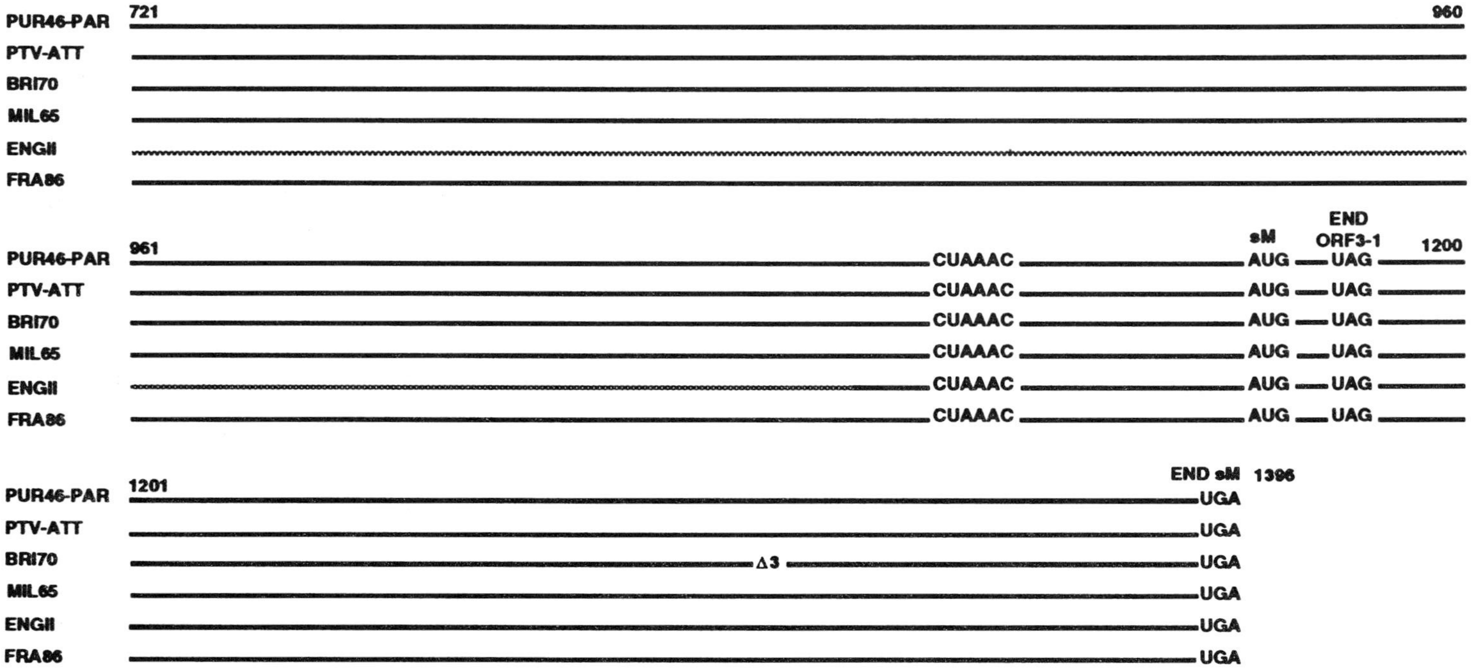


FIGURE 6. Simplified alignment of ORFs 3, 3-1, and sM of enteric and respiratory coronavirus of the TGEV group. Sequences of the indicated ORF were aligned using the Pileup program of Genetics Computer Group (Wisconsin University). Sequence starts after S gene termination codon. The approximate position and sequence of start and stop codons and of the consensus sequence are shown. The number of deleted nucleotides is shown after Δ. Solid and dotted bars represent determined or undetermined sequences. The sequences were obtained from: PUR46-PAR (Rasschaert *et al.*, 1987); PTV-ATT (Purdue type virus, attenuated, formerly named NEB72) (Sánchez *et al.*, 1992); BRI70 (Britton *et al.*, 1991); MIL65 (Wesley *et al.*, 1989); ENGII (Britton *et al.*, 1990b, 1991); FRA86 (Rasschaert *et al.*, 1990); IND89 (Wesley *et al.*, 1991a).

1993; C. M. Sánchez and L. Enjuanes, unpublished results). Between the 3′ end of the S gene and the start of gene 3, there are deletions in six different positions, in the seven viral isolates. In contrast, in ORFs 3-1 and 4 (translated and nontranslated areas), deletions were only observed in two and one position, respectively. The presence of a deletion (ENGII, FRA86) or a mutation (IND89) has altered the ORF 3 consensus sequence CUAAAC in PRCV isolates, preventing its expression. In contrast, ORF 3-1, which is expressed in PRCVs, is not functional in TGEVs isolates: PUR46, PTV-ATT, and BRI70 due to alterations in the consensus sequence. Finally, TGEV MIL65 strain expresses both 3 and 3-1 ORFs. In ORFs 3 and 3-1 amino acid identity was higher than 95%. The lack of expression of ORF3 in PRCVs and ORF3-1 in some TGEVs indicates that they are not essential for *in vitro* or *in vivo* replication. Although, alterations in these ORFs might be involved in the pathogenicity of these isolates.

CCV and FIPV have ORFs equivalent to the TGEV ORFs 3 and 3-1. In addition, CCV has an extra ORF (named 3x) that, to date, has not been detected in this group of viruses and could potentially encode a 71 amino acid protein overlapping with ORFs 3a and 3b of CCV. Due to the context of the ORF, it is very unlikely that it is expressed and probably represents an evolutionary redundant sequence that is no longer required by the virus (Horsburgh *et al.*, 1992). Amino acid identity in ORFs 3 and 3-1, between TGEV and CCV, is high: 83.5% and 92.7%, respectively.

The genomic organization at the 3′ end of TGEV differs from that of CCV and FIPV in that the latter viruses contain additional ORFs (Fig. 1) (de Groot *et al.*, 1988; Vennema *et al.*, 1992; Horsburgh *et al.*, 1992). At the 3′ end, TGEV has ORF 7, CCV has ORFs 7a and 7b, while FIPV has ORFs 6a and 6b. In addition, TGEV ORF 7 has a 69 nt deletion. These nucleotides are present in all FIPV 6a ORFs. The presence rather than absence of ORF 6b appears to be the common theme in FIPV and CCV, strongly suggesting that these nucleotides have been lost in TGEV (Vennema *et al.*, 1992). Paired comparison of the colinear parts of the amino acid sequences of the TGEV and CCV ORF 7, with FIPV ORF 6, have shown an identity of 96% and 77%, respectively (Vennema *et al.*, 1992).

The antigenic and genetic homology among the isolates of the TGEV cluster strongly suggests that they have diverged from a common ancestor. Differences in the antigenic and genomic maps of coronaviruses, including TGEV and FIPV, show that in their divergence certain transcription units have been lost, gained, or translocated (de Groot *et al.*, 1987, 1988; Sánchez *et al.*, 1990, 1992). There are three regions where deletions can occur at a higher frequency: within the S gene, between the S and M genes, and downstream of the N gene (Wesley *et al.*, 1990, Rasschaert *et al.*, 1990; Britton *et al.*, 1991; Sánchez *et al.*, 1992; Horsburgh *et al.*, 1992; Vennema *et al.*, 1992). The relevance of RNA recombination to the evolution of these viruses has been suggested by the occurrence of recombination during TGEV infections (Ballesteros *et al.*, 1994). The porcine, feline, and canine coronaviruses can infect the same cell type in the gut of their natural host species (Reynolds *et al.*, 1980; Stoddart and Scott, 1989; Woods *et al.*, 1981), which would facilitate recombination between them. More specifically, CCV (Woods and Wesley, 1986), FIPV (Woods *et al.*, 1981), and obviously TGEV can infect swine.

## IV. CLASSIFICATION OF TGEV AND RELATED CORONAVIRUSES

Based on the antigenic cross-reaction among coronaviruses, four antigenic groups have been defined (Siddell *et al.*, 1982; Sturman and Holmes, 1983; Wege *et al.*, 1982; see also Chapter 1, this volume), two of which include mammalian viruses and the other two, avian viruses. In this grouping several isolates, not all of which are yet recognized as coronavirus species, remain unclassified: namely, hemagglutinating encephalomyelitis virus (HEV) (Mengeling *et al.*, 1972); human enteric necrotic coronavirus (HENCV) (Resta *et al.*, 1985); rabbit coronavirus (RbCV) (Small and Woods, 1987); mink coronavirus [which has been tentatively related to TGEV and PEDV (Have *et al.*, 1992), although these results have not been confirmed], and cheetah coronavirus (ChCV), which is antigenically related to the feline coronaviruses by immunofluorescence (Evermann *et al.*, 1989). The RbCV could be related to TGEV based on *in vivo* protection experiments (Small and Woods, 1987), but there is evidence that this virus may have a gp65 glycoprotein and a proteolytically cleaved peplomer (Descoteaux *et al.*, 1985), both of which are not characteristic for the TGEV group.

TGEV and related viruses can be classified on the basis of their protein composition and antigenic properties. The protein composition of TGEV and related coronaviruses (Table II) shows that these viruses have three families of abundant proteins with molecular weight 160–220 kDa (S protein), 47–56 kDa (N protein), and 22–36 kDa (M protein) (Laude *et al.*, 1986; Jiménez *et al.*, 1986a,b; Cavanagh *et al.*, 1990, 1994). In addition, TGEV has another structural protein, sM (Godet *et al.*, 1992). Some coronaviruses have yet another structural glycoprotein (HE, gp65), which forms dimers through disulfide bridges. These dimers form a second "crown" below the one formed by the S protein and have hemagglutinating activity. TGEV and the related coronaviruses PRCV, FIPV, FECV, and CCV (Sánchez *et al.* 1990) do not have gp65. Other coronaviruses, including rat coronavirus (RCV), human coronavirus (HCV) OC43, HENCV, bovine coronavirus (BCV), and diarrhea virus of infant mice (DVIM) have gp65 as an abundant component. MHV and HEV contain variable amounts of the hemagglutinin esterase protein, depending on the particular isolate or the host cell.

Another characteristic useful for classification of coronaviruses is the cleavage of the spike protein into two halves. The peplomer protein of TGEV and related isolates (PRCV, FIPV, FECV, CCV, and HCV 229E) is not cleaved, in contrast to the antigenically unrelated coronaviruses: turkey coronavirus (TCV), HCV OC43, BCV, MHV, IBV, and, tentatively, HEV. These later viruses have a protease cleavage site on the peplomer protein that is absent on the uncleaved S proteins (Table II and references therein).

The binding of 42 MAbs, which recognized at least 25 epitopes, to enteric isolates of TGEV, collected during a period of 41 years in three distinct geographical areas (America, Europe, and Asia), PRCV isolates, and other coronaviruses, showed a strong antigenic homology among the corresponding viral proteins (S, M, and N) of TGEV, PRCV, FIPV, FECV, and CCV (Fig. 3). In contrast, no cross-reactivity with PEDV, HCV 229E, HCV OC43, HECV, BCV, or MHV

TABLE II. Coronavirus Proteins

| Protein | Virus | | | | | | | | | | | | | | | |
|---|---|---|---|---|---|---|---|---|---|---|---|---|---|---|---|---|
| | TGEV | PRCV | PEDV | CCV | FECV | FIPV | RbCV | TCV | HCV229E | HCVOC43 | HENCV | HEV | BCV | DVIM | MHV | IBV |
| S | 160–220 | <220 | 200 | 204 | 180–200 | 180–200 | | 180 | 180 | 190 | 190 | 180[c] | | 180 | 180 | 128 |
| S1 | — | — | | — | — | — | (82)[c] | | 107[d,e] | 110 | | (100)[c] | 120 | — | 90 | 90 |
| S2 | — | — | | — | — | — | (81)[c] | (95)[c] | 92[d,e] | 90 | | | 100 | — | 90 | 84 |
| | | | 70[b] | | | | 76[b] | 75[b] | | | | | | | | |
| | | | | | | | 71[b] | 72[b] | | | | | | | | |
| HE | — | — | — | — | — | — | (65)[c] | (66)[c] | ±[d,e] | 65 | 60 | ±[d] | 65 | 69 | ±[d] | — |
| N | 47–56 | 47–56 | 58 | 50 | 45–50 | 45–50 | 54 | 52 | 50 | 55 | 50 | 56 | 52 | 58 | 50 | 50–54 |
| | | | | | | | | | 39[e] | | | | | | | |
| M | 23–33 | 23–33 | 27–32 | 22–32 | 25–30 | 25–30 | 34 | 27 | 21–25 | 26 | 23 | 26 | 26 | 25 | 23 | 23–36 |
| References[a] | 1 | 2 | 3 | 4 | 5 | 6 | 7 | 8 | 9 | 10 | 11 | 12 | 13 | 14 | 15 | 16 |

[a]References: 1. Garwes and Pocock, 1975; Jiménez *et al.*, 1986a,b. An additional protein of 21 kDa has been reported by other authors (Horzinek *et al.*, 1982). 2. Callebaut *et al.*, 1988. 3. Egberink *et al.*, 1988; Knuchel *et al.*, 1992; Utiger *et al.*, 1993. 4. Garwes and Reynolds, 1981; Horzinek *et al.*, 1982. 5. Boyle *et al.*, 1984; Fiscus and Teramoto, 1987b. 6. Evermann *et al.*, 1981; Boyle *et al.*, 1984; Fiscus and Teramoto, 1987a,b. 7. Descoteaux *et al.*, 1985. 8. Dea and Tijssen, 1988. 9. Kemp *et al.*, 1984; Schmidt and Kenny, 1982; Spaan *et al.*, 1988. 10. Hogue and Brian, 1986. 11. Resta *et al.*, 1985. 12. Callebaut and Pensaert, 1980; Pocock and Garwes, 1977. 13. King and Brian, 1982; Hogue *et al.*, 1989; Parker *et al.*, 1989. 14. Sugiyama *et al.*, 1986. 15. Sturman and Holmes, 1977; Sturman, 1977. 16. Cavanagh, 1981.

[b]Possibly degradation products of higher-molecular-weight proteins.

[c]Bracketed numbers were tentatively assigned to a protein family.

[d]Cleavage occurs only at low rate or varies depending on the virus strain and its host cell (Spaan, 1990; Sturman *et al.*, 1985; Sugiyama *et al.*, 1986).

[e]Variable. Not confirmed.

was detected (Sánchez *et al.*, 1990). Interestingly, all TGEV, PRCV, FIPV, FECV, and CCV shared the antigenic subsite Ac, defined by three MAbs that neutralized all these isolates (Sánchez *et al.*, 1990). The presence of the antigenic subsite Ac in a coronavirus could be taken as the basis to define an antigenic cluster, which groups TGEV, PRCV, FIPV, FECV, and CCV with all members having: (1) common epitopes in the three structural proteins; (2) no conventional cleavage site in the peplomer protein; and (3) no gp65 glycoprotein.

HCV 229E was classified into the TGEV serological group (Pedersen *et al.*, 1978) based on the binding of polyvalent antibodies, as determined by immunofluorescence. In contrast, no reactivity was detected by radioimmunoassay (RIA) or ELISA (Sánchez *et al.*, 1990), with a panel of 42 MAbs (Correa *et al.*, 1988) or with a second collection of independently derived MAbs (Laude *et al.*, 1986; P. Talbot, personal communication). Also, the binding of TGEV- or HCV 229E-specific polyvalent antisera adsorbed with noninfected cells to these viruses in a RIA was absent or slight (Sánchez *et al.*, 1990). Reynolds *et al.* (1980) did not detect neutralizing activity for CCV or TGEV in the antisera specific to the 229E strain of human coronavirus. This suggests that the cross-reactivity observed by Pedersen *et al.* (1978) could be due to nonstructural antigenic determinants (Garwes, 1982). No cross-hybridization between the HCV 229E RNA and a cDNA probe complementary to the N and M genes of TGEV has been reported, while a strong hybridization was observed with TGEV, FIPV, and CCV (Shockley *et al.*, 1987).

Sequence data, in contrast to antigenicity and cross-hybridization data, have shown significant sequence homology in three genes coding for the main structural proteins of HCV 229E, PEDV, and other TGEV-related viruses. On the basis of S gene sequences, there is a closer relationship of PEDV with the TGEV-related subset than with the MHV-related subset and IBV (Duarte and Laude, 1994), confirming the reported sequence homology between these viruses in their N and M genes (Raabe and Siddell, 1989; Schreiber *et al.*, 1989; Bridgen *et al.*, 1993; Duarte *et al.*, 1994). A multiple alignment of the S protein sequences of PEDV, TGEV, FIPV, CCV, and HCV 229E has been constructed (Duarte and Laude, 1994), and a significant sequence relationship between PEDV and HCV 229E was found, with level of identity of 60% (S2 region) and 37% (S1 region). The percentage identity of the PEDV N protein with these coronaviruses ranged from 12 to 19% with MHV, IBV, HCV OC43, and BCV and from 32 to 37% with FIPV, CCV, PRCV, TGEV, and HCV 229E (Bridgen *et al.*, 1993). A comparison of the amino sequences of the M proteins showed that the HCV 229E M protein has a higher sequence similarity to the homologous protein of TGEV (HCV/TGEV, 68%; HCV/MHV, 58%; HCV/BCV, 57%; and HCV/IBV, 52%) (Raabe and Siddell, 1989).

Some cross-reactivity, as detected by immunoblotting, has been reported among the N proteins of various coronaviruses previously considered antigenically unrelated: FIPV, PEDV, and HEV (Yaling *et al.*, 1988); TGEV, HCV 229E, MHV 3, and HEV (Yassen and Johnson-Lussenburg, 1978); and MHV-A59 and IBV-M41 (H. G. M. Niesters, unpublished observations). More work is needed to determine whether all coronaviruses are antigenically related at the N-protein level.

In spite of the sequence homology between PEDV and viruses related to

TGEV, no antigenic cross-reaction has been detected using both polyvalent or MAbs with this group of viruses (Pensaert and De Bouck, 1978; Garwes and Reynolds, 1981; Callebaut *et al.*, 1988; Enjuanes *et al.*, 1990; C. M. Sánchez and L. Enjuanes, unpublished results). Also, no antigenic relationship was detected between TGEV and HEV. These results are interesting, in the case of PEDV, as it causes a disease similar to the one produced by TGEV. In summary, using sequencing data there is a cluster of viruses formed by TGEV, PRCV, CCV, FIPV, FECV, PEDV, and HCV 229E. Using serological criteria, an antigenic cluster including TGEV, PRCV, CCV, FIPV, and FECV is differentiated.

## V. DIAGNOSIS OF TGEV, PRCV, AND RELATED CORONAVIRUSES

A rapid diagnosis of TGE is important to discriminate it from enteritis caused by enteropathogenic *E. coli*, in order to determine if treatment with antibiotics is required. Several techniques have been developed for TGEV diagnosis: immunofluorescence, reversed passive hemagglutination, ELISA, RIA, and hybridization with DNA probes. TGEV antigens can be specifically detected by immunofluorescence. Coronaviruses of the TGEV cluster can be distinguished with type-, group- and interspecies-specific MAbs (Sánchez *et al.*, 1990). Reversed passive hemagglutination, based on the agglutination of erythrocytes coated with TGEV-specific antibodies, is more sensitive than immunofluorescent staining of primary cultures of porcine kidney cells inoculated with the specimens (Asagi *et al.*, 1986). It is also simple and rapid.

PRCV induces a serological response that originally could not be distinguished from that of TGEV-infected swine. This was a considerable drawback, since the movement of pigs between countries is frequently restricted until evidence that the stock is specifically free from TGEV infection has been obtained. This requires a diagnostic procedure capable of differentiating the two viruses. For this purpose, competitive RIA (Sánchez *et al.*, 1990) and ELISA (Callebaut *et al.*. 1989) with type- and group-specific MAbs have been developed. These assays are sensitive tests for the detection of epitope-specific antibodies, and provide no false-positive results. As a small percentage of false-negative results may occur with this test, a negative result must be confirmed, using several serum samples from the same farm. A cDNA clone containing 396 base pairs (bp) from the 5′ end of the TGEV S gene of MIL65 virus has been used to differentiate TGEV from PRCV. The probe also hybridizes to CCV but not to FIPV (Wesley *et al.*, 1991b).

To differentiate coronaviruses from other agents that cause similar disease and to determine the mechanism by which coronaviruses perpetuate enzootic or epizootic outbreaks, cloned cDNA probes, representing 2 kb from the TGEV genome, have been used in dot blot hybridization assays to detect viral RNA from cell culture and from fecal specimens. The cloned sequence encompasses the 3′-noncoding region, the nucleocapsid protein gene, and a large portion of the membrane protein gene. $^{32}$P-labeled cDNA probes prepared from these clones detected as little as 23 pg of homologous RNA, but did not detect RNA from the nonrelated virus, even when amounts of up to 10 ng per dot were used. In cell culture fluids, these probes detected TGEV, FIPV, and CCV, but not HCV 229E, HCV OC43, BCV, HEV, and MHV A59 (Shockley *et al.*, 1987).

To differentiate PEDV from TGEV and other porcine coronaviruses, an ELISA blocking test, based on crude virus preparations (Callebaut *et al.*, 1982), or an ELISA test that uses purified PEDV adapted to grow on Vero cells (Hofmann and Wyler, 1990) have been developed.

## VI. IMMUNE PROTECTION

The main economic losses caused by TGEV result from high mortality rates in newborn piglets, under the age of 10 days. These piglets do not have a fully mature immune system, and the time is too short to elicit a protective immune response. Passive immunity from colostrum and postcolostral milk is crucial in providing protection to neonates against TGEV infection (Abou-Youssef and Ristic, 1975; Bohl and Saif, 1975). Key experiments (Stone *et al.*, 1977; Wesley *et al.*, 1988; De Diego *et al.*, 1992) showed that both the IgG and sIgA immunoglobulin fractions of colostrum and postcolostral milk from immune sows confer protection when fed to susceptible piglets. Thus, newborn animals can be protected by natural lactogenic immunity provided by sows or by artificial lactogenic immunity using serum or protective MAbs.

To induce mammary antibodies, the sows have to be immunized 2 weeks before delivery, preferentially in the intestinal tract (Saif and Wesley, 1992). Stimulation of the mucosal humoral immune system against TGEV also can be induced by priming of the bronchus-associated lymphoid tissue (Cox *et al.*, 1993; De Diego *et al.* 1992; Wesley and Woods, 1993). Immunization in the mammary glands gave variable results. Some authors claimed protection (Bohl *et al.*, 1975; Saif and Bohl, 1983; Woods, 1984), while others (Aynaud *et al.*, 1986) did not find the intramammary route a valid alternative to the oral route. The available data agree on the low efficacy of the intramuscular route (Saif and Wesley, 1992; Moxley and Olson, 1989; Moxley *et al.*, 1989).

Three types of viruses have been used to develop lactogenic immunity: (1) virulent TGEV; (2) attenuated TGEV; and (3) TGEV-related coronaviruses (PRCV, FIPV, and CCV). The oral administration of nonattenuated TGEV in sows has generally resulted in protective levels of immunity for the sow and passive (lactogenic) immunity for suckling pigs (Haelterman, 1965; Bohl *et al.*, 1972; Saif and Wesley, 1992; Moxley and Olson, 1989; De Diego *et al.*, 1992). Because vaccination with nonattenuated TGEV has the obvious risk of spreading pathogenic virus (Bohl, 1982), vaccine research has been focused on attenuated and variant strains of TGEV (Chen and Kahn, 1985; Aynaud *et al.*, 1985, 1988; Fitzgerald *et al.*, 1986; Nguyen *et al.*, 1987; De Diego *et al.*, 1992; Cox *et al.*, 1993, Wesley and Woods, 1993) or on the use of related coronaviruses (Woods, 1984; Hooyberghs *et al.*, 1988). Attenuated TGEV vaccines have only limited efficacy. The problems were inconsistent results in their experimental evaluation, occurrence of epizooties, and persistence of TGEV in vaccinated herds (Saif and Wesley, 1992; Moxley and Olson, 1989).

Comparative studies of virulent and attenuated strains of TGEV revealed that the former are more stable to pancreatic enzymes (Chen, 1985; Furuuchi *et al.*, 1975, 1976), acidity (Chen, 1985; Hess and Bachmann, 1976; Laude, 1981), and porcine intestinal fluids (Chen, 1985). These properties may protect virulent TGEV from inactivation during its passage through the upper part of the gastrointestinal tract and permit viral replication in the small intestine, where it stimulates enteric immunity. A variant of TGEV (derived from the attenuated Purdue strain), resistant to trypsin and α-chymotrypsin, induced lactogenic immunity (Chen and Kahn, 1985). Using a survivor selection process in gastric juice, a mutant of TGEV, selected from a low passage strain (D-52), acquired simultaneous resistance to acidity and pepsin or trypsin cleavage (Aynaud *et al.*, 1985; Nguyen *et al.*, 1987). This mutant induced protective immunity when administered by the oral route but not by the intramuscular or intramammary routes (Aynaud *et al.*, 1988). Molecular characterization of attenuated vaccine strains of TGEV has shown alterations in mRNAs 2 and 3 of these viruses, affecting the spike and the nonstructural protein 3 (Register and Wesley, 1994).

Protection by the TGEV-related coronavirus PRCV has been studied by several groups. Some of them have found no protection (Hooyberghs *et al.*, 1988; Van Nieuwstadt *et al.*, 1989; Paton and Brown, 1990), while others have described a significant level of protection (Bernard *et al.*, 1989; Cox *et al.*, 1993; De Diego *et al.*, 1992; Wesley and Woods, 1993). The poor protection observed by the first group might be due to the lack of or insufficient antigenic stimulation of the gut-associated lymphoid tissue during PRCV infection of the sows. PRCV is known to have a respiratory tropism. An intermediate degree of protection induced by PRCV has been described. Infection of pigs with this virus primes the systemic and mucosal humoral immune system against TGEV, so that subsequent challenge with TGEV results in a secondary antibody response and in a decreased duration of infectious TGEV excretion. Experimental vaccination of seronegative naive gilts with PRCV induced lactogenic immunity against TGEV. The overall survival rate ranged from 47 to 70%, but was variable from sow to sow (Bernard *et al.*, 1989; De Diego *et al.*, 1992; Wesley and Woods, 1993).

These results suggest that there is a link between respiratory infection with PRCV and secreted protective antibody in the mammary glands of postparturient gilts. The level of virus-neutralizing antibody in serum and colostrum that was induced by PRCV vaccination did not correlate with piglet survival (Wesley and Woods, 1993; Bernard *et al.*, 1989). This has also been recognized as a consistent feature of TGEV vaccination and challenge experiments (Saif and Wesley, 1992), which suggests that the immunodominant neutralizing epitopes for PRCV and TGEV are probably not the major contributors to passive protection. Pigs previously infected with PRCV develop a rapid secondary immune response upon infection with field strains of TGEV. The duration of the TGE outbreak is shortened and the loss of piglets is substantially reduced. This situation may open new ways for vaccination against TGEV. These results are consistent with the apparent correlation between dissemination of PRCV and the reduction in incidence of TGEV observed by Jestin *et al.* (1987a,b).

Vaccination against TGE with heterologous coronaviruses such FIPV

(Woods and Pedersen, 1979; Woods, 1984) and CCV (Woods and Wesley, 1986) only provided partial protection against TGEV. The reverse experiments, that is, vaccination to protect against FIPV with heterologous live virus vaccines (TGEV, CCV, and HCV 229E), also did not provide satisfactory results (Barlough *et al.*, 1984, 1985; Toma *et al.*, 1979; Woods and Pedersen, 1979; Scott, 1987).

## VII. NEW TRENDS IN VACCINE DEVELOPMENT

An effective vaccine against TGE should protect newborn piglets through lactogenic immunity. At least two approaches could be undertaken: (1) the production of noninfectious antigens that are targeted to the gut, or (2) the use of live vectors with enteric tropism. Both approaches require the definition of the B- and T-cell epitopes involved in protection and the search for molecules promoting IgA responses. In addition, the first approach requires the incorporation of molecules with affinity for enteric gut cells. B-cell epitopes involved in protection are, most frequently, those involved in the induction of neutralizing antibodies, although viral proteins that do not have this activity have been shown to induce protection against herpes simplex virus (Chan *et al.*, 1985), cytomegalovirus (Reddehase *et al.*, 1987), and other viral systems (Whitton *et al.*, 1989; Klavinskis *et al.*, 1989). At least five antigenic sites are involved in the induction of TGEV-neutralizing antibodies: sites A, B, and D, and the sites on the S protein defined by MAbs 5G1 and 5D5 from R. D. Woods' and R. Wesley's laboratory (personal communication). Of these five domains, site A is the major inducer of neutralizing antibodies. Sites A and B are complex, conformational, and glycosylation-dependent. Site D can be represented by synthetic peptides, although glycosylation has a minor effect on its conformation. Site C is continuous and glycosylation-independent. A peptide from site D, which includes residues 379 to 386 (SFFSYGEI) from the S protein of TGEV, induced neutralizing antibodies. This peptide induced antibodies with a higher titer in neutralization when coupled to a second S protein derived peptide, which includes residues 1160 to 1180 (Posthumus *et al.*, 1990a,b). The combination of these peptides could be the first candidate for a subunit vaccine. Unfortunately, the neutralizing epitopes selected in these studies were defined using swine testicle cells and intestinal porcine epithelial cells in culture (C. M. Sánchez and L. Enjuanes, unpublished results), and may differ from the epitopes that induce protection *in vivo*. This is a real possibility, since no correlation has been found between neutralization titer and protection by some authors (Bernard *et al.*, 1989; Saif and Wesley, 1992; Wesley and Woods, 1993).

The T-cell epitopes potentially involved in protection against TGEV are being defined by using virus-specific T-cell hybridomas (Bullido *et al.*1989) and polyclonal T cells. T-cell epitopes have been identified on the three major structural proteins of TGEV: S, M, and N. A dominant T-helper epitope defined in the N protein helps the synthesis of TGEV-neutralizing antibodies specific for the S protein *in vitro* (Antón *et al.*, 1995).

IgA isotype immunoglobulin is more stable in the gut than those of the IgG isotype (Porter and Allen, 1972). Several types of T-cell factors have been implicated in the promotion of IgA responses: interleukin 5 (Harriman *et al.*, 1988; Lebman and Coffman, 1988; Strober and Harriman, 1989), transforming growth factor β (Coffman *et al.*, 1989; Sonoda *et al.*, 1989), and a factor that binds to the Fc portion of the IgA immunoglobulin on the surface of B cells (Mestecky and McGhee, 1987). Genes coding for these factors should be cloned and expressed in combination with viral antigens in order to study their role in the induction of secretory immune responses against TGEV.

Two types of molecules have shown to be effective as carriers of unrelated antigens in the stimulation of gut-associated lymphoid tissue after oral administration: cholera toxin B subunit and the pili of the enterotoxic K88 strain of *E. coli*. Most nonviable antigens are often inefficient in promoting antibody responses. They require large (milligram) quantities of immunogen and yield, if any, modest antibody responses (Czerkinsky *et al.*, 1987, 1989; Mestecky, 1987). A notable exception is cholera toxin (CT), which is a potent immunogen (Pierce, 1978). CT and *E. coli* labile toxin (LT) have been shown to exert potent adjuvant effects in mice on gut-immune responses to unrelated antigens presented orally (Elson and Ealding, 1984; Lycke and Holmgren, 1986; Nedrud *et al.*, 1987). There has been a debate on the requirements of the holotoxin (A and B subunits of CT or LT). The best adjuvant effect is induced by the complete toxin. This adjuvant effect appears to be closely linked to the ADP-ribosylating action of CT and LT, associated with enhanced cyclic AMP formation in the affected cells, and thus it may prove difficult to eliminate the enterotoxic activity without loss of adjuvant activity. However, as an antigen carrier system providing specific binding to epithelium, including the M cells of the intestinal Peyer's patches, both CT and its nontoxic binding subunit moiety have been shown to markedly enhance the mucosal immune response to various foreign antigens or epitopes covalently linked to these molecules (Holmgren *et al.*, 1993). An antigenic determinant representing multimers of site D from the peplomer protein of TGEV has been expressed as a fusion protein on the carboxy-terminus of the LT-B subunit. The recombinant product induced TGEV-neutralizing antibodies (Smerdou *et al.*, 1994). The LT-B-site D antigen was expressed using attenuated forms of *Salmonella typhimurium*, which have a tropism for the Peyer's patches (Curtiss *et al.*, 1986). Double mutants of this recombinant *Salmonella*, defective in the synthesis of cAMP and the cAMP receptor, persisted in the gut lymphoid organs for about 3 weeks and induced TGEV antibodies (Smerdou *et al.*, 1994). Its role in protection is being evaluated. Other attempts to develop genetically engineered vaccines using prokaryotic vectors have failed. Most of the S protein gene has been expressed at high levels in *E. coli*. Subcutaneous immunizations with the recombinant antigen did not induce neutralizing antibodies (Hu *et al.*, 1984, 1987) nor induce protection *in vivo*.

A 23-kDa subunit immunogen obtained from purified TGEV by sonification, isopycnic centrifugation, and gel filtration through Sephadex G200 was administered intramuscularly to gilts prior to farrowing. It protected piglets suckling on the vaccinated gilts (Gough *et al.*, 1983a,b). These results were unexpected, as intramuscular inoculation of inactivated TGEV did not provide protection (Saif and Wesley, 1992) and have not been confirmed.

Expression in eukaryotic vectors is required for those TGEV antigenic determinants that are dependent on glycosylation, such as sites A and B, of the S glycoprotein. Two types of live eukaryotic vectors have been considered to induce protection against TGEV: poxvirus and adenovirus. Porcine poxviruses and adenoviruses induced systemic and secretory IgA responses in pigs (Tuboly *et al.*, 1993). Attempts to immunize against TGEV with TGEV–vaccinia recombinants expressing most of the peplomer protein have resulted in the induction of TGEV-neutralizing antibodies (Hu *et al.*, 1987) but not in protection. Adenoviruses have the advantage of their tropism for gut- and bronchus-associated lymphoid tissue. Also, these viruses have a DNA genome that facilitates their use as vector (Graham and Prevec, 1992; Prevec *et al.*, 1989; Schneider *et al.*, 1989; Lubeck *et al.*, 1989; Bett *et al.*, 1993). Using human adenoviruses-based vectors, the whole spike protein and truncated fragments of this glycoprotein have been expressed. The Ad5–TGEV recombinant induced TGEV-neutralizing antibodies when administered to hamsters (Sánchez *et al.*, 1994). Ad5–TGEV recombinants infected swine. The tropism of these recombinants is being determined using a recombinant Ad5-luc vector (Mittal *et al.*, 1993), expressing luciferase activity. Both respiratory and enteric tissues were infected and the enzymatic activity expressed (J. M. Torres, C. M. Sánchez, F. Graham, and L. Enjuanes, unpublished results). Porcine adenoviruses are being developed as vectors to induce protection against TGEV. The genome of porcine adenovirus type 3 (Seshidar-Reddy *et al.*, 1993) and type 4 (Kleibocker *et al.*, 1993) have been cloned and partially sequenced and deletions on the E3 gene introduced to facilitate the cloning of heterologous genes, including the S gene of TGEV.

Expression of TGEV proteins under the control of the baculovirus polyhedrin promoter or in filamentous fungi (as *Aspergillus nidulans* or the Mucor system) may be an economic way of producing large amounts of glycosylated antigens (Van Brunt, 1986). The S, M, and N proteins of TGEV have been expressed using baculovirus (Britton *et al.*, 1990a; Pulford *et al.*, 1990; Tuboly *et al.*, 1993). The recombinant baculovirus developed by Tuboly *et al.* (1993) contained inserts ranging from 1.6 kb, encoding sites C, B, and D, to 3.3 kb, encoding all the major antigenic sites (C, B, D, and A). Piglets immunized with the recombinants developed a strong antibody response, but only recombinants expressing at least part of antigenic site A induced *in vitro* neutralizing antibodies (T. Tuboly and J. B. Derbyshire, personal communication).

There is an increasing interest in the use of RNA viruses that do not replicate through a DNA intermediate as vectors for the expression of heterologous genes. This is the case of alphaviruses (Schlesinger, 1993; Bredenbeeck *et al.*, 1993; Liljeström and Garoff, 1991). Coronaviruses have a genome of 30 kb, which is too large to be easily manipulated. This has prompted the isolation and characterization of defective subgenomic RNAs, which are easily generated in the murine coronavirus MHV. Identification of the minimum requirements to replicate these subgenomic RNAs (Makino *et al.*, 1988; Makino and Joo, 1993), the packaging signal (Makino *et al.*, 1990; van der Most and Bredenbeek, 1991; Fosmire *et al.*, 1992), and the possibility of engineering cDNAs coding for these RNAs (Koetzner *et al.*, 1992; van der Most *et al.*, 1992; Masters, 1992; Masters *et al.*, 1994) are helping the development of vectors based in coronaviruses.

TGEV subgenomic RNAs have been identified. A cDNA has been derived from a 10-kb defective RNA of TGEV. This cDNA has been cloned and sequenced (Mendez *et al.*, 1994). The subgenomic RNA is packaged into capsids and it can be used to develop a TGEV-based vector to induce secretory immunity in swine (C. Smerdou, A. Mendez, M. L. Ballesteros, C. M. Sánchez, and L. Enjuanes, unpublished results).

Anti-idiotypic antibodies are in theory an interesting source of antigen, especially those mimicking the complex antigenic site A, the major inducer of neutralizing antibodies in TGEV. The use of antibodies as antigen may have the advantage of their adsorption into the gut, particularly in the first days after birth. Induction of neutralizing antibodies to TGEV by porcine anti-idiotypic antibodies generated against a murine-neutralizing MAb has been shown in mice (Hariharan *et al.*, 1989). These antibodies neutralized TGEV *in vitro*, but no protection studies *in vivo* have been reported. Monoclonal anti-idiotypic antibodies of the γ and β (internal-image) type, which induced and Ab3 TGEV-neutralizing, are now being tested *in vivo* for protection (Suñé *et al.*, 1991).

ACKNOWLEDGMENTS. We acknowledge C. Smerdou and I. M. Antón for critical reading of the manuscript. The studies on which this review was based were supported by grants to L. E. from the Consejo Superior de Investigaciones Científicas, the Commisión Interministerial de Ciencia y Tecnología, La Consejería de Educación y Cultura de la Comunidad de Madrid from Spain, and the European Communities (Projects Science and Biotech). B. A. M. v/d Z. was supported by Solvay-Duphar b.v., Weesp, The Netherlands.

## VIII. REFERENCES

Abou-Youssef, M. H., and Ristic, M., 1975, Protective effect of immunoglobulins in serum and milk of sows exposed to transmissible gastroenteritis virus, *Can. J. Comp. Med.* **39**:41.

Anonymous, 1989, Virus de la gastroenteritis porcina transmisible, Consejería de Agricultura, Ganadería y Pesca de la Región de Murcia, Spain.

Antón, I. M., González, S., Bullido, M. J., Suñé, C., Meloen, R. H., Borrás-Cuesta, F., and Enjuanes, L., 1995, Transmissible gastroenteritis coronavirus nucleoprotein specific T-helper epitope collaborates in the in vitro antibody synthesis to the three major structural viral proteins, in press.

Asagi, M., Ogawa, T., Minetoma, T., Sato, K., and Inaba, Y., 1986, Detection of transmissible gastroenteritis virus in feces from pigs by reverse passive hemagglutination, *Am. J. Vet. Res.* **47**:2161.

Aynaud, J. M., Nguyen, T. D., Bottreau, E., Brun, A., and Vannier, P., 1985, Transmissible gastroenteritis (TGE) of swine: Survivor selection of TGE virus mutants in stomach juice of adult pigs, *J. Gen. Virol.* **66**:1911.

Aynaud, J. M, Salmon, H., Bottreau, E., Bernard, S., and Lantier, I., 1986, Transmissible gastroenteritis: Immunization of the pregnant sow with the 188-SG strain of TGE coronavirus (Nouzilly strain) using the intramammary route, in: Proceedings of the 9th Congress of the International Pig Veterinary Society, p 202.

Aynaud, J. M., Bernard, S., and Shirai, J., 1988, Les enterites virales du porcelet: Données recentes sur l'immunisation de la truie contre la gastroenterite transmissible TGE en vue de la proteccion passive du porcelet, in: Proceedings of the 10th Congress of the International Pig Veterinary Society, p 32, Rio de Janeiro.

Ballesteros, M. L., Sánchez, C. M., Méndez, A., and Enjuanes, L., 1995, Recombination between transmissible gastroenteritis coronavirus isolates which differ in tropism, in press.

Barlough, J. E., Stoddart, C. A., Sorresso, G. P., Jacobson, R. H., and Scott, F. W., 1984, Experimental inoculation of cats with canine coronavirus and subsequent challenge with feline infectious peritonitis virus, *Lab. Anim. Sci.* **34:**592.

Barlough, J. E., Johnson-Lussenburg, C. M., Stoddart, C. A., Jacobson, R. H., and Scott, F. W., 1985, Experimental inoculation of cats with human coronavirus 229E and subsequent challenge with feline infectious peritonitis virus, *Can. J. Comp. Med.* **49:**303.

Bernard, S., Bottreau, E., Aynaud, J. M., Have, P., and Szymansky, J., 1989, Natural infection with the porcine respiratory coronavirus induces protective lactogenic immunity against transmissible gastroenteritis, *Vet. Microbiol.* **21:**1.

Bett, A. J., Prevec, L., and Graham, F. L., 1993, Packaging capacity and stability of human Adenovirus type 5 vectors, *J. Virol.* **67:**5911.

Bohl, E. H., 1982, Vaccination against transmissible gastroenteritis (TGE) pigs, pros and cons, in: Proceedings of the 23rd Annual George A. Young Conference, p. 77, Nebraska.

Bohl, E. H., and Saif, L. J., 1975, Passive immunity in transmissible gastroenteritis of swine: Immunoglobulin characteristics of antibodies in milk after inoculating virus by different routes, *Infect. Immun.* **11:**23.

Bohl, E. H., Gupta, R. K. P., Olquin, M. Y. F., and Saif, L., 1972, Antibody responses in serum, colostrum and milk of swine after infection or vaccination with transmissible gastroenteritis virus, *Infect. Immun.* **6:**289.

Bohl, E. H., Frederick, G. T., and Saif, L. J., 1975, Passive immunity in transmissible gastroenteritis of swine: Intramuscular injection of pregnant swine with a modified live-virus vaccine, *Am. J. Vet. Res.* **36:**267.

Boyle, J. F., Pedersen, N. C., Evermann, J. F., McKeirman, A. J., Ott, R. L., and Black, J. W., 1984, Plaque assay, polypeptide composition and immunochemistry of feline infectious peritonitis virus and feline enteric coronavirus isolates, *Adv. Exp. Med. Biol.* **173:**133.

Bredenbeek, P. J., Frolov, I., Rice, C. M., and Schlesinger, S., 1993, Sindbis virus expression vectors: Packaging of RNA replicons by using defective helper RNAs, *J. Virol.* **67:**6439.

Bridgen, A., Duarte, M., Tobler, K., Laude, H., and Ackermann, M., 1993, Sequence determination of the nucleocapsid protein gene of the porcine epidemic diarrhoea virus confirms that this virus is a coronavirus related to human coronavirus 229E and porcine transmissible gastroenteritis virus, *J. Gen. Virol.* **74:**1795.

Britton, P., and Page, K. W., 1990, Sequence of the S-gene from a virulent British field isolate of transmissible gastroenteritis virus, *Virus Res.* **18:**71.

Britton, P., Cármenes, R. S., Page, K. W., and Garwes, D. J., 1988a, The integral membrane protein from a virulent isolate of transmissible gastroenteritis virus: Molecular characterization, sequence and expression in *E. coli*, *Molec. Microbiol.* **2:**497.

Britton, P., Cármenes, R. S., Page, K. W., Garwes, D. J., and Parra, F., 19883b, Sequence of the nucleoprotein from a virulent British field isolate of transmissible gastroenteritis virus and its expression in *Saccharomyces cerevisiae*, *Molec. Microbiol.* **2:**89.

Britton, P., López Otín, C., Martín Alonso, J. M., and Parra, F., 1989, Sequence of the coding regions from the 3.0 kb and 3.9 kb mRNA subgenomic species from a virulent isolate of transmissible gastroenteritis virus, *Arch. Virol.* **105:**165.

Britton, P., Garwes, D. J., Page, K. W., and Stewart, F., 1990a, Molecular aspects of the relationship of the transmissible gastroenteritis virus with porcine respiratory coronavirus, *Adv. Exp. Med. Biol.* **276:**441.

Britton, P., Page, K. W., Mawditt, K., and Pocock, D. H., 1990b, Sequence comparison of porcine transmissible gastroenteritis virus (TGEV) with porcine respiratory coronavirus in: *Proceedings of the VIIIth International Congress of Virology*, p. P6., IUMS, Berlin.

Britton, P., Mawditt, K. L., and Page, K. W., 1991, The cloning and sequencing of the virion protein genes from a British isolate of porcine respiratory coronairus: Comparison with transmissible gastroenteritis virus genes, *Virus Res.* **21:**181.

Brown, J., and Cartwright, S. F., 1986, New porcine coronavirus? *Vet. Rec.* **119:**282.

Bullido, M. J., Correa, I., Jiménez, G., Suñé, C., Gebauer, F., and Enjuanes, L., 1989, Induction of transmissible gastroenteritis coronavirus-neutralizing antibodies *in vitro* by virus-specific T helper cell hybridomas, *J. Gen. Virol.* **70:**659.

Callebaut, P. E., and Pensaert, M. B., 1980, Characterization and isolation of structural polypeptides in hemagglutinating encephalomyelitis virus, *J. Gen. Virol.* **48:**193.

Callebaut, P., Debouck, P., and Pensaert, M., 1982, Enzyme-linked immunosorbent assay for the detection of the coronavirus-like agent and its antibodies in pigs with porcine epidemic diarrhea, *Vet. Microbiol.* **7:**295.

Callebaut, P., Correa, I., Pensaert, M., Jiménez, G., and Enjuanes, L., 1988, Antigenic differentiation between transmissible gastroenteritis virus of swine and a related porcine respiratory coronavirus, *J. Gen. Virol.* **69:**1725.

Callebaut, P. E., Pensaert, M. B., and Hooyberghs, J., 1989, A comparative inhibition ELISA for the differentiation of serum antibodies from pigs infected with transmissible gastroenteritis virus (TGEV) or with the TGEV-related porcine respiratory coronavirus, *Vet. Microbiol.* **20:**9.

Cavanagh, D., 1981, Structural polypeptides of coronavirus IBV, *J. Gen. Virol.* **53:**93.

Cavanagh, D., Davis, P. J., and Mockett, A. P. A., 1988, Amino acids within hypervariable region 1 of avian coronavirus IBV (Massachusetts serotype) spike glycoprotein are associated with neutralizing epitopes, *Virus Res.* **11:**141.

Cavanagh, D., Brian, D. A., Enjuanes, L., Holmes, K. V., Lai, M. M. C., Laude, H., Siddell, S. G., Spaan, W., Taguchi, F., and Talbot, P., 1990, Recommendations of the coronavirus study group for the nomenclature of the structural proteins, mRNAs and genes of coronavirus, *Virology* **176:**306.

Cavanagh, D., Brian, D. A., Enjuanes, L., Holmes, K. V., Lai, M. M. C., Laude, H., Siddell, S. G., Spaan, W., Taguchi, F., and Talbot, P., 1994, Revision of the taxonomy of the *Coronavirus*, *Torovirus*, and *Arterivirus* genera, *Arch. Virol.* **135:**227.

Chan, L., Lukig, M. L., and Liew, F. Y., 1985, Helper T cells induced by an immunopurified Herpes simplex virus type I (HSV-I) 115 kilodalton glycoprotein (gB) protect mice against HSV-I infection, *J. Exp. Med.* **162:**1304.

Chasey, D., and Cartwright, S. F., 1978, Virus-like particles associated with porcine epidemic diarrhea, *Res. Vet. Sci.* **25:**255.

Chen, K. S., 1985, Enzymatic and acidic sensitivity profiles of selected virulent and attenuated transmissible gastroenteritis viruses of swine, *Am. J. Vet. Res.* **46:**632.

Chen, K. S., and Kahn, D. E., 1985, A double-protease-resistant variant of transmissible gastroenteritis virus and its ability to induce lactogenic immunity, *Am. J. Vet. Res.* **46:**1632.

Chu, R. M., Glock, R. D., and Ross, R. F., 1982, Changes in gut associated lymphoid tissues of the small intestine of eight-week-old pigs infected with transmissible gastroenteritis virus, *Am. J. Vet. Res.* **43:**67.

Coffman, R. L., Lebman, D. A., and Shrader, B., 1989, Transforming growth factor β specifically enhances IgA production by lipopolysaccharide-stimulated murine B lymphocytes, *J. Exp. Med.* **170:**1039.

Concellón Martínez, A., 1960, Gastroenteritis epizootica transmisible de los cerdos, *Bol. Inf. Con. Gen. Col. Vet. Esp.* **7:**479.

Correa, I., Jiménez, G., Suñé, C., Bullido, M. J., and Enjuanes, L., 1988, Antigenic structure of E2-glycoprotein of transmissible gastroenteritis coronavirus, *Virus Res.* **10:**77.

Correa, I., Gebauer, F., Bullido, M. J., Suñé, C., Baay, M. F. D., Zwaagstra, K. A., Posthumus, W. P. A., Lenstra, J. A., and Enjuanes, L., 1990, Localization of antigenic sites of the S glycoprotein of transmissible gastroenteritis coronavirus, *J. Gen. Virol.* **71:**271.

Cox, E., Pensaert, M. B., Callebaut, P., and van Deun, K., 1990b, Intestinal replication of a porcine respiratory coronavirus closely related antigenically to the enteric transmissible gastroenteritis virus, *Vet. Microbiol.* **23:**237.

Cox, E., Pensaert, M. B., and Callebaut, P., 1993, Intestinal protection against challenge with transmissible gastroenteritis virus of pigs immune after infection with the porcine respiratory coronavirus, *Vaccine* **11:**267.

Cox, E., Hooyberghs, J., and Pensaert, M. B., 1990a, Sites of replication of a porcine respiratory coronavirus related to transmissible gastroenteritis virus, *Res. Vet. Sci.* **48:**165.

Cubero, M. J., León, L., Contreras, A., and Astorga, R., 1990, Epidemiological enquire by serological survey (ELISA) of transmissible gastroenteritis virus (TGEV) and porcine respiratory coronavirus (PRCV) in the region of Murcia (Spain), in: Proceedings of the XIth Congress of the International Pig Veterinary Society, p. 264. Lausanne, Switzerland.

Cubero, M. J., León, L., Contreras, A., Lanza, I., Zamora, E., and Caro, M. R., 1992, Seroepidemiological survey of porcine respiratory coronavirus (PRCV) infection in breeding herds in Southeastern Spain, *J. Vet. Med.* **39**:1.

Cubero, M. J., Bernard, S., Leon, L., Lantier, I., and Contreras, A., 1993a, Comparative study of different immunoserological techniques for the detection of antibodies against transmissible gastroenteritis (TGE) coronavirus, *Vet. Res.* **24**:47.

Cubero, M. J., León, L., Contreras, A., Astorga, R., Lanza, I., and Garcia, A., 1993b, Transmissible gastroenteritis in pigs in South East Spain—prevalence and factors associated with infection, *Vet. Rec.* **132**:238.

Curtiss, R., Goldschmidt, R., Pastian, R., Lyons, M., Michalek, S. M., Mestecky, L., 1986, Cloning virulence determinants from *S. mutans* and the use of recombinant clones to construct bivalent oral vaccine strains to confer protective immunity against *S. mutans*-induced dental caries, in: *Molecular Microbiology and Immunobiology of Streptococcus mutans* (S. Hamada, ed.), pp. 173–180, Elsevier Science Publishers, New York.

Czerkinsky, C., Prince, S. J., Michalek, S. M., Jackson, S., Moldoveanu, Z., Russell, M. W., McGhee, J. R., and Mestecky, J., 1987, IgA antibody-producing cells after antigen ingestion: Evidence for a common mucosal immune system in humans, *Proc. Natl. Acad. Sci. USA* **84**:2449.

Czerkinsky, C., Russell, M. W., Lycke, N., Lindblad, M., and Holmgren, J., 1989, Oral administration of a streptococcal antigen coupled to cholera toxin B subunit evokes strong antibody responses in salivary glands and extramucosal tissues, *Infect. Immun.* **57**:72.

Dea, S., and Tijssen, P., 1988, Identification of the structural proteins of turkey enteric coronavirus, *Arch. Virol.* **99**:173.

De Diego, M., Laviada, M. D., Enjuanes, L., and Escribano, J. M., 1992, Epitope specificity of protective lactogenic immunity against swine transmissible gastroenteritis virus, *J. Virol.* **66**:6502.

de Groot, R. J., Maduro, J., Lenstra, J. A., Horzinek, M. C., van der Zeijst, B. A. M., and Spaan, W. J. M., 1987, cDNA cloning and sequence analysis of the gene encoding the peplomer protein of feline infectious peritonitis virus, *J. Gen. Virol.* **68**:2639.

de Groot, R. J., Andeweg, A. C., Horzinek, M. C., and Spaan, W. J. M., 1988, Sequence analysis of the 3′ end of the feline coronavirus FIPV79-1146 genome: Comparison with the genome of porcine coronavirus TGEV reveals large insertions, *Virology* **167**:370.

Delmas, B., Godet, M., Gelfi, J., Raschaert, D., and, Laude, H., 1990a, Enteric coronavirus TGEV: Mapping of four major antigenic determinants in the amino-terminal half of peplomer protein S, *Adv. Exp. Med. Biol.* **276**:151.

Delmas, B., Rasschaert, D., Godet, M., Gelfi, J., and Laude, H., 1990b, Four major antigenic sites of the coronavirus transmissible gastroenteritis virus are located on the amino-terminal half of spike protein, *J. Gen. Virol.* **71**:1313.

Delmas, B., Gelfi, J., and Laude, H., 1986, Antigenic structure of transmissible gastroenteritis virus. II. Domains in the peplomer protein, *J. Gen. Virol.* **67**:1405.

Delmas, B., Gelfi, J., L'Haridon, R., Vogel, L. K., Norén, O., and Laude, H., 1992, Aminopeptidase N is a major receptor for the enteropathogenic coronavirus TGEV, *Nature* **357**:417.

Delmas, B., Gelfi, J., Sjöström, N., O., and Laude, H., 1993, Further characterization of aminopeptidase N as a receptor for coronaviruses, *J. Exp. Med. Biol.* **342**:293.

Descoteaux, J. P., Lussier, G., Berthiaume, L., Alain, R., Seguin, C., and Trudel, M., 1985, An enteric coronavirus of the rabbit: Detection by immunoelectron microscopy and identification of structural polypeptides, *Arch. Virol.* **84**:241.

Doyle, L. P., and Hutchings, L. M., 1946, A transmissible gastroenteritis in pigs, *J. Am. Vet. Med. Assoc.* **108**:257.

Duarte, M., and Laude, H., 1994, The porcine epidemic diarrhoea virus genome encodes an uncleaved, large type coronavirus spike protein, *J. Gen. Virol.* **75**:1195.

Duarte, M., Tobler, K., Bridgen, A., Rasschaert, D., Ackermann, M., and Laude, H., 1994, Sequence analysis of the porcine epidemic diarrhoea virus genome between the nucleocapsid and spike protein genes reveal a polymorphic ORF, *Virology* **198**:466.

Duret, C., Brun, A., Guilmoto, H., and Dauvergne, M., 1988, Isolement, identification et pouvoir pathogéne chez le porc d'un coronavirus apparenté au virus de la gastro-entérite transmissible, *Rec. Méd. Vét.* **164:**221.

Egan, I. T., Harris, D. L., and Hill, H. T. 1982. Prevalence of swine dysentery, transmissible gastroenteritis, and pseudorabies in Iowa, Illinois and Missouri. Proceedings of the 86th Annual Meeting of United States Animal Health Association, p. 497.

Egberink, H. F., Ederveen, J., Callebaut, P., and Horzinek, M. C., 1988, Characterization of the structural proteins of porcine epidemic epizootic diarrhea virus, strain CV 777, *Am. J. Vet. Res.* **49:**1320.

Elson, C. O., and Ealding, W., 1984, Cholera toxin feeding did not induce oral tolerance in mice and abrogated oral tolerance to unrelated protein antigen, *J. Immunol.* **33:**2892.

Enjuanes, L., Gebauer, F., Correa, I., Bullido, M. J., Suñé, C., Smerdou, C., Sánchez, C. M., Lenstra, J. A., Posthumus, W. P. A., and Meloen, R., 1990, Location of antigenic sites of the S-glycoprotein of transmissible gastroenteritis virus and their conservation in coronavirus, *Adv. Exp. Biol. Med.* **276:**159.

Evermann, J. F., Baumgartner, L., Ott, R. L., Davis, E. V., and McKeirnan, A. J., 1981, Characterization of a feline infectious peritonitis virus isolate, *Vet. Pathol.* **18:**256.

Evermann, J. F., Heeney, J. L., McKeirnan, A. J., and O'Brien, J. S., 1989, Comparative features of a coronavirus isolated from a cheetah with feline infectious peritonitis, *Virus Res.* **13:**15.

FAO, WHO, OIE, 1984, *Animal Health Yearbook 1983* (V. Kouba, ed.), International Office of Epizootics, Rome, Italy.

Fazakerley, J. K., Parker, S. E., Bloom, F., and Buchmeier, M. J., 1992, The V5A13.1 envelope glycoprotein deletion mutant of mouse hepatitis virus type-4 is neuroattenuated by its reduced rate of spread in the central nervous system, *Virology* **187:**178.

Fiscus, S. A., and Teramoto, Y. A., 1987a, Antigenic comparison of feline coronavirus isolates: Evidence for markedly different peplomer glycoproteins, *J. Virol.* **61:**2607.

Fiscus, S. A., and Teramoto, Y. A., 1987b, Functional differences in the peplomer glycoproteins of feline coronavirus isolates, *J. Virol.* **61:**2655.

Fitzgerald, G. R., Welter, M. W., and Welter, C. J., 1986, Improving the efficacy of oral TGE vaccination, *Vet. Med.* **81:**184.

Fleming, J. O., Trousdale, M. D., El-Zaatari, F. A., Stohlman, S. A., and Weiner, L. P., 1986, Pathogenicity of antigenic variants of murine coronavirus JHM selected with monoclonal antibodies, *J. Virol.* **58:**869.

Fosmire, J. A., Hwang, K., and Makino, S., 1992, Identification and characterization of a coronavirus packaging signal, *J. Virol.* **66:**3522.

Frana, M. F., Behnke, J. N., Sturman, L. S., and Holmes, K. V., 1985, Proteolytic cleavage of the E2 polyprotein of murine coronavirus: Host-dependent differences in proteolytic cleavage and cell fusion, *J. Virol.* **56:**912.

Furuuchi, S., Shimizu, Y., and Kumagai, T., 1975, Comparison of properties between virulent and attenuated strains of transmissible gastroenteritis virus, *Natl. Inst. Anim. Health Q.* **15:**159.

Furuuchi, S., Shimizu, Y., and Kumagai, T., 1976, Vaccination of pigs with an attenuated strain of transmissible gastroenteritis virus, *Am. J. Vet. Res.* **37:**1401.

Furuuchi, S., Shimizu, M., and Shimizu, Y., 1978, Field trials on transmissible gastroenteritis live virus vaccine in newborn piglets. *Natl. Inst. Anim. Health Q.* **18:**135.

Furuuchi, S., Shimizu, Y., and Kumagai, T., 1979, Multiplication of low and high cell culture passaged strains of transmissible gastroenteritis virus in organs of newborn piglets, *Vet. Microbiol.* **3:**169.

Garwes, D. J., 1982, Coronavirus in animals, in: *Virus Infections of the Gastrointestinal Tract* (D. A. J. Tyrell and A. Z. Kapikian, eds.), p. 319, Marcel Dekker, New York.

Garwes, D. J., and Pocock, D. H., 1975, The polypeptide structure of transmissible gastroenteritis virus, *J. Gen. Virol.* **29:**25.

Garwes, D. J., and Reynolds, D. J., 1981, The polypeptide structure of canine coronavirus and its relationship to porcine transmissible gastroenteritis virus, *J. Gen. Virol.* **52:**153.

Gebauer, F., Posthumus, W. P. A., Correa, I., Suñé, C., Smerdou, C., Sánchez, C. M., Lenstra, J. A., Meloen, R. H., and Enjuanes, L., 1991, Residues involved in the antigenic sites of transmissible gastroenteritis coronavirus S glycoprotein, *Virology* **183:**225.

Godet, M., L'Haridon, R., Vautherot, J. F., and Laude, H., 1992, TGEV coronavirus ORF4 encodes a membrane protein that is incorporated into virions, *Virology* **188:**666.

Gough, P. M., Ellis, C. H., Frank, C. J., and Johnson, C. J., 1983a, A viral subunit immunogen for porcine transmissible gastroenteritis, *Antiviral Res.* **3**:211.

Gough, P. M., Frank, C. J., Moore, D. G., Sagona, M. A., and Johnson, C. J., 1983b, Lactogenic immunity to transmissible gastroenteritis virus induced by a subunit immunogen, *Vaccine* **1**:37.

Graham, F. L., and Prevec, L., 1992, Adenovirus-based expression vectors and recombinant vaccines, in: *Vaccines: New Approaches to Immunological Problems* (R. W. Ellis, ed.), p. 363, Butterworth-Heinemann, Stoneham, MA.

Haelterman, E. O., 1965, Lactogenic immunity to transmissible gastroenteritis of swine, *J. Am. Vet. Med. Assoc.* **147**:1661.

Halbur, P. G., Paul, P. S., Vaughn, E. M., and Andrews, J. J., 1992, Porcine respiratory coronavirus, *Am. Assoc. Swine Prod.* March/April, 21.

Halbur, P. G., Paul, P. S., Vaughn, E. M., and Andrews, J. J., 1993, Experimental reproduction of pneumonia in gnotobiotic pigs with porcine respiratory coronavirus isolate AR310, *J. Vet. Diag. Invest.* **5**:184.

Harada, K., Furuuchi, S., Kumagai, T., and Sasahara, J., 1969, Pathogenicity, immunogenicity and distribution of transmissible gastroenteritis virus in pigs, *Natl. Inst. Anim. Health Quart.* **9**:185.

Hariharan, K., Srikumara, S., Moxley, R. A., Osorio, F. A., and Arevalo-Morales, A., 1989, Induction of neutralizing antibodies to transmissible gastroenteritis virus by anti-idiotypic antibodies, *Viral Immunol.* **2**:133.

Harriman, G. R., Kunimoto, D. Y., Elliot, J. F., Paetkau, V., and Strober, W., 1988, The role of IL-5 in IgA B cell differentiation, *J. Immunol.* **140**:3033.

Have, P., 1991, Infection with a new porcine respiratory coronavirus in Denmark. Serologic differentiation from transmissible gastroenteritis virus using monoclonal antibodies, *Adv. Exp. Med. Biol.* **276**:435.

Henningsen, A. D., Mousing, J., and Aalund, O., 1988, Porcine coronavirus (PCV) in Denmark: An epidemiological study based on questionaire data from screening districts, *Dansk Vet. Tidsskrift* **71**:1168.

Hess, R. G., and Bachmann, P. A., 1976, *In vitro* differentiation and pH sensitivity of field and cell culture-attenuated strains of transmissible gastroenteritis virus, *Infect. Immun.* **13**:1642.

Hill, H. T., 1989, Preventing epizootic TGE from becoming enzootic TGE, *Vet. Med.* April:432.

Hoefling, D., 1989, Tracking the culprits behind diarrhea in neonatal pigs, *Vet. Med.* April:426.

Hofmann, M., and Wyler, R., 1990, Enzyme-linked immunosorbent assay for the detection of porcine epidemic diarrhea coronavirus antibodies in swine sera, *Vet. Microbiol.* **21**:263.

Hogue, B. G., and Brian, D. A., 1986, Structural proteins of human respiratory coronavirus OC43, *Virus Res.* **5**:131.

Hogue, B. G., Kienzle, T. E., and Brian, D. A., 1989, Synthesis and processing of the bovine enteric coronavirus haemagglutinin protein, *J. Gen. Virol.* **70**:345.

Hohdatsu, T., Eiguchi, Y., Tsuchimoto, M., Ide, S., Yamagishi, H., and Matumoto, M., 1987, Antigenic variation of porcine transmissible gastroenteritis virus detected by monoclonal antibodies, *Vet. Microbiol.* **14**:115.

Holmes, K. V., Doller, E. W., and Behnke, J. N., 1981, Analysis of the functions of coronavirus glycoproteins by differential inhibition of synthesis with tunicamycin, *Adv. Exp. Med. Biol.* **142**:133.

Holmes, V., Williams, R. K., Stephensen, C. B., Compton, R., Cardellichio, C. B., Hay, C. M., Knobler, R. L., Weismiller, D. G., and J. F., Boyle, 1989, Coronavirus receptors, in: *Cell Biology of Virus Entry, Replication, and Pathogenesis* (R. W. Compans, A. Helenius and M. B. A. Oldstone, eds.), p. 85, Alan R. Liss, New York.

Holmgren, J., Lycke, N., and Czerkinsky, C., 1993, Cholera toxin and cholera B subunit as oral-mucosal adjuvant and antigen vector systems, *Vaccine* **11**:1179.

Hooyberghs, J., Pensaert, M. B., and Callebaut, P., 1988, Transmissible gastroenteritis: Outbreaks in swine herds previously infected with a TGEV-like porcine respiratory coronavirus, Proceedings of the 10th International Pig Veterinary Society Congress, 1988, Rio Janeiro, p. 200.

Horsburgh, B. C., Brierley, I., and Brown, T. D. K., 1992, Analysis of a 9.6 kb sequence from the 3′ end of canine coronavirus genomic RNA, *J. Gen. Virol.* **73**:2849.

Horzinek, M. C., Lutz, H., and Pedersen, N. C., 1982, Antigenic relationship among homologous structural polypeptides of porcine, feline, and canine coronaviruses, *Infect. Immun.* **37:**1148.

Hu, S., Bruszewski, J., Boone, T., and Souza, L., 1984, Cloning and expression of the surface glycoprotein of porcine transmissible gastroenteritis virus, in: *Modern Approaches to Vaccines* (R. Chanock and R. Lerner, eds.), p. 219, CSHL, New York.

Hu, S., Bruszewski, J., Smallig, R., and Browne, J. K., 1987, Studies of TGEV spike protein GP195 expressed in *E. coli* and by a TGE-vaccinia virus recombinant, in: *Immunobiology of Proteins and Peptides. II. Viral and Bacterial Antigens* (M. Zouhair Attasi and H. L. Bachrach, eds.), p. 63, Plenum Press, New York.

Jabrane, A., and Elazhary, Y., 1993, Pathogenicity of porcine respiratory coronavirus isolated in Quebec, *Can. Vet. J.* **15:**16.

Jacobs, L., van der Zeijst, B. A. M., and Horzinek, M., 1986, Characterization and translation of transmissible gastroenteritis virus mRNAs, *J. Virol.* **57:**1010.

Jacobs, L., de Groot, R., van der Zeijst, B. A. M., Horzinek, M. C., and Spaan, W., 1987, The nucleotide sequence of the peplomer gene of porcine transmissible gastroenteritis virus (TGEV): Comparison with the sequence of the peplomer protein of feline infectious peritonitis virus (FIPV), *Virus Res.* **8:**363.

Jestin, A., Leforban, Y., and Vannier, P., 1987a, Porcine coronavirus, *Rec. Med. Vet.* **163:**583.

Jestin, A., LeForban, Y., Vannier, P., Madec, F., and Gourreau, J. M., 1987b, Un nouveau coronavirus porcin. Etudes sero-épidémilogiques retrospectives dans les élévages de Bretagne, *Rec. Méd. Vét.* **163:**567.

Jiménez, G., Castro, J. M., Del Pozo, M., Correa, I., De la Torre, J., and Enjuanes, L., 1986a, Identification of a coronavirus inducing porcine gastroenteritis in Spain, Proceedings of the 9th International Pig Veterinary Society, p. 186, Barcelona, Spain.

Jiménez, G., Correa, I., Melgosa, M. P., Bullido, M. J., and Enjuanes, L., 1986b, Critical epitopes in transmissible gastroenteritis virus neutralization, *J. Virol.* **60:**131.

Kapke, P. A., and Brian, D. A., 1986, Sequence analysis of the porcine transmissible gastroenteritis coronavirus nucleocapsid protein gene, *Virology* **151:**41.

Kapke, P. A., Tung, F. Y. C., Brian, D. A., Woods, R. D., and Wesley, R., 1987, Nucleotide sequence of the porcine transmissible gastroenteritis coronavirus matrix protein, *Adv. Exp. Med. Biol.* **218:**117.

Kemeny, L. J., 1978, Isolation of transmissible gastroenteritis virus from pharyngeal swabs obtained from sows at slaughter, *Am. J. Vet. Res.* **39:**703.

Kemeny, L. J., and Woods, R. D., 1977, Quantitative transmissible gastroenteritis virus shedding patterns in lactating sows, *Am. J. Vet. Res.* **38:**307.

Kemeny, L. J., Wiltsey, V. L., and Riley, J. L., 1975, Upper respiratory infection of lactating sows with transmissible gastroenteritis virus following contac exposure to infected piglets, *Cornell Vet.* **65:**352.

Kemp, M. C., Hierholzer, J. C., Harrison, A., and Burks, J. S., 1984, Characterization of viral proteins synthesized in 229-E infected cells and effect (s) of inhibition of glycosylation and glycoprotein transport, in: *Molecular Biology and Pathogenesis of Coronaviruses*, Vol. 173 (P. J. M. Rottier, B. A. M. van der Zeijst, W. J. M. Spaan, and M. C. Horzinek, eds.), p. 65, Plenum Press, New York.

Kenny, A. J., Stephenson, S. L., Turner, A. J., 1987, Cell surface peptidases, in: *Mammalian ecto-enzymes* (Kenny, A. J., Turner, A. J., eds.), p 169, Elsevier, New York.

King, B., and Brian, D. A., 1982, Bovine coronavirus structural proteins, *J. Virol.* **42:**700.

Klavinskis, L. S., Lindsay, Whitton, J., and Oldstone, M. B. A., 1989, Moleculary engineered vaccine which expresses an immunodominant T-cell epitope induces cytotoxic T lymphocytes that confer protection from lethal virus infection, *J. Virol.* **63:**4311.

Kleibocker, S. B., Seal, B. S., and Mengeling, W. L., 1993, Genomic cloning and restriction site mapping of a porcine adenovirus isolate: Demonstration of genomic stablility in porcine adenovirus, *Arch. Virol.* **133:**357.

Knuchel, M., Ackermann, M., Muller, H., and Kihm, H., 1992, An ELISA for detection of antibodies against porcine epidemic diarrhoea virus (PEDV) based on the specific solubility of the viral surface glycoprotein, *Vet. Microbiol.* **32:**117.

Koetzner, C. A., Parker, M. M., Ricard, C. S., Sturman, L. S., and Masters, P. S., 1992, Repair and mutagenesis of the genome of a deletion mutant of the coronavirus mouse hepatitis virus by targeted RNA recombination, *J. Virol.* **66:**1841.

Kusters, J. G., Niesters, H. G. M., Lenstra, S. A., Horzinek, M. C., and van der Zeijst, B. A. M., 1989, Phylogeny of antigenic variants of avian coronavirus IBV, *Virology* **169:**217.
Lanza, I., Brown, I., and Paton, D. J., 1992, Pathogenicity of concurrent infection in pigs with porcine respiratory coronavirus and swine influenza virus, *Res. Vet. Sci.* **53:**309.
Lanza, I., Rubio, P., Enjuanes, L., Callebaut, P., and Carmenes, P., 1990, Improvement of an ELISA for the detection of IgG anti-TGEV/PRCV in swine area, Proceedings of the 11th International Pig Veterinary Society, p 213, Lausanne, Switzerland.
Lanza, I., Rubio, P., Fernández, M., Muñoz, M., and Cármenes, P., 1993a, Seroprevalence of porcine respiratory coronavirus infection in Spanish breeding sows, *Prev. Vet. Med.* **17:**263.
Lanza, I., Rubio, P., Muñoz, M., and Cármenes, P., 1993b, Comparison of a monoclonal antibody capture ELISA (MACELISA) to indirect ELISA and virus neutralization test for the serodiagnosis of transmissible gastroenteritis virus, *J. Vet. Diagn. Invest.* **5:**21.
Laude, H., 1981, *In vitro* properties of low- and high-passaged strains of transmissible gastroenteritis coronavirus of swine, *Am. J. Vet. Res.* **42:**447.
Laude, H., Charley, B., and Gelfi, J., 1984, Replication of transmissible gastroenteritis coronavirus (TGEV) in swine alveolar macrophages, *J. Gen. Virol.* **65:**327.
Laude, H., Chapsal, J. M., Gelfi, J., Labiau, S., and Grosclaude, J., 1986, Antigenic structure of transmissible gastroenteritis virus. I. Properties of monoclonal antibodies directed against virion proteins, *J. Gen. Virol.* **67:**119.
Laude, H., Rasschaert, D., and Huet, J. C., 1987, Sequence and N-terminal processing of the transmembrane protein E1 of the coronavirus transmissible gastroenteritis virus, *J. Gen. Virol.* **68:**1687.
Laude, H., Vanreeth, K., and Pensaert, M., 1993, Porcine respiratory coronavirus—Molecular features and virus host interactions, *Vet. Res.* **24:**125.
Laviada, M. D., Marcotegui, M. A., and Escribano, J. M., 1988, Diagnóstico e identificación de un brote de gastroenteritis porcina transmisible en España, *Med. Vet.* **5:**63.
Lebman, D. A., and Coffman, R. L., 1988, The effects of IL-4 and IL-5 on the IgA response by murine Peyer's patch B cell subpopulations, *J. Immunol.* **141:**2050.
Liljeström, P., and Garoff, H., 1991, A new generation of animal cell expression vectors based on the Semliki forest virus replicon, *Biotechnology* **9:**1356.
Look, A. T., Ashmun, R. A., Shapiro, L. H., and Peiper, S. C., 1989, Human myeloid plasma membrane glycoprotein CD13 (gp150) is identical to aminopeptidase, *N. J. Clin. Invest.* **83:**1299.
Lubeck, M. D., Davis, A. R., Chengalvala, M., Naatuk, R. J., Morin, J. E., Molnar-Kimber, K., Moson, B. B., Bhat, B. M., Mizutani, S., Hung, P. P., and Purcell, R. H., 1989, Immunogenicity and efficacy testing in chimpanzees of an oral hepatitis B vaccine based on live recombinant adenovirus, *Proc. Natl. Acad. Sci. USA* **86:**6763.
Luytjes, W., Sturman, L. S., Bredenbeek, P. J., Charité, J., van der Zeijst, B. A. M., Horzinek, M. C., and Spaan, W. J. M., 1987, Primary structure of the E2 glycoprotein of coronavirus MHV-A59 and identification of the trypsin cleavage site, *Virology* **161:**479.
Lycke, N., and Holmgren, J., 1986, Strong adjuvant properties of cholera toxin on gut mucosal immune responses to orally presented antigens, *Immunology* **59:**301.
Makino, S., and Joo, M., 1993, Effect of intergenic consensus sequence flanking sequences on coronavirus transcription, *J. Virol.* **67:**3304.
Makino, S., Shieh, C.-K., Soe, L. H., Baker, S. C., and Lai, M. C., 1988, Primary structure and translation of a defective interfering RNA of murine coronavirus, *Virology* **166:**550.
Makino, S., Yokomori, K., and Lai, M. M. C., 1990, Analysis of efficiently packaged defective interfering RNAs of murine coronavirus–localization of a possible RNA-packaging signal, *J. Virol.* **64:**6045.
Martin-Alonso, J. M., Balbin, M., Garwes, D. J., Enjuanes, L., Gascon, S., and Parra, F., 1992, Antigenic structure of transmissible gastroenteritis virus nucleoprotein, *Virology* **188:**168.
Masters, P. S., 1992, Repair and mutagenesis of the genome of a deletion mutant of the coronavirus mouse hepatitis virus by targeted RNA recombination, *J. Virol.* **66:**1841.
Masters, P. S., Koetzner, C. A., Kerr, C. A., and Heo, Y., 1994, Optimization of targeted RNA recombination and mapping of a novel nucleocapsid gene mutation in the coronavirus mouse hepatitis virus, *J. Virol.* **68:**328.
Méndez, A., Smerdou, C., and Enjuanes, L., 1995, Primary structure of a defective interfering RNA of transmissible gastroenteritis coronavirus (in preparation).

Mengeling, W. L., Boothe, A. D., and Ritchie, A. E., 1972, Characteristics of a coronavirus (strain 67N) of pigs, *Am. J. Vet. Res.* **33**:297.
Mestecky, J., 1987, The common mucosal immune system and current strategies for induction of immune responses in external secretions, *J. Clin. Immunol.* **7**:265.
Mestecky, J., and McGhee, J. R., 1987, Immunoglobulin A (IgA): Molecular and cellular interactions involved in IgA biosynthesis and immune response, *Adv. Immunol.* **40**:153.
Mittal, S. K., McDermott, M. R., Johnson, D. C., Prevec, L., and Graham, F. L., 1993, Monitoring foreign gene expression by a human adenovirus based vector using the firefly luciferase as a reporter gene, *Virus Res.* **28**:67.
Moxley, R. A., and Olson, L. D., 1989, Clinical evaluation of transmissible gastroenteritis virus vaccines and vaccination procedures for inducing lactogenic immunity in sows, *Am. J. Vet. Res.* **50**:111.
Moxley, R. A., Olson, L. D., and Solorzano, R. F., 1989, Relationship among transmissible gastroenteritis virus antibody titers in serum, colostrum, and milk from vaccinated sows, and protection in their suckling pigs, *Am. J. Vet. Res.* **50**:119.
National Animal Health Monitoring Systems (NAMHS), 1992. Advisory Group Report. United States Department of Agriculture. Veterinary Services, Fort Collins, Colorado
Nedrud, J. G., Liang, X., Hague, N., and Lamm, M. E., 1987, Combined oral/nasal immunization protects mice from Sendai virus infection, *J. Immunol.* **139**:3484.
Nguyen, T. D., Bernard, S., Botreau, E., Lantier, I., and Aynaud, J. M., 1987, Etude comparée de trois souche du coronavirus de la gastroentérite transmissible: Conditiona de la réplication virale et de ls synthésis des antigánes estructuraux, *Ann. Inst. Pasteur/Virol.* **138**:315.
Niesters, H. G. M., Lenstra, J. A., Spaan, W. J. M., Zijderveld, A. J., Bleumink-Pluym, N. M. C., Hong, F., Van Scharrenburg, G. J. M., Horzinek, M. C., and van der Zeijst, B. A. M., 1986, The peplomer protein sequences of the M41 strain of coronavirus IBV and its comparison with Beaudette strains, *Virus Res.* **5**:253.
Norén, O., Sjöström, H., Danielsen, E. M., Cowell, G. M., and Skovbjerg, H. (eds.), 1986, *The Enzymes of the Enterocyte Plasma Membrane,* Elsevier/North-Holland Biomedical Press, Amsterdam.
O'Toole, D., Brown, I., Bridges, A., and Cartwright, S. F., 1989, Pathogenicity of experimental infection with "pneumotropic" procine coronavirus, *Res. Vet. Sci.* **47**:23.
Parker, M. D., Cox, G. J., Deregt, D., Fitzpatrick, D. C., and Babiuk, L. A., 1989, Cloning and *in vitro* expression of the gene for the E3 haemagglutining glycoprotein of bovine coronavirus, *J. Gen. Virol.* **70**:155.
Paton, D. J., and Brown, I. H., 1990, Sows infected in pregnancy with porcine respiratory coronavirus show no evidence of protecting their suckling piplets against transmissible gastroenteritis, *Vet. Res. Commun.* **14**:329.
Pedersen, N. C., Ward, J., and Mengeling, W. L., 1978, Antigenic relationship of the feline infectious peritonitis virus to coronaviruses of other species, *Arch. Virol.* **58**:45.
Pensaert, M. B., and Debouck, P., 1978, A new coronavirus-like particle associated with diarrhea in swine, *Arch. Virol.* **58**:243.
Pensaert, M., Callebaut, P., and Vergote, J., 1986, Isolation of a porcine respiratory, non-enteric coronavirus related to transmissible gastroenteritis, *Vet. Quart.* **8**:257.
Pensaert, M., Callebaut, P., and Hooyberghs, J., 1987, Transmissible gastroenteritis virus in swine: Old and news, in: Proceedings of the 9th International Pig Veterinary Society Congress, p 40, Barcelona, Spain.
Pensaert, M., Cox, E., Deun, V., and Callebaut, P., 1993, A seroepizootiological study of the porcine respiratory coronavirus in the Belgian swine population, *Vet. Quart.* **65**:16.
Pierce, N. F., 1978, The role of antigen form and function in the primary and secondary intestinal immune response to cholera toxin and toxoid in rats, *J. Exp. Med.* **148**:195.
Plana, J., Vayreda, M., and Marull, L., 1982, Diagnosis of a deadly outbreak of transmissible gastroenteritis in Spain, Proceedings of the 7th International Symposium of World Association of Veterinary Microbiologists, Immunologists and Specialists in Infectious Diseases (WAVMI), p162. Barcelona, Spain.
Pocock, D. H., and Garwes, D. J., 1977, The polypeptides of haemagglutinating encephalomyelitis virus and isolated subviral particles, J. Gen. Virol. *37* **:**487.

Popischil, A., Cox, E., and Pensaert, M., 1990, Localization of porcine respiratory coronavirus in the small intestine of experimental infected piglets, Proceedings of the 11th International Pig Veterinary Society, p219. Lausanne, Switzerland.
Porter, P., and Allen, W. D., 1972, Classes of immunoglobulins related to immunity in the pig: A review, *J. Am. Vet. Med. Assoc.* **160:**511.
Posthumus, W., Meloen, R. H., Enjuanes, L., Correa, I., Van Nieuwstadt, A. P., Koch, G., de Groot, R. J., Kusters, J. G., Luytjes, W., Spaan, W. J., van der Zeijst, B. A. M., and Lenstra, J. A., 1990a, Linear neutralizing epitopes on the peplomer protein of coronaviruses, *Adv. Exp. Med. Biol.* **276:**181.
Posthumus, W. P. A., Lenstra, J. A., Schaaper, W. M. M., van Nieuwstadt, A. P., Enjuanes, L., and Meloen, R. H., 1990b, Analysis and simulation of a neutralizing epitope of transmissible gastroenteritis virus, *J. Virol.* **64:**3304.
Prevec, L., Schneider, M., Rosenthal, K. L., Belbeck, L. W., Derbyshire, J. B., and Graham, F. L., 1989, Use of human adenovirus-based vectors for antigen expression in animals, *J. Gen. Virol.* **70:**429.
Pritchard, G. C., 1987, Transmissible gastroenteritis in endemically infected breeding herds of pigs in East Anglia, 1781–85, *Vet. Rec.* **120:**226.
Pritchard, G. C., and Cartwright, S. F., 1982, TGE of pigs, *Vet. Rec.* **111:**512.
Pulford, D. J., Britton, P., Page, K. W., and Garwes, D. J., 1990, Expression of transmissible gastroenteritis virus structural genes by virus vectors, *Adv. Exp. Med. Biol.* **276:**223.
Raabe, T., and Siddell, S. G., 1989, Nucleotide sequence of the gene encoding the membrane protein of human coronavirus 229E, *Arch. Virol.* **107:**323.
Rasschaert, D., and Laude, L., 1987, The predicted primary structure of the peplomer protein E2 of the porcine coronavirus transmissible gastroenteritis virus, *J. Gen. Virol.* **68:**1883.
Rasschaert, D., Gelfi, J., and Laude, H., 1987, Enteric coronavirus TGEV: Partial sequence of the genomic RNA, its organization and expression, *Biochemie* **69:**591.
Rasschaert, D., Duarte, M., and Laude, H., 1990, Porcine respiratory coronavirus differs from transmissible gastroenteritis virus by a few genomic deletions, *J. Gen. Virol.* **71:**2599.
Reddehase, M. J., Mutter, W., Münch, K., Bühring, H. J., and Koszinowski, U. H., 1987, CD8 positive T lymphocytes specific for murine cytomegalovirus immediate-early antigens mediate protective immunity, *J. Virol.* **61:**3102.
Redman, D. R., Bohl, E. H., and Cross, R. F., 1978, Intrafetal inoculation of swine with transmissible gastroenteritis virus, *Am. J. Vet. Res.* **39:**907.
Register, K. B., and Wesley, R. D., 1994, Molecular characterization of attenuated vaccine strains of transmissible gastroenteritis virus, *J. Vet. Diagn. Invest.* **6:**16.
Resta, S., Luby, J. P., Rosenfeld, C. R., and Siegel, J. D., 1985, Isolation and propagation of a human enteric coronavirus, *Science* **229:**978.
Reynolds, D. J., Garwes, D. J., and Lucey, S., 1980, Differenciation of canine coronavirus and porcine transmissible gastroenteritis virus by neutralization with canine, porcine and feline sera, *Vet. Microbiol.* **5:**283.
Rubio, P., Alvarez, M., and Carmenes, P., 1987, Estudio epizootiológico de la gastroenteritis transmisible en Castilla y León, in: 8th Symposium Asociación Nacional de Porcinocultura Científica), p40. Barcelona, Spain.
Saif, L. J., and Bohl, E. H., 1983, Passive immunity to transmissible gastroenteritis virus: Intramammary viral inoculation of sows, *Ann. NY Acad. Sci.* **409:**708.
Saif, L. J., and Wesley, R. D., 1992, Transmissible gastroenteritis, in: *Diseases of Swine* (A. D. Leman, B. Straw, W. L. Mengeling, S. D' Allaire, and D. J. Taylor, eds.), p. 362, Iowa State University Press, Ames.
Sánchez, C. M., Jiménez, G., Laviada, M. D., Correa, I., Suñé, C., Bullido, M. J., Gebauer, F., Smerdou, C., Callebaut, P., Escribano, J. M., and Enjuanes, L., 1990, Antigenic homology among coronaviruses related to transmissible gastroenteritis virus, *Virology* **174:**410.
Sánchez, C. M., Gebauer, F., Suñé, C., Méndez, A., Dopazo, J., and Enjuanes, L., 1992, Genetic evolution and tropism of transmissible gastroenteritis coronaviruses, *Virology* **190:**92.
Sasahara, J., Harada, K., Hayashi, S., and Watanabe, M., 1958, Studies on transmissible gastroenteritis pigs in Japan, *Jap. J. Vet. Sci.* **20:**1.
Schlesinger, S., 1993, Alphaviruses-vectors for the expression of heterologous genes, *Trends Biotechnol.* **11:**18.

Schmidt, O. W., and Kenny, G. E., 1982, Polypeptides and functions of antigens from human coronaviruses 229 E and OC43, *Infect. Immun.* **35:**515.

Schneider, M., Graham, F. L., and Prevec. L., 1989, Expression of the glycoprotein of vesicular stomatitis virus by infectious adenovirus vectors, *J. Gen. Virol.* **70:**417.

Schreiber, S., Kamahora, T., and Lai, M. M. C., 1989, Sequence analysis of the nucleocapsid protein gene of human coronavirus 229E, *Virology* **169:**142.

Scott, F. W., 1987, Immunization against feline coronaviruses, *Adv. Exp. Med. Biol.* **218:**569.

Seshidhar-Reddy, P., Nagy, E., and Derbyshire, J. B., 1993, Restriction endonuclease analysis and molecular cloning of porcine Adenovirus type3, *Intervirology* **36:**161.

Sethna, P. B., Hung, S.-L., and Brian, D. A., 1989, Coronavirus subgenomic minus-strand RNAs and the potential for mRNA replicons, *Proc. Natl. Acad. Sci. USA* **86:**5626.

Shockley, L. J., Kapke, P. A., Lapps, W., Brian, D. A., Potgieters, L. N. D., and Woods, R., 1987, Diagnosis of porcine and bovine enteric coronavirus infections using cloned DNA prober, *J. Clin. Microbiol.* **25:**1591.

Siddell, S. G., Wege, H., and Ter Meulen, V., 1982, The structure and replication of coronaviruses, *Curr. Top. Microbiol. Immunol.* **99:**131.

Small, J. D., and Woods, R. D., 1987, Relatedness of rabbit coronavirus to other coronaviruses, *Adv. Exp. Med. Biol.* **218:**521.

Smerdou, C., Antón, I. M., Plana, J., Curtiss, R., and Enjuanes, L., 1995, Expression of a continuous epitope from transmissible gastroenteritis coronavirus S protein fused to *E. coli* heat-labile toxin B subunit in attenuated *Salmonella* for oral immunization (in preparation).

Smith, H. C., 1956, Advances made in swine practice, *Vet. Med.* **51:**425.

Söderberg, C., Giugni, T. D., Zaia, J. A., Larsson, S., Wahlberg, J. M., and Möller, E., 1993, CD13 (human aminopeptidase N) mediates human cytomegalovirus infection, *J. Virol.* **67:**6576.

Sonoda, E., Matsumoto, R., Hitoshi, Y., Ishii, T., Sugimoto, M., Araki, S., Tominaga, A., Yamaguchi, N., and Takatsu, K., 1989, Transforming growth factor β induces IgA production and acts additively with interleukin 5 for IgA production, *J. Exp. Med.* **170:**1415.

Spaan, W. J. M., 1990, Towards a coronavirus recombinant DNA vaccine, *Adv. Exp. Med. Biol.* **276:**201.

Spaan, W., Cavanagh, D., and Horzinek, M. C., 1988, Coronaviruses: Structure and genome expression, *J. Gen. Virol.* **69:**2939.

Stoddart, C. A., and Scott, F. W., 1989, Intrinsic resistance of feline peritoneal macrophages to coronavirus infection correlates with *in vivo* virulence, *J. Virol.* **63:**436.

Stone, S. S., Kemeny, L. J., Woods, R. D., and Jensen, M. T., 1977, Efficacy of isolated colostral IgA, IgG, and IgM (A) to protect neonatal pigs against the coronavirus of transmissible gastroenteritis, *Am. J. Vet. Res.* **38:**1285.

Strober, W., and Harriman, G. R., 1989, The role of cells and cytokines in IgA isotype differentiation, Proceedings of the International Congress on Mucososal Immunolology, p 8A. London

Sturman, L. S., 1977, Characterization of a coronavirus. I. Structural proteins: Effects of preparative conditions on the migration of polyacrylamide gels, *Virology* **77:**637.

Sturman, L. S., and Holmes, K. V., 1977, The molecular biology of coronaviruses. II. Glycoproteins of the viral envelope: Tryptic peptide analysis, *Virology* **77:**650.

Sturman, L. S., and Holmes, K. V., 1983, The molecular biology of coronaviruses, *Adv. Virus. Res.* **28:**36.

Sturman, L. S., Ricard, C. S., and Holmes, K. V., 1985, Proteolytic cleavage of E2 glycoprotein of murine coronavirus: Activation of cell-fusing activity of virions by trypsin and separation of two different 90K cleavage fragments, *J. Virol.* **56:**904.

Sugiyama, K., Ishikama, R., and Fukuhara, N., 1986, Structural polypeptides of the murine coronavirus DVIM, *Arch. Virol.* **89:**245.

Suñé, C., Jiménez, G., Correa, I., Bullido, M. J., Gebauer, F., Smerdou, C., and Enjuanes, L., 1990, Mechanisms of transmissible gastroenteritis coronavirus neutralization, *Virology* **177:**559.

Suñé, C., Smerdou, C., Antón, I. M., Abril, P., Plana, J., and Enjuanes, L., 1991, A conserved coronavirus epitope, critical in virus neutralization, represented by internal image monoclonal anti-idiotypic antibodies, *J. Virol.* **65:**6979.

Toma, B., Duret, C., Chappuis, G., and Labadie, J., 1979, Péritonite infectieuse féline: étude des anticorps antivirus de la gastroentérite transmissible du porc par séroneutralisation et hémagglutionation passive, *Rec. Méd. Vét.* **155:**541.

Torres-Medina, A., 1975, Adult pigs carry TGE virus, in: *Nebraska Swine Report*, University of Nebraska. Lincoln Institute of Agriculture and Natural Resources. E. C. 75–219.

Tuboly, T., Nagy, E., and Derbyshire, J. B., 1993, Potential viral vectors for the stimulation of mucosal antibody responses against enteric viral antigens in pigs, *Res. Vet. Sci.* **54:**345.

Tung, F. Y. T., Abraham, S., Sethna, M., Hung, S. L., Sethna, P., Hogue, B. G., and Brian, D. A., 1992, The 9-kDa hydrophobic protein encoded at the 3′ end of the porcine transmissible gastroenteritis coronavirus genome is membrane-associated, *Virology* **186:**676.

Underdahl, N. R., Mebus, C. A., Stair, E. L., Rhodes, M. B., McGill, L. D., and Twiehaus, M. J., 1974, Isolation of transmissible gastroenteritis virus from lungs of market-weight swine, *Am. J. Vet. Res.* **35:**1209.

Underdahl, N. R., Mebus, C. A., and Torres-Medina, A., 1975, Recovery of transmissible gastroenteritis virus from chronically infected experimental pigs, *Am. J. Vet. Res.* **36:**1473.

Utiger, A., Rosskpf, M., Guscetti, F., and Ackermann, M., 1993, Preliminary characterization of a monoclonal antibody specific for a viral 27 kD glycoprotein family synthesized in porcine epidemic diarrhoea virus infected cells, in: *Coronaviruses: Molecular Biology and Virus–Host interactions* (H. Laude and J. F. Vautherot, eds.), p. 197, Plenum Press, New York.

Van Brunt, J., 1986, Fungi: The perfect host? *Biotechnology* **12:**1057.

Vancott, J. L., Brim, T. A., Simkins, R. A., and Saif, L. J., 1993, Isotype-specific antibody-secreting cells to transmissible gastroenteritis virus and porcine respiratory coronavirus in gut-associated and bronchus-associated lymphoid tissues of suckling pigs, *J. Immunol.* **150:**3990.

van der Most, R. G., and Bredenbeek, P. J., 1991, A domain at the 3′ end of the polymerase gene is essential for encapsidation of coronavirus defective interfering RNAs, *J. Virol.* **65:**3219.

van der Most, R. G., Heijnen, L., Spaan, W. J. M., and Degroot, R. J., 1992, Homologous RNA recombination allows efficient introduction of site-specific mutations into the genome of coronavirus MHV-A59 via synthetic coreplicating RNAs, *Nucleic Acids Res.* **20:**3375.

Van Nieuwstadt, A. P., and Pol, J. M. A., 1989, Isolation of a TGE-virus-related respiratory coronavirus causing fetal pneumonia in pigs, *Vet. Rec.* **124:**43.

Van Nieuwstadt, A. P., Cornelissen, J. B. W. J., and Zetstra, T., 1988, Comparison of two methods for detection of transmissible gastroenteritis virus in feces of pigs with experimentally induced infection, *Am. J. Vet. Res.* **49:**1836.

Van Nieuwstadt, A. P., Zetstra, T., and Boonstra, J., 1989, Infection with porcine respiratory coronavirus does not fully protect pigs against intestinal transmissible gastroenteritis virus, *Vet. Rec.* **125:**58.

Vaughn, E. M., Halbur, P. G., and Paul, P. S., 1994, Three new isolates of porcine respiratory coronavirus with various pathogenicities and spike gene deletions, *J. Clin. Microbiol.* **32:**1809.

Vennema, H., Rossen, J. W. A., Wesseling, J., Horzinek, M. C., and Rottier, P. J. M., 1992, Genomic organization and expression of the 3′ end of the canine and feline enteric coronaviruses, *Virology* **191:**134.

Wagner, J. E., Beamer, P. D., and Ristic, M., 1973, Electron microscopy of intestinal epithelial cells of piglets infected with a transmissible gastroenteritis virus, *Can. J. Comp. Med.* **37:**177.

Wege, M., Siddell, S. G., and Ter Meulen, V., 1982, The biology and pathogenesis of coronaviruses, *Curr. Top. Microbiol. Immunol.* **99:**165.

Weingartl, H. M., and Derbyshire, J. B., 1993a, Binding of porcine transmissible gastroenteritis virus by enterocytes from newborn and weaned piglets, *Vet. Microbiol.* **35:**23.

Weingartl, H. M., and Derbyshire, J. B., 1993b, Cellular receptors for porcine transmissible gastroenteritis virus, in: *74th Annual Meeting Conference of Research Workers in Animal Disease* p. 12, Chicago.

Wesley, R. D., 1990, Nucleotide sequence of the E2-peplomer protein gene and partial nucleotide sequence of the upstream polymerase gene of transmissible gastroenteritis virus (Miller strain), *Adv. Exp. Med. Biol.* **276:**301.

Wesley, R. D., and Woods, R. D., 1993, Immunization of pregnant gilts with PRCV induces lactogenic immunity for protection of nursing piglets from challenge with TGEV, *Vet. Microbiol.* **38:**40.

Wesley, R. D., Woods, R. D., Correa, I., and Enjuanes, L., 1988, Lack of protection *in vivo* with neutralizing monoclonal antibodies to transmissible gastroenteritis virus, *Vet. Microbiol.* **18:**197.

Wesley, R. D., Cheung, A. K., Michael, D. D., and Woods, R. D., 1989, Nucleotide sequence of coronavirus TGEV genomic RNA: Evidence for 3 mRNA species between the peplomer and matrix protein genes, *Virus Res.* **13:**87.

Wesley, R. D., Woods, R. D., and Cheung, A. K., 1991a, Genetic analysis of porcine respiratory coronavirus, an attenuated variant of transmissible gastroenteritis virus, *J. Virol.* **65:**3369.

Wesley, R. D., Wesley, I. V., and Woods, R. D., 1991b, Differentiation between transmissible gastroenteritis virus and porcine respiratory coronavirus using a cDNA probe, *J. Vet. Diagn. Invest.* **3:**29.

Whitton, J. L., Tishon, A., Lewicki, H., Gebhard, J., Cook, T., Salvato, M., Joly, E., and Oldstone, M. B. A., 1989, Molecular analysis of a five-amino acid cytotoxic T-lymphocyte (CTL) epitope: An immunodominant region which induces nonreciprocal CTL cross-reactivity, *J. Virol.* **63:**4303.

Witte, K. H., and Walther, C., 1976, Age-dependent susceptibility of pigs to infection with the virus of transmissible gastroenteritis. Proceedings of the 4th International Congress of Pig Veterinary Society. Iowa State University, p. K3.

Woode, G. N., 1969, Transmissible gastroenteritis of swine, *Vet. Bull.* **39:**239.

Woods, R. D., 1984, Efficacy of vaccination of sows with serologically related coronaviruses for control of transmissible gastroenteritis in nursing pigs, *Am. J. Vet. Res.* **45:**1726.

Woods, R. D., and Pedersen, N. C., 1979, Cross-protection studies between feline infectious peritonitis and porcine transmissible gastroenteritis viruses, *Vet. Microbiol.* **4:**11.

Woods, R. D., and Wesley, R. D., 1986, Immune response in sows given transmissible gastroenteritis virus or canine coronavirus, *Am. J. Vet. Res.* **47:**1239.

Woods, R. D., Cheville, N. F., and Gallagher, J. E., 1981, Lesions in the small intestine of newborn pigs inoculated with procine, feline and canine coronaviruses, *Am. J. Vet. Res.* **42:**1163.

Yaling, Z., Ederveen, J., Egberink, H., Pensaert, M., and Horzinek, M. C., 1988, Porcine epidemic diarrhea virus (CV777) and feline infectious peritonitis virus (FIPV) are antigenically related, *Arch. Virol.* **102:**63.

Yassen, S. A., and Johnson-Lussenburg, C. M., 1978, Comparative antigenic studies on coronaviruses, *Int. Virol.* **4:**451.

Yeager, C. L., Ashmun, R. A., Williams, R. K., Cardellichio, C. B., Shapiro, L. H., Look, A. T., and Holmes, K. V., 1992, Human aminopeptidase N is a receptor for human coronavirus 229E, *Nature* **357:**420.

Yokomori, K., Asanaka, M., Stohlman, S. A., and Lai, M. M. C., 1993, A spike protein-dependent cellular factor other than the viral receptor is required for mouse hepatitis virus entry, *Virology* **196:**45.

CHAPTER 17

# Pathogenesis of the Porcine Coronaviruses

DAVID J. GARWES

## I. GENERAL INTRODUCTION

Of all the animal species that the coronaviruses have evolved to grow in, the pig appears to have become host to more members of the family than any other. The filterable agent identified as the cause of transmissible gastroenteritis (TGE) in American swine in the 1960s was subsequently confirmed as a coronavirus, TGEV, shortly after the family was formally identified in 1969. During the following decade, the causal agent of an encephalomyelitic disease that was associated with vomiting and wasting in swine in North America and Europe was shown to be a second coronavirus, antigenically unrelated to TGEV, named hemagglutinating encephalomyelitis virus (HEV).

A third porcine coronavirus was identified in association with a TGE-like disease of pigs in England and Europe during the 1970s. This virus, named porcine epidemic diarrhea virus (PEDV), had a pathogenesis similar to that of TGEV but was quite unrelated to either TGEV or HEV. The fourth porcine coronavirus to be identified was isolated from mild respiratory infections of European pigs that had serum antibodies to TGEV but showed no evidence of gastroenteric infection. Recent work has confirmed that this virus is closely related antigenically and genetically to TGEV but has a greatly reduced ability to grow in the intestinal mucosa. It is clear that the virus, named porcine respiratory coronavirus (PRCV), is a mutant strain of TGEV, but, as its pathogenesis is quite distinct from that of TGEV, it will be described separately.

---

DAVID J. GARWES • Ministry of Agriculture, Fisheries & Food, London SW1P 3JR, England.

*The Coronaviridae*, edited by Stuart G. Siddell, Plenum Press, New York, 1995.

## II. TRANSMISSIBLE GASTROENTERITIS VIRUS

### A. History of the Disease

Transmissible gastroenteritis of swine was probably first described in the 1930s (Smith, 1956), although the demonstration of a viral etiology was not confirmed until the report of Doyle and Hutchings in the United States was published in 1946. They clearly showed that a filterable agent could be transmitted to piglets with the subsequent induction of a frequently fatal gastroenteritis. The growth of the science of virology following this allowed further demonstrations of the viral cause of TGE, its isolation in cell culture, and the subsequent identification of TGEV in Canada, Europe, Taiwan, and Japan (reviewed by Woode, 1969). Following the formal acceptance of the family Coronaviridae (Tyrrell *et al.*, 1968), the causal agent of TGE was recognized as a coronavirus by Tajima (1970)

### B. Clinical Signs

The severity of disease caused by TGEV is related to the age of the animal. Although pigs of all ages can become infected and sows can suffer from TGE to such an extent that agalactia occurs, it is the young piglet that usually shows the most severe clinical signs. Infection of the piglet during the first 2 weeks of life results in vomiting within 18 to 24 hr. This is usually followed by a profuse watery diarrhea that continues for several days and results in obvious dehydration and frequently death 2 to 5 days after infection. In pigs over 2 weeks of age, the initial response to infection is the same as that of the neonatal piglet, but the mortality rate drops as the body weight of the animal at the time of infection increases. This probably reflects the ability of the animal to withstand dehydration during induction of an immune response that leads to subsequent recovery.

### C. Pathology

The usual route of infection is oral, following ingestion of contaminated material. The virus is resistant to the low pH of the stomach and passes to the small intestine where it infects the columnar epithelial cells covering the distal portion of the villi in the jejunum and ileum. The virus rarely colonizes the epithelium of the duodenum, supporting a hypothesis that the virus is actively taken up by the microcanalicular–vesicular system found in villus epithelium of the jejunum and ileum in the young piglet but not in the duodenum after the second day of life (Wagner *et al.*, 1973). The cells covering the tips of the villus are the first to be infected, and the virus then spreads down the villus while the infected cells at the tips are shed. The virus is never found in the cells at the base of the villi or in the crypt cells, and it may be that these cells are refractory to infection. As the infected cells are shed from the villous tips, cells migrate up from the crypt to replace them and this terminates the infection, usually 5 to 7 days after infection, as the new cells are not reinfected (Moon *et al.*, 1976). Villous atrophy, detected as a measurable shortening of the villi, is associated with TGEV infection as with other forms of gastroenteritis (Hooper and Haelterman, 1966).

In a comparative study with attenuated and virulent strains of the virus, an apparent correlation was seen between virulence and tissue tropism. While the Purdue-115 and Gep II strains, which both induce clinical signs in week-old piglets, could be demonstrated in the enterocytes of the jejunum and ileum, Peyer's patches, and the mesenteric lymph nodes, the attenuated Nouzilly vaccine strain multiplied only in the ileum and mesenteric lymph nodes (Cubero *et al.*, 1992).

TGEV has not been detected in the cells of the large intestine, and the clinical signs appear to result from an increase in osmotic pressure in the lumen of the small intestine. The villous epithelium lost during infection carries the enymes responsible for the digestion of milk lactose, resulting in an inability to absorb the sugar (Hooper and Haelterman, 1966). Additional studies also suggested that the cells that migrate from the crypt to repopulate the denuded villi retain a secretory role, producing an abnormal level of sodium secretion into the lumen (Butler *et al.*, 1974)) and loss of blood albumin (Prochazka *et al.*, 1975).

TGEV can be isolated from other organs of the infected pig (Cartwright, 1967) including the kidneys and the lung. Several studies have shown that virus replication in the lung may be important for transmission and persistence of the disease (Underdahl *et al.*, 1974; Kemeny *et al.*, 1975). Lung infection, with TGEV antigen being detected in bronchiolar epithelial cells and alveolar cells, does not appear to be associated with respiratory disease, but may explain the emergence of the porcine respiratory coronavirus, discussed in Section III.

A study of the thymus in TGEV-infected conventional and gnotobiotic piglets suggested that, over a 4-day period, infection accelerated thymus involution, reduced thymocyte density, and resulted in hypertrophy of Hassall's bodies in the medulla (Kvachev and But, 1990).

## D. Immune Response

Since the target cell for TGEV is the apical epithelium of the intestinal villi, host immunity requires the secretion of antibodies into the lumen of the gut. For this reason, parenteral immunization with TGEV antigens, while providing good levels of circulating serum antibodies to the virus, provides no protection against the disease. During TGE infection the immune response results in the secretion of protective antibodies into the lumen of the small intestine and this plays a part in limiting the infection in the older pig (Bohl *et al.*, 1972a). The rapid progression of the disease to death in newborn piglets infected with TGEV probably provides insufficient time for immunity to be induced. It is well documented, however, that sows that have recovered from TGE can protect their offspring by passive transfer of immunity (Bay *et al.*, 1953). Following infection of the pregnant sow, B lymphocytes secreting IgA antibodies to TGEV circulate from the gut-associated lymphoid tissue and relocate in the mammary gland. For the first 3 days after piglets are born, the colostrum secreted by the mammary gland contains antibodies derived directly from the serum. Following this, however, the only immunoglobulin present to any extent in porcine milk is secretory IgA, which is synthesized in the mammary gland. In the convalescent sow the IgA secreting cells relocated from the gut-associated lymphoid tissue can provide protective antibodies to the piglets for several weeks (Bohl *et al.*, 1972b).

More recently, however, research findings have suggested that there may be other factors involved in the ability of milk from immune sows to protect piglets. The levels and classes of anti-TGEV antibodies were measured in milk from sows naturally infected with TGEV or immunized with the Nouzilly vaccine strain (Bernard *et al.*, 1990). No correlation was demonstrated between the level of protection afforded and the titers of the various classes of antibodies to the virus in the colostrum and milk for the first 10 days after farrowing, although an inverse correlation was seen after 10 days.

The role of cellular immunity in recovery from TGE has not been clearly demonstrated, although studies have shown lymphocyte responses in the Peyer's patches and mesenteric lymph nodes of young piglets infected with virulent TGEV but not with attenuated virus (Welch *et al.*, 1988). These authors suggested that this may in part explain why attenuated virus vaccines have proved to be less than fully successful for TGE. When used in the pregnant sow, vaccines have had limited success in generating lactogenic immunity, since most have failed to induce a secretory IgA response and relocation of active B lymphocytes in the mammary gland (Bohl *et al.*, 1975; Saif and Bohl, 1979).

## III. PORCINE RESPIRATORY CORONAVIRUS

### A. History of the Disease

During routine serological surveillance of pig herds in Great Britain, Belgium, Holland, and France in the mid-1980s, an increase in the number of herds with antibodies to TGEV was noted but with no concomitant increase in clinical enteric disease. A coronavirus, PRCV, was isolated in 1986 from respiratory tissue of affected pigs in Belgium (Pensaert *et al.*, 1986), Great Britain (Brown and Cartwright, 1986), and subsequently from other parts of Europe and North America (Wesley *et al.*, 1990). The virus was shown to be indistinguishable from TGEV by conventional serological tests, although monoclonal antibodies to TGEV could be used as the basis for a differential diagnostic immunoassay (Garwes *et al.*, 1988).

By the end of the 1980s, the virus had spread to infect almost 100% of Belgium pig farms (Pensaert and Cox, 1989). It is likely that the virus is spread by the oronasal route and by aerosol. This is supported by data from a study with experimentally infected pigs in which airborne virus could be recovered 1 to 6 days after infection and at levels that correlated with those in the nasal mucosa (Bourgueil *et al.*, 1992).

Several possible sources for the origin of this respiratory coronavirus have been discussed (Jestin *et al.*, 1987). These include recombination between TGEV and another coronavirus, mutation of TGEV into a nonenteric strain, and adaptation of a TGEV-related coronavirus from another host species or its release as an attenuated TGEV vaccine. Comparative sequence analysis of the genomes from PRCV and TGEV have shown that there are high levels of homology (Britton *et al.*, 1991; Wesley *et al.*, 1991). The major difference between the two viruses was determined to be a 600–700 base deletion in the gene encoding the surface peplomer, suggesting that the difference in tissue tropism for the two viruses might reside in the absence of a specific cell receptor in PRCV (see Chapter 16, this volume).

## B. Clinical Signs

Most observers have noted little or no sign of respiratory disease associated with infection of pigs of all ages with this virus. Experimental studies to examine the disease have given somewhat contradictory results, although differences in the PRCV isolate and the age of the pigs used may explain the findings.

O'Toole and co-workers (1989) infected colostrum-deprived piglets aged 6 to 7 days with the UK isolate of PRCV. The animals remained clinically healthy and developed specific antibodies 1 week later. Van Nieuwstadt and Pol (1989), however, used a Dutch isolate of the virus to infect specific pathogen-free pigs aged 5 weeks and recorded a severe illness in all infected animals after 1 to 2 days. The disease was characterized by anorexia, lethargy, labored respiration, and fever. Coughing was observed after 8 days and the condition improved from 10 days on, although two of the animals died of pneumonia 4 and 7 days after inoculation. Vannier (1990) noted a transient hyperthermia, mild to severe dyspnea, polypnea, and an obvious cessation of growth in colostrum-derived piglets infected by the intratracheal route.

A clinical disease, with fever, growth retardation, and lung lesions, was induced by PRCV in 6- to 8-week-old specific pathogen-free pigs and this was not exacerbated by concurrent infection with swine influenza viruses (Lanza *et al.*, 1992).

## C. Pathology

In the UK study, O'Toole and his co-workers (1989) examined the pathology produced by PRCV and compared it with that produced by TGEV. They found that both viruses caused a mild cranioventral bronchointerstitial pneumonia involving between 5 and 20% of the lung parenchyma. Both viruses were shown to replicate in the nonciliated cuboidal bronchiolar cells, and these cells then bulged into the lumen and subsequently detached. There was an accumulation of fibrin, macrophages, and cellular debris in the bronchioles, alveolar ducts, and alveoli. A minority of the infected piglets showed other pathology, including mild laryngitis, tracheitis, and rhinitis. While TGEV was identified in association with intestinal pathology, PRCV produced no intestinal lesions, although low levels of virus were isolated from the small intestine, suggesting limited replication.

The pigs that died in the study by Van Nieuwstadt and Pol (1989) were shown to have a catarrhal lobular disseminated bronchopneumonia involving all lobes of the lung. An American strain of PRCV, isolated from the intestine of a pig from a herd with endemic TGE, was capable of causing 60% consolidation of the lung 10 days after oronasal infection (Halbur *et al.*, 1993). There was a necrotizing and proliferative bronchointerstitial pneumonia characterized by necrosis, squamous metaplasia, dysplasia, proliferation of airway epithelium, mononuclear cell infiltration of alveolar septa, mild type II pneumocyte proliferation, and lymphohistiocytic alveolar exudation. The lesions were not associated with clinical disease and were resolved by 15 days after infection.

In addition to PRCV replication in the respiratory tract, Pensaert's group demonstrated that viremia occurred and some virus reached the gastrointestinal tract (Cox *et al.*, 1990a). There was limited replication of the virus in the jejunum, spreading to the ileum and the duodenum, in a few cells located in the subepithelial layers of the villi and crypts. Following direct inoculation into the lumen of the gut of 1-week-old piglets, PRCV replication was similarly limited to these sites and virus could be isolated from feces for several days (Cox *et al.*, 1990b).

## D. Immune Response

Infection of pigs with PRCV results in the production of serum antibodies that react with PRCV and with TGEV in virus neutralization, immune fluorescence, and enzyme-linked immunosorbent assay (ELISA) tests. It is likely that these antibodies confer some degree of protection on the pigs, although the mildness of the disease and the widespread occurrence of the virus make this difficult to assess. The antigenic relatedness of PRCV and TGEV suggested the possibility that seroconversion to PRCV would protect pigs against subsequent infection with TGEV. Protection against subsequent challenge with TGEV was seen in 10-week-old pigs that had been infected with PRCV by aerosol (Cox *et al.*, 1993). However, pig herds that are known to be seropositive to PRCV have become infected with TGEV, and no protection has been afforded to nursing piglets in an experimental study (Paton and Brown, 1990). Whether this absence of cross-protection is due to critical, yet minor, antigenic differences between the two viruses or to the absence of lactogenic immunity following PRCV infection of sows has yet to be determined, although a degree of protective lactogenic immunity was claimed by Bernard *et al.* (1989).

## IV. PORCINE EPIDEMIC DIARRHEA VIRUS

### A. History of the Disease

A TGE-like disease was observed in young and old pigs in England at the end of 1976 and was termed "epidemic diarrhea type II" (Wood, 1977). The affected animals did not seroconvert to TGEV, however, and TGEV antigen could not be detected in the infected intestinal epithelium. Examination of gut contents from affected pigs in England (Chasey and Cartwright, 1978) and Belgium (Pensaert and Debouck, 1978) showed particles with characteristic coronavirus morphology, and the agent was recognized as a distinct coronavirus and named porcine epidemic diarrhea virus (PEDV). The disease was subsequently identified in other parts of Europe, including Czechoslovakia, Hungary, and Germany (Pensaert, 1981). Attempts to characterize the virus were hindered by its inability to replicate in cell culture until Hofmann and Wyler (1988) showed that the use of trypsin overcomes this block. The agent was subsequently shown to have physicochemical characteristics that confirmed its classification as a coronavirus (Hofmann and Wyler, 1989; Kusanagi *et al.*, 1992)

### B. Clinical Signs

The disease resembles TGE very closely and it is difficult to distinguish between them in the field. Mortality is generally restricted to piglets aged 1 week or less, but clinical signs and morbidity are frequently seen in the older animal, with diarrhea, depression, and anorexia. Experimental infection of pigs with an isolate of the PEDV from Germany, V215/78 (Witte *et al.*, 1981), showed that a proportion of animals aged 12 weeks or older displayed severe symptoms of depression and lassitude.

### C. Pathology

The replication of PEDV closely resembles that of TGEV, with viral antigens being detected in the epithelial cells of the villi in the small intestine shortly after infection, followed by desquamation and villous stunting. Unlike TGE, however, PEDV antigens have been detected in the epithelium of the colon. There have been no reports to date of replication in the lung.

Macroscopic lesions appear to be limited to the small intestine. The damage to the villi results in impaired intestinal function, as with TGE, but the extent of intestinal degeneration is reported to be less than that seen with TGEV.

### D. Immune Response

Serum of pigs recovering from infection with PEDV contain antibodies to the virus detectable by immune fluorescence and virus neutralization tests.

These antibodies do not cross-react with TGEV, and further evidence for the lack of serological relationship between the two viruses was provided by the absence of cross-protection in pigs that had recovered from infection with one virus and were subsequently infected with the other (Debouck and Pensaert, 1980). It is very likely that protection of suckling piglets against PEDV would be based on lactogenic immunity and, as discussed for TGEV, suitable vaccines have yet to be developed.

## V. HEMAGGLUTINATING ENCEPHALOMYELITIS VIRUS

### A. History of the Disease

A disease of neonatal pigs was described in Canada in 1958 in which the clinical signs were vomiting and wasting (Roe and Alexander, 1958). A similar disease, reported in Ontario the following year, developed into an acute encephalomyelitis (Alexander *et al.*, 1959), and subsequent studies showed that the disease could be transmitted. The causal agent was not identified until 1962, however, when a virus was isolated in cell culture from piglets with encephalomyelitis (Greig *et al.*, 1962). The virus produced syncytia in monolayer cultures of infected pig kidney cells, and the supernatant fluid from these cultures could hemagglutinate chick erythrocytes, prompting the name hemagglutinating encephalomyelitis virus (HEV).

Several years later in England, a virus was isolated in cell culture from the brain of a piglet exhibiting the vomiting and wasting disease described earlier in Canada (Cartwright *et al.*, 1969). This virus also had the ability to produce syncytia and hemagglutinate red cells, and a serological relationship with the Canadian HEV isolate was shown, indicating that the two viruses were closely related if not identical. The virus was recognized from its morphology to be a member of the coronavirus group in 1971, and was subsequently identified in association with diseased pigs from many parts of the world (reviewed by Greig, 1981).

### B. Clinical Signs

The two diseases induced by HEV, vomiting and wasting or encephalomyelitis, both become established in piglets in the first few days of their lives but are quite distinct clinically. The vomiting and wasting form starts with anorexia, depression, and inappetence. This is followed by retching and, in some piglets, vomiting of undigested milk (Lai *et al.*, 1992). The ability to drink appears to be impaired and the piglets rapidly lose weight. There is no evidence of diarrhea, although constipation has been recorded. Three to four days after the first clinical signs are seen, infected piglets may have distended abdomens. Affected animals may die within 1 to 2 weeks but frequently survive for several weeks before dying of starvation or secondary infection.

The encephalomyelitic form of the disease often starts with clinical signs similar to those described for the vomiting and wasting form, but the coat becomes staring, the extremities cyanotic, and signs of CNS involvement are soon manifest. These include some respiratory distress, hyperesthesia, and slight paralysis of the hind legs. In advanced stages of the disease the piglets have trouble breathing, become blind, and progress into terminal coma. The clinical signs develop over a period of about 10 days, but occasionally pigs survive and recover to show no aftereffects.

## C. Pathology

Infection occurs via the oronasal route, and the primary site of virus replication appears to be the respiratory tract and the pharyngeal tonsils. There is little or no replication of the virus in the alimentary tract, and the vomiting produced is not related to a gastrointestinal infection as seen with TGEV and PEDV. The elegant studies of Andries and Pensaert (1980a,b) clearly demonstrated the spread of the virus by way of the peripheral nervous system to the CNS. Viral antigens were identified in the early stages of infection in the trigeminal ganglion, the inferior vagal ganglion, the superior cervical ganglion, the intestinal nervous plexuses, the solar ganglion, and the dorsal root ganglia of the lower thoracic region. Infection of the brain stem was shown to occur first in the medulla oblongata and then spread to other parts of the brain, including the cerebrum and cerebellum in later stages of infection. Virus was detected in the nervous plexus of the stomach after the incubation period, but it was not clear whether vomiting resulted from damage to these nerve plexuses or was controlled by CNS.

## D. Immune Response

Circulating antibodies to HEV are found in serum from infected animals and these have been used to diagnose the disease by hemagglutination inhibition and virus neutralization tests. The epidemiology of HEV-induced disease suggests that herd immunity plays a part in limiting the spread of the virus (Appel *et al.*, 1965). These authors conducted experimental infections that indicated that piglets are protected by maternal antibodies present in colostrum.

In a serological survey of pig herds in Belgium, Pensaert and co-workers (1980) found that piglets suckled by immune dams acquired neutralizing antibodies to HEV that persisted for 11 to 12 weeks. This passive immunity was replaced by active immunity between 8 to 16 weeks as the piglets became infected by virus circulating within the herd, but there were no clinical signs associated with this seroconversion, probably because the piglets were protected by the maternal antibodies.

## VI. CONCLUSIONS

It is not obvious why the pig should provide such a suitable host for coronaviruses. Intensive rearing focuses attention on the health of a species and provides an environment that allows rapid spread of a newly emerged virus and this may provide a partial explanation. The factors that determine host species range are not clear for most virus groups, although the presence of closely related coronaviruses in pigs (TGEV), dogs (canine enteric coronavirus), and cats (feline infectious peritonitis) suggests that coronaviruses may move readily between species. HEV is antigenically related to murine, bovine, and human coronaviruses and may have evolved from one of these. The evolution of PRCV is readily explained by mutation of TGEV which is well established in pig herds in all parts of the world, but the emergence of PEDV as a pig pathogen is more difficult to explain. It is most likely that it arose from a coronavirus infecting a species of wild life in close contact with pigs. If this is so, then the conclusion may be drawn that other coronaviruses remain to be isolated from species that have not yet been studied in depth, thereby providing a reservoir from which new diseases of commercially important animals may arise.

## VII. REFERENCES


Alexander, T. J. L., Richards, W. P. C., and Roe, C. K., 1959, An encephalomyelitis of suckling pigs in Ontario, *Can. J. Comp.* **23**:316.

Andries, K., and Pensaert, M., 1980a, Virus isolation and immunofluorescence in different organs of pigs infected with hemagglutinating encephalomyelitis virus, *Am. J. Vet. Res.* **41**:215.

Andries, K., and Pensaert, M., 1980b, Immunofluorescence studies on the pathogenesis of hemagglutinating encephalomyelitis virus infection in pigs after oronasal inoculation, *Am. J. Vet. Res.* **41**:1372.

Appel, M., Greig, A. S., and Corner, A. H., 1965, Encephalomyelitis of swine caused by a hemagglutinating virus. IV. Transmission studies, *Res. Vet. Sci.* **6**:482.

Bay, W. W., Doyle, L. P., and Hutchings, L. M., 1953, Transmissible gastroenteritis in swine. A study of immunity, *J. Am. Vet. Med. Assoc.* **122**:200.

Bernard, S., Bottreau, E., Aynaud, J. M., Have, P., and Szymansky, J., 1989, Natural infection with the porcine respiratory coronavirus induces protective lactogenic immunity against transmissible gastroenteritis, *Vet. Microbiol.* **21**:1.

Bernard, S., Shirai, J., Lantier, I., Bottreau, E., and Aynaud, J. M., 1990, Lactogenic immunity to transmissible gastroenteritis (TGE) of swine induced by the attenuated Nouzilly strain of TGE virus: Passive protection of piglets and detection of serum and milk antibody classes by ELISA, *Vet. Immunol. Immunopathol.* **24**:37.

Bohl, E. H., Gupta, R. K. P., McCloskey, L. W., and Saif, L. J., 1972a, Immunology of transmissible gastroenteritis, *Am. J. Vet. Med. Assoc.* **160**:543.

Bohl, E. H., Gupta, R. K. P., Olquin, M. V. F., and Saif, L. J., 1972b, Antibody responses in serum, colostrum and milk of swine after infection or vaccination with transmissible gastroenteritis virus, *Infect. Immun.* **6**:289.

Bohl, E. H., Frederick, G. T., and Saif, L. J., 1975, Passive immunity in transmissible gastroenteritis of swine: Intramuscular injection of pregnant swine with a modified live-virus vaccine, *Am. J. Vet. Res.* **36**:267.

Bourgueil, E., Hutet, E., Cariolet, R., and Vannier, P., 1992, Experimental infection of pigs with the porcine respiratory coronavirus (PRCV): Measure of viral excretion, *Vet. Microbiol.* **31**:11.

Britton, P., Mawditt, K. L., and Page, K. W., 1991, The cloning and sequencing of the virion protein genes of porcine respiratory coronavirus: Comparison with transmissible gastroenteritis virus genes, *Virus Res.* **21:**181.

Brown, I., and Cartwright, S. F., 1986, New porcine coronavirus? *Vet. Rec.* **119:**282.

Butler, D. G., Gall, D. G., Kelly, M. H., and Hamilton, J. R., 1974, Transmissible gastroenteritis: Mechanisms responsible for diarrhea in an acute enteritis in piglets, *J. Clin. Invest.* **53:**1335.

Cartwright, S. F., 1967, Recovery of virus and copro-antibody from piglets infected experimentally with transmissible gastroenteritis, in: Proceedings of the 18th World Veterinary Congress (Paris), Organising Committee of the 18th World Veterinary Congress (eds.), 2:565, published by National Syndicate of French Veterinary Surgeons and Conseil National de l'Ordre, Paris.

Cartwright, S. F., Lucas, M., Cavill, J. P., Gush, A. F., and Blandford, T. B., 1969, Vomiting and wasting disease of piglets, *Vet. Rec.* **84:**175.

Chasey, D., and Cartwright, S. F., 1978, Virus-like particles associated with porcine epidemic diarrhoea, *Res. Vet. Sci.* **25:**255.

Cox, E., Hooyberghs, J., and Pensaert, M., 1990a, Sites of replication of a porcine respiratory coronavirus related to transmissible gastroenteritis, *Res. Vet. Sci.* **48:**165.

Cox, E., Pensaert, M., Callebaut, P., and van Deun, K., 1990b, Intestinal replication of a porcine respiratory coronavirus closely related antigenically to the enteric transmissible gastroenteritis virus, *Vet. Microbiol.* **23:**237.

Cox, E., Pensaert, M., and Callebaut, P., 1993, Intestinal protection against challenge with transmissible gastroenteritis virus of pigs immune after infection with the porcine respiratory coronavirus, *Vaccine* **11:**267.

Cubero, M. J., Bernard, S., Leon, L., Berthon, P., and Contreras, A., 1992, Pathogenicity and antigen detection of the Nouzilly strain of transmissible gastroenteritis coronavirus in 1-week-old piglets, *J. Comp. Pathol.* **106:**61.

Debouck, P., and Pensaert, M., 1980, Experimental infection of pigs with a new porcine enteric coronavirus, CV 777, *Am. J. Vet. Res.* **41:**219.

Doyle, L. P., and Hutchings, L. M., 1946, A transmissible gastroenteritis in pigs, *Am. J. Vet. Med. Assoc.* **108:**257.

Garwes, D. J., Stewart, F., Cartwright, S. F., and Brown, I., 1988, Differentiation of porcine coronavirus from transmissible gastroenteritis virus, *Vet. Rec.* **122:**86.

Greig, A. S., 1981, Hemagglutinating encephalomyelitis, in: *Diseases of Swine* (A. D. Leman *et al.*, eds.), pp. 246–253, The Iowa State University Press, Ames.

Greig, A. S., Mitchell, D., Corner, A. H., Bannister, G. L., Meads, E. B., and Julian, R. J., 1962, A hemagglutinating virus producing encephalomyelitis in baby pigs, *Can. J. Comp. Med.* **26:**49.

Halbur, P. G., Paul, P. S., Vaughn, E. M., and Andrews, J. J., 1993, Experimental reproduction of pneumonia in gnotobiotic pigs with porcine respiratory coronavirus isolate AR310, *J. Vet. Diag. Invest.* **5:**184.

Hofmann, M., and Wyler, R., 1988, Propagation of the virus of porcine epidemic diarrhea in cell culture, *J. Clin. Microbiol.* **26:**2235.

Hofmann, M., and Wyler, R., 1989, Quantitation, biological and physicochemical properties of cell culture-adapted porcine epidemic diarrhea coronavirus (PEDV), *Vet. Microbiol.* **20:**131.

Hooper, B. E., and Haelterman, E. O., 1966, Concepts of pathogenesis and passive immunity in transmissible gastroenteritis of swine, *Am. J. Vet. Med. Assoc.* **149:**1580.

Jestin, A., Leforban, Y., and Vannier, P., 1987, Les coronavirus du porc, *Rec. Med. Vet.* **163:**583.

Kemeny, L. J., Wiltsey, V. L., and Riley, J. L., 1975, Upper respiratory infection of lactating sows with transmissible gastroenteritis virus following contact exposure to infected piglets, *Cornell Vet.* **65:**352.

Kusanagi, K., Kuwahara, H., Katoh, T., Nunoya, T., Ishikawa, Y., Samejima, T., and Tajima, M., 1992, Isolation and serial propagation of porcine epidemic diarrhea virus in cell cultures and partial characterisation of the isolate, *J. Vet. Med. Sci.* **54:**313.

Kvachev, V. G., and But, V. I., 1990, Thymus involution in piglets with transmissible gastroenteritis, *Veterinariya (Moskva)* **3:**29.

Lai, S. S., Ho, W. C., and Li, N. J., 1992, A preliminary report of haemagglutinating encephalomyelitis virus infections in Taiwan, *J. Chinese Soc. Vet. Sci.* **17:**183.

Lanza, I., Brown, I. H., and Paton, D. J., 1992, Pathogenicity of concurrent infection of pigs with porcine respiratory coronavirus and swine influenza virus, *Res. Vet. Sci.* **53:**309.

Moon, H. W., Kemeny, L. J., and Lambert, G., 1976, Effects of epithelial cell kinetics on age dependent resistance to transmissible gastroenteritis of swine, in: *Proceedings of the 4th International Pig Veterinary Society, K12*, Programme Committee of the 4th International Pig Veterinary Society (eds.), published by the American Association of Swine Practitioners, Ames, Iowa.

O'Toole, D., Brown, I., Bridges, A., and Cartwright, S. F., 1989, Pathogenicity of experimental infection with "pneumotropic" porcine coronavirus, *Res. Vet. Sci.* **47:**23.

Paton, D. J., and Brown, I. H., 1990, Sows infected in pregnancy with porcine respiratory coronavirus show no evidence of protecting their sucking piglets against transmissible gastroenteritis, *Vet. Res. Commun.* **14:**329.

Pensaert, M. B., 1981, Porcine epidemic diarrhea, in: *Diseases of Swine* (A. D. Leman *et al.*, eds.), pp. 344–346, The Iowa State University Press, Ames.

Pensaert, M., and Cox, E., 1989, Porcine respiratory coronavirus related to transmissible gastroenteritis virus, *Agri-Practice* **10:**17.

Pensaert, M. B., and Debouck, P., 1978, A new coronavirus-like particle associated with diarrhea in swine, *Arch. Virol.* **58:**243.

Pensaert, M., Andries, K., and Callebaut, P., 1980, A seroepizootiologic study of vomiting and wasting disease in pigs, *Vet. Quart.* **2:**142.

Pensaert, M., Callebaut, P., and Vergote, J., 1986, Isolation of a porcine respiratory, non-enteric coronavirus related to transmissible gastroenteritis, *Vet. Quart.* **8:**257.

Prochazka, Z., Hampl, J., Sedlacek, M., Masek, J., and Stepanek, J., 1975, Protein loss in piglets infected with transmissible gastroenteritis virus, *Zentralbl. Veterinaermed. [B]* **22:**138.

Roe, C. K., and Alexander, T. J. L., 1958, A disease of nursing pigs previously unreported in Ontario, *Can. J. Comp. Med.* **22:**305.

Saif, L. J., and Bohl, E. H., 1979, Passive immunity in transmissible gastroenteritis of swine: Immunoglobulin classes of milk antibodies after oral-intranasal inoculation of sows with a live low cell culture-passaged virus, *Am. J. Vet. Res.* **40:**115.

Smith, H. C., 1956, Advances made in swine practice. IX. Transmissible gastroenteritis, *Vet. Med.* **51:**425.

Tajima, M., 1970, Morphology of transmissible gastroenteritis virus of pigs, *Arch. Ges. Virusforsch.* **29:**105.

Tyrrell, D. A. J., Almeida, J. D., Berry, D. M., Cunningham, C. H., Hamre, D., Hofstad, M. S., Mallucci, L., and McIntosh, K., 1968, Coronaviruses, *Nature* **220:**650.

Underdahl, N. R., Mebus, C. A., Stair, E. L., Rhodes, M. B., McGill, L. D., and Twiehaus, M. J., 1974, Isolation of transmissible gastroenteritis virus from lungs of market-weight swine, *Am. J. Vet. Res.* **35:**1209.

Vannier, P., 1990, Disorders induced by the experimental infection of pigs with the porcine respiratory coronavirus (PRCV), *J. Vet. Med. Ser. B.* **37:**177.

Van Nieuwstadt, A. P., and Pol, J. M. A., 1989, Isolation of a TGE virus-related respiratory coronavirus causing fatal pneumonia in pigs, *Vet. Rec.* **124:**43.

Wagner, J. E., Beamer, P. D., and Ristic, M., 1973, Electron microscopy of intestinal epithelial cells of piglets infected with a transmissible gastroenteritis virus, *Can. J. Comp. Med.* **37:**177.

Welch, S-K. W., Saif, L. J., and Ram, S., 1988, Cell-mediated immune responses of suckling pigs inoculated with attenuated or virulent transmissible gastroenteritis virus, *Am. J. Vet. Res.* **49:**1228.

Wesley, R. D., Woods, R. D., Hill, H. T., and Biwer, J. D., 1990, Evidence for a porcine respiratory coronavirus, antigenically similar to transmissible gastroenteritis virus, in the United States, *J. Vet. Diag. Invest.* **2:**312.

Wesley, R. D., Woods, R. D., and Cheung, A. K., 1991, Genetic analysis of porcine respiratory coronavirus, an attenuated variant of transmissible gastroenteritis, *J. Virol.* **65:**3369.

Witte, K. H., Prager, D., Ernst, H., and Nienhoff, H., 1981, Die epizootische virusdiarrhoe (EVD), *Tierarztl. Umschau.* **36:**235.

Wood, E. N., 1977, An apparently new syndrome of porcine epidemic diarrhoea, *Vet. Rec.* **100:**243.

Woode, G. N., 1969, Transmissible gastroenteritis of swine, *Vet. Bull.* **39:**239.

1

CHAPTER 18

# Human Coronavirus Infections

STEVEN H. MYINT

## I. INTRODUCTION AND HISTORY

The first report of a human coronavirus was in 1965 when Tyrrell and Bynoe (1965) isolated a virus from the nasal washings of a male child. The child had typical symptoms and signs of a common cold and the washing was found to be able to induce common colds in volunteers challenged intranasally. The virus, termed B814 (after the number of the nasal washing), could be cultivated in human embryo tracheal organ tissue but not in cell lines used at that time for growing other known etiologic agents of the common cold. At the same time, Hamre and Procknow (1966) were characterizing five "new" agents isolated from the respiratory tract of medical students with colds. One of these agents, strain 229E, was adapted to grow in WI-38 cells. Subsequently, Almeida and Tyrrell (1967) showed that these isolates were morphologically identical to the viruses of avian bronchitis and mouse hepatitis. McIntosh and colleagues (1967a), working at the National Institutes of Health in Bethesda, then isolated six morphologically related viruses that could not be adapted to cell monolayer culture but would grow in organ cultures. Two of these isolates, OC (for organ culture) 38 and 43 were then adapted to grow in suckling mice brain. The term "coronavirus," which described the characteristic morphology of these agents, was accepted in 1968 (Tyrrell *et al.*, 1968a).

## II. HUMAN RESPIRATORY CORONAVIRUSES

There are a number of human respiratory coronaviruses described in the literature, but few have been well characterized (Table I). On the basis of

STEVEN H. MYINT • Department of Microbiology, University of Leicester, Leicester LE1 9HN, England.

*The Coronaviridae*, edited by Stuart G. Siddell, Plenum Press, New York, 1995.

TABLE I. Classification of Human Coronaviruses

| Serogroup | Prototype | Virus |
|---|---|---|
| A | 229E | 229E |
| | | LP |
| | | PR |
| | | TO |
| | | KI |
| | | PA |
| | | AD |
| | | Linder |
| | | Others |
| B | OC43 | OC43 |
| | | OC38 |
| | | OC44 |
| Unclassified | | B814 |
| | | 692 |
| | | OC16 |
| | | OC37 |
| | | OC48 |
| | | HO |
| | | GI |
| | | RO |

serological cross-reactivity, however, it is possible to classify most of them (Reed, 1984; Bradburne, 1970; McIntosh *et al.*, 1969). The two main serogroups are 229E-related and OC43-related and it is these prototype viruses that will be discussed in the rest of this chapter. Coronaviruslike particles have been seen in the stools of humans, but, as they have not been characterized, they will not be discussed in this chapter.

## III. EPIDEMIOLOGY

Most epidemiological surveys of these viruses as agents of respiratory tract illness have been based on serology, using either complement fixation or hemagglutination-inhibition tests for the two prototype viruses 229E and OC43. As it is clear that serologically unrelated coronaviruses exist, the prevalence of these viruses is likely to be underestimated. Moreover, it is also certain that not all human respiratory coronaviruses have yet been adapted to tissue or organ culture. This conclusion is supported by a study based in London, England, in which coronaviruses were isolated, by tissue or organ culture, from 18.4% of patients with common colds. However, a further 13% of the isolates were able to induce colds in inoculated volunteers, although they could not be identified as any of the known "common cold" viruses (Larson *et al.*, 1980).

The most extensive epidemiological survey of OC43 infection has been the study of the community of Tecumseh, Michigan (Monto and Lim, 1974). A 4-year study involved looking for serological evidence of infection with OC43 in 910 persons in 269 families. A mean of 17.1% of individuals showed evidence of infection with OC43 in any one year. There was, however, a cycling of the frequency of occurrence: OC43 infections occurred in most years but there were peaks of infection every 3 years. Over 80% of infections occurred despite preexisting antibody with infections gradually diminishing with age. The peaks of infection took place in the winter–spring months. A study of 229E infections in the same community showed a mean annual rate of 7.7% infected persons (Cavallaro and Monto, 1970). This figure was half of that recorded in a 6-year study of Chicago medical students, in which there was also marked year-to-year variation with peaks of 35% incidence (Hamre and Beem, 1972). Nearly all 229E infections also occurred in the winter and spring months. The Tecumseh study showed that either 229E or OC43 was dominant in any one year.

Studies of coronaviruses as causes of clinical illness have shown that coronaviruses are second only to rhinoviruses as the causes of the common cold. In the United States and England, 229E and OC43 are responsible for 1 to 30% of all clinical cases (McIntosh *et al.*, 1970a,b; Isaacs *et al.*, 1983; Wenzel *et al.*, 1974; Kaye *et al.*, 1971; Owen-Hendley *et al.*, 1972; Bradburne *et al.*, 1967), with approximately an equal number of subclinical infections.

## IV. DISEASE MANIFESTATIONS

Human respiratory coronaviruses are now well accepted as causes of upper respiratory tract illness and, in particular, the common cold. The viruses have been isolated from patients with the common cold and they produce common colds when inoculated into volunteers intranasally (Andrewes, 1962). There is a mean incubation period of 3 days (range 2–5 days) followed by an illness that lasts a mean of 6–7 days (range 2–18 days). The classical clinical illness is well known to all of us and consists of general malaise, headache, nasal discharge, sneezing, and a mild sore throat (Tyrrell *et al.*, 1993). Approximately one tenth of patients will also have a fever and one fifth will have a cough. Table II contrasts the clinical features of rhinovirus type 2-induced common colds, those induced by coronavirus 229E, and the illness caused by the influenza A virus (compiled from Lowenstein and Parrino, 1987; McIntosh *et al.*, 1973, 1974). It is not possible on an individual basis to distinguish rhinovirus colds from coronavirus colds. Although earlier studies suggested that there may be differences in the clinical symptomatology with different coronaviruses, this has not been substantiated.

Less well documented are lower respiratory tract infections associated with coronaviruses. A seroepidemiological study, using a complement fixation test, found that coronaviruses were less likely to be found in hospitalized children with lower respiratory tract infection than in controls with non-respiratory tract disease. From 565 children with lower respiratory tract infection there was evidence of infection with OC38 or OC43 in 3.5%, compared to 8.2% of 245 children in the group with nonrespiratory tract disease (McIntosh *et al.*, 1970a,b). In a later study, however, there was serological evidence of either 229E or OC43 infection in 8.2% of 417 hospitalized children under 18

TABLE II. Clinical Features of Rhinovirus, Coronavirus, and Influenza A Respiratory Tract Infection

| Clinical feature | Coronavirus 229E (%) | Rhinovirus 2 (%) | Influenza A (%) |
|---|---|---|---|
| Fever | 9–23 | 7–16 | 98 |
| Nasal discharge and/or obstruction | 94–100 | 64–100 | 20–30 |
| Headache | 32–85 | 28–50 | 85 |
| General malaise | 46–47 | 28–43 | 80 |
| Sneezing | 85 | 50 | 30 |
| Sore throat | 54–68 | 87–93 | 50–60 |
| Cough | 21–31 | 64–68 | 90 |
| Hoarseness | 12 | 57 | 10 |
| Myalgia | 9 | 21 | 60–75 |
| Watery/sore eyes | 29 | 43 | 60–70 |
| Chills | 18 | 21 | 90 |

months of age with lower respiratory tract disease (McIntosh *et al.*, 1974). The incidence of coronavirus infection was higher than that of other respiratory viruses, except parainfluenza virus type 3 and respiratory syncytial virus. It is unlikely that there is a direct infection of the lower respiratory tract with coronaviruses and any association is most probably through secondary phenomena. Though these secondary phenomena are yet to be defined, a link between coronavirus infection of the upper respiratory tract and wheezing attacks has been shown in several studies over the last 20 years (Isaacs *et al.*, 1983; McIntosh *et al.*, 1973; S. L. Johnston *et al.*, unpublished data). At particular risk are asthmatic children, in which up to 30% of acute wheezing episodes may be due to coronavirus infection. Increased airways resistance has been shown to occur in the upper respiratory tract in nonatopic individuals with colds (Bende *et al.*, 1989; Akerlund, 1993) and asthmatic individuals are likely to be at greater risk.

Nonrespiratory tract illnesses have also been associated with coronaviruses and include multiple sclerosis, pancreatitis, thyroiditis, pericarditis, nephropathy, and infectious mononucleosis (Riski and Hovi, 1980; Apostolov and Spasic, 1975; Arnold *et al.*, 1981). The association with multiple sclerosis has been of particular interest since coronaviruslike particles were seen in the postmortem brain of a patient who died with the disease (Tanaka *et al.*, 1976). Subsequently, OC43-related coronaviruses were isolated from the brain material of two multiple sclerosis patients (Burks *et al.*, 1980), although it is likely that these isolates were murine coronaviruses present in the mice used for cultivation (Weiss, 1983; Gerdes *et al.*, 1981; Fleming *et al.*, 1988). Seroepidemiological studies trying to ascertain an association of coronaviruses with multiple sclerosis have been conflicting (Madden *et al.*, 1981; Salmi *et al.*, 1982; Hovanec and Flanagan, 1983; Leinikki *et al.*, 1981). Gene detection has failed to detect OC43 (Sorensen *et al.*, 1986) but has suggested a neurotropism for 229E (Stewart *et al.*, 1992). It is clear that further research in this area is needed.

The role of coronaviruses in causing diseases outside the respiratory tract has been doubted, in part, because there has been no evidence that the virus can spread from the nasal mucosa. Using the polymerase chain reaction, however, it has recently been possible to show that there is a short viremic phase of 229E, in at least some experimentally inoculated volunteers (S. Myint, unpublished data). The identification of the 229E receptor as a metalloprotease that is found on the surface of cells in many tissue types would also suggest a possible involvement of human coronaviruses in diseases other than just the common cold.

## V. PATHOGENESIS AND IMMUNE RESPONSE

Little is known about the detailed pathogenic mechanisms in human coronavirus infection, principally because there is no animal model of infection. Even the predominant mode of transmission is uncertain. Infection can be induced experimentally by direct inoculation of virus into the nose, but this is unlikely to be the natural route. By analogy with rhinoviruses it is likely that infection is either by aerosols or fomites (Editorial, 1988). In support of this view is the finding that 229E survives well in an atmosphere of high humidity and low temperature (Ijaz *et al.*, 1985).

Once in the nose, the virus is thought to enter the cell via a specific receptor, aminopeptidase N. Replication is optimal at 32–33 °C, the temperature in the superficial layers of the nasal mucosa. This results in sloughing of the superficial nasal epithelium and a proteinaceous exudate. Serum antibody levels rise after about a week, but it is not clear whether it is this response or cell-mediated mechanisms that clear the infection (Callow *et al.*, 1990). Certainly, there is some correlation of the severity and likelihood of disease with preexisting serum antibody, but the mere presence of such antibody is not protective (Callow, 1985). Serum antibody levels peak about 2 weeks after infection and decline to low or undetectable levels at 12–18 months. Reinfections are common.

## VI. DIAGNOSIS

Because of the trivial and temporary nature of common colds, the detection of coronavirus infections has not been attempted in routine diagnostic laboratories. This situation is unlikely to change unless antiviral therapy becomes available. Although most techniques available to the diagnostic virologist have been used to detect human respiratory coronaviruses, even in research laboratories the range of tests employed by any one center tends to be limited. This chapter will give a synopsis of the range of methods used, but the reader is referred elsewhere for details of methods (Myint and Tyrrell, 1994).

### A. Organ Cultures

The method, or modifications of it, developed by Tyrrell and Bynoe (1965) to isolate B814 is still used by some laboratories but is hampered by the difficulty of obtaining human embryonic tracheal tissue. It is the best method available for primary isolation of the broadest range of respiratory coronaviruses.

Tracheal tissue is taken from 14- to 24-week embryos and planted in sterile plastic Petri dishes containing 199 medium. The tissue is immersed with cilia uppermost. Virus can be inoculated onto the cilia and then incubated at 33 °C for up to 10 days. Viral replication is indicated by cessation of ciliary activity and confirmed by interference with another virus (echovirus, parainfluenza, or Sendai). Electron microscopy was used originally to confirm the isolates as coronavirus but other tests such as virus neutralization are now employed.

Trachea obtained from 5- to 9-month-old fetuses have also been shown to support the growth of coronaviruses, and different media recipes can also be used (McIntosh *et al.*, 1967a,b).

## B. Mouse Brain Culture

OC38 and OC43 (but not B814, OC16, OC37, OC44, or OC48) have been adapted to grow in suckling mice brains (McIntosh *et al.*, 1967a,b, 1970; Tyrrell *et al.*, 1968b). The mice can be inoculated intracerebrally or via the peritoneum, with encephalitis occurring in some several days later. In the initial description of McIntosh and colleagues, CD-1 Swiss mice were used, with encephalitis occurring 11–15 days after inoculation with virus that had been passaged several times in organ culture. After the fourth passage, the time to illness was reduced to 40–60 hr. Evidence of infection was not found in other organs (e.g., liver, heart, or lungs). Virus can be prepared from brain suspensions by clarification through low-speed centrifugation and then adsorption to and elution from group O erythrocytes. Virus may be visualized by electron microscopy. Although this is not used for primary isolation of virus, this method is still a commonly used means of preparing OC43 antigen for serological assays.

## C. Cell Culture

Cell cultures have proved to be unreliable for the primary isolation of all human respiratory coronaviruses, but certain strains have been adapted to growth in them. The 229E and related strains grow well in a continuous heteropoid cell line termed C16, as they were the 16th clone of MRC-C cells that was selected (Philpotts, 1983). In the original description of these cells from the MRC Common Cold Unit in Salisbury, the morphology showed a mixture of fibroblastic and epithelioid cells. The former constituted three quarters of the cell population. These cells were contaminated with organisms detected by Hoechst stain 33258, presumably *Mycoplasma* spp. Thus, C16 cells would arguably be the cell line of choice for the isolation of 229E, but frequent passage of these cells results in a increasing proportion of the epithelioid content and a consequent reduction in the ability of the cell line to sustain replication of virus.

Many laboratories have continued to utilize a cell line that was originally used in the work of Hamre and Procknow (1966): a human diploid cell strain from Wistar Institute, so-called WI-38 cells. Although 229E was readily adapted to this cell line, primary isolation was in human kidney cells, and the authors noted that WI-38 cells may not be ideal for primary isolation.

Apart from C16 and Wi-38 cells, many other cell lines have been used to grow individual virus strains (Hamre and Procknow, 1966; Reed, 1984; Larson *et al.*, 1980; Hamre *et al.*, 1967; Kapikian *et al.*, 1969; Bradburne, 1969, 1972; Schmidt *et al.*, 1979; Schmidt and Kenny, 1982; Tyrrell *et al.*, 1979; Chaloner-Larsson and Johnson-Lussenberg, 1981; Bruckova *et al.*, 1970). These are summarized in Table III. These viruses do not grow well in cell types commonly used for the isolation of other respiratory viruses such as HEp-2 or Rhesus monkey kidney, which makes routine identification unlikely. It is, moreover, clear that the ideal cell line for isolation and propagation is not yet available and other methods of diagnosis have to be applied.

## D. Electron Microscopy and Immune Electron Microscopy

Electron microscopy of nasal washings is impractical as the virus load is usually below the level of sensitivity of standard methods. It has been used, however, to detect virus in tissue sections such as in mouse brain culture of OC43. It is also the means by which human enteric coronaviruses (HECVs) have been detected. Negative staining with tungsten has usually been the method of choice (Almeida and Tyrrell, 1967; Tyrrell and Bynoe, 1965), but molybdenum and uranium salts have also been used. The size of the particles appear greater if uranium salts are used in place of tungsten (Davies and MacNaughton, 1979).

TABLE III. Cell Lines and Strains Used for the Cultivation of Human Coronaviruses

| Cell type | Virus | Primary isolation or adaptation |
|---|---|---|
| Human embryonic kidney | 229E | Primary |
| C16 (see text) | 229E | Adaptation ≥ primary |
| Human embryonic lung fibroblast, WI-38 | 229E | Adaptation ≥ primary |
| Human embryonic lung fibroblast, MRC-c | 229E and OC43 | Adaptation > primary (OC43, adaptation only) |
| Human embryonic lung epithelium, L132 | 229E and OC43 | Adaptation > primary |
| Human embryonic intestinal fibroblast, MA177 | 229E | Primary isolation of some strains |
| Human type II pneumocytes | 229E | Primary > adaptation |
| Human fetal tonsil fibroblast | 229E and OC43 | Adaptation |
| Human embryonic rhabdomyosarcoma | 229E and OC43 | Adaptation |
| Primary monkey kidney | OC43 | Adaptation |
| Rhesus monkey kidney epithelium, LLC-MK2 | OC43 | Adaptation |
| Continuous green monkey kidney epithelioid, BSC-1 | OC43 | Adaptation |

An attempt to enhance electron microscopy by utilizing antibody concentration of cultured virus has been used successfully to detect 692 virus in washings from an adult with upper respiratory tract infection (Kapikian *et al.*, 1973). Nasal washings were passaged through both cell culture and tracheal organ culture, and the resulting supernatant was then incubated with convalescent serum from the same patient. After centrifugation the pellet was then examined on a Formvar-carbon-coated grid. Aggregates of virus were clearly discernible. Supernatant that had been incubated with phosphate-buffered saline, instead of convalescent serum, was also examined but virus particles were not seen.

## E. Immunofluorescence

An immunofluorescence method has been developed and applied to the detection of 229E and OC43 in nasopharyngeal secretions and washings (McIntosh *et al.*, 1967a). Sera were raised in rabbits against mouse brain-derived OC43 and cell culture-grown 229E and used in an indirect fluorescence assay. This test was able to detect homologous coronavirus antigens in nasal washings from infected volunteers, though cross-reactivity was noted with the 229E antiserum in washings from volunteers who had been inoculated with OC43 and OC44. No nasopharyngeal aspirates from 106 children who were hospitalized with respiratory tract infection had detectable coronavirus antigen by this method. It is difficult to ascertain whether this was due to a lack of sensitivity, as paired sera collected from 66 children during the study period did not show evidence of coronavirus infection.

## F. Enzyme-Linked Immunoassay

An enzyme-linked immunosorbent assay (ELISA) method based on purified 229E and HECV CV-Paris (which has cross-reactivity with OC43) has been used to diagnose infections in children (Isaacs *et al.*, 1983; MacNaughton *et al.*, 1983). The ELISA method is a modification of that described for antibody detection using rabbit antisera (see Section VI.I). In a study of 30 children aged 6 months to 6 years, 159 samples were collected: 111 nose swabs, 11 throat swabs, and 55 nasopharyngeal aspirates. Of these, 34.2% of the nose swabs, but only 18.2% of the throat swabs and nasopharyngeal aspirates, were positive for either 229E or OC43. No comparison with serology was attempted, but the positivity rate would suggest that this ELISA was a sensitive test.

## G. Nucleic Acid Hybridization

The first application of gene detection methods to detecting human coronaviruses was developed for 229E using Northern hybridization (Myint *et al.*, 1990). A cDNA that encoded the entire nucleocapsid gene for 229E was ligated into a Riboprobe (Promega) vector, pGEM-1, from which $^{32}P$-labeled full-length transcripts could be generated. These transcripts could be made as sense or antisense, depending on whether an SP6 or T7 promoter as used. This method has been applied to the detection of 229E in nasal washings from inoculated volunteers (Myint *et al.*, 1989) and has been shown to be at least as sensitive as culture. There is also the advantage of a diagnostic result being available within 48 hr. An interesting observation was that the probe method was able to detect virus for longer than cell culture in sequential samples from the volunteers. This probe method will not detect OC43 and attempts to remove the radioactive-labeling by incorporating biotin or digoxigein into transcripts have led to significant loss of sensitivity (S. Myint, unpublished data).

## H. Reverse Transcription–Polymerase Chain Reaction

With the advent of gene and probe amplification strategies, it was to be expected that these methods would be seen as advantageous for viruses that are difficult to cultivate. Gene amplification methods based on "nested" priming have been shown to be a sensitive and specific means of detecting both 229E and OC43 (Myint *et al.*, 1994). Serotype-specific nested primers were designed from the known sequences of the nucleocapsid genes of 229E and OC43. The inner primers were, in particular, chosen to produce a small fragment of about 100 base pairs for maximum sensitivity. RNA is extracted using an acid-phenol/guanidinium isothiocyanate procedure followed by reverse transcription using murine Moloney leukemia virus reverse transcription. Two 20-cycle amplification steps are then used with the outer and inner sets of primers, respectively. The sensitivity of the assay appears to be much greater than that of cell culture or probe methods and each primer pair appears to be either 229E or OC43 specific. The use of this method has greatly enhanced the diagnostic yield of coronaviruses in clinical material from asthmatic children (S. L. Johnston *et al.*, unpublished data). The assay is, at least, as sensitive as a combination of culture and serology for diagnosing infection and is more specific. Reverse transcription–polymerase chain reaction is likely to become the method of choice for direct virus detection.

## I. Serological Methods

Because of the lack of reliable detection methods prior to the development of those based on gene detection, most epidemiological studies have used serological assays to determine evidence of coronavirus infection. The most widely used and sensitive format is the enzyme-linked immunoassay. The assay was first described for strain 229E (Kraaijeveld *et al.*, 1980; MacNaughton, 1982) but has since been adapted for detection of antibodies to OC43 (Schmidt, 1984). The 229E assay uses antigen that is grown in cell monolayers and then clarified. The OC43 test uses mouse-brain-derived antigen. Rabbit antisera have been used for both 229E and OC43 tests. The specificities of the assays are similar to that of counterimmunoelectrophoresis, neutralization, and complement fixation assays but sensitivity is over 1000-fold greater. In volunteer studies, the 229E assay has shown a close correlation between clinical illness and virus shedding. It has been the principal method for determining the occurrence and frequency of coronaviruses infections in serological surveys, but recent data suggest that some false-positive and false-negative reactions occur (Myint *et al.*, 1994). The 229E assay also detects antibody rises to some 229E-like viruses (PR, KI, and TO).

Other serological test formats have also been used: indirect hemagglutination (Kaye *et al.*, 1972) and immune-adherence hemagglutination (Gerna *et al.*, 1978) for 229E antibody; rapid microneutralization (Gerna *et al.*, 1979) and plague-reduction (Gerna *et al.*, 1980) for OC43 antibody; and immunofluorescence (Monto and Rhodes, 1977), complement fixation (Hovi, 1978), and single radial hemolysis (Hierholzer and Tannock, 1977; Riski *et al.*, 1977) for both 229E and OC43 antibody. These test formats have been superseded by the ELISA test.

## VII. REFERENCES


Akerlund, A., 1993, Nasal pathophysiology in the common cold, PhD thesis, University of Lund, Sweden.

Almeida, J. D., and Tyrrell, D. A. J., 1967, The morphology of three previously uncharacterised human respiratory viruses that grow in organ culture, *J. Gen. Virol.* **1**:175.

Andrewes, C. H., 1962, The Harben Lectures: The common cold, *J. Roy. Inst. Public Health Hyg.* (Suppl.).

Apostolov, K., and Spasic, P., 1975, Evidence of a viral aetiology in endemic (Balkan) nephropathy, *Lancet* **2**:1271.

Arnold, W., Klein, M., Wang, J. B., Schmidt, W. A. K., and Trampisch, H. J., 1981, Coronavirus-associated antibodies in nasopharyngeal carcinoma and infectious mononucleosis, *Arch. Otorhinolaryngol.* **232**:165.

Bende, M., Barrow, G. I., Heptonstall, J., Higgins, P. G., Al- Nakib, W., Tyrrell, D. A. J., and Akerlund, A., 1989, Changes in human nasal mucosa during experimental coronavirus commun colds, *Acta Otolaryngol.* **107**:262.

Bradburne, A. F., 1969, Sensitivity of L 132 cells to some "new" respiratory viruses, *Nature* **221**:85.

Bradburne, A. F., 1970, Antigenic relationships amongst coronaviruses, *Arch. Ges. Virusforschung* **31**:352.

Bradburne, A. F., 1972, An investigation of the replication of coronaviruses in suspension cultures of L132 cells, *Arch. Gesamte Virusforsch.* **34**:297.

Bradburne, A. F., Bynoe, M. L., and Tyrrell, D. A. J., 1967, Effects of a "new" human respiratory virus in volunteers, *Br. Med. J.* **3**:767.

Bruckova, M., McIntosh, K., Kapikian, A. Z., and Chanock, R. M., 1970, The adaptation of two human coronavirus strains (OC38 and OC43) to growth in cell monolayers, *Proc. Soc. Exp. Biol. Med.* **135**:431.

Burks, J. S., DeVald, B. L., Jankovsky, L. D., and Gerdes, J. C., 1980, Two coronaviruses isolated from central nerveous system tissue of two multiple sclerosis patients, *Science* **209**:933.

Callow, K. A., 1985, Effect of specific humoral immunity and some non-specific factors on resistance of volunteers to respiratory coronavirus infection, *J. Hyg.* **95**:173.

Callow, K. A., Parry, H. F., Sergeant, M., and Tyrrell, D. A. J., 1990, The time course of the immune response to experimental coronavirus infection of man, *Epidemiol. Infect.* **105**:435.

Cavallaro, J. J., and Monto, A. S., 1970, Community-wide outbreak of infection with a 229E-like coronavirus in Tecumseh, Michigan, *J. Infect. Dis.* **122**:272.

Chaloner-Larsson, G., and Johnson-Lussenberg, C. M., 1981, Establishment and maintenance of a persistent infection of L132 cells by human coronavirus strain 229E, *Arch. Virol.* **69**:117.

Davies, H. A., and MacNaughton, M. R., 1979, Comparison of the morphology of three coronaviruses, *Arch. Virol.* **59**:25.

Editorial, 1988, Splints don't stop colds—surprising! *Lancet* **1**:277.

Fleming, J. O., El Zaatari, F. A. K., Gilmore, W., Berne, J. D., Burks, J. S., Stohlman, S. A., Tourtellote, W. W., and Weiner, L. P., 1988, Antigenic assessment of coronaviruses isolated from patients with multiple sclerosis, *Arch. Neurol.* **45**:629.

Gerdes, J. C., Klein, I., DeVald, B. L., and Burks, J. S., 1981, Coronavirus isolates SK and SD from multiple sclerosis patients are serologically related to murine coronaviruses A59 and JHM and human coronavirus OC43, but not to human coronavirus 229E, *J. Virol.* **38**:231.

Gerna, G., Achilli, G., Cattaneo, E., and Cereda, P., 1978, Determination of coronavirus 229E antibody by an immune-adherence haemagglutination method, *J. Med. Virol.* **2**:215.

Gerna, G., Cereda, P. M., Revello, M. G., Torsellini Gerna, M., and Costa, J., 1979, A rapid microneutralisation test for antibody determination and serodiagnosis of human coronavirus OC 43 infections, *Microbiologica* **2**:331.

Gerna, G., Cattaneo, E., Cereda, P. M., Revelo, M. G., and Achilli, G., 1980, Human coronavirus serum inhibitor and neutralizing antibody by a new plaque-reduction assay, *Proc. Soc. Exp. Biol. Med.* **163**:360.

Hamre, D., and Beem, M., 1972, Virologic studies of acute respiratory disease in young adults. V. Coronavirus 229E infections during six years of surveillance, *Am. J. Epidemiol.* **96**:94.

Hamre, D., and Procknow, J. J., 1966, A new virus isolated from the human respiratory tract, *Proc. Soc. Exp. Biol.* **121**:190.

Hamre, D., Kindig, D. A., and Mann, J., 1967, Growth and intracellular development of a new respiratory virus, *J. Virol.* **1**:810.

Hierholzer, J. C., and Tannock, G. A., 1977, Quantitation of antibody to nonhaemagglutinating viruses by single radial haemolysis: Serological test for human coronaviruses, *J. Clin. Microbiol.* **5**:613.

Hovanec, D. L., and Flanagan, T. D., 1933, Detection of antibodies to human coronaviruses 229E and OC43 in the sera of multiple sclerosis patients and normal subjects, *Infect. Immun.* **41**:426.

Hovi, T., 1978,Nonspecific inhibitors of coronavirus OC43 haemagglutination in human sera, *Med. Microbiol. Immunol.* **166**:1773.

Ijaz, M. K., Brunner, A. H., Sattar, S. A., Nair, R. C., and Johnson-Lussenberg, C. M., 1985, Survival characteristics of airborne human coronavirus 229E, *J. Gen. Virol.* **66**:2743.

Isaacs, D., Flowers, D., Clarke, J. R., Valman, H. B., and MacNaughton, M. R., 1983, Epdemiology of coronavirus respiratory infections, *Arch. Dis. Child.* **58**:500.

Kapikian, A. Z., James, H. D., Kelly, S. J., Dees, J. H., Turner, H. C., Mcintosh, K., Kim, H. W., Parrott, R. H., Vincent, M. M., and Chanock, R., 1969, Isolation from man of "avian infectious bronchitis virus-like" viruses (coronaviruses) similiar to 229E virus with some epidemiological observations, *J. Infect. Dis.* **119**:282.

Kapikian, A. Z., James, H. D., Kelly, S. J., and Vaughn, A. L., 1973, Detection of coronavirus strain 692 by immune electron microscopy, *Infect. Immun.* **7**:111.

Kaye, H. S., Marsh, H. B., and Dowdle, W. R., 1971, Seroepidemiologic survey of coronavirus (strain OC43) related infections in a children population, *Am. J. Epidemiol.* **94**:43.

Kaye, H. S., Ong, S. B., and Dowdle, W. R., 1972, Detection of coronavirus 229E antibody by indirect haemagglutination, *Appl. Microbiol.* **24**:703.

Kraaijeveld, C. A., Reed, S. E., and MacNaughton, M. R., 1980, Enzyme-linked immunosorbent assay for detection of antibody in volunteers experimentally infected with human coronavirus strain 229E, *J. Clin. Microbiol.* **12**:493.

Larson, H. E., Reed, S. E., and Tyrrell, D. A. J., 1980, Isolation of rhinoviruses and coronaviruses from 38 colds in adults, *J. Med. Virol.* **5**:221.

Leinikki, P. O., Holmes, K. V., Shekarchi, I., Iivainen, M., Madden, D., Sever, J. L., 1981, Coronavirus antibodies in patients with multiple sclerosis, *Adv. Exp. Med. Biol.* **142**:323.

Lowenstein, S. R., and Parrino, T. A., 1987, Management of the common cold, *Adv. Intern. Med.* **32**:207.

MacNaughton, M. R., 1982, Occurrence and frequency of coronavirus infections in humans as determined by enzyme-linked immunosorbent assay, *Infect. Immun.* **38:**419.
MacNaughton, M. R., Flowers, D., and Isaacs, D., 1983, Diagnosis of human coronavirus infections in children using enzyme-linked immunosorbent assay, *J. Med. Virol.* **11:**319.
Madden, D. L., Wallen, W. C., Houff, S. A., Leinikki, P. A., Sever, J. L., Holmes, K. A., Castellano, G. A., and Shekarchi, I. C., 1981, Coronavirus antibodies in sera from patients with multiple sclerosis and matched controls, *Arch. Neurol.* **38:**209.
McIntosh, K., Dees, J. H., Becker, W. B., Kapikian, A. Z., and Chanock, R. M., 1967a, Recovery in tracheal organ cultures of novel viruses from patients with respiratory disease, *Proc. Natl. Acad. Sci. USA* **57:**933.
McIntosh, K., Becker, W. B., and Chanock, R. M., 1967b, Growth in suckling mice brain of IBV-like viruses from patients with upper respiratory tract disease, *Proc. Natl. Acad. Sci. USA* **58:**2268.
McIntosh, K., Kapikian, A. Z., Hardison, K. A., Hartley, J. W., and Chanock, R. M., 1969, Antigenic relationships among the coronaviruses of man and between human and animal coronaviruses, *J. Immunol.* **102:**1109.
McIntosh, K., Bruckova, M., Kapikian, A. Z., Chanock, R., and Turner, H., 1970a, Studies on new virus isolates recovered in tracheal organ culture, *Ann. NY Acad. Sci.*, **174:**983.
McIntosh, K., Kapikian, A. Z., Turner, H. C., Hartley, J. W., Parrott, R. H., and Chanock, R. M., 1970b, Seroepidemiologic studies of coronavirus infection in adults and children, *Am. J. Epidemiol.* **91:**585.
McIntosh, K., Ellis, E. F., Hoffmann, L. S., Lybass, T. G., Eller, J. J., and Fulginiti, V. A., 1973, The association of viral and bacterial respiratory infections with exacerbations of wheezing in young asthmatic children, *J. Paediatr.* **82:**579.
McIntosh, K., Chao, R. K., Krause, H. E., Wasil, R., Mocega, H. E., and Mufson, M. A., 1974, Coronavirus infections in lower respiratory tract disease of infants, *J. Infect. Dis.* **130:**502.
Monto, A. S., and Lim, S. K., 1974, The Tecumseh study of respiratory illness. VI. Frequency of and relationship between outbreaks of coronavirus infection, *J. Infect. Dis.* **129:**271.
Monto, A. S., and Rhodes, L. M., 1977, Detection of coronavirus infection of man by immunofluorescence, *Proc. Soc. Exp. Biol. Med.* **155:**143.
Myint, S., and Tyrrell, D. A. J., 1994, Coronaviruses, in: *Diagnostic Procedures for Viral, Rickettsial and Chlamydial Infections* (E. H. Lenette, E. T. Lennette, and D. A. Lennette, eds.), pp. 709–723. American Public Health Association, Berkeley, CA.
Myint, S., Siddell, S., and Tyrrell, D., 1989, Detection of human coronavirus 229E in nasal washings using RNA:RNA hybridisation, *J. Med. Virol.* **29:**70.
Myint, S., Harmsen, D., Raabe, T., and Siddell, S. G., 1990, Characterisation of a nucleic acid probe for the diagnosis of human coronavirus 229E infections, *J. Med. Virol.* **31:**165.
Myint, S., Johnstone, S., Sanderson, G., and Simpson, H., 1994, The evaluation of "nested" RT-PCR for the detection of human coronaviruses 229E and OC43 in clinical specimens, *Mol. Cell. Probes.* **8:**357.
Owen-Hendley, J. O., Fishburne, H. B., and Gwaltney, J. M., 1972, Coronavirus infections in working adults, *Am. Rev. Resp. Dis.* **105:**805.
Philpotts, R., 1983, Clones of MRC-c cells may be superior to the parent line for the culture of 229E-like strains of human respiratory coronavirus, *J. Virol. Methods* **6:**267.
Reed, S. E., 1984, The behaviour of recent isolates of human respiratory coronavirus *in vitro* and in volunteers: Evidence of heterogeneity among 229E-related strains, *J. Med. Virol.* **13:**179.
Riski, H., and Hovi, T., 1980, Coronavirus infections of man associated with diseases other than the common cold, *J. Med. Virol.* **6:**259.
Riski, H., Hovi, T., Vaananen, P., and Penttinen, K., 1977, Antibodies to human coronavirus OC 43 measured by radial haemolysis in gel, *Scand. J. Infect. Dis.* **9:**75.
Salmi, A., Ziola, B., Hovi, T., and Reunanen, M., 1982, Antibodies to coronaviruses OC43 and 229E in multiple sclerosis patients, *Neurology* **32:**292.
Schmidt, O. W., 1984, Antigenic characterisation of human coronaviruses 229E and OC43 by enzyme-linked immunosorbent assay, *J. Clin. Microbiol.* **20:**175.
Schmidt, O. W., and Kenny, G. E., 1982, Polypeptides and functions of antigens from human coronaviruses 229E and OC 43, *Infect. Immun.* **35:**515.
Schmidt, O. W., Cooney, M. K., and Kenny, G. E., 1979, Plaque assay and improved yield of human

coronaviruses 229E and OC43 in a human rhabdomyosarcoma cell line, *J. Clin. Microbiol.* **9**:722.

Sorensen, O., Collins, A., Flintoff, W., Ebers, G., and Dales, S., 1986, Probing for the human coronavirus OC43 in multiple sclerosis, *Neurology* **35**:1604.

Stewart, J. N., Mounir, S., and Talbot, P. J., 1992, Human coronavirus gene expression in the brains of multiple sclerosis patients, *Virology* **191**:502.

Tanaka, R., Iwasaki, Y., and Koprowski, H., 1976, Intracisternal virus-like particles in brain of a multiple sclerosis patient, *J. Neurol. Sci.* **28**:121.

Tyrrell, D. A. J., and Bynoe, M. L., 1965, Cultivation of a novel type of common cold virus in organ culture, *Br. Med. J.* **1**:1467.

Tyrrell, D. A. J., Almeida, J. D., Berry, D. M., Cunningham, C. H., Hamre, D., Hofstad, M. S., Malluci, L., and McIntosh, K., 1968a, Coronaviruses, *Nature* **220**:650.

Tyrrell, D. A. J., Bynoe, M. L., and Hoorn, B., 1968b, Cultivation of "difficult" viruses from patients with common colds, *Br. Med. J.* **1**:606.

Tyrrell, D. A. J., Mika-Johnson, M., Philips, G., Douglas, W. H. J., Chapple, P. J., 1979, Infection of cultured human type II pneumocytes with certain respiratory viruses, *Infect. Immun.* **26**:621.

Tyrrell, D. A. J., Cohen, S., and Schlarb, J. E., 1993, Signs and symptoms in common colds, *Epidemiol. Infect.* **111**:143.

Weiss, S. R., 1983, Coronaviruses SD and SK share extensive nucleotide homology with murine coronavirus MHV-A59, more than that shared between human and murine coronaviruses, *Virology* **126**:669.

Wenzel, R. P., Hendley, J. O., Davies, J. A., and Gwaltney, J. M., 1974, Coronavirus infections in military recruits, *Annu. Rev. Resp. Dis.* **109**:621.

CHAPTER 19

# The Pathogenesis of Torovirus Infections in Animals and Humans

MARION KOOPMANS AND MARIAN C. HORZINEK

## I. INTRODUCTION

In 1992, the new genus torovirus was added to the family Coronaviridae (Pringle, 1992; Cavanagh and Horzinek, 1993), ending a period of controversy about the assignment of this taxonomic cluster. Toroviruses are enveloped, positive-stranded RNA viruses that may cause enteric infections in animals and humans. The first descriptions mentioned superficial morphological resemblances with coronaviruses, both being 80–120 nm, enveloped, peplomer-bearing particles with a pleomorphic appearance as seen by electron microscopy (EM) (Weiss *et al.*, 1983; Woode *et al.*, 1982). However, further studies revealed morphological and antigenic differences, sparking discussions about the place toroviruses should occupy in taxonomy (Horzinek *et al.*, 1984, 1985, 1986, 1987; Horzinek and Weiss, 1984; Koopmans *et al.*, 1986; Weiss *et al.*, 1984; Weiss and Horzinek, 1986, 1987; Woode *et al.*, 1982, 1985). The matter was settled when studies of the replication mechanism and genomic sequence showed fundamental similarities between toro- and coronaviruses (reviewed by Snijder and Horzinek, 1993; Bredenbeek *et al.*, 1990; den Boon *et al.*, 1991; Snijder *et al.*, 1988, 1989, 1990 a-c; see also Chapter 11, this volume).

The torovirus prototype Berne virus (BEV) was isolated in Berne, Switzer-

MARION KOOPMANS • Virology Section, National Institute of Public Health and Environmental Protection, 3720 BA Bilthoven, The Netherlands. MARIAN C. HORZINEK • Virology Division, Department of Infectious Diseases and Immunology, University of Utrecht, 3584 CL Utrecht, The Netherlands.

*The Coronaviridae*, edited by Stuart G. Siddell, Plenum Press, New York, 1995.

land from a rectal swab taken from a horse with severe gastroenteritis 1 week before it died (Weiss *et al.*, 1983). *Salmonella lille*, isolated from the same swab, was considered to be the cause of the disease in this animal, and BEV has remained a virus in search of a disease. However, since it grows in cell culture and has therefore been extensively studied by the Utrecht group, BEV has been designated the torovirus prototype.

The situation is quite different for another torovirus that had been described 1 year before; it had been found by EM in feces from calves in a dairy herd in Breda, Iowa, that had severe diarrhea (Woode *et al.*, 1982). Breda virus (BRV) does not replicate in cell or tissue culture, but has been identified as a pathogen causing gastroenteritis in calves and possibly in older cattle (Koopmans *et al.*, 1990, 1991c; Saif *et al.*, 1981; Woode *et al.*, 1982, 1985). In the years following the discoveries of BEV and BRV, serological evidence of torovirus infection has been obtained in all ungulates that were tested using a BEV neutralization test (horses, cattle, sheep, goats, pigs), and in rats, rabbits, and some species of feral mice (Weiss *et al.*, 1984). Also, toroviruslike particles have been detected in stool specimens from pigs (Penrith and Gerdes, 1992; Durham *et al.*, 1989; Scott *et al.*, 1987; L. Saif, personal communication), humans (Beards *et al.*, 1984, 1986; Brown *et al.*, 1987; Koopmans *et al.*, 1991a, 1993b), cats (Muir *et al.*, 1990), and dogs (Hill and Yang, 1984). There is little doubt that solitary torovirions have been seen by electron microscopists, but their pleomorphism precluded their identification as viruses since confirmatory testing was unavailable (Koopmans *et al.*, 1991a, 1993b; Liebler *et al.*, 1992). No antibodies have been found in the sera of cats and humans (Brown *et al.*, 1988; Weiss *et al.*, 1984).

The purpose of this chapter is to highlight epidemiological and clinical studies with an emphasis on the pathogenesis of toroviruses. We will focus on bovine toroviruses since most information on the infection *in vivo* has been obtained from studies in cattle. For an update on the structural and morphological properties of the virus, on its replication strategy, and on diagnosis of torovirus infections in animals and humans, the reader is referred to other recent reviews (Koopmans and Horzinek, 1994; Snijder and Horzinek, 1993) and to Chapter 11, this volume.

## II. INFECTION IN CATTLE

### A. Enteric Infections

Three strains of BRV have been used to study the course of infection in its natural host. Besides the original "isolate," two additional strains of bovine enteric toroviruses have been reported; the three strains were assigned to two groups (BRV1 and BRV2) based on antigenic comparisons: BRV1 refers to the first Iowa strain, and BRV2 comprises the two strains that had been detected in feces from a 5-month-old diarrheal calf in Ohio and from a 2-day-old calf in Iowa (Saif *et al.*, 1981; Woode *et al.*, 1983, 1985). All three strains are pathogenic for newborn gnotobiotic and nonimmune conventional calves, aged 1 hour to 10 weeks (Woode *et al.*, 1985).

Although the respiratory system may be sporadically involved (see Section II.C), BRV infections are usually limited to the gut. Between 1 and 3 days after oral inoculation, calves typically develop a watery diarrhea that persists for 4 to 5 days (Woode *et al.*, 1982, 1983, 1985). The most severe symptoms occur within 2 days after onset of the diarrhea, with dehydration, weakness, and depression (Woode, 1987). In the presence of a normal intestinal flora (but in the absence of maternal antibodies), diarrhea generally is more severe than in gnotobiotic calves. Both crypt and villus epithelial cells are infected from the midjejunum through to the large intestine (Woode *et al.*, 1982; Woode, 1987).

Infections with BRV seem to be ubiquitous, as evidence of infection has been obtained in every country where serological and/or virological studies were done: Belgium, Great Britain, France, Germany, India, Italy, The Netherlands, Switzerland, and the United States (Vanopdenbosch *et al.*, 1992a; Brown *et al.*, 1987, 1988; Lamouliatte *et al.*, 1987; Liebler *et al.*, 1992; Koopmans *et al.*, 1989; Weiss *et al.*, 1984; Woode *et al.*, 1985). In addition, unconfirmed torovirus-like particles have been found in cattle from South Africa (Penrith and Gerdes, 1992) and New Zealand (Horner, personal communication).

The infections are quite common in dairy cattle; by 1 year of age, 85–95% of the animals have antibodies to BRV (Koopmans *et al.*, 1989; Weiss *et al.*, 1984; Woode *et al.*, 1985). In a study among dairy cattle in The Netherlands, most seroconversions took place between 6 and 12 months of age, after the waning of maternal immunity (Koopmans *et al.*, 1986).

The presence of maternal antibodies in calves (in 95% of the animals) did not prevent infection, but may have modified its outcome, since diarrhea generally was mild in BRV-excreting calves (Koopmans *et al.*, 1990, 1991c). BRV infections accounted for 4% of cases of diarrhea in calves in this study. BRV-associated diarrhea under farm conditions was clinically indistinguishable from that caused by the most common enteropathogens (rota- or coronavirus), although it lasted slightly longer (average 9.2, 6.8, and 6.8 days, respectively, for the three viruses) and affected slightly older calves (average 12.7, 7.7, and 8.3 days, respectively) (Koopmans *et al.*, 1991c). Calves 3–4 months of age showed very mild diarrhea or no diarrhea at all in association with torovirus shedding (Koopmans *et al.*, 1991c).

Besides its association with gastroenteritis in young calves, BRV is a possible cause of diarrhea in adult dairy cows (Koopmans *et al.*, 1991c).

## B. Pathogenesis of Enteric Torovirus Infections

The exact pathogenic pathways that typify torovirus infections in cattle are not known. However, results from experimental infections of calves suggest a mechanism similar to that of bovine coronaviruses (BCV) (Clark, 1993). Calves may be infected with BRV by the oral route and possibly by the respiratory route (Woode *et al.*, 1982, 1985; Vanopdenbosch *et al.*, 1991). The epithelial cells lining the small and large intestine become infected, with progression from areas of the midjejunum down through the ileum and colon (Fagerland *et al.*, 1986). Within the small intestine, not only epithelial cells at the top of intestinal villi are infected but also cells in the upper third of the crypts (Fagerland *et al.*, 1986; Pohlenz *et al.*, 1984). This has important consequences for the outcome of the infection as can be deduced from the physiology of the gastrointestinal tract: in an uninfected gut there is a well-attuned balance between the sequestration of cells at the top of the intestinal villi and the production of new enterocytes in the crypts. Crypt epithelial cells rapidly divide and are pushed up toward the top of the intestinal villus. During this process the new cells mature into absorbing cells by the development of fingerlike (brush border) projections that increase the cell surface many times and by the production of brush border-associated enzymes. Following a toroviral infection, the turnover rate of epithelial cells increases because infected cells die and detach (Fagerland *et al.*, 1986; Pohlenz *et al.*, 1984), leading to an efficient spread of the infection. In fecal preparations from experimentally BRV infected calves, hemagglutination (HA) titers up to $3 \times 10^7$ units/ml have been measured (Woode *et al.*, 1983), corresponding to virus titers of $10^{11}$ to $10^{12}$ (Zanoni *et al.*, 1986). With such high titers the infection will spread rapidly. The lost epithelium is replaced by immature cells that do not yet possess a mature brush border, thereby presenting insufficient absorptive surface and digestive enzymes (15–65% reduction in D-xylose resorption in BRV-infected calves) (Woode *et al.*, 1982, 1985). Since BRV infects both crypt cells and the villus epithelium, restoration of the normal structure and function may be slower than after infections that affect only the villous epithelium (e.g., by rotaviruses and BCV). A longer duration of diarrhea can therefore be expected (Koopmans *et al.*, 1991c) with more severe complications. The decrease in digestive and absorptive capacities leads to accumulation of lactose in the gut lumen, which in turn results in water and electrolyte retention leading to diarrhea and occasionally to dehydration, acidosis, hypoglycemia, and death.

## C. Respiratory Infection

Recently, Vanopdenbosch *et al.* (1992a,b) reported the isolation of a torovirus from the respiratory tract of calves with pneumonia [bovine respiratory torovirus (BRTV)]. If confirmed, this finding would suggest that bovine toroviruses are both entero- and pneumotropic, as has been clearly demonstrated for BCV (Reynolds, 1983; Reynolds *et al.*, 1985; Saif *et al.*, 1986). Most likely different strains are involved, because respiratory tract symptoms were rarely observed in calves after experimental BRV infection that suffered from diarrhea, whereas in the calves described by Vanopdenbosch *et al.* (1992a,b) respiratory symptoms were predominant (Koopmans *et al.*, 1990; Woode *et al.*, 1982, 1985). Also, when testing RNA extracts from cells infected with the BRTV isolate in a diagnostic polymerase chain reaction assay with BEV/BRV consensus primers, no detectable amplification product was obtained (M. Koopmans, unpublished results). The "respiratory" isolate is the only bovine torovirus that has been adapted to cell culture, but its presence still needs to be confirmed. Using infected cells in an immunofluorescence test, Vanopdenbosch *et al.* (1992b) found high levels of antibody in all calves ($n = 50$) that persisted for 5 months.

These authors also detected antibodies in 20 batches of fetal bovine serum, in contrast with earlier findings (Koopmans *et al.*, 1990; Woode *et al.*, 1982).

## D. Infection of Other Organ Systems

The unexpected finding of antibodies in all commercially available batches of fetal bovine serum and in 7 of 13 precolostral calf sera (Vanopdenbosch *et al.*, 1992b) indicated the possibility of transplacental torovirus infections in cattle; this concept was underscored by the demonstration of toroviral antigen in placental cotyledons of spontaneously aborted calves using an immunofluorescence assay. The same authors suggested a possible role in central nervous disturbances, sudden death, and in a syndrome resembling mucosal disease (Vanopdenbosch *et al.*, 1992a).

## E. BRV Infection and the Immune System

In addition to crypt and villus epithelial cells of the small and large intestine, BRV also infects the dome epithelium overlying Peyer's patches. The dome epithelial cells, including the M cells, show the same cytopathic changes that occur in the absorptive villous cells (Woode *et al.*, 1984; Pohlenz *et al.*, 1984). In addition, the germinal centers in Peyer's patches of BRV-infected calves were found depleted of lymphocytes and occasionally showed fresh hemorrhage (Woode *et al.*, 1982). The M cells are specialized gut epithelial cells that endocytose macromolecules and microorganisms from the gut lumen and present them to underlying lymphoid tissue; they play a key role in the development of an immune response against enteric pathogens. Direct or IgA-mediated adherence to M cells has been observed for other viruses that infect them (bovine astrovirus) or use their transepithelial transport capacity to cross the epithelium and to invade neural and lymphoid tissues (reovirus and poliovirus) (Pearson *et al.*, 1978; Sicinski *et al.*, 1990; Weltzin *et al.*, 1989; Wolf *et al.*, 1981). Infection of M cells by BRV and bovine astrovirus results in a cytopathic effect, while poliovirus and reovirus infections are noncytopathic. It is unknown whether BRV (or astroviruses) pass the epithelial lining via the M cell pathway; no such evidence was obtained in clinical and pathological studies (Fagerland *et al.*, 1986; Woode *et al.*, 1982, 1985).

The tropism of BRV for dome cells may have consequences for the (mucosal) immune response. Development of BRV-specific IgM antibodies interestingly occurred very late after primary infection of sentinel calves ($<$ 1 month of age), and no memory response was seen in them after a second BRV infection at 10 months of age (Koopmans *et al.*, 1990). An alternative explanation for these findings may be the suppressive effect of maternal antibodies on infection: In experimental BCV infections the immune response was delayed in calves that had been fed colostrum with high titers of specific antibody compared with the response in calves fed low-titered colostrum (Heckert *et al.*, 1991).

## F. Chronic Infection

Chronic torovirus infections also may occur, as data from a recent epidemiological study suggest. In a closed herd of 10 dairy calves repeated torovirus shedding was observed in several animals, with intervals of several weeks to 4 months. When the calves were introduced into a herd of adult cows, they all had an episode of diarrhea in association with BRV shedding, followed by seroconversion. The adult cows showed no evidence of active BRV infection in the 2 weeks preceding the arrival of the calves, but they all had high levels of preexisting antibodies (Koopmans *et al.*, 1990). These observations indicate the presence of carriers that shed low levels of virus or undergo recurrent subclinical infections. Virus persistence and shedding may be an important source of virus in epizootics of neonatal torovirus infections. Infections with BCV may serve as a model for understanding the situation in bovine torovirus infections, given the close resemblance in other aspects of the pathogenesis of these viruses. Evidence of coronavirus shedding has been found in up to 75% of clinically normal cows (Collins *et al.*, 1987; Crouch and Acres, 1984), but only 5% were shedding free virus that could be detected in regular enzyme-linked immunosorbent assay (ELISA); the remaining 70% were identified using an ELISA for the detection of immune complexes (Crouch and Acres, 1984). In a follow-up study, free virus could be detected intermittently in some animals, but immune-complexed virus was present in the feces throughout the 12 weeks of the study (Crouch *et al.*, 1985). Chronic shedders can be a source of infection as shown by Bulgin *et al.* (1989): calves from carrier cows had a 60% chance of developing clinical illness, whereas those from noncarriers had only a 22% chance.

With the development of highly sensitive, polymerase chain reaction-based detection methods for toroviruses, the role of carriers in the epidemiology and pathogenesis of torovirus infections can now be addressed (Koopmans *et al.*, 1993a).

# III. INFECTION IN HORSES

## A. A Virus in Search of a Disease

Similar to the situation for bovine toroviruses, infection of horses is quite common in the populations that have been examined; 81% of the adult horses in Switzerland and The Netherlands possess neutralizing antibodies to BEV (Weiss *et al.*, 1983, 1984), 35% in Germany, and 38% in India (Liebermann, 1990).

Equine toroviruses may cause similar disease pictures as BRV in calves, but epidemiological studies in populations most likely at risk have not been done so far. Diarrhea in young foals is a big problem, and epidemiological studies should be aimed at this age group (<1 month of age). Infection experiments with BEV have been very limited: two yearlings that had been injected intravenously with $10^7$ $TCID_{50}$ of tissue culture grown BEV seroconverted without clinical symptoms (Weiss *et al.*, 1984). Virus shedding was not monitored in these horses. A

3-day-old gnotobiotic foal was inoculated orally, and again no symptoms were seen, although virus shedding and seroconversion occurred. Attempts to infect a 3-month-old foal were unsuccessful (Dr. F. Scott, Moredun Research Institute, Edinburgh, England, personal communication, 1992). Tissue culture-passaged virus was used in these experiments, which may be of low virulence as a result of the adaptation.

## B. The Torovirus Mutant BEV

Irrespective of many attempts (Weiss, unpublished observations), isolation of BEV has been a unique event and could only be repeated with the field sample from the same horse. This observation suggests that the Berne isolate is a mutant.

Most parts of the BEV genome have been sequenced; it would be interesting to obtain sequence information from pathogenic toroviruses and to look for explanations for the difference in pathogenicity. Of special interest is a presumed pseudogene [open reading frame 4 (ORF 4)] that has been identified in the BEV genome. The predicted amino acid sequence of the ORF 4 product bears similarities to the C-terminal part of the hemagglutinin esterase (HE) of BCV and of human influenza C virus; however, the 5′ two-thirds of the HE gene are missing, and an ORF 4 product has not been identified in BEV virions or in lysates of infected cells (Snijder *et al.*, 1991). The BCV HE (like the influenza C virus HE) uses *N*-acetyl-9-*O*-acetylneuraminic acid as a receptor to initiate the infection of cultured cells and probably plays a role in infection of host cells *in vivo* (Clark, 1993; Schultze and Herrler, 1992). Recently, 2 kb of the 3′-end of BRV has been sequenced; it was demonstrated that BRV has a complete HE gene that is probably functional (L. A. H. M. Cornelissen and R. J. de Groot, unpublished observation; Koopmans *et al.*, 1986).

## C. Host Range

Infection with BEV *in vitro* is limited to cells of equine origin. Neutralizing antibodies have been found in sera from cattle, goats, sheep, pigs, rabbits, and feral mice, indicating a close antigenic relationship between toroviruses of these species or cross-species infections with one or a few related toroviruses. The latter option is not likely. Mouse immune sera raised against BRV2 show very little cross-reactivity with BRV, except at the level of the peplomers; the sera recognized the polypeptides of the homologous virus and the two highest-molecular-weight proteins (105 kDa and 85 kDa) of BRV1 in radioimmune precipitation. The same sera inhibited hemagglutination of the heterologous serotype and efficiently neutralized the infectivity of BEV (Koopmans *et al.*, 1986).

No evidence of viral replication was obtained in rats, mice, or lambs that had been experimentally infected with BRV (Woode *et al.*, 1982; Woode, 1987).

In contrast, the respiratory bovine toroviruses reportedly replicate in cells from a wide range of host species (Vanopdenbosch *et al.*, 1992b).

## IV. TOROVIRUS INFECTIONS IN OTHER SPECIES

Toroviruslike particles and torovirus antibodies have been found in species other than the cattle and horse, indicating that these viruses may infect a broad range of animal hosts (Muir *et al.*, 1990; Scott *et al.*, 1987; Weiss *et al.*, 1984). Although in most cases stool specimens from animals with diarrhea were examined, the pathogenic role of these toroviruses remains unclear, and epidemiological studies are needed to study the causal relationship between virus presence and disease (Durham *et al.*, 1989; Hill and Yang, 1984; Muir *et al.*, 1990; Scott *et al.*, 1987).

The toroviruslike particles found in humans (Beards *et al.*, 1984, 1986) cross-react antigenically with BRV (Koopmans *et al.*, 1993b); preliminary data from an ongoing epidemiological study in Brazil indicate an association with diarrhea (Koopmans and Guerrant, submitted for publication).

## V. REFERENCES

Beards, G. M., Green, J., Hall, C., Flewett, T. H., Lamouliatte F., and Du Pasquier, P., 1984, An enveloped virus in stools of children and adults with gastroenteritis that resembles the Breda virus of calves, *Lancet* **2:**1050.

Beards, G. M., Brown, D. W. G., Green, J., and Flewett, T. H., 1986, Preliminary characterization of torovirus-like particles of humans: Comparison with Berne virus of horses and Breda virus of calves, *J. Med. Virol.* **20:**67.

Bredenbeek, P. J., Snijder, E. J., Noten, A. F. H., den Boon, J. A., Schaaper, W. M. M., Horzinek, M. C., and Spaan, W. J. M., 1990, The polymerase gene of corona- and toroviruses: Evidence for an evolutionary relationship, *Adv. Exp. Med. Biol.* **276:**307.

Brown, D. W. G., Beards, G. M., and Flewett, T. H., 1987, Detection of Breda virus antigen and antibody in humans and animals by enzyme immunoassay, *J. Clin. Microbiol.* **25:**637.

Brown, D. W. G., Selvakumar, R., Daniel, D. J., and Mathan, V. I., 1988, Prevalence of neutralizing antibodies to Berne virus in animals and humans in Vellore, South India, *Arch. Virol.* **98:**267.

Bulgin, M. S., Ward, A. C. S., Barrtett, D. P., and Lane, V. M., 1989, Detection of rotavirus and coronavirus shedding in two beef cow herds in Idaho, *Can. Vet. J.* **30:**235.

Cavanagh, D., and Horzinek, M. C., 1993, Genus *Torovirus* assigned to the *Coronaviridae*, *Arch. Virol.* **128:**395.

Clark, M. A., 1993, Bovine coronavirus, *Br. Vet. J.* **149:**51.

Collins, J. K., Riegel, B. S., Olson, J. D., and Fountain, A., 1987, Shedding of enteric coronavirus in adult cattle, *Am. J. Vet. Res.* **48:**361.

Crouch, C. F., and Acres, S. D., 1984, Prevalence of rotavirus and coronavirus antigens in the feces of normal cows, *Can. J. Comp. Med.* **48:**340.

Crouch, C. F., Bielefeldt Ohmann, H., Watts, T. C., and Babiuk, L. A., 1985, Chronic shedding of bovine enteric coronavirus antigen-antibody complexes by clinically normal cows, *J. Gen. Virol.* **66:**1489.

den Boon, J. A., Snijder, E. J., Krijnse-Locker, J., Horzinek, M. C., and Rottier, P. J. M., 1991, Another triple-spanning envelope protein among intracellularly budding RNA viruses: The torovirus E protein, *Virology* **182:**655.

Durham, P. K. J., Hassard, L. E., Norman, G. R., and Yemen, R. L., 1989, Viruses and virus-like particles detected during examination of feces from calves and piglets with diarrhea, *Can. Vet. J.* **30:**876.

Fagerland, J. A., Pohlenz, J. F. L., and Woode, G. N., 1986, A morphologic study of the replication of Breda virus (proposed family Toroviridae), *J. Gen. Virol.* **67:**1293– 1304.

Heckert, R. A., Saif, L. J., and Myers, G. W., 1991, Mucosal and systemic isotype-specific antibody responses to bovine coronavirus structural proteins in naturally infected dairy calves, *Am. J. Vet. Res.* **52:**852.

Hill, D. L., and Yang, T. J., 1984, Virus-like particles in a positive case of canine parvovirus enteritis, *Micron Microscop. Acta* **15:**207.

Horzinek, M. C., and Weiss, M., 1984, Toroviridae: A taxonomic proposal, *Zentrlbl. Vet. Med. B* **31:**649.

Horzinek, M. C., Weiss, M., and Ederveen, J., 1984, Berne virus is not "coronavirus-like," *J. Gen. Virol.* **65:**645.

Horzinek, M. C., Ederveen, J., and Weiss, M., 1985, The nucleocapsid of Berne virus, *J. Gen. Virol.* **66:**1287.

Horzinek, M. C., Ederveen, J., Kaeffer, B., de Boer, D., and Weiss, M., 1986, The peplomers of Berne virus, *J. Gen. Virol.* **67:**2475.

Horzinek, M. C., Flewett, T. H., Saif, L. F., Spaan, W. J. M., Weiss, M., and Woode, G. N., 1987, A new family of vertebrate viruses: Toroviridae, *Intervirology* **27:**17.

Koopmans, M., and Horzinek, M. C., 1994, Toroviruses of animals and humans: A review, *Adv. Virus Res.* **43:**233.

Koopmans, M., Ederveen, J., Woode, G. N., and Horzinek, M. C., 1986, Surface proteins of Breda virus, *Am. J. Vet. Res.* **47:**1896.

Koopmans, M., van den Boom, U., Woode, G. N., and Horzinek, M. C., 1989, Seroepidemiology of Breda virus in cattle using ELISA, *Vet. Microbiol.* **19:**233.

Koopmans, M., Cremers, H., Woode, G. N., and Horzinek, M. C., 1990, Breda virus (Toroviridae) infection and systemic antibody response in sentinel calves, *Am. J. Vet. Res.* **51:**1443.

Koopmans, M., Herrewegh, A., and Horzinek, M. C., 1991a, Diagnosis of torovirus infection, *Lancet* **337:**85.

Koopmans, M., Snijder, E. J., and Horzinek, M. C., 1991b, cDNA probes for the diagnosis of bovine torovirus (Breda virus) infection, *J. Clin. Microbiol.* **29:**493.

Koopmans, M., van Wuijckhuise-Sjouke, L., Cremers, H., and Horzinek, M. C., 1991c, Association of diarrhea in cattle with torovirus infections on farms, *Am. J. Vet. Res.* **52:**1769.

Koopmans, M., Monroe, S. S., Coffield, L. M., and Zaki, S. R., 1993a, Optimization of extraction and PCR amplification of RNA from paraffin-embedded tissue in different fixatives, *J. Virol. Meth.* **43:**189.

Koopmans, M., Petric, M., Glass, R. I., and Monroe, S. S., 1993b, ELISA reactivity of torovirus-like particles, *J. Clin. Microbiol.* **31:**2738.

Lamouliatte, F., du Pasquier, P., Rossi, F., Laporte, J., and Lose, J. P., 1987, Studies on bovine Breda virus, *Vet. Microbiol.* **15:**261.

Liebermann, H., 1990, Für die DDR neuartige Virusinfektionen der Haustiere. 1. Mitteilung: Serologische Übersichtsuntersuchungen über die Verbreitung equiner Torovirusinfektionen in der DDR, *Arch. Exp. Vet. Med.* **44:**251.

Liebler, E. M., Kluever, S., Pohlenz, J., and Koopmans, M., 1992, Zur Bedeutung des Bredavirus als Durchfallerreger in niedersächsischen Kälberbeständen, *Dtsch. Tierärztl. Wschr.* **99:**195.

Muir, P., Harbour, D. A., Gruffydd-Jones, T. J., Howard, P. E., Hopper, C. D., Gruffydd-Jones, E. A., Broadhead, H. M., Clarke, C. M., and Jones, M. E., 1990, A clinical and microbiological study of cats with protruding nictitating membranes and diarrhoea: Isolation of a novel virus, *Vet. Rec.* **127:**324.

Pearson, G. R., Logan, E. F., and Brennan, G. P., 1978, Scanning electron microscopy of the small intestine of a normal unsuckled calf and a calf with enteric colibacillosis, *Vet. Pathol.* **15:**400.

Penrith, M. L., and Gerdes, G. H., 1992, Breda virus-like particles in pigs in South Africa, *J. S. Afr. Vet. Assoc.* **63:**102.

Pohlenz, J. F. L., Cheville, N. F., Woode, G. N., and Mokresh, A. H., 1984, Cellular lesions in intestinal mucosa of gnotobiotic calves experimentally infected with a new unclassified bovine virus (Breda virus), *Vet. Pathol.* **21:**407.

Pringle, C. R., 1992, Committee pursues medley of virus taxonomic issues, *ASM News* **58:**475.
Reynolds, D. J., 1983, Coronavirus replication in the intestinal and respiratory tracts during infection of calves, *Ann. Rech. Vet.* **14:**445.
Reynolds, D. J., Debney, T. G., Hall, G. A., Thomas, L. H., and Parson, K. R., 1985, Studies on the relationships between coronaviruses from the intestinal and respiratory tracts of calves, *Arch. Virol.* **85:**71.
Saif, L. J., Redman, D. R., Theil, K. W., Moorhead, P. D., and Smith, C. K., 1981, Studies of an enteric "Breda" virus in calves, in: 62nd Annual Meeting of the Confederation of Research Workers in Animal Diseases, Abstract 236, Chicago.
Saif, L. J., Redman, D. R., Moorhead, P. D., and Theil, K. W., 1986, Experimental coronavirus infections in calves: Viral replication in the respiratory and intestinal tracts, *Am. J. Vet. Res.* **47:**1426.
Schultze, B., and Herrler, G., 1992, Bovine coronavirus uses *N*-acetyl-9-*O*-acetylneuraminic acid as a receptor determinant to initiate the infection of cultured cells, *J. Gen. Virol.* **73:**901.
Scott, A. C., Chaplin, M. J., Stack, M. J., and Lund, L. J., 1987, Porcine torovirus? *Vet. Rec.* **120:**583.
Sicinski, P., Rowinski, J., Warchol, J. B., Jarzabek, Z., Gut, W., Szczygiel, B., Bielecki, K., and Koch, G., 1990, Poliovirus type 1 enters the human host through intestinal M cells, *Gastroenterology* **98:**56.
Snijder, E. J., and Horzinek, M. C., 1993, Toroviruses: Replication, evolution and comparison with other members of the coronavirus-like superfamily, *J. Gen. Virol.* **74:**2305.
Snijder, E. J., Ederveen, J., Spaan, W. J. M., Weiss, M., and Horzinek, M. C., 1988, Characterization of Berne virus genomic and messenger RNAs, *J. Gen. Virol.* **69:**2135– 2144.
Snijder, E. J., den Boon, J. A., Verjans, G. M. G. M., Spaan, W. J. M., and Horzinek, M. C., 1989, Identification and primary structure of the gene encoding the Berne virus nucleocapsid protein, *J. Gen. Virol.* **70:**3363.
Snijder, E. J., Horzinek, M. C., and Spaan, W. J. M., 1990a, A 3′-coterminal nested set of independently transcribed mRNAs is generated during Berne virus replication, *J. Virol.* **64:**331.
Snijder, E. J., den Boon, J. A., Spaan, W. J. M., Weiss, M., and Horzinek, M. C., 1990b, Primary structure and post-translational processing of the Berne virus peplomer protein, *Virology* **178:**355.
Snijder, E. J., den Boon, J. A., Bredenbeek, P. J., Horzinek, M. C., Rijnbrand, R., and Spaan, W. J. M., 1990c, The carboxyl-terminal part of the putative Berne virus polymerase is expressed by ribosomal frameshifting and contains sequence motifs which indicate that toro- and coronaviruses are evolutionary related, *Nucleic Acids Res.* **18:**4535.
Snijder, E. J., den Boon, J. A., Horzinek, M. C., and Spaan, W. J. M., 1991, Comparison of the genome organization of toro- and coronaviruses: Both divergence from a common ancestor and RNA recombination have played a role in Berne virus evolution, *Virology* **180:**448.
Vanopdenbosch, E., Wellemans, G., and Petroff, K., 1991, Breda virus associated with respiratory disease in calves, *Vet. Rec.* **31:**203.
Vanopdenbosch, E., Wellemans, G., Oudewater, J., and Petroff, K., 1992a, Prevalence of torovirus infections in Belgian cattle and their role in respiratory, digestive and reproductive disorders, *Vlaams Diergeneesk. Tijdschr.* **61:**1.
Vanopdenbosch, E., Wellemans, G., Charlier, G., and Petroff, K., 1992b, Bovine torovirus: Cell culture propagation of a respiratory isolate and some epidemiological data, *Vlaams Diergeneesk. Tijdschr.* **61:**45.
Weiss, M., and Horzinek, M. C., 1986, The morphogenesis of Berne virus (proposed family Toroviridae), *J. Gen. Virol.* **67:**1305.
Weiss, M., and Horzinek, M. C., 1987, The proposed family Toroviridae: Agents of enteric infections, *Arch. Virol.* **92:**1.
Weiss, M., Steck, F., and Horzinek, M. C., 1983, Purification and partial characterization of a new enveloped RNA virus (Berne virus), *J. Gen. Virol.* **64:**1849.
Weiss, M., Steck, F., Kaderli, R., and Horzinek, M. C., 1984, Antibodies to Berne virus in horses and other animals, *Vet. Microbiol.* **9:**523.
Weltzin, R., Lucia-Jandris, P., Michetti, P., Fields, B. N., Kraehenbuhl, J. P., and Neutra, M. R., 1989, Binding and transepithelial transport of immunoglobulins by intestinal M cells: Demonstration using monoclonal antibodies against enteric viral proteins, *J. Cell Biol.* **108:**1673.

Wolf, J. L., Kauffman, R. S., Finberg, R., Dambrauskas, R., Fields, B. N., and Trier, J. S., 1981, Intestinal M cells: A pathway for entry of reovirus into the host, *Science* **212**:471.
Woode, G. N., 1987, Breda and Breda-like viruses: Diagnosis, pathology and epidemiology, in: *Novel Diarrhea Viruses* (CIBA Foundation Symp. 128) (G. Bock and J. Whelan, eds.), pp. 175–191, Wiley & Sons, Chichester, England.
Woode, G. N., Reed, D. E., Runnels, P. L., Herrig, M. A., and Hill, H. T., 1982, Studies with an unclassified virus isolated from diarrheal calves, *Vet. Microbiol.* **7**:221.
Woode, G. N., Mohammed, K. A., Saif, L. J., Winand, N. J., Quesada, M., Kelso, N. E., and Pohlenz, J. F., 1983, Diagnostic methods for the newly discovered "Breda" group of calf enteritis inducing viruses, in: pp. 533.
Woode, G. N., Pohlenz, J. F., Kelso-Gourley, N. E., and Fagerland, J., 1984, Astrovirus and Bredavirus infections of dome cell epithelium of bovine ileum, *J. Clin. Microbiol.* **19**:623.
Woode, G. N., Saif, L. J., Quesada, M., Winand, N. J., Pohlenz, J. F., and Kelso Gourley, N., 1985, Comparative studies on three isolates of Breda virus of calves, *Am. J. Vet. Res.* **46**:1003.
Zanoni, R., Weiss, M., and Peterhans, E., 1986, The hemagglutinating activity of Berne virus, *J. Gen. Virol.* **67**:2485.

# Index